C0-AUZ-844

HANDBOOK OF TRANSPORT SYSTEMS AND TRAFFIC CONTROL

HANDBOOKS IN TRANSPORT

3

Series Editors

DAVID A. HENSHER

KENNETH J. BUTTON

HANDBOOK OF TRANSPORT SYSTEMS AND TRAFFIC CONTROL

Edited by

KENNETH J. BUTTON
The School of Public Policy,
George Mason University

DAVID A. HENSHER
Institute of Transport Studies,
University of Sydney

2001
PERGAMON
An Imprint of Elsevier Science
Amsterdam – London – New York – Oxford – Paris – Shannon – Tokyo

ELSEVIER SCIENCE Ltd
The Boulevard, Langford Lane
Kidlington, Oxford OX5 1GB, UK

First edition 2001

Library of Congress Cataloging in Publication Data
A catalog record from the Library of Congress has been applied for.

British Library Cataloguing in Publication Data
A catalogue record from the British Library has been applied for.

ISBN: 0 08 043595 5
Series ISSN: 1472-7889

∞ The paper used in this publication meets the requirements of ANSI/NISO Z39.48-1992 (Permanence of Paper).
Printed in The Netherlands.

INTRODUCTION TO THE SERIES

Transportation and logistics research has now reached maturity, with a solid foundation of established methodology for professionals to turn to and for future researchers and practitioners to build on. Elsevier is marking this stage in the life of the subject by launching a landmark series of reference works: *Elsevier's Handbooks in Transport*. Comprising specially commissioned chapters from the leading experts of their topics, each title in the series will encapsulate the essential knowledge of a major area within transportation and logistics. To practitioners, researchers and students alike, these books will be authoritative, accessible and invaluable.

David A. Hensher
Kenneth J. Button

This Handbook is supported by The Institution of Highways & Transportation (IHT).

The Institution is the foremost learned society in the U.K. concerned specifically with the design, construction, maintenance and operation of sustainable transport systems and infrastructure.

It aims to provide a forum for the exchange of technical information and views on highways and transport policy; to produce practical technical publications; to provide specialist advice to government and other bodies; to make roads safer for the travelling public; and to encourage training and professional development to meet today's requirements.

Membership, which is currently 10,400 world-wide, is open to all those working in Highways and Transportation whatever their discipline.

The Institution of Highways & Transportation
6 Endsleigh Street
London WC1H 0DZ
U.K.

Tel: (0)207 387 2525
Fax: (0)207 387 2808
E-mail: iht@iht.org
Website: www.iht@iht.org.uk

CONTENTS

Chapter 1

INTRODUCTION

KENNETH J. BUTTON
George Mason University, Fairfax, VA

DAVID A. HENSHER
University of Sydney

1. Introduction

Transport is an important element in economic development and affords the social and political interactions that most people take for granted. This series of Handbooks is intended to provide readers with up-to-date information on how the transport system functions, and how the various components that constitute transport systems interact with each other. This volume focuses on matters pertaining to transport systems. It complements the previous volumes in the series on transport modeling, and on logistics and supply chain management.

Transport is provided across a range of networks. These have grown with time as technology improvements and innovation have allowed individuals to develop new networks and refine the systems that exist. Earlier transport by foot or pack animal involved lanes, tracks, and pathways, but the system was extended as boats became available for inland navigation and ships for maritime movement. The advent of the sail led to further developments on both water and land (the "Chinese wheelbarrow"). The past two hundred years have seen further transformation with steam power, the internal combustion engine, and jet power revolutionizing transport. More recently telecommunications has provided an alternative to travel, and sometimes a complementary infrastructure that leads to increased travel.

In all these cases, and at any period of time, transport was provided as a system over a network of links and nodes. In a previous volume in this series – *Handbook of Transport Modelling* – papers were published that were largely concerned with looking at how one may develop models to assist in public and private decision-making. In Volume 1 of the series we emphasized the theoretical aspects of transport models with limited attention given to the working of the transport system. These are of immense importance in policy formulation but they say little

Handbook of Transport Systems and Traffic Control, Edited by K.J. Button and D.A. Hensher

about the ways transport systems are developed, planned, managed, and regulated. They take as axiomatic that policy instruments are available and may be deployed to influence future events.

A second volume – *Handbook of Logistics and Supply Chain Management* – was concerned with how those operating transport services make use of transport systems with some limited information on the public sector's role. This volume looks at transport systems more directly. It is also much more public sector focused in the sense that it is government and government agencies that provide much of the key underlying infrastructure that allows transport systems to operate, and that control access to that infrastructure and the ways in which it is used.

The current volume is more concerned with the broader nature of transport systems and with the implementation of public policy across transport systems. These are no small issues at a time when problems such as urban traffic congestion, environmental intrusion, transport safety, and budgetary constraints are exercising the minds of policy makers. But first, what exactly do we mean by a transport system?

2. What are transport systems?

Transport systems are complex and subject to continual change. Furthermore, there is no simple definition of what constitutes a transport system. Much depends on the eye of the beholder.

From one perspective there are the various modes of transportation (ships, airplanes, trucks, and so on), each of which may be seen as a system. From another perspective one can separate out different infrastructure systems (rail networks, seaports, roads, and so on) and again each may be seen as a system although in some cases they may be used by several different modes of transport (roads by trucks, buses, bicycles, and cars). Again, another slice of the pie may produce systems seen as ways of meeting a particular transport objective (e.g., the various modes and route alternatives that allow a package to move from origin A to destination B).

A rearrangement of this idea can lead to transport systems being differentiated by distance into urban, interurban and international transport. From a transport operator's point of view (e.g., a distributor such as the Post Office or UPS), the vocabulary of collection, trunk haul systems, and distribution systems also comes very close to this concept. This same type of geographical separation takes on a slightly different form, however, when it comes to public policy and terms such as urban transport and rural transport systems become the more common parlance.

More recent distinctions may be between physical transport systems and those concerned with the transport of information through cables or by wireless means.

Some transport systems may be perceived as the combination of physical and informational inputs that allow a transport activity to take place (e.g., the combination of telecommunications and transport required in modern just-in-time production).

Much also depends on who is concerned with the system. The consumer of transport services normally views the transport system that is available in a rather narrow way and is only concerned with those aspects of any transport system that are immediately useful. A national transport planner, in contrast, by necessity takes a much wider, generally multimodal view that embraces the interaction of infrastructure and the use made of that infrastructure.

3. Transport systems from the user's perspective

Few individuals are happy with the transport systems that exist. And if one goes back and reads the stories of travelers in the past when one learns of dirty, uncomfortable, delayed, and expensive trips, this feeling is not certainly one peculiar to the current populace. Transportation is part of everyone's life and a lot is demanded of the transport systems that provide it. The difficulty is in part that users are not homogeneous in their demands, the transport system is not infinitely flexible (particularly in the short term), and there are other parties to consider who live along transport links or would be displaced if the transport network were to be expanded.

Users of transport systems represent a variety of diverse groups. These groups seek different things from the systems and have different degrees of access to them. There are, for example, gender differences and differences in the perceptions of able-bodied and disabled people. The commercial parts of the overall system are allocated on market principles with profits and losses determining the scale and nature of supply. This means that those willing and able to buy space on any system get priority in its use. Others either have to use it at different times (e.g., when they can afford or are willing to pay off-peak-period airfares), use different systems, or cannot use it at all.

Large parts of the overall transport system are not, however, provided on a commercial basis. They are often subsidized on the argument that these parts meet a social need and should be provided free or at reduced cost. Other parts of the system, most notably roads, are not provided commercially in most countries but users have to pay a variety of charges and fees to access them. The user's perspective at the time of using the transport system is that the monetary costs of infrastructure use are virtually zero, and the result is inevitably excessive use, with high levels of congestion the outcome. Equally, there are costs that are borne by non-users, most notably environmental costs that again affect the perspective of users and can enhance their proclivity to overuse.

The situation is also not static, and sometimes not logical. The users of transport systems have a tendency to use increasingly higher benchmarks. Air travel is far safer now than in the 1930s but the quest is for even safer flights. The safe speeds attained by cars now far exceeds those of earlier vehicles and they are much more comfortable. But road users seek even faster travel. Comparisons between systems are also often not based on objective fact but rather on perception – many people fear flying, and every time there is a plane crash there is a major inquiry and often considerable sums are then spent on various forms or remedial action, despite the fact that flying is safer than driving.

4. Transport systems and transport policy

Transport has always been under some degree of control from public policy. It serves important military needs that the state has always sought to control and is important for the successful governing of internal affairs. But it also serves important commercial purposes. The exchange economy could never have developed if farmers and craftsmen could not have got their products to market. The exchange economy is underpinned by at least a very basic transport system. Crossroads have traditionally been centers of commerce and production and provided the focal points for the first great civilizations. Maritime transport for two thousands years carried the bulk of long-distance commerce and led to the economic power of large port cities. Even in the 21st century, high-technology companies make extensive use of air and road transport to meet the needs of customers and to conduct their own business-to-business dealings.

At the broad macro level, the links between transport and development are fairly clear but when moving to more detail, complexities emerge. There is also the policy question of how government can best develop positive links between transport and the economy.

An initial question concerns the old problem of the "chicken and the egg." Whilst there is correlation between transport and economic change, it is not altogether clear what the direction of causality is. On the one hand, there is a school of thought, amply backed by historical evidence, showing transport to be the stimulating factor. The 18th century Industrial Revolution in the U.K., for instance, is often depicted as being preceded and stimulated by a Transport Revolution involving improved shipping and the construction of canals and turnpikes. But equally, the evidence has been marshaled to show that much of the new transport infrastructure building chronologically followed enhanced industrial performance and could only have been financed from surplus revenues that enhanced industrial performance generated. Similar debates have emerged in recent years regarding the social and economic returns from public infrastructure investment. The topic is still largely unsolved but there has been a move away from looking at this sort of macro question, with a greater focus now on micro issues concerning individual systems, modes, or technologies.

Nevertheless, the large matter of how transport policy in general is best formulated still remains. Two broad schools of thought dominate the policy debate, each in a way viewing the causality matter in a different way. On one side is the "Continental philosophy" that has much of its basis in concepts underpinning the *Code Napoléon*. Transport systems, within this approach, treat transport as input into a wider sociopolitical–economic framework. Transport is seen as a device for achieving a range of policy objectives and as such should be heavily regulated and controlled. Whether transport infrastructure is provided and operated in a narrow economic manner is less important than its role in meeting wider aims and goals of government. The level of governmental intervention, at various levels, is considerable with state ownership of transport infrastructure and a large part of operations being the norm.

This view of the transport system is in contrast to the "Anglo-Saxon philosophy," which sees transport as just another sector in the economy that should be provided as efficiently as possible in its own right. If all sectors of the economy are involved in market processes then people will be able to consume goods and services that yield them the greatest benefits, and producers will provide what the consumers want at lowest resource costs. This is more akin to the classical economic view of the world. Markets are normally seen to provide a more efficient way of enhancing welfare than excessive government intervention that, either because of lack of adequate information or through manipulation of the system to serve the interest of policy makers, results in losses to society. Policies based on this view of the role of a transport system involve private ownership of the system with access to it determined largely by commercial criteria.

No public policy perspective regarding transport systems falls exactly into one of these schools. They are extremes. The public perception of how to treat transport systems, however, can be characterized as leaning toward one or other of these positions. There are also shifts in the way transport systems have been viewed over time. Although some countries, such as the U.S.A. and U.K., have always had a tendency towards the Anglo-Saxon school, the intensity of this position has varied. The 1930s to the 1970s, for example, saw extensive regulation in the ways that transport systems could be provided and used but more recently measures of deregulation, privatization, and the encouragement of more competitive markets have seen a more laissez-faire approach to transport. France has always been much more of an advocate of the Continental philosophy but even in that country the current trend is towards allowing markets more scope to function freely.

The involvement of government means that somewhat different decision-making tools are required from those used for commercial decisions. This involves modeling (much of which was covered in Volume 1 in this series) but also embraces such things as cost–benefit analysis, issues of environmental protection, and matters of safety. The lack of markets, or efficiently operating markets, also means that there is a need to control the transport system. This may be done through fiscal measures such as road pricing and subsidies or through physical controls over traffic flows and, in the longer term, in the way the transport system is planned.

5. Traffic control

Traffic has probably never been free of controls and regulations. Certainly in Roman times there were rules about when carts could be moved around Rome, but even prior to that, evidence from archaeological sites of real systematic rutting of primitive roads indicates a controlled use of infrastructure. Much of the control was as much for reasons of efficient military and political control as for facilitating commerce and personal activities.

The rules applied to controlling traffic are only partially to do with simple efficiency of the network. In many cases there are also issues of equity and also the meeting of wider social goals. Traffic light sequencing, for example, is seldom designed purely to maximize traffic throughput. If it were then vehicles entering from a side street would have very little priority. Instead the rules represent a trade-off between getting the greatest flow possible and allowing "reasonable" priority for those entering from secondary links.

Recent advances in information technology have provided new opportunities in traffic control. The notion of an intelligent transportation system has become much more of a possibility. This applies not only to road networks, where there are new methods of conveying traffic information, directions, and warnings, but also to areas such as air traffic control and activities such as public transport scheduling. The very way in which information is disseminated across transport system users influences traffic patterns. These advances also offer opportunities for applying control measures that have in the past been largely ignored, in part because of the difficulties of implementation. For example, the tolling of roads can now be done without large numbers of collection points, and electronic congestion charging, whatever the political barriers, is now a technical possibility.

These developments in traffic control have come at an opportune time as many countries face mounting pressures on their existing transportation infrastructure and, for a variety of fiscal, environmental, and practical reasons, have limited scope for capacity expansion. The battle cry of transport policy-makers, although not entirely constraining the construction of new systems or the expansion of existing ones, is that of "managing traffic growth." Consequently traffic control, which was once thought of primarily in terms of making efficient use of an optimal transport system or network, is increasingly being seen as a mechanism for expanding the use of whatever is available irrespective of whether the system is optimal or not.

6. The Handbook

This book is the third in the series of *Handbooks in Transport*. The earlier volumes, as we noted above, have been concerned with transport modeling and with transport logistics and supply-chain management. As with the others in the series the coverage here is meant to be neither comprehensive nor always excessively deep; it is neither a textbook nor a research monograph. It aims to follow the *Oxford Dictionary*'s definition of a handbook, namely a "guidebook" or a "manual." The objective is to furnish the reader with information and concepts in a concise manner. Accessibility is seen as an important requirement of any handbook and every effort has been made to make the material contained here as accessible as possible.

The collection does not contain all of the material that one might anticipate in a volume dealing with transport systems and traffic control. Some topics are not covered in this handbook quite simply because articles on them can be found in companion volumes (e.g., several forms of systems modeling, such as that relating to demand and signal systems, are covered in Volume 1 of the series as well as in this book). Such material was put in these other volumes because they had as much right to be there as to be in this collection. The lines that have been drawn are pragmatic and open to question, but lines did have to be drawn to make the project manageable.

The material here has been organized with chapters grouped to reflect the important elements of modern transport systems and approaches to traffic control. There are more general chapters at the beginning of the volume, where there is more emphasis on transport systems. The later part of the volume is concerned with matters of traffic control.

The chapters are all original and the international collection of authors were selected for both their knowledge of a subject area and, of equal importance, their ability to put on paper in a fairly few words the core of that subject. They have not all gone about their task in exactly the same way but that adds to the richness of the material and reflects the diversity of approaches that can be adopted. It also shows the individuality of the various contributors and their ways of thinking.

The common denominator extending across the contributions is that they try to be up to date in their treatment of topics. The topics themselves were selected to embrace most elements of modern transport systems and to cover the important aspects of traffic control. The topics included in the volume extend beyond domestic issues and the coverage embraces many matters, such as cross-border issues, that are important for efficient international trade.

The authors have approached their task using a variety of tools including modeling, synthesis, and case studies but have aimed at ensuring a reader should be provided with contemporary material, set within a contextual background when that is important. Equally, some of the topics and their coverage are more abstract while others take a much more pragmatic bent. Again this is by design. Further, there is a degree of interlinkage and overlap between some contributions that reflects the artificial boundaries that often have to be drawn in our quest for understanding. Humans are creatures who habitually compartmentalize things but the resultant divisions should be seen as porous.

The readership that contributors have been asked to target is far from homogeneous. Handbooks are aimed at serving both the purpose of keeping individuals informed and of updating those already in the field, whilst offering an opening for those who wish to gain an introductory acquaintance with it. This has by necessity limited the technical components of some papers, although not all, but equally has drawn forth some very articulate verbal accounts of topics. Overall, we think that the contributors have done an admirable job in meeting the challenges that we posed them.

Chapter 2

TRANSPORT PLANNING

DAVID BANISTER
University College London

1. The three great changes

Transport planning has undergone radical change in the last forty years, and it is now barely recognizable from its origins in the highway-building movement and its concerns over increasing the capacity of the system to meet the expected levels of demand. The focus in this chapter is more modest as it concentrates on the recent past and on land-based passenger transport, with a heavy emphasis on strategic and urban planning issues. It reviews the most recent developments in transport planning (1990–2000), and it describes the new agenda and the key role of transport planners in promoting sustainable development. Over this period, there have been three great changes in the requirements placed on transport planning, namely the huge growth in car ownership and congestion, the withdrawal of the state from the provision of transport services through regulatory reform and privatization, and the new environmental debates.

The 1990s have been remembered for the huge growth in the numbers of cars and drivers. The costs of acquisition and use of the car were significantly reduced, with extensive systems of subsidies being provided to many motorists through their employers and through the pricing of transport at levels substantially lower than its full social and environmental costs. Traffic levels in many car-dependent countries have doubled (1975–1995), but the expansion of the infrastructure has been more modest, typically a 10–15% increase in the road network (mainly the motorways). These two underlying trends have resulted in the inevitable increase in congestion and this situation is well illustrated with data from Great Britain (Table 1).

Important conclusions have been reached by most governments in developed countries. Congestion is going to get worse, as the capacity of the network will never increase at a level to match the increase in demand. Even if it was possible to invest in expanding the infrastructure, this is not seen to be desirable for financial and environmental reasons. There are other higher priorities for public expenditure (e.g., health and education), and investment in new roads has often

Handbook of Transport Systems and Traffic Control, Edited by K.J. Button and D.A. Hensher

Table 1
Measures of congestion and traffic in Great Britain (a)

	1977	1987	1997
Cars per km of road	39.4	49.4	58.6
Vehicles per km of road	52.2	62.8	72.9
Passenger km per km of road (thousands)	1 375	1 710	1 939
Distance travelled per person per year (km)	7 536	8 507	10 512
Percentage of all distance by car	70.5	75.6	81.9

Note: (a) Length of road includes trunk and classified roads (about 45%) and unclassified roads (about 55%).

proved to be politically unpopular for many governments. The role of the transport planner has changed from the provider of roads and additional capacity to exploring the means by which the existing capacity can be better used and allocated to priority users (Dunn, 1998).

This links into the second great change. In the past governments have always played a major interventionist role in transport decisions. The underlying philosophy was that transport should be made available to meet the needs of the population and that everyone could expect a minimum level of mobility, almost as a right. To this end, once the basic road network had been established, investment was switched to public transport in the form of capital and revenue expenditure. Very few roads or public transport services were operated by the private sector. This underlying philosophy was common across most countries, perhaps with the exception of the U.S.A., where the market was allocated a much stronger role (Roth, 1996). This has all changed as the market-based approach has now been extended to many other countries and the role of government has been significantly reduced, with market forces determining both the quantity and the quality of transport services.

The new government policy on transport means that services should be provided by the private sector wherever possible, service levels should be determined competitively, not in a co-ordinated fashion, and fares should be market priced. Coupled with these fundamental changes is a move towards a greater precision in defining the objectives of public transport enterprises, particularly financial performance objectives and quality-of-service standards. The traditional role of the transport planner has been radically modified as transport services are now provided by transport professionals rather than planners.

The third great change is the environmental impact of transport and the key role that transport should play in the achievement of targets to reduce global warming, to reduce dependence on non-renewable energy sources, and to minimize the local pollution and adverse social impacts. In the past, it was accepted that transport had an impact on the local environment through unacceptable levels of road accidents, through increases in noise levels, through community severance, through visual intrusion, through vibration from lorries, and through some pollution effects (e.g., carbon monoxide and carbon emissions from the incomplete combustion of fuel). But now, the environmental agenda is much wider and includes the global emissions (principally carbon dioxide), an extended list of local emissions (including small particles which may cause breathing difficulties), the consumption of non-renewable resources (fuel), the use of land, and the impact on the local ecology and ecosystems (Royal Commission on Environmental Pollution, 1994).

These three fundamental changes mean that the traditional role of the transport planner must be reinterpreted so that the new requirements can be accommodated. In the next section the changes are further developed through an assessment of the reactions to the new requirements and the means by which the transport planning process has responded.

2. The new agenda

Transport planning has been completely transformed from an essentially technical activity based on the simple assumption of predict and provide to a much more complex approach that attempts to place limits on mobility through pricing, regulation, and other control strategies. The underlying principle is one of demand management. In addition to tackling congestion in cities and along many interurban routes, transport planning has to address questions of meeting environmental standards, identifying pollution hotspots, and setting and achieving traffic reduction targets, but at the same time ensuring that all people have appropriate levels of accessibility to jobs, services, and facilities. The underlying question here is whether transport planning should do more than just respond to these fundamentally new requirements, and whether it is prepared to take a lead in developing the new agenda.

Transport planning is essential at both the strategic and the local levels, but the nature of the tasks to be tackled has fundamentally changed. In the recent past, policy has been dictated more by ideological concerns about the appropriate roles for the state and the market, but even here a strategic planning framework is still required. This framework is very different from that used in the 1970s and 1980s which was based on the production of large-scale proposals to meet expected shortfalls in capacity. The planning process has become more holistic with the transport elements being linked to housing allocations and the need to maintain regional competitiveness. The process is also more broad based, involving a wider range of stakeholders and affected parties. It has been democratized and become more normative, principally through the introduction of complex objectives such as those related to sustainable development. This contrasts with the more traditional positive approach based on the "scientific" method and the belief that through careful quantitative analysis one could understand the complexity of cities and evaluate a range of alternative strategies to meet expected levels of traffic demand. Transport planning has recognized the limitations of the quantitative approach and accepted the fact that decisions in transport are essentially political, requiring a wider range of both quantitative and qualitative analysis. In addition, there is a maturity in approach which acknowledges the necessity both to explain what is happening in the methods used and to recognize the limitations of these methods (Banister, 1994).

Whether investment is funded by the public or private sector (or jointly), some stability is required in the decision-making framework so that decisions can be made with some certainly, otherwise only low-risk strategies will be adopted. This means that for transport investments, only those projects which are underwritten by the government will actually proceed. If the private sector is to be fully involved in the new generation of investment, then a planning framework is required so that the risk and return levels on investments are reasonably well known.

Predicted increases in travel demand cannot be met. The expectation that new capacity could be provided, either through new construction or through traffic management, to meet that growth has now been firmly rejected. Even if it were possible, it would be undesirable. The land required for new infrastructure is at a premium in many countries, and the environmental and social costs involved make it unacceptable, particularly if a solution is only temporary. In many transport systems working close to capacity, additional increases in that capacity will be immediately taken up with "latent" demand with previous or even worse levels of congestion being quickly reestablished (Standing Advisory Committee on Trunk Road Assessment, 1999).

The new realism in transport planning recognizes that there is no possibility of increasing road supply to a level which approaches the forecast increases in traffic. Whatever road construction policy is followed, the amount of traffic per unit of road will increase, not reduce (Table 1). In effect, all available road construction policies differ only in the speed at which congestion gets worse, either in its intensity or in its spread. The only way forward here is to make demand management the central feature of all transport strategies, independent of the ideological or political stance. Transport planning has always been an intensely political activity, but there does now seem to be some agreement on what needs to be done both politically and professionally. This means that radical action could take place provided that public support can be obtained, and that the necessary skills and organizational framework are in place so that the traditional mold of transport planning can be broken.

A third element in the new picture is a vision about the future of the city in terms of the quality of life and its "livability." In the past, the central concern has been over increasing the quantity of travel, the acquisition of the car, and the notion of the freedom to use that car. As affluence increases, other factors related to the quality of travel, the quality of life, and environmental responsibility become important – values change. Important economic concerns, such as those relating to employment and local economic competitiveness, are still central to policymaking, but other factors relating to the environment and social justice have also become key elements in the search to promote the sustainable city (Williams et al., 2000).

The fairly narrow conceptual framework used by many countries (the Anglo-Saxon approach) allocates a market role to transport with efficiency and productivity being the prime policy objectives. Intervention only takes place where the market is seen to fail or where there need to be adjustments for social reasons. A more Continental approach (the French approach) to transport planning assigns transport the status of an intermediate activity that requires direct control to achieve wider social, industrial, regional, and national objectives. The state still has a central role to play, but policy is driven much more by national objectives (usually normative) that can embrace a wider range of interests, including quality of life.

3. The new challenge

Having made a strong argument for new approaches to transport planning and one that places it firmly in the context of economic, environmental, and social policy objectives of sustainable development, the next step is to determine what the main challenges will be in the new century. The underlying rationale for transport planning has not changed, as its primary objectives have always been to facilitate access to and participation in activities, but the means to achieve these objectives have radically changed. Moreover, these traditional roles have been extended to ensure that all groups within society benefit (i.e., avoiding exclusivity), that broader objectives are met (e.g., on the environment), and that the quality of life for all is maintained and enhanced (e.g., livability of cities). Some of these objectives are likely to be in conflict with each other, and in these cases planners should assess all the evidence before advising decision-makers what to do.

One clear lesson from the "market experiment" in the 1980s has been that the market works well in certain situations, but there is still a need for a clear overall strategic framework within which the market can operate. It is in the creation of this framework that planning has a new role to play. This is really in helping to define the social market and the means by which planning can work alongside the market. Transport planning has in the past tended to limit itself to a narrow conceptual base, looking at problems from the transport perspective and presenting solutions only in terms of transport options. Even within transport itself, the range of options considered has been mainly restricted to pricing and physical measures. Increasingly, problems, analysis, and solutions must be examined holistically as all these questions are part of the same process.

At the strategic scale, there is the market–state relationship which has now shifted substantially towards the market. This shift is likely to remain, but the role of the state has to be redefined as supporting the private sector. New forms of transport planning have evolved, with regulatory organizations ensuring that the newly privatized transport operators are promoting the public interest as well as their own profit- and shareholder-related interests. The role of the state is not only one of control, but also to ensure fair competition and value for money, particularly where public investment is still involved.

Part of the new agenda places environmental issues at the center of planning, which means that economic objectives need to be matched by clear environmental improvement to meet global and local targets. There are tremendous opportunities here for creating solutions that improve both the environment and economic performance. For planning, the role is somewhat more difficult as moves towards sustainable development strategies may result in economic benefits being moderated. For example, in transport there should be clear environmental targets, with charges being related to the full environmental and social costs of each journey made. Environmental audits and best-practice guidelines are now part of the responsibilities placed on companies to promote the minimization of total resource consumption, to encourage recycling, and to reduce the levels of environmental damage (Maddison et al., 1996).

Underlying much of the debate at the strategic scale is the public acceptability of the proposals. To have a real effect on the behavior of firms and individuals, there is the considerable task of convincing them that change is necessary – commitment often falls short of real action. No matter how attractive public transport is, no matter how expensive petrol is, no matter how close facilities are located to the home, people will still use their cars. Policy levers such as pricing and control may have only a limited effect on actual behavior. To hold any expectations that reality is different misinterprets the dependence of current lifestyles on the car and the perceived freedom it provides. One real challenge to the transport planner is to convincingly argue for cars not to be used and for people to accept that argument and to leave their cars at home. It is widely recognized that the acquisition of the car is the most important single factor in changing travel patterns – from the destinations used, to when the trip is taken, to the trip lengths, and to the forms of transport used. The new issue here is that where car ownership levels are universally high, how can transport planners actually influence travel patterns (Banister and Marshall, 2000)?

One interesting possibility is the notion of traffic degeneration. As discussed earlier, we have argued that additional capacity is likely to increase demand. The counterargument is that if we reduce capacity, some travel will be lost. If the use of the car is restricted in city centers (e.g., by traffic calming) or through land use and development policies (e.g., by concentrating on city center development), or through the use of technology (e.g., by telecommuting), will travel be reduced? Similarly, in the past the concern has primarily focused on the amount of travel, but now quality of travel is also important. So if that quality declines, will people reassess the necessity to make a journey? Underlying these key questions is an understanding both of public acceptability and of receptiveness to radical change, and their behavioral response.

Radical alternatives are required. Transport systems management was used in the 1970s to increase the capacity of the road network through low-cost schemes such as area traffic control, restrictions on parking, and extensive one-way systems. This was followed by demand management in the 1980s to promote high-occupancy vehicles, new public transport systems, parking controls and pricing, and extensive pedestrianization and calming schemes. In the 1990s, there have been more demand management schemes (e.g., park and ride, bicycle priority, central area management, access restrictions) and a reliance on technological solutions (e.g., route guidance, parking guidance). But there is an institutional reluctance to implement a road pricing policy or to use the planning system to limit the growth of traffic at source (e.g., through clear strategies on sustainable development).

As a result of this inertia at the strategic level, most of the action has taken place at the city and local levels, where responsibilities are clear. It is also at this level that many of the real effects of congestion, inefficiency and poor environmental quality are actually felt. Greater car dependence and higher levels of mobility result in increased congestion as the capacity of the transport system fails to respond. Cities become less attractive places in which to live as decentralization takes people and jobs to peripheral car-accessible locations. To reverse these trends means that cities must become attractive locations for investment, with affordable housing and high-quality facilities and amenities. They must be seen as safe, secure, and pleasant places to live. Transport plays a pivotal role in achieving such a city, and although the car may still be essential, it must be seen as promoting cities rather than as one of the main reasons why cities have become hostile environments (Tolley, 1997).

What then are the options for transport planners in cities? The car is an inefficient user of road space and this must be realized and accepted by all, through charging for parking and for using road space. Smaller city vehicles (including clean fuels) offer some potential, but this does not solve the problem of space. Parking controls are the main means available to limit the use of the car in urban areas. The bus (or tram) is the most efficient user of city road space, yet sharing that space with the car reduces efficiency. The urban road network could be designated for particular uses. For example, in the city center 30% of roads could be allocated to pedestrians and cyclists, a further 30% to public transport and access only, and the remaining 40% to general use (Banister, 1994). The proportions would not be uniform across the city, but would relate to the dominant land user. The implementation of such a scheme could take place immediately, and the capacity and quality of the bus system would be dramatically increased as scheduled operating speeds would be improved. There would be fewer cars in the city center and the environmental benefits would also be significant. Segregated bus networks would then become a reality.

Traditionally, the cases for both pedestrian areas and traffic calming have been argued on safety and (more recently) environmental criteria. More extensive schemes would be established in residential areas and quality neighborhoods could be created where there was no access for polluting vehicles. All travel would be by soft modes (e.g., walk and cycle) or through electric delivery vehicles or clean public transport. In each case schemes should be seen as being a part of a more general area-wide strategy, as the concern is to reduce traffic overall (degeneration) rather than to divert traffic elsewhere.

Apart from actions on individual forms of transport, there needs to be co-ordination between all forms of transport. Included here are parkway stations and park and ride facilities at peripheral sites or at suitable interchange points, where car users can switch to regional and suburban rail, tram, or bus services. Car sharing can be encouraged through high-occupancy vehicle lanes, and all employers should develop company transport plans to reduce car dependence (e.g., for firms, schools, universities, hospitals, local authorities, etc.). Investment is still required to upgrade and expand the existing public transport infrastructure. There is no lack of ideas here and many cities have addressed these problems, some with very innovative solutions. Similarly, a new optimism can be seen in investments in the high-speed rail networks (Whitelegg, 1993). But in many situations, there is an absence of co-ordination between the different schemes and in the financing of all schemes. It has proved almost impossible to get the private sector to invest substantial capital in transport infrastructure, even in partnership with the public sector. In addition, the availability of traditional forms of public sector finance have been limited and are unlikely to be renewed, at least to the same levels as before. New financing mechanisms are required in the form of low-interest government-backed loans (as in France) or through public bonds (as in New York), or through creative forms of partnership finance with the public and private sectors. This again is a real challenge to transport planning.

4. Into the new millennium

In the longer term, the most important contribution from the transport planner is in the form of a real integration of land use and development decisions, together with an assessment of their transport impacts. Decisions now being made on the allocation of new housing to meet the demands for smaller households, single-person households, and an ageing population will determine levels of traffic generation in the future. Similarly, the location of workplaces, shops, services, and facilities will all influence the journey lengths and forms of transport used in the future. This integration will actually tackle the transport problems at their source, namely where the travel is generated, and it has a substantial contribution to make to the sustainable-development agenda.

Location strategy, together with clear analysis of density, settlement size, facilities, and services, and mixed-use development, would all enhance accessibility and provide a choice for individuals not to use their cars. The solutions for transport planning in cities, particularly city centers, are clear, through the creation of high-quality environments at intermediate densities with efficient and attractive public transport – there is no reason to own or use the car. But the solutions in the suburbs, in rural areas, on the interurban networks, and to the growth in international travel are much harder. In each case (except the last) the role of the car is crucially important and some people seem to be prepared to pay a high price for transport so that they can live in low-density locations (Williams et al., 2000).

Transport planning must take the lead in understanding and analyzing these problems, and placing them in the context of environmental and social concerns. Planning should seek to reduce the environmental impact of transport, both at source (through technological innovation) and on the road (through pricing and regulation). Equally important is the necessity to provide access to jobs, services, and facilities for all people, particularly those on lower incomes without access to a car. But accessibility is not just a matter of transport. Accessibility is having local facilities, a strong social network, and the means to communicate with others, as well as the means to get there and the necessary resources. Closure of local facilities often results in savings to the producer, but the costs are passed onto the users as they have to travel further (by car) to get to these centralized and specialized facilities. Transport planning must balance the requirements of the market with those of the users, particularly those who are likely to have problems of accessibility. There is a substantial potential here for the new technology to break down many of these barriers, but in reality more barriers may be raised as the inaccessible are also those with the lowest (potential) access to the internet. This may provide the greatest new challenge to transport planners, namely to establish the means by which technology can actually replace car-based travel, but at the same time ensuring high levels of access by all to the new technology.

For example, in rural areas where the quality of public transport is limited, there are substantial opportunities for transport innovation through community transport, social car schemes, and flexible, shared taxis. Technology now provides additional flexibility in the scheduling of vehicles and in being able to respond to requests almost instantaneously. There is a substantial potential synergy between transport and technology in rural areas that needs to be exploited. Even though local facilities have been closed and many forms of rural services are under threat, it may be the new technology that comes to the rescue. Goods can now be ordered over the internet, but one of the main problems is the distribution of those goods to people's homes, particularly when people are out at work. The local shop or post office could become the local internet distribution center where you could collect your goods when it is convenient. For the supplier, this offers the opportunity for less distribution and a guaranteed point of delivery. This is a win–win situation and may allow uneconomic rural services to have new life.

As we move into the new millennium, how will our three great changes continue to evolve? There will also be a reinterpretation of priorities as little new infrastructure will be built, due to the political unpopularity of roads and the need to use available resources to replace much of the existing infrastructure. Most investment will be in the public transport infrastructure and in the means by which technology can be used to maximum effectiveness in squeezing more capacity out of the available transport system. Transport planning will be transformed and take a leading role in providing a link between the different public and private sector organizations so that the full implications of decisions taken in any particular sector can be interpreted. There must be a balance between the narrow market-driven economy, and the broader values assigned to environmental and social priorities. There is a huge opportunity for transport planning to become a key player in the sustainable-development debate, as transport continues to be a major user of non-renewable resources and a polluter of the environment. Through the use of technology and planning controls, cities can be made clean, desirable, and livable. There is also scope for new methods of environmental evaluation and auditing to become central in all planning, together with new responsibilities for monitoring and achievement of challenging environmental targets – this is the vision of the sustainable city.

Transport needs basically depend on where people live and work. But more importantly, with the new patterns of household structures, with the aging population, with the growth in leisure-based activities, and with the known and unknown effects of new technology, this will all change. Added complexity arises out of the interaction effects of how quickly or conveniently people can travel even over short distances, as this influences where and how they work and live, and this in turn is affected by the other factors mentioned above. If people change how and where they work and live, and what they do in their leisure time, different transport needs will arise. This interdependence continues to become more complex with time, providing a range of exciting new challenges to transport planning.

References

Banister, D. (1994) *Transport planning: In the UK, USA and Europe*. London: Spon.
Banister, D. and S. Marshall (2000) *Encouraging transport alternatives: Good practice in reducing travel*. London: The Stationery Office.
Dunn, J.A. (1998) *Driving forces: The automobile, its enemies and the politics of mobility*. Washington: Brookings Institution.
Maddison, D., D. Pearce, O. Johansson, E. Calthorp, T. Litman, and E.Verhoef (1996) *The true costs of road transport*. London: Earthscan.
Roth, G. (1996) *Roads in a market economy*. Aldershot: Avebury Technical.
Royal Commission on Environmental Pollution (1994) "Transport and the environment", HMSO, London, Cm 2674.
Standing Advisory Committee on Trunk Road Assessment (1999) "Transport and the economy", Department of the Environment, Transport and the Regions, London.
Tolley, R. ed. (1997) *The greening of urban transport*. Chichester: Wiley.
Whitelegg, J. (1993) *Transport for a sustainable future: The case of Europe*. London: Belhaven.
Williams, K., E. Burton, and M. Jenks, eds. (2000) *Achieving sustainable urban form*. London: Spon.

Chapter 3

TRAFFIC REDUCTION

PHILLIP B. GOODWIN
University College London

1. Introduction: why did the idea of traffic reduction become important?

During the 20th century, car ownership and use grew from virtually nothing to become the dominant method of personal transport, with many benefits to those enjoying the greater mobility, speed, comfort, and convenience which cars could provide. Similar developments occurred for freight transport. In turn, these trends led to changes in the structure and functioning of cities (encouraging a tendency to "sprawl," with the main growth mainly in suburban areas and a more diffuse pattern of origins and destinations), and to major commitments of funds, mostly public, as networks of roads were constructed and extended to deal with the extra traffic.

During this period, the most influential thinking among transport planners and political bodies was influenced by two main presumptions:

(1) Such a growth in mobility was, in general terms, to be encouraged rather than discouraged, but was in any case largely outside the scope of any policy intervention: it was viewed as an autonomous and inevitable trend, driven primarily by the free choices of individuals, and enabled by the steady growth of real income that all countries sought to provide.
(2) Therefore the main objective of policy in national and local government was to accommodate this growth in as civilized and efficient a manner as possible, providing sufficient road capacity to guarantee reasonably free movement, and control systems which would manage the resulting traffic at acceptable standards of safety and impact.

However, the further the trends proceeded – which they did, to a greater or lesser extent, in all countries – the more weight was given to an alternative view, which tended to emphasize the mounting negative consequences of the same trends, especially concerning congestion and its effects on economic efficiency, environmental damage, safety, and social inclusion. In some countries (notably in Europe – Goodwin, 1999) such a shift of emphasis was expressed in national

Handbook of Transport Systems and Traffic Control, Edited by K.J. Button and D.A. Hensher

policy debates, while in others it was confined to metropolitan areas where traffic congestion was most extensive, or to historic cities where it was most apparent (Department of the Environment, Transport and the Regions, 1998). In both cases, policy advocacy developed in parallel with changes in some underlying technical assumptions by professionals, changes in public opinion as expressed in polls, elections or grass-roots campaigning, and changes in the thinking of some politicians both about strategic policies and also for electoral advantage.

In summary, there were four factors which underpinned this change of view. First, the growth in traffic always seemed to outpace the provision of road capacity, no matter how extensive (and expensive) the programs of public works. So individuals and businesses were spending more and more of their resources stuck in traffic jams – or so it seemed to them – which resulted in increased business costs and reduced quality of life, especially when public transport systems, faced with falling demand, reduced their quality of service. There was a question about whether it would even be *possible* to provide sufficient road capacity to keep pace with the traffic, particularly when, after a technical dispute lasting over 20 years, it was finally professionally accepted* that the provision of improved infrastructure can itself generate extra traffic (see Standing Advisory Committee on Trunk Road Assessment, 1994, and Chapter 9).

Second, the transport sector became identified as one of the major causes of environmental damage, both in terms of land take for road building and also in terms of emissions, with local, regional, and global effects, and use of fossil fuel, noise, and disruption of communities (see, e.g., Royal Commission on Environmental Pollution, 1994).

Third, traffic accidents (although rarely increasing at the same pace as the traffic itself) nevertheless, at a world level, became higher and higher in the lists of sources of violent death or disablement, greater in importance than many of the world's diseases. In local areas, even where mounting congestion reduced the speed and therefore severity of accidents, there was an increasing sensitivity resulting from rising expectations, especially in residential areas.

And fourth, there was an unstable but significant shift in the public mood. In the early stages the opening of a new motorway had been a unifying and popular symbol of progress, welcomed perhaps not by absolutely everybody, but certainly by majorities large enough to give politicians confidence that this was a way to re-election. But increasingly, new motorways became symbols of dissent and division, with demonstrations, and sometimes with quite bitter election battles, especially where roads involved demolition of houses or loss of green space.

Not all these factors were of equal importance in all countries, but to some degree they occurred everywhere. International bodies also increasingly

*There are still some professionals in the U.S.A. who do not accept this proposition.

considered and gave weight to these considerations. At the heart of a developing new agenda was the question: *is it really true that traffic growth is an inevitable and unstoppable trend – or is it possible that policy intervention to stabilize, or even reverse, the trend, could be successful and desirable?* If this could be done, then it might be possible to achieve a better quality of travel, more efficient use of resources, and less damage to the environment, life, and limb – which might then be more politically popular than the present controversies.

It should be said that not all the reconsideration resulted in substantial policy change. A number of countries have from time to time discussed the possibility of seeking traffic reductions at the level of the whole country, notably in Scandinavia, the Netherlands, and the U.K. But at the level of an entire nation, no country has yet achieved a lasting and large-scale downturn in the total volume of traffic, and indeed none have yet actually tried to do so, in a sufficiently determined manner, sustained for long enough, to achieve unambiguous results. Such discussions have typically evolved into intentions to "reduce the rate of growth of traffic," which is obviously in principle easier to achieve, though it is much more difficult to prove that it has occurred (since it depends on demonstrating "what would have happened otherwise," which is inherently uncertain). But there is much more experience at local level, including many successes and a developing theory and practice. For these reasons, the issue of "traffic reduction" is much better established as a local question than as a national one, at the present time.

2. The instruments and objectives of traffic reduction

There have been five main developments, mostly at local or regional levels, relevant to achieving reductions in traffic.

2.1. Town center pedestrianization

Although some cities, especially in the 1950s to the 1970s, undertook major new road building right to and into the central areas, these were often controversial (unlike the almost unchallenged popularity of interurban roads in the early days), with early signs, for example in London and in some German cities, of demonstrations that later became much larger. This was partly because the destruction of the existing urban fabric was so obvious and so extensive (the phrase "the planners have done more damage than the bombing in the last war" was ubiquitous in many European countries), and partly because the technical merit of building roads near the center of large cities was more dubious: theoretical work by traffic scientists had proved that the larger the city, the less successful would be attempts to provide for car-based mobility (e.g., Smeed, 1968), and theoretical work by economists had suggested these were the areas where markets were most distorted and therefore revealed demand to be the least reliable signal for beneficial investment (e.g., Solow, 1973).

In the event, however, these arguments were sidelined by a powerful movement in civic planning, most strongly established in Germany and its neighboring countries (Hass-Klau, 1990). This was a movement at city level, especially, in the early stages, in those cities with a strong and culturally important historic city center, a medieval or Renaissance street pattern incapable of dealing with heavy traffic, and an attractive urban environment of squares, beautiful buildings, and untouchable monuments. The view developed that such centers would be more attractive if, instead of providing for traffic growth, traffic was simply banned.

In most cases, such ideas were controversial, with opposition mainly from two sources, namely from local traders who feared that restrictions on traffic would lead to loss of trade, and from some traffic engineers who feared that restricting traffic in some streets would cause insupportable stress in other streets, often described as "traffic chaos." (Resistance was especially strong among those traffic engineers who relied on a form of computer model that assumed that changing the choice of route was the only, or main, response that drivers could make to changes in road conditions – a fundamentally misleading but widely applied assumption.) In addition, there were practical problems – what to do about deliveries of goods to shops, what to do about the cars owned by people who actually lived within the restricted areas, where to draw the boundaries, etc.

There is no particularly persuasive theory to solve these problems, but by now there is more than a quarter of a century of practice, and at least some of them are now solved (Hass-Klau, 1993, Carley and Donaldsons, 1997). The provision of a good-quality pedestrian-only space in the heart of a city center is now so widespread and so popular that it can no longer be treated as an experiment: these places manifestly work, deliver commercial and cultural success, and win votes. In the largest areas, there are usually special arrangements to enable public transport vehicles to enter the restricted streets; delivery lorries are allowed in at specified times of the day, usually early mornings (and sometimes cars are allowed in during the evenings). Some cities have provided inner ring roads to accommodate a proportion of the displaced traffic; other cities have decided not to. Both approaches seem to work.

For these cases, it is quite clear that traffic can be reduced substantially in a specified area, with desirable and popular consequences, and no impossibly difficult side effects.

There has been some argument about whether such schemes reduce traffic overall, or just redistribute it. A review of a large number of such schemes by Cairns et al. (1998) suggested that while some of the displaced traffic did reappear on other streets outside the restricted area, not all of it did. Taking the town center and the surrounding inner-area streets together, there was to some degree an overall reduction in traffic, although the amount varied widely according to the local circumstances. Considering the town as a whole, the evidence is unclear: town centers are very important for the cultural and economic life of a city, but they usually take a relatively small proportion of the total traffic in a town, and even quite large changes may not be visible when averaged over the whole town and its hinterland.

Figure 1 (Cairns et al., 1998) shows the range of results seen in a sample of case study areas, the evidence being supported by other cases where the reduction in capacity was not an act of policy but resulted from a natural disaster, extended road works, or other causes. In summary, of the cases studied, there was an average reduction in traffic on the affected road or area of 41%, of which rather under a half reappeared on neighboring streets or alternative routes, the remainder "disappearing." Thus the overall average reduction in traffic was 25%. This figure is influenced by a few extreme cases: the median result showed that 50% of cases showed traffic reductions of over 14%. In should be stressed that in every case of this sort, there are always special local factors and caveats, spelled out in detail in the source references.

2.2. Residential-area traffic calming

Alongside interest in banning traffic from town centers, there has been increasing pressure to "tame" traffic in residential areas, using a wide variety of engineering and psychological measures (speed humps of different sizes and designs, chicanes, culs-de-sac, road surfaces, street furniture, signs, trees and plants, etc.). These are usually intended to reduce through traffic, especially due to people using the residential area as a rat-run to avoid congestion on the main roads, and to "shift the balance of power" between the remaining vehicles and pedestrians, especially by reducing vehicle speeds, sometimes even to walking pace. The prime motivation is usually safety.

Apart from the through traffic, this approach is usually more concerned with the behavior of vehicles than the number, in the interests of a safe and pleasant living environment. But, especially where the road network is changed with one-way systems, and no-entry or access-only streets, a side effect is usually to reduce the volume of traffic within the treated area, as well as to calm it down. Again, there is a technical argument about whether all the displaced traffic simply reappears on other roads or not. The evidence is less clear-cut than in the case of town center pedestrianization (because the schemes are usually more modest in scale, and fewer detailed monitoring studies have been carried out), but there are some indications of a small reduction in traffic overall.

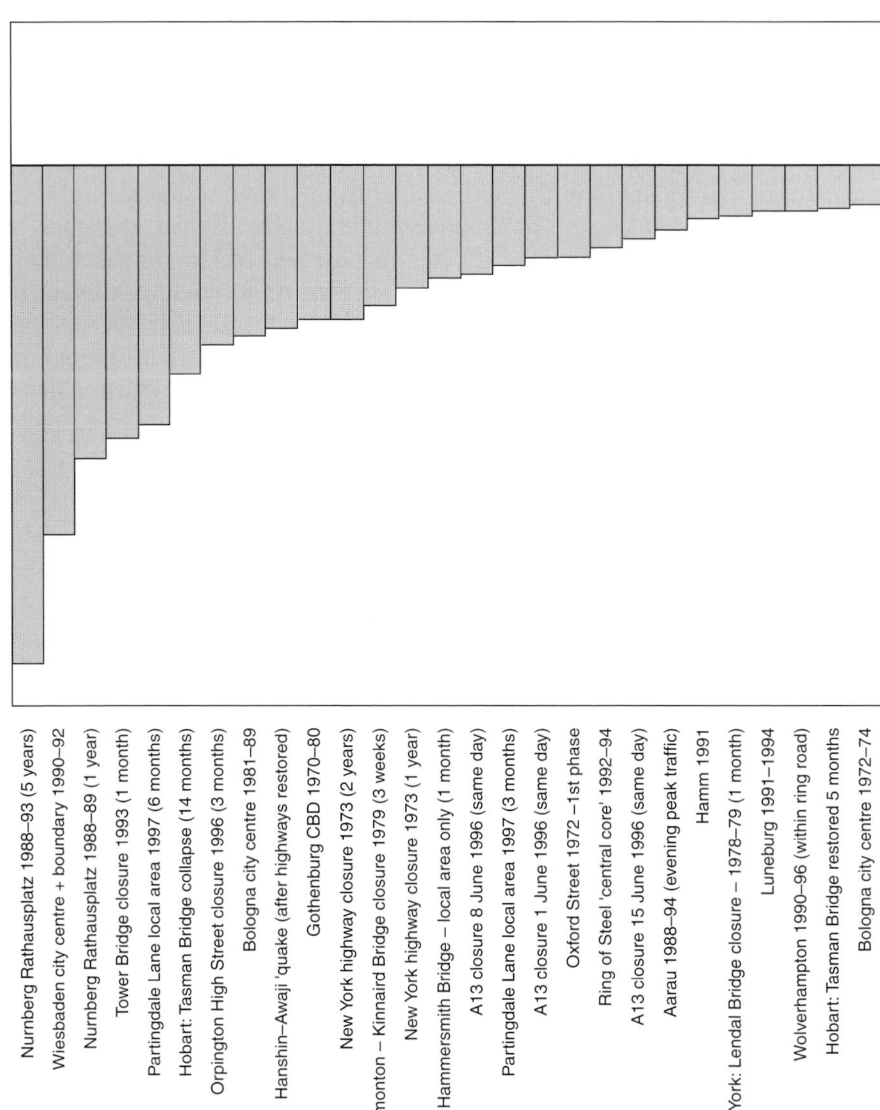
Nurnberg Rathausplatz 1988–93 (5 years)
Wiesbaden city centre + boundary 1990–92
Nurnberg Rathausplatz 1988–89 (1 year)
Tower Bridge closure 1993 (1 month)
Partingdale Lane local area 1997 (6 months)
Hobart: Tasman Bridge collapse (14 months)
Orpington High Street closure 1996 (3 months)
Bologna city centre 1981–89
Hanshin–Awaji 'quake (after highways restored)
Gothenburg CBD 1970–80
New York highway closure 1973 (2 years)
Edmonton – Kinnaird Bridge closure 1979 (3 weeks)
New York highway closure 1973 (1 year)
Hammersmith Bridge – local area only (1 month)
A13 closure 8 June 1996 (same day)
Partingdale Lane local area 1997 (3 months)
A13 closure 1 June 1996 (same day)
Oxford Street 1972 –1st phase
Ring of Steel 'central core' 1992–94
A13 closure 15 June 1996 (same day)
Aarau 1988–94 (evening peak traffic)
Hamm 1991
York: Lendal Bridge closure – 1978–79 (1 month)
Luneburg 1991–1994
Wolverhampton 1990–96 (within ring road)
Hobart: Tasman Bridge restored 5 months
Bologna city centre 1972–74

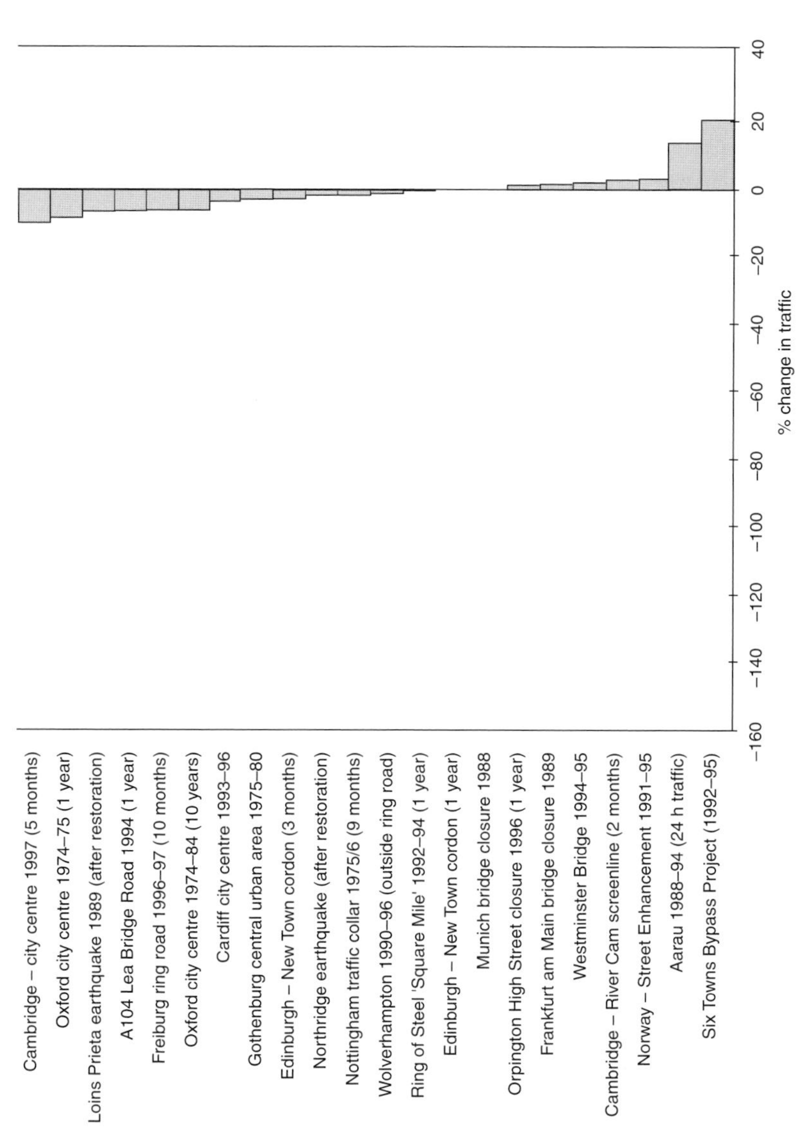

Figure 1. Changes in traffic flows after capacity reductions, shown as a percentage of previous traffic on the affected roads (Cairns et al., 1998).

2.3. National parks and tourist attractions

With the rapid increase of car-based tourism, particular pressure has often been placed on the most attractive areas such as parks, forests, lakes, sea shores, and mountains, whose attractiveness typically resides in precisely the attributes of tranquillity, cleanliness, visual beauty, plants, and animals which are threatened by the visitors themselves, if there are too many of them. Such schemes have sometimes been popular even in countries where traffic reduction in general is not thought of highly, such as schemes in the Grand Canyon National Park, and proposals in the Yosemite National Park, in the U.S.A.

A number of different strategies have been adopted to cope with this problem, including very large-scale development which, at least for a period, is financially attractive.

Some areas have deliberately sought to preserve the attractive features of the area by reducing the number of visitors to a sustainable level, and encouraging those that come to use environmentally less damaging means of transport. This usually involves restricted parking availability (and therefore control of illegal parking), restrictions on use of particularly intrusive vehicles such as beach buggies, four-wheel drive off-road vehicles, etc., encouragement of access by tourist coaches rather than cars, and provision of tourist centers from which guided tours or individual walking is promoted. Discreetly marked networks of hiking routes are common. Some islands and national parks restrict vehicle access entirely. There have even been plans to create substitute destinations (artificial monuments, for example) to protect the fragile real ones.

Currently, forecasts of the growth of international air travel, especially for tourism, are among the highest growth rates forecast for any area of economic activity (one recent forecast speaks of a 28-fold growth internationally in air travel over the next 50 years, for which it is difficult to imagine how an infrastructure of transport and destination management could possibly be created). Therefore the interest in controlling traffic levels in these honey pot sites is also growing quickly, often using metaphors of tourism as a plague of locusts moving from site to site and destroying each in turn. While such language may be seen as political rhetoric rather than helpful analysis, the underlying problem to which it relates is entirely genuine.

2.4. Transport pricing: parking, fuel taxes, and road user charges

From economic theory, we know that if prices are not correctly adjusted to costs, there will be some distortions in the resulting choices which reduce economic efficiency and overall welfare. This has been one of the technical criticisms of the "traffic growth is inevitable" school of thought: the argument is that traffic growth has been higher than it should have been, because people have not been faced with paying for the full costs (such as congestion and environmental damage) resulting from their choices.

For this reason, there has been much interest in using prices as an instrument of transport policy (see Dargay and Goodwin, 2000), the approach occurring in one of two forms, i.e., either "get the prices right: in cases where travel is currently undercharged, this will result in a reduction in traffic" or "let's decide how much traffic we want, and then use prices to achieve it." One of the great advantages of using price as a mechanism in this way, compared with physical restrictions, is that to do so provides considerable amounts of revenue, which can then be used to make whatever other improvements are technically merited or politically popular. A disadvantage, however, is that as in any use of prices, there are problems of fairness and equity: a ban which applies equally to everybody is often perceived as fairer than using prices, which will affect some people more than others.

The use of pricing, combined with regulation of the number of parking spaces available, is now very widespread in busy areas. In most cases, authorities decide (within the powers open to them) how much parking should be allowed for a building, street, or area, restrict the number of parking spaces to this, and then choose a price which brings about a comfortable level of utilization of those spaces. There is scope for trial and error in setting the charges, necessary because drivers' sensitivity to the price level is very variable between areas. It is generally thought that the combination of parking control and pricing is a very powerful tool indeed for influencing the number of vehicles attracted to an area, and with rapid effects on reducing traffic where desired. But there are limitations, especially because this does not affect through journeys (which may indeed increase, if a reduction in parking vehicles releases some road space for other traffic), and because there is usually a proportion of parking spaces not subject to public authority control, both legal and in some cases illegal. Even in those dense urban areas where parking is scarce and expensive, direct or indirect employer subsidies can encourage greater car use than would otherwise be warranted.

Fuel prices also have an influence on traffic levels, whether this is an intended result of deliberate policies or an accidental side effect of price changes due to aims of revenue collection or simply the operation of the international oil market. When there are policy discussions about traffic volumes at the national level, fuel price is often the focus of attention, because it will cover the whole country and because it is within the control of governments. But it is difficult to sustain large price differences between the price levels in neighboring countries, due to legal or illegal cross-border purchases.

It used to be the case that the effect of fuel price was thought to be very insignificant. This was partly because the dominant trend was for increases in traffic, and this seemed to overwhelm the effects of changes in fuel price. However, in recent years there has been much research on this question (see, e.g.,

Glaister and Graham, 2000). It is clear that there are small, but very significant effects of changes in fuel price – a sustained 10% real increase in fuel price will reduce traffic levels by about 3% below what they would otherwise be (and reduce fuel consumption by about 7%). But one of the problems is that this effect takes some years to build up, only half or less of the eventual full effect being seen in the first year or so after a change. It is this early period during which most attention is usually focused on the results, hence the widespread feeling that there is little useful effect.

The pricing tool which is now widely discussed, but so far with relatively little practical experience, is road-user charging, either using permits to enter a designated area, or using technically more advanced payments systems analogous to taxi meters within cars (see Chapter 7). Theoretical studies indicate that such systems would have a significant effect on reducing traffic within the priced area (figures between 10% and 30% are common, depending on assumptions about the price level, etc.), but do not agree on the effects on traffic levels outside this area.

2.5. Improvements in alternatives

A very widespread practice is to seek to reduce traffic levels, especially by car, by making the alternatives to car use more attractive, notably by bus priority systems, investment in new light rail networks, facilities for cyclists such as cycle lanes and safe lock-up storage areas, wider pavements or more favorable traffic signals for pedestrians, park-and-ride services, and price reductions supported by subsidy either for all users or for particular groups. A similar logic applies to objectives of shifting a proportion of freight movement from road to rail.

These approaches may be divided into two types, "carrots," and "carrots and sticks." The carrots are cases where the alternative is improved but without any deterioration in the attractiveness of car use. These cases, if well done, usually do find a significant increase in use of the improved method of transport, but little or no detectable reduction in traffic levels – partly because the increased use is not brought about by diversion from the car, and partly because even when it is, the road space freed up by such diversion is then filled by other car users or traffic growth generally. The "carrots and sticks" are those cases where the act of improving public transport, for example, is simultaneously connected with making car use less attractive. This especially applies to a whole range of techniques of reallocation of existing road capacity among the competing users, for example in bus lanes, or street-running tram systems. In these cases there is often an overall reduction of traffic level, usually rather small because the amount of road space reallocated is itself rather small (with the exception of large-scale pedestrian areas as discussed above).

3. Overall assessment

Policy instruments designed to reduce traffic levels, or doing so as a side effect of other objectives, are now firmly established as an important strand of transport planning. The degree of success is limited by the strength of trends leading to increased traffic, and by the fairly early days in developing tools and experience.

Success is very dependent on the specific context, being most advanced in developed countries with long-established historic cities, but also of relevance in many developing countries, especially those with rapid economic growth and very swiftly expanding cities – and even discussed from time to time in areas where plentiful supply of space and low-density development create the possibility of larger-scale car use. At the present stage, the most successful examples are the best town center schemes, where all except a small, manageable minimum of service and residents' traffic can be removed with great benefits to the commercial and social success of the center. Most planners involved in this activity emphasize that one measure on its own is very unlikely to be successful, no matter how stringent: best practice invariably talks of a combination or "package" of traffic bans, public transport improvements, supporting land use policies, and consistent price signals.

At the level of a whole city, current discussion often talks of the possibility of reducing traffic volumes by 10–30% for the largest cities, with smaller reductions for the smaller cities and maintaining current traffic levels in other towns. While it is agreed that this could not be achieved without sustained, long-term, multistranded policies, there is not yet agreement on the appropriate level to aim at, and the political and electoral consequences of doing so. Both professional and public attitudes are changing rapidly, and it is certain that this will be a controversial, important, and illuminating area of transport practice. The current discussion has also highlighted the importance of taking a wide view of how people respond to new traveling conditions: resulting choices are more complex, and take longer, than is often suggested.

References

Cairns, S., C. Hass-Klau and P. Goodwin (1998) *Traffic effects of highway capacity reduction*. London: Landor.

Carley, M. and Donaldsons (1997) "Sustainable transport & retail vitality: State of the art for towns & cities", Historic Burghs Association of Scotland and Transport 2000.

Dargay, J. and P. B. Goodwin (2000) *Changing prices: The role of pricing in travel behaviour and transport policy*. London: Landor.

Department of the Environment, Transport and the Regions (1998) *A new deal for transport: better for everyone*. London: TSO.

Glaister, S. and D. Graham (2000) *The effects of fuel prices on motorists*. Basingstoke Automobile Association.

Goodwin, P. B. (1999) "Transformation of transport policy in Great Britain", *Transportation Research A*, 33:655–699.

Hass-Klau, C. (1990) *The pedestrian and city traffic*. London: Bellhaven.

Hass-Klau, C. (1993) "Impact of pedestrianisation and traffic calming on retailing: A review of the evidence from Germany and the UK", *Transport Policy*, 1 (1):21–31.

Royal Commission on Environmental Pollution (1994) "Transport and the environment", The Stationery Office, London, Cm2674.

Smeed, R. J. (1968) "Traffic studies and urban congestion", *Journal of Transport Economics and Policy*, 2(1):33–70.

Solow, R. M. (1973) "Congestion and the use of land for streets", *Bell Journal of Economics and Management Science* 4(2):602–618.

Standing Advisory Committee on Trunk Road Assessment (1994) *Trunk roads and the generation of traffic*. London: HMSO.

Chapter 4

EQUITY VERSUS EFFICIENCY IN TRANSPORT SYSTEMS

CHRIS NASH
University of Leeds

1. Introduction

Most textbooks on welfare economics see the issue of equity as being a central issue (e.g., Johanson, 1991). However, this issue is less often explicitly recognized when it comes to applied studies of transport economic issues. Yet it is clearly a central feature of many debates over transport pricing and investment decisions.

In this chapter we will first discuss the two concepts of efficiency and equity, and then consider their application to transport pricing and appraisal decisions. We will then consider explicitly the concept of social exclusion and its relevance for transport decisions before reaching our conclusions.

2. Efficiency versus equity – key concepts

Before discussing the conflicts between equity and efficiency in the transport sector, it is necessary to clarify these concepts. To do so it is necessary to introduce a little economics. For those with an economic background, this section will be very straightforward, but for others it is important to make the effort to understand these principles.

The concept of efficiency in general relates to the idea of maximizing output per unit of input. In economics it has a more precise meaning. Specifically it refers to the achievement of a situation in which it is impossible, with the resources available, to make one person better off without making another worse off. This situation is known to economists as being a "Pareto" optimum, after the Italian economist Vilfredo Pareto. More practically an improvement in economic efficiency is said to occur whenever a change is made, the benefits from which could be redistributed in such a way as to make at least one person better off, without making anyone worse off – a "so-called" potential Pareto improvement in welfare. Clearly this occurs whenever those benefiting from the change are willing

Handbook of Transport Systems and Traffic Control, Edited by K.J. Button and D.A. Hensher

to pay enough for the benefits to be able to fully compensate the losers and still be better off themselves. This is the compensation test that forms a basic principle of cost–benefit analysis.

Figure 1 illustrates these concepts for a very simple economy comprising two individuals (or identical groups of individuals). The curves show the maximum amount of utility that person 1 can obtain for any given level of utility for person 2. The inner curve shows this for the existing situation, whereas the outer curve shows it for a situation in which a transport project is implemented that improves economic efficiency. Anywhere on the outer curve is a Pareto optimum, since person 1 can only be made better off at the expense of person 2 (we are assuming that there are no other projects which could be undertaken and which would improve economic efficiency). On the other hand a point on the inner curve, such as point A, is not a Pareto optimum, since by moving to point C both people can be made better off – the move from A to C is a Pareto improvement in welfare. The move from point A to point B is not a Pareto improvement in welfare, since it makes person 2 worse off, but by implementing the project and redistributing some of the benefits from person 1 to person 2, a position such as C can be reached and both people can be made better off. The move from A to B is therefore a potential Pareto improvement in welfare.

These concepts translate directly into simple rules for pricing and project appraisal, if economic efficiency is derived. Regarding pricing, the implication is that goods should be supplied as long as those receiving them are willing to pay at least the marginal social cost of providing them, where marginal social cost

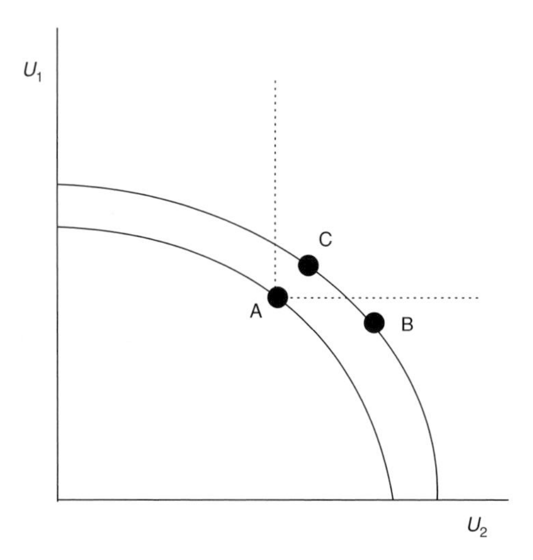

Figure 1. Pareto optimality.

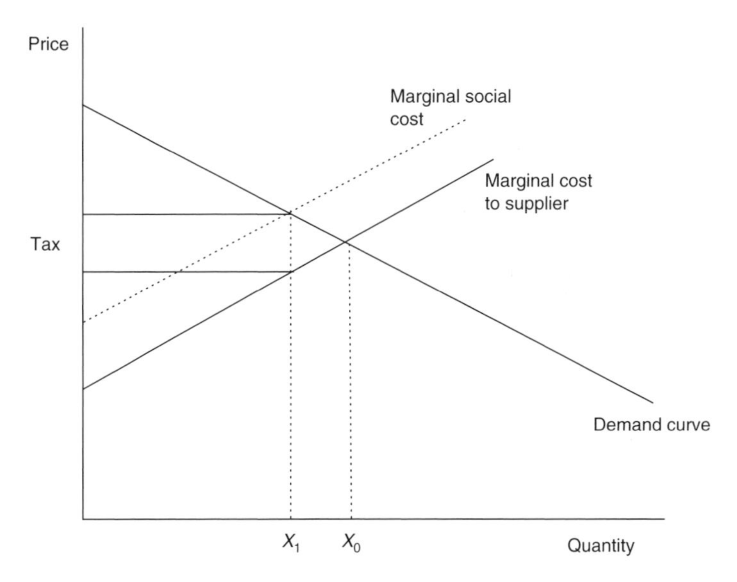

Figure 2. Marginal-cost pricing.

includes both those costs borne directly by the supplier and external costs imposed upon third parties by their use (for instance, environmental costs such as air pollution and noise). The optimal output of a good is the amount for which consumers are willing to pay these costs (X_1 in Figure 2). However, in the absence of monopoly power, the supplier will be willing to supply more of the good as long as they receive at least their own marginal cost. In the absence of some form of intervention then the output of the good will settle down to X_0. This is optimal for the producers and consumers but not for society as a whole, since it ignores the external costs. Consumers would not be willing to pay for the final units of the good sufficient to compensate those bearing the external costs as well as the producers for their costs. The usually advocated intervention is for the government to impose a tax equal to the amount of the external costs created by the good, so that an equilibrium is now reached at X_1.

In terms of project appraisal, the appropriate approach is to examine the costs and benefits of the proposed course of action, valuing the benefits in terms of the willingness to pay of the beneficiaries and the costs in terms of the compensation required by the losers. In this way we can determine economically efficient levels of service provision, investment, and prices.

There is an obvious and long-standing objection to this analysis, however. That is that it ignores the issue of who enjoys the benefits and who bears the costs. There is not one Pareto optimum but many. All the points on the outer curve in Figure 1 are Pareto optima, since it is only possible to improve the utility of one person by lowering that of the other. Which point the economy will end up at depends on the initial distribution of resources between the groups. As explained above, a move such as that from point A to point B represents a potential Pareto improvement in welfare. Whilst notionally it should be possible to redistribute the benefits of the policy in such a way that everyone is better off than under any alternative transport pricing and appraisal system (e.g., at point C in the diagram), in practice that is not often achieved. The tax and income supplementation system is too coarse to do so even if decision-takers wished to; it is usually impossible to identify specifically the gainers and losers and identify taxes and subsidies which achieve the desired redistribution. Moreover, there are political constraints on the extent to which particular taxes may be raised. In any case taxes and income supplementation introduce their own inefficiencies into the system, by distorting relative prices and the incentive to earn more income. But as long as it is accepted that optimal pricing and appraisal policies will not be accompanied by optimal distribution policies, then the distributional aspects of transport systems decisions must be addressed.

Unfortunately, whilst there is consensus at least amongst economists as to what constitutes efficiency, there is no such consensus about equity. Indeed at least three different views of equity may commonly be found amongst those commenting on transport policy.

The common view amongst economists is that equity is essentially concerned with income distribution. There is an obvious sense in which the above approach to transport pricing and appraisal may be seen as unfair. This is that a willingness to pay of one pound, or a willingness to accept compensation of one pound, will have the same weight in pricing and appraisal decisions whatever the income of the person it relates to (Pearce and Nash, 1981). It is a standard assumption, though one that is hard to prove, that as incomes rise the marginal utility of income falls. Thus decisions based solely on willingness to pay or willingness to accept compensation will not maximize the sum of the utilities of the members of society (in the absence of appropriate redistribution). For instance, a rich group of motorists may be willing to pay enough to use a new motorway to more than compensate the poor residents whose homes it pollutes, but this does not mean to say that the utility it yields to the motorists exceeds what it costs the residents. Moreover, even if it did, it might still be argued that as the motorists already have a higher level of utility than the residents, the change in their utility should count for less in social terms.

This approach to welfare economics essentially goes beyond pure efficiency considerations to seek to judge changes in terms of a social welfare function, where social welfare is seen as the sum of the utility of the individual members of society, weighted by factors which represent the relative social weight attached to a change in the utility of the person in question. This approach makes the strong assumption that in principle utility may be measured on a scale which permits comparability between individuals, and many economists reject it for that reason. Amongst proponents it is often argued that the weights should be anonymous, relating only to some relevant aspect of the individual concerned – in this case their income.

Suppose that we seek to maximize such a welfare function but assume that we do not have policy instruments available which can redistribute the benefits and costs of our decisions. The result is a set of pricing and appraisal rules that are as clear as those relating solely to efficiency, but which have the added difficulty of requiring all costs and benefits to be traced through to see who ultimately bears them, and of requiring a set of weights to be introduced whose values are pure value judgment and may therefore vary from decision-maker to decision-maker. Unlike in the case of economic efficiency, there is no single set of pricing, service, and investment decisions that can be argued to be "right" (strictly, even in the case of efficiency these values will vary with the distribution of income, but that is often not recognized).

But for many commentators this remains a very narrow view of equity. People differ in much more than their incomes and it is argued that allowance should be made for other factors which might lead to them being disadvantaged. Age and physical or mental disability are obvious factors to consider, but in many societies other factors such as race or gender may also be important. This approach may still be seen as consistent with the social welfare maximization approach, but where factors other than income influence the weights. However, other people would argue that the whole idea of a compensatory welfare function, in which a large increase in utility to the already well off could compensate for any reduction in utility to the disadvantaged, is anathema. They would tend to see the provision for the disadvantaged more in terms of recognizing their right to a decent standard of living (Neuburger and Fraser, 1993).

There is yet a third view of equity which plays an important part in transport planning. This is a view which says that equity demands that those who benefit from the provision of a facility should pay for it. This is a form of equity which is hard to justify in terms of any reasonable social welfare function, since it seems to rest on the assumption that in the absence of the facility the distribution of income would be fair, so that all that is needed to preserve that fairness is that provision of the facility does not change the distribution of income or utility. Indeed many people would recognize this shortcoming of the rule, to the extent of exempting facilities provided specifically for the poor or otherwise disadvantaged from its application. But the rule that equity demands that the users of each facility should pay its cost remains extremely forceful in practical politics and amongst public opinion (Newbery, 1998).

A somewhat more sophisticated version of this argument recognizes the existence of joint costs, and argues that each group of users should pay at least their avoidable cost, but not more than the "stand-alone" cost of providing the facility for themselves alone. In other words, they should benefit rather than lose from the fact that they share the facility with others (Brown and Sibley, 1986). For instance suppose a rail route carries both passenger and freight traffic. The existing cost of the route K comprises the cost of providing the facility for freight traffic, S_F, plus the avoidable cost of adding passenger traffic to it, A_p. The same figure K may also be derived as the stand-alone cost of providing passenger service, S_p, plus the avoidable cost of freight service, A_F. It is argued that passenger traffic should collectively pay something between A_p and S_p, and freight traffic between A_F and S_F.

We now proceed to consider the implications of these alternative views of equity for the derivation of pricing and appraisal rules.

3. Transport pricing

The concept of marginal-cost pricing for transport has been long debated but little implemented. Marginal social cost of infrastructure use may be defined (Jansson, 1997) as the sum of the following marginal costs:

(1) costs borne directly by the user (provision of vehicle and fuel, users' time);
(2) costs imposed on the infrastructure provider (provision and maintenance of the infrastructure);
(3) costs imposed on other infrastructure users (delays, increased risk of accidents);
(4) costs imposed on society in general (air pollution, noise, global warming, etc.).

A distinction must be made between short-run marginal cost, where no additional infrastructure is provided and thus the infrastructure capital costs are irrelevant, and long-run marginal cost, where the infrastructure is optimally expanded as a result of additional traffic but consequently the additional impact of the traffic in terms of congestion and perhaps accidents and environment is reduced. It is generally accepted that short-run marginal cost is the appropriate pricing concept but that long-run marginal cost is relevant to the decision as to whether to provide additional capacity.

The economically efficient charge for the use of infrastructure is then the sum of these costs except those costs already borne by the infrastructure user.

In the case of transport services, similar considerations apply. The marginal social cost is the sum of the costs imposed by an additional user on the transport operator, on other transport users, and on society at large. In the short run, where the service level does not adjust, the extra costs on the transport operator will be confined to any delays caused by the passenger boarding and alighting, and trivial additions to fuel and wear and tear. But the costs of crowding and possibly being left behind by a full vehicle or train imposed on other passengers may be severe, at least at peak times. In the longer run, when service levels adjust, the operator will bear a higher marginal cost but other passengers will actually gain from an improved level of service – the so-called Mohring effect (Mohring, 1972). Given the speed at which services can be adjusted it is typically considered that this slightly longer-term marginal social cost is the appropriate basis for pricing of transport services.

An increasing number of case studies has been undertaken to illustrate the effects of marginal-social-cost pricing. Generally the outcome is that it would lead to road users paying more than the total costs of provision of the road system in congested urban areas but less than the total cost outside those areas. Public transport would generally justify subsidy, particularly over shorter distances where the Mohring effect is more significant as a proportion of total cost (Nash et al., 2001).

The implication of moving from a purely efficiency-oriented approach to transport pricing to one based on maximization of a social welfare function incorporating income weights can be seen using the framework put forward in a surprisingly little known paper by Feldstein (1972). He puts forward the concept of the distributional characteristic of a good. This is defined as the weighted average of the marginal social utility of all consumers of the good, where the weight is that consumer's (or group of consumers') share of consumption of the good. Essentially this enables us to weight the willingness to pay of users by the appropriate equity weight. The benefit or cost of a change in the price of the good may thus be multiplied by the distributional characteristic of the good to assess its social value. The effect of such a pricing system will be to lower systematically the prices of goods where their users tend to be poor, and to raise the prices of goods used by the rich. There is no guarantee that in general this will move prices towards full cost recovery. Similar considerations may be introduced for other motives for weighting in terms of other aspects of disadvantage, so that lower prices for facilities used by the elderly or disabled are justified, or indeed specifically lower tariffs for them in the form of concessionary fares.

One important point should be noted. This approach to pricing is very much a second best one; it would be possible to make everyone better off by implementing marginal-cost pricing and redistributing the benefits. Thus wherever implementing such a pricing policy involves the direct input of government funding, it should be examined whether there is a more efficient way of directing that funding towards the people it is intended to help. For instance, if the only reason for introducing concessionary fares is to help people because they are poor, it may be more efficient to supplement their incomes in some other way. The case for concessionary fares then rests on a supplementary argument concerning their right to access to public transport facilities.

By contrast, a view of equity which requires users of a facility to pay its total cost will lead to a very different approach to pricing. If this is to be achieved in the way which compromises efficiency the least then it will result in some combination of Ramsey pricing and multipart tariffs. Ramsey pricing minimizes the distorting effect of charging more than marginal cost by increasing prices more in those markets where demand is least sensitive to price. If markets are unrelated then the proportionate deviation between price and marginal cost will be inversely proportional to the price elasticity of demand. Multipart tariffs charge different prices according to the amount of the good purchased. For instance a simple two-part tariff comprises a fixed charge for the supply of the good plus a variable component according to the amount of the good supplied. Multipart tariffs will be most efficient where the fixed component has (virtually) no effect on consumption of the good. This might be true of a small annual vehicle ownership tax, for instance, or – perhaps less likely – a small fixed charge for access to the public transport system. Where there is a distorting effect on use from such a fixed charge, Ramsey pricing, which raises charges more the less sensitive demand is to price, is needed. Ramsey pricing will thus charge people more the more captive they are; for instance, commuters will be charged more than shoppers; business travelers more than leisure. It should be noted, however, that this approach to pricing has itself often been regarded as unfair as it exploits market power to raise price to the captive user (Jansson, 1984). A poor person who needs to use a transport service to earn a living may be charged more than a rich person using it for an inessential leisure journey. If the view of equity is that all users should contribute to the cost of the facility in proportion to their use of it, then some form of average-cost pricing is the only admissible pricing policy.

It should be noted that in what we have said about the desire for users to cover the total costs of the facilities they use we have referred here only to equity considerations. Other justifications for such a rule may be given, in terms of public sector budget constraints or the need to give suppliers incentives for efficiency. However, none of these alternative arguments gives a clear case for total cost coverage for each mode or facility. Meeting a budget constraint in the most efficient way implies applying Ramsey pricing and/or multipart tariffs over the whole set of prices under government control (including those influenced by taxation) rather than for a specific good, and is therefore likely to involve cross subsidy. Incentives for efficiency may be separated from the need to cover costs by competitive tendering for the required subsidy.

In practice, the number of studies of the distributional consequences of transport pricing is limited. There has, however, been a lively debate over the distributional effects of the introduction of road pricing and of public transport subsidies. In both cases the conclusion seems to be that the result depends very much on the circumstances. In the U.S.A., with high levels of car ownership and low use of public transport, it appears that road pricing may be regressive (Small, 1983), whilst in London poorer people benefit because of their heavy use of the bus system, which gains substantially from the reduced congestion resulting from introduction of road pricing (MVA Consultancy, 1995). By the same token subsidies to bus services may be progressive, whereas – in European circumstances at least – subsidies to rail and air are more likely to be concentrated on the better off (Gwilliam, 1987).

4. Appraisal

Transport appraisal is most generally viewed as a way of taking decisions on investment projects, but the same techniques may be applied to a range of decisions, including whether to subsidize a public transport service, and if so what service level should be provided. Indeed it can also be applied to pricing decisions, particularly to assess whether the benefits of a more expensive system to implement, such as electronic road pricing, are worthwhile (MVA Consultancy, 1995). The common feature is that all these decisions require comparison of a number of discrete options.

The pure efficiency approach to transport appraisal is the application of social cost–benefit analysis to assess the benefits and costs in terms of willingness to pay or to accept compensation, and then to compute the net present value of these items. The decision as to the worthiness of the proposal is then based on a single numerical indicator. Again, the decision can be based on a weighting of costs and benefits based on the incomes or other characteristics of the recipients, at the expense of tracing through the benefits and costs to see who the ultimate recipients are. This process is no simple task, as changes in other markets, such as property prices or rents, may mean that the direct beneficiaries end up little or no better off, and third parties – such as property owners or landlords – gain the benefits.

However, this pure approach to transport appraisal is not often used in practice for a number of reasons. First there is the real or apparent difficulty in placing monetary values on some of the costs or benefits. Research into the valuation of environmental effects means that these problems are now less acute than they were, but they are still seen as preventing total valuation of all relevant items by many commentators (Nash, 1997). But secondly there is a belief that decisions should not be based on economic efficiency (or even efficiency and equity) alone. Many appraisal methods see a whole range of objectives as relevant. For instance, the New Approach to Transport Appraisal (NATA) (Department of the Environment, Transport and the Regions,1998) in Britain sees as objectives safety, economy, accessibility, environmental protection, and integration. That these different objectives are all relevant to the proper measure of economic efficiency is not often appreciated. Thirdly there is the wish to avoid attaching explicit weights to certain items, including distributive weights. Thus decision-takers often prefer to see explicit information on who gains and who loses from projects, rather than defining a weight to be built into the appraisal.

The result has been popularity of the use of matrices showing a range of costs and benefits in monetary and/or physical or descriptive dimensions and often also showing the incidence group bearing them, along the lines of the "planning balance sheet" approach popularized by Lichfield in the 1960s (Lichfield, 1968). In Britain, there has been an implicitly multicriterion approach to road investment appraisal since the introduction of the "Leitch framework" in the 1970s (Advisory Committee on Trunk Road Assessment, 1977). This framework put forward measures of the economic and environmental performance of schemes, although these have only recently been explicitly linked to the objectives of transport policy. The Leitch framework and its NATA successor also explicitly recognize different incidence groups for transport projects, including infrastructure providers, operators, users, residents, those concerned with heritage, etc. However, there has never been any explicit tracing through of the income distribution of such groups. This would be interesting given that the beneficiaries of transport investment are often the better off, whereas the external costs may be borne by the poor, particularly where investment attracts traffic to inner city areas (Dalvi and Nash, 1977).

5. Transport and social exclusion

The approach taken to issues of equity above is one that sees equity very much in terms of the overall distribution of income or utility. Transport is relevant simply as one contributor to that whole. However, some studies see transport as having a more fundamental role than that. To them, transport is one of the basic rights that, along with food, clothing, and housing, make it possible to participate in society. Access to facilities such as work, shops, medical care, education, friends and relatives, and entertainment forms an essential element of an acceptable standard of living, and without it one is socially excluded. On this argument, accessibility should be provided to all regardless of cost and ability to pay.

One clear manifestation of this point of view is in legislation which requires all public transport facilities to be accessible to the disabled. This legislation has existed in the U.S.A. for some time and is being progressively introduced in Europe. More generally, this approach suggests a need to provide at least a minimum level of public transport for all. This has long been a principle behind local authority subsidies for lightly used bus services in Great Britain, where criteria are frequently defined in terms of minimum levels of service according to the type of settlement served (Bristow et al., 1992). A more accurate measure of the adequacy of the transport system may be in terms of accessibility criteria. These may be defined in terms of simple contours, such as the proportion of the population within x minutes' travel of each relevant type of facility, or more complex measures based on trip distribution models (Ingram, 1971).

It has long been argued that transport systems, and particularly those based on the private car, lead to an intrinsically unfair distribution of accessibility (Hillman et al., 1973). The argument may be broken down into a number of propositions. Firstly it is often argued that transport systems are typically planned and managed by middle class males who have no understanding of the needs of other segments of the population and that the mix of price and quality is based on average needs, which pay no respect to the needs of poorer sectors of the community. Secondly it may be argued that even with the expansion of the number of cars, car ownership and availability are heavily concentrated on adult males and that the growth of transport systems based on the car favors these individuals to the exclusion of the young and the old, and even of adult women, who in countries such as Britain still have much lower levels of access to the car than do males (for instance, in Britain, in the National Travel Survey 1997/9 82% of adult males but only 59% of adult women possessed a driving licence). Thirdly, as car travel expands, so the quality of public transport worsens, so that the distribution of accessibility across the population becomes more and more unequal. At the same time walking and cycling become unpleasant and dangerous, and facilities more remote as out-of-town sites are promoted. Whilst all this may be true, it is also the case that even non-car-owners make extensive use of car transport as passengers. The counterargument is that the car has brought big increases in accessibility for wide sections of the community even though they rely on other members of the household to do the driving.

That the development of transport systems leads to social exclusion and inequality of access to facilities is a point that seems to be widely accepted. Transport is a necessary element for the consumption of many other important goods, including healthcare and education; indeed its absence is a major problem in the provision of these services in developing countries (Howe and Richards, 1984). It is also subject to economies of scale and must be produced where it is consumed, with the result that there may be many locations where it is simply not available to those without access to a car. Equity issues in the provision of transport services are therefore of particular importance. However, this clearly cannot mean that high-quality transport systems can be available at all locations. Economic efficiency cannot be totally neglected, and it is in dealing with these sorts of issues that the trade-off between equity and economic efficiency is most pronounced.

6. Conclusions

We have seen how a pure efficiency-based approach to planning transport systems ignores all equity issues. Equity in transport planning may be taken to mean a wide variety of different things, ranging from attaching weights in pricing and transport appraisal to different groups according to their incomes or other factors such as disability, through a view that users should pay the total costs of the facilities they use, to a belief that accessibility should be provided for all regardless of ability or willingness to pay. Given the clear impossibility of achieving optimal redistribution of costs and benefits of transport systems through the tax system, there is no doubt that equity considerations should play a role in transport planning. Pricing, service provision, and investment decisions all need to be accompanied by an analysis of who gains and who loses by them in order to ensure that decision-makers have all the relevant information available to them. This remains surprisingly rarely the case at present.

REFERENCES

Advisory Committee on Trunk Road Assessment (1977) *Report*. London: HMSO.

Bristow, A.L., P.J. Mackie and C.A. Nash (1992) "Evaluation criteria in the allocation of subsidies to bus operations", in: *Transport investment appraisal and evaluation criteria: Proceedings of Seminar C, 20th PTRC Summer Annual Meeting*. London: PTRC Education and Research Services Ltd.

Brown, S.J. and D.S. Sibley (1986) *The theory of public utility pricing*. Cambridge: Cambridge University Press.

Dalvi, M.Q. and C.A. Nash (1977) "The redistributive impact of road investment", in: P. Bonsall, M.Q. Dalvi and P.J. Hills, eds., *Urban transportation planning*. London: Abacus Press.

Department of the Envioronment, Transport and the Regions (1998) *Guidance on the new approach to appraisal*. London: HMSO.

Feldstein, M.S. (1972) "Distributional equity and the optimal structure of public prices", *American Economic Review*, 62:32–36.

Gwilliam, K.M. (1987) "Market failures, subsidy and welfare maximisation", in: S. Glaister, ed. *Transport subsidy*. Newbury: Policy Journals.

Hillmann, M, I. Henderson and A. Whalley (1973) "Personal mobility and transport policy", Political and Economic Planning Broadsheet 542, London.

Howe, J. and P. Richards, eds. (1984) *Rural roads and poverty alleviation*". London: Intermediate Technology Publications.

Ingram, D.R. (1971) "The concept of accessibility: A search for an operational form", *Regional Studies*, 5:101–107.

Jansson, J.O. (1984) *Transport system optimisation and pricing*. Chichester: Wiley.

Jansson, J.O. (1997) "Theory and practice of transport infrastructure and public transport pricing", in: G. de Rus and C. Nash, eds., *Recent developments in transport economics*. Aldershot: Ashgate.

Johansson, P.O. (1991) *An introduction to modern welfare economics*. Cambridge: Cambridge University Press.

Lichfield, N. (1968) "Economics in town planning", *Town Planning Review*, 38:5–20.

Mohring, H. (1972) "Optimisation and scale economies in urban bus transportation", *American Economic Review, Papers and Proceedings*, 591–604.

MVA Consultancy (1995) *The London congestion charging research programme*. London: HMSO.

Nash, C.A. (1997) "Transport externalities: Does monetary valuation make sense?", in G. de Rus and C. Nash eds., *Recent developments in transport economics*. Aldershot: Ashgate.

Nash, C.A., T. Sansom and B. Still (200i) "Modifying transport prices to internalise externalities – evidence from European case studies", *Regional Science and Urban Economics* (in press).

Neuburger, H. and N. Fraser (1993) *Economic policy analysis: A rights based approach*. Aldershot: Avebury.

Newbery, D. (1998) *Fair payment from road-users. A review of the evidence on social and environmental costs.* Basingstoke: Automobile Association.
Pearce, D.W. and C.A. Nash (1981) *The social appraisal of projects: A text in cost–benefit analysis*. London: Macmillan.
Small, K. (1983) "The incidence of congestion tolls on urban highways", *Journal of Urban Economics*, 13:90–111.

47-59

Chapter 5

THE CONCEPT OF OPTIMAL TRANSPORT SYSTEMS

ROGER VICKERMAN
The University of Kent at Canterbury

R41
R48

1. Introduction

Optimality in any system implies the most efficient use of resources in that system. Efficiency refers to a relationship between inputs and outputs, such that either a given set of inputs is being used to create the maximum possible output or a given output is being produced with the minimum set of inputs. Technical efficiency usually refers to the physical relationship between inputs and outputs. Economic efficiency introduces the prices of both inputs and outputs so that we can obtain some idea of value. Efficiency thus carries with it a notion of welfare: in an economically efficient situation we cannot make someone better off without simultaneously making someone else worse off (see Chapter 4).

The concept of optimality in transport systems has to bring together these two elements. On the one hand we can look at transport systems and networks in terms of their technical efficiency: how good they are at moving traffic, goods, and people from one point to another with the least "cost" in terms of the physical units of distance traveled or time taken. On the other hand there is a more complex concept of optimality which seeks to define the optimal amount of resources devoted to transport within the economy as a whole. Transport is in most cases only a means to an end; it does not create welfare on its own, that is done by the final activities for which the transport is used. Hence we would typically seek to minimize the amount of transport consistent with any given level of economic activity. On the other hand it can be argued that improved transport will lead to an increase in economic activity, indeed, in certain circumstances, to an increase in the rate of economic growth.

However, it is not just a simple relationship of transport input to economic output; transport is used for both inputs and outputs, for individuals as suppliers of labor and as consumers of final goods and activities, and as both a service which is supplied and demanded in the open market and a non-marketed service provided by individuals and firms for their own use. This requires us to look at all the linkages between transport and all the other markets in the economy, for

Handbook of Transport Systems and Traffic Control, Edited by K.J. Button and D.A. Hensher

goods, labor, and capital, and see how they are brought into a general equilibrium of the economy as a whole. Then we would need to ask whether some reallocation of resources would produce a more efficient economy which would be more competitive and achieve a higher rate of growth.

There are five main sections to the discussion. First, we consider the notion of optimal networks. Secondly, we explore the relationships between the transport sectors and the overall efficiency of the economy. Thirdly, we look in more detail at the labor market, which has a critical role in this link. Fourthly, we look at the role of transport in a general equilibrium view of the economy. Fifthly, we explore some implications for the appraisal and evaluation of transport investments in the light of this discussion.

2. Optimal networks

Transport systems are networks. Networks are systems which involve *nodes* which are joined together by *links* or *arcs*. Either the nodes can be given externally to the network, for example they are settlements, cities, etc., or they are endogenous to the network, for example when a network operator identifies a location where it is efficient to arrange transshipment of goods, or interchange for passengers. Each node, other than a transshipment point, is either the origin or the destination of a flow. Flows occur along the arcs of a network and each arc has associated with it a cost per unit of flow. The cost will be determined by the distance and the operating characteristics of the transport system, but it will also be related to the amount of flow along that arc. Each arc has a capacity which defines the maximum flow along that arc. This set of factors determines the characteristics of the network (see Kansky, 1963; Haggett and Chorley, 1972).

The optimizing problem is then first to determine the shortest, or least-cost, path through the network from any one point to any other, and secondly to determine the minimum cost of the potential flows in the network as a whole, given the pattern of origins (supply) and destinations (demand). This allows for interaction between the flows on particular arcs such that when the flow on one arc approaches capacity, other flows are rerouted to satisfy the minimum-cost rule. For an introduction to the problem of network optimization see Ahuja et al. (1995), and for one to the location of facilities on networks see Labbé et al. (1995).

There are two sets of interests in optimal networks. One is the optimal use of the network by the user, the logistics operator, or the solver of the classic traveling-salesman problem, who needs to know how to find the optimal way of linking nodes and minimizing the costs of transport. The second is the network designer or transport operator, who needs to establish the most efficient network to operate given the existing pattern of demands. Sometimes the most efficient network for the operator will not be the network which maximizes the users' welfare. For example, the establishing of hub and spoke networks by airlines often deprives users of the most direct (shortest time cost) route between two nodes, although the argument can be made that the savings to the operator result in lower fares or more frequent services such that the effect on overall welfare is ambiguous.

In some cases, such as the airline example given above, it is the service operator who establishes the network (except for the need to take the location of the major nodes as given). In the development of railways this was also the pattern. The eventual service provider constructed the physical network according to expected patterns and levels of demand. In the development of road transport, however, a rather different situation evolved in which the basic network was established (originally as a publicly provided infrastructure) and service operators then have used the physical network to develop their own service network. This pattern is now becoming more common in the provision of rail services, in which competing rail operators have to bid for usage of a network provided by a separate infrastructure provider.

In the case of a privately provided network, the main objective of the provider is to maximize profits (return on assets employed). No concern is therefore given to the social welfare of potential users. Networks can therefore be developed which are efficient or optimal to the provider, but not optimal from the point of view of society as a whole since they may exclude certain groups or geographical areas, serving whom is not profitable to the operator. It is for this reason that many transport networks have been provided by the public sector. In these cases the full costs of an entire network could be spread over the network as a whole (average cost pricing) such that the users in a densely trafficked part of the network would cross-subsidize those in lightly trafficked parts, who, if charged full costs, would not be able to afford the network and, if only charged the marginal costs of the service, would see the network not provided or rapidly falling into disrepair.

There is, however, also an efficiency argument for preserving unprofitable parts of networks, where the existence of such links promotes a wider efficiency in the economy. There are two elements to this, network economies which exist only when a network is complete, but which might not be able to be realized by the individual operator or infrastructure provider, and external effects on the rest of the economy. We deal with the second of these in the following section, but network economies are recognized to be an increasingly important issue, though one which is less developed than it needs to be in the context of the move to greater privatization of both transport services and infrastructure.

Network economies to the operator comprise economies of scale, scope, and density (Keeler, 1983; Caves et al., 1984). It is important to separate out the potential reductions in unit cost which relate just to the size of the firm (independent of the nature of the network operated), those which relate to the range of different services provided by a multiproduct operator, and those which relate to the characteristics of the network. Multiproduct outputs are common in transport; the same operator serves many different markets, defined both by the nature of the service provided (passenger, freight, scheduled, charter, business, leisure, etc.) and by the origins and destinations served. This makes it important to specify output correctly in order to assess the presence of scale economies. Jara-Diaz and Cortes (1996) identify the correct building block for output in transport cost functions as "a vector of commodity-specific O-D flows." Even where such information is not directly available it should not preclude the use of the concept in interpreting the data which is available.

Network density economies are of two types. One relates to the density of the network, such that an operator can provide services which link conveniently together, thus lowering idle time of rolling stock and maximizing the number of passengers who face less disrupted journeys. The other relates to the average length of haul within a network of given density, since the longer the average journey length the greater the economy from spreading the fixed terminal costs. It is particularly important to consider these different cost elements in the case of the privatization of transport operators where large natural monopolies (e.g., railway or airline companies) are broken into smaller units. What the analysis of network economies suggests is the importance of considering the competitive structure of the industry alongside the structure of the relevant networks. Thus, for example, most economies can probably be found for local bus operators at a relatively small size, and for regional rail operations at a rather larger network size, but national intercity services depend more heavily on wider national networks. Where the local or regional services act as an important feeder to the intercity network then this will affect the optimal organization of the network, but the answer is essentially an empirical one.

Firms and other organizations also employ these concepts when establishing the location of their facilities (Labbé et al., 1995). For example, logistics organizations, wholesalers, retailers, etc. all have to determine the optimal number and location of depots and stores. Changes in the transport network can lead to the pressure to change both of these parameters. Given the cost of such changes, the response may not be a continuous one. A degree of inertia is likely to build up whilst the organization operates with a suboptimal network until the benefits from changing exceed the costs. This is one of the reasons why it may be difficult to observe the wider impacts of transport changes; the improvement of individual links is of no benefit to users, since they are unable to adjust their operations to save from the improvement, but a cumulation of network improvements does provide such an opportunity.

Similar considerations also apply to the location of various public facilities; schools, hospitals, libraries, etc. are all network facilities, whose location interacts with the transport system, and accessibility to which directly affects consumer welfare. Planners can both identify the best locations for facilities for a given transport network, and consider how improvements to the transport network, for a given set of facilities, may lead to an improvement in access times. Such factors may be of particular importance when considering the distribution of accessibility between different groups to help minimize the degree of social exclusion caused by poor transport to already disadvantaged groups.

The shape of the network in a region will affect the likely characteristics of a region's economy, through the determination of where firms and facilities will locate (Kansky, 1963; Haggett and Chorley, 1972). Rather different patterns of location and particularly of the concentration and dispersion of facilities arise in the case of radial networks, circumradial networks (i.e., radial networks, but with peripheral routes), and grid networks (Peeters et al., 1998). Thus additions to networks which seek to improve the accessibility of one location may have complex impacts on accessibility to all locations on the network, thus leading to a change in the location of other facilities which could either reinforce the original objective or work against it (Blum et al., 1992; Vickerman et al., 1999). Moreover, as Peeters et al. (1998) show, where adjacent regions have developed different types of network, any reduction of borders between these regions (for example through a customs union or common market) can have differential effects on the two regions. A region with a grid network is less likely to experience a reduction in facilities than one with a radial network. Shape of the network is identified as being more important than network density in determining this outcome.

So far we have only considered direct links between transport and various measures of efficiency, of the firm, of the accessibility of certain individuals or groups, etc. What we now need to turn to is the relationship between the efficiency of the transport network and the overall efficiency of the economy.

3. Transport and the efficiency of the economy

Transport is treated in the optimal network situation as a derived demand. Demands and supplies exist at nodes, between which goods and people have to be moved. In practice, however, different networks will lead to different total transport costs and hence different delivered costs of the goods (or people) moved around the network. These different total costs will affect prices, and hence ex post or revealed demands at the destinations. Thus the quality of the network will have an effect on the volume of activity. This is the most direct way in which transport can affect both the overall volume of economic activity and the efficiency of the economy.

If transport costs change as a result of an improvement in the network, then users will use this to substitute cheaper transport for other inputs to the production process. Transport improvements also enable firms to adopt just-in-time production techniques, using the efficiency of the transport network to reduce the level of inventories and thus save cost. Increasingly firms use electronic means both to order and then to track goods along the transport chain; thus there is a mapping of a virtual network onto a physical network. Thus supermarkets, for example, take advantage of the completion of highway networks to reduce the number of supply depots they need to serve all their stores. The improved reliability of delivery means that they can reduce the level of stocks and deliver as the store needs supplies.

There is a paradox in this for the evaluation of transport. On the one hand transport contributes to the overall level of economic activity through its contribution as one of the sectors of the economy in terms of employment, value added, etc. Thus the more transport in the economy the greater the overall level of activity. However, transport is also a consumer of resources which may be able to be used more efficiently in other sectors of the economy; thus there is an incentive to reduce the transport input to any activity in order to increase the level of aggregate activity. Improvements in transport which lower transport costs may thus also reduce the size of the transport sector, at least in terms of employment. In an increasingly global economy, some countries may have a comparative advantage in the provision of transport services and thus will specialize in their production, selling such services to others in order to increase the level of economic activity in their own economy. This can be seen in the increasing global networks in airlines and shipping, but also in the emergence of major pan-European road haulage carriers within the European Union's single transport market.

4. Transport, labor markets and efficient cities

In the previous section we have discussed the way in which transport interacts with the production process as an input, in order to contribute to the overall efficiency of the economy. In this section we shall look in more detail at the interaction between transport and another input to the production process, labor. The transport network will define the size of a labor market for any particular process in any particular location. Workers have to use the transport network to get to and from the place of work. The more efficient is the network, the larger will be the potential labor market and thus the lower will be the labor costs for the employer. This may be in part because competition in an enlarged labor market keeps wages down and in part because the more efficient transport system brings workers more quickly to work and allows them to be more productive.

This idea goes back to the basic model of the city as developed by Alonso (1964), in turn based on the original model of von Thünen (1826). This model develops a trade-off between transport costs and land rents which helps determine the size of a market area or city. As we move away from the center of the city or market area, transport costs rise and thus the amount of rent which potential users of land are prepared to bid will fall. In a perfectly competitive situation it can be shown that this trade-off keeps the total payment constant. The simplest model just looks at a single arc of a transport network, a spoke on a radial transport network for the city. In more complex models, however, the network is allowed to interact with the activities of the city (Mirrlees, 1972; Dixit, 1973; Solow, 1973). As the city grows, the potential profitability from activities at the center rises and the maximum rent which these activities will bid for the use of land will increase. Thus the city can also grow in terms of its physical size and the maximum transport cost from the edge of the city will increase. Moreover, as the city increases in size the number of people trying to gain entry to the center of the city will increase, causing congestion on the network. There is pressure therefore for more land to be devoted to transport, thus restricting its availability for other uses with a consequent effect on its rental value. Since more people are traveling, it becomes profitable for the city to invest in more efficient transport networks.

We can observe this in the historical growth of cities (Schaeffer and Sclar, 1975). As they have grown there has been pressure to invest in urban rail and transit networks. Nineteenth-century cities in Europe and North America grew particularly as the street tramway provided the means for an increasing urban labor force to move to and from factories in densely populated areas. As cities grew further there was increasing pressure on land, with the potential for transport to need an excessive share of land at the urban center. This led first to the development of underground metro systems which need much less surface land for the same capacity, but later to pressure for decentralization of activities within cities and deconcentration from large cities to smaller cities. The latter moves have been enabled by the development of the automobile and its mass ownership. However, this makes mass transit a less efficient operation and an increasing burden on the city's economy. Thus different stages in city growth imply different types of transport network for their efficient operation, but transport network developments have led to cities growing in different ways. Very large cities are increasingly facing a burden which relates in part to the maintenance of a historic transport network which is no longer efficient. This problem of financing a more efficient, but unprofitable, public transit system whilst facing the external costs of a rapidly growing, less efficient, private-car-based urban transport system is central to our understanding of optimality in urban transport (see Jansson, 1984, 1993).

This leads to consideration of the optimal form of the city to increase environmental sustainability. This is not the place to deal with the full issue of sustainable transport, but there is one important link to the optimal network, the concept of the compact city (Breheny, 1992). Much of this has followed the attempt to establish a comparative framework for urban efficiency in relation to transport use (Newman and Kenworthy, 1989a,b; Hall, 1995). However, others have been much less willing to accept the view that increasing the density of cities will lead to greater efficiency, at least in terms of energy consumption (Gordon and Richardson, 1989).

At the more microeconomic level the effect of the interaction between transport networks and labor markets can be measured in terms of time savings. Time savings can be thought of as equivalent to an increase in productivity by those individuals traveling in the course of work and as the source of a potential improvement in productivity by those traveling to and from work. This is the theoretical basis on which the value of time savings used in transport appraisal is based (Jara-Diaz, 2000). The implication of this can be explored at different levels. For the individual we know that this is equivalent to an increase in welfare, since it effectively means a reduction in the price of time spent on an activity which is directly unproductive. For the individual's employer, the employee can reach work quicker and more reliably and hence may be seen as being potentially more productive on arrival. But for the employer there is also the advantage that the potential labor market supply has increased in size, which will lead either to competitive downward pressure on wages or to an improvement in skill availability, or to both. For the economy as a whole the increase in productivity is a source of growth. If this enables firms to restructure their operations and/or reduce operational slack this could lead not just to an increase in the level of economic growth, but also to an ongoing increase in the rate of economic growth – so-called endogenous growth (Aghion and Howitt, 1998).

5. Transport in a general equilibrium model of the economy

In the discussion so far we have taken what is usually referred to as a partial equilibrium view of the transport sector. In a partial equilibrium model we look at each sector in turn, assuming that all other sectors remain unchanged. Thus we focused, in turn, on the transport sector and then on each sector with which the transport sector interacts. For example, we can observe the effect of a change in price, caused by an investment in transport, on the use of transport assuming at first that there is no change in the pattern of activities using transport. Secondly, we can look at the effect of the change in the price of transport on the pattern of activity in each sector in turn. However, we have not considered the feedback from this induced change in the pattern of activity on the use of transport, nor have we taken account of the impacts from the transport-induced change in one sector on other sectors, such as the links between product markets and the labor markets explored in the previous section.

In general equilibrium all of these linkages are considered simultaneously. Thus both the feedbacks onto the sector where the initial change takes place and the linkages between other sectors are considered. There are two advantages of this approach. First, by allowing for feedbacks and other linkages, the full net impact on the economy of a transport change can be estimated more accurately. Secondly, the general equilibrium approach allows us to consider interaction between sectors with varying degrees of imperfect competition.

Where only a partial equilibrium approach is followed, there is a danger that the full economic impact of a transport change can be either over- or underestimated. The most obvious case of overestimation is where the transport change induces either an increase in, or a relocation of, activity and there is an induced increase in traffic over and above that originally estimated. Such induced traffic will lead to any extra capacity being used, thus leading to a higher level of congestion than predicted. This could, in certain circumstances, eliminate all the estimated benefits (Standing Advisory Committee on Trunk Road Assessment, 1994: Venables, 1999). Underestimation of benefits could occur where the transport improvement leads to a general increase in economic activity in an area. These wider benefits are often asserted by proponents of a scheme, but are rarely subjected to rigorous economic analysis.

What such analysis needs to show is that economic activity is raised by the improvement to a level which is higher than measured by the direct improvement in welfare attributable to the improvement itself. In a world of perfect competition in the sectors which use transport, this would not be possible since transport users would evaluate the improvement at a level which ensured their willingness to pay (reflecting the full benefits to them) was exactly equal, at the margin, to the cost of using the facility. Thus there would be no additional benefits; all users would pay for transport exactly the value to their activity of using that amount of transport. The conventional appraisal of the improvement would capture all of the economic benefits. Any additional benefits included would involve double counting of transfers to rents or profits of factor owners (Dodgson, 1973; Jara-Diaz, 1986). This would suggest that wider benefits are rarely attributable, and where there is an increase in economic activity observed at one location as a result of a transport improvement, this will only be a transfer from another location. This might involve an equity consideration, but does not signify an increase in efficiency which can be said to optimize the transport system relative to the economy.

But suppose that the sectors which use transport are not in perfect competition, such that they charge prices which deviate from marginal cost. In this case the user's valuation of the transport service will not be accurately reflected in the willingness to pay and there will be a deviation of overall benefits from those measured in the transport sector. Furthermore, a firm in a market-dominating position in a region may be in a position to force down the price which it pays to its transport suppliers and force up the price of the finished product or service to its customers. The market will thus not result in an optimum use of transport. In the situation above it may overconsume transport and any improvement of transport facilities may simply add to the firm's profit. The net impact of the improvement could thus be to reduce the efficiency of the transport sector even further and transfer welfare towards the firm.

However, transport improvements could benefit the situation if they increase the competition which a firm faces. Monopoly power often depends on the spatial monopoly which firms are able to create because of the cost which competitors face in entering the market, of which the cost of transport is a major element. In this case the improvement of a link between two regions which introduces additional competition can reduce the excess of price over marginal cost and the firm's monopoly rent. In this case there will be an additional net welfare benefit to the economy, which is not measured in the direct benefit to users of the transport facility. Of course, if one of the firms in one of the regions has lower costs of production, for example because of scale economies enjoyed in a larger "home" market, then this firm can use the lower transport costs to expand even further, gaining more monopoly rents, reducing production and employment in the other region, and in the long run leading to a potential reduction in the net economic impact (see Krugman, 1991, 1998; Venables and Gasiorek, 1999).

This is the partial equilibrium outcome with imperfectly competitive firms in the transport-using sector. The general equilibrium outcome looks at the impacts of the initial change in one transport-using sector on all other sectors in the local economy. How significant this will be will depend on the linkages between firms in the local economy. In an economy where the initial impact is on a sector which has little or no forward or backward linkages in the local economy, there will be very little wider impact other than that on the initial sector. Where there are strong linkages the local impact will be much greater (Venables, 1995). This is logical since it suggests that where an economy depends on a lot of efficient linkages between firms, improvement of transport will improve the efficiency of those linkages and lead to a greater impact on the economy as a whole.

The size of this impact will depend on the relative elasticities of demand, the degree of scale economies, and both the level and change in transport costs. There is no a priori outcome to this problem; in some cases, depending on the values of these parameters, there will be quite large additional net benefits (up to 40% or more of the initial user benefit estimates, according to Venables and Gasiorek, 1999). In other cases the impact is such that it can reduce the net impact to below that of the initial user benefits.

6. Implications for appraisal and evaluation of transport investments

If the optimality of a transport network is determined, at least in part, by its contribution to the overall efficiency of the economy then it is important that these wider impacts are incorporated into the appraisal framework. The extent of such impacts will be case dependent: it depends on both the responsiveness of transport demand to any change in the generalized price of transport and the responsiveness of the demand for, and production of, goods and services to these changes in the transport sector. This responsiveness will depend in part on the degree of competition in each sector. It is not therefore possible to make a standard adjustment to a conventional transport cost–benefit analysis to allow for these effects. As we have seen in the preceding sections, however, the complexity of the linkages which produce these impacts will make it extremely difficult to model all these effects for every possible investment (Standing Advisory Committee on Trunk Road Assessment, 1999). In the case of large infrastructure investments or the introduction of major policy initiatives, such as the imposition of congestion tolls or general road pricing, it is desirable to model the impacts of such changes on the wider economy.

This can be achieved through the use of computable general equilibrium models. Such models (e.g., Venables and Gasiorek, 1999) use available information on the relevant demand and elasticity of substitution elasticities in each sector, allowing for imperfectly competitive markets, and compute an equilibrium response to a change in one particular parameter, such as the cost of transport. Such models allow for space by allocating resource and production to individual regions and can thus reallocate production and employment as a result of any change in the cost of transport. This gets us as near as possible to a practical application of a method of optimizing a transport system with respect to its role in the economy as a whole. There is much to be done to refine such models, and in particular to make them a practical proposition, given the large data requirements, but only with such progress can we hope to be able to understand the critical nature of transport in the working of the economy.

7. Conclusion

We have dealt here with a number of complex issues. The purpose has been to set transport in its wider economic context so that we can make some assessment of its role and significance in the economy as a whole. We have shown that a number of critical linkages exist such that transport has to be considered both as part of aggregate output in its own right, its demand being derived from the activities for which it is used, and as an input to the process of production which can be substituted for other inputs. Here the critical link is seen to be with the labor market since this helps to determine the size and shape of cities. Such a complex set of relationships needs fairly sophisticated modeling approaches to be able to trace through the full effects of any change. Currently this is being achieved with computable general equilibrium models which may offer some scope for development into tools which can assist with the appraisal process.

References

Aghion, P. and P. Howitt (1998) *Endogenous growth theory*. Cambridge, MA: MIT Press.
Ahuja, R.K., T.L. Magnati, J.B. Orlin, and M.R. Reddy (1995) "Applications of network optimization", in M.O. Ball, T.L. Magnati, C.L. Monma and G.L. Nemhauser eds., *Handbooks in Operations Research and Management Science*, vol. 7. *Network Models*. Amsterdam: Elsevier.
Alonso, W. (1964) *Location and land use*. Cambridge, MA: Harvard University Press.
Blum, U., H. Gercek and J. Viegas (1992) "High-speed railway and the European peripheries: Opportunities and challenges", *Transportation Research A*, 26A:211–221.
Breheny, M. (1992) "The contradictions of the compact city: A review", in M. Breheny, ed., *European Research in Regional Science*, vol. 2. *Sustainable development and urban form*. London: Pion.
Caves, D.W., L.R. Christensen and M.W. Tretheway (1984) "Economies of density versus economies of scale: Why trunk and local service airline costs differ", *Rand Journal of Economics*, 15:471–489.
Dixit, A. (1973) "The optimal factory town", *Bell Journal of Economics and Management Science*, 4:637–651.
Dodgson, J. (1973) "External effects and secondary benefits in road investment appraisal", *Journal of Transport Economics and Policy*, 7:169–185.
Gordon, P. and H.W. Richardson (1989) "Gasoline consumption and cities: A reply", *Journal of the American Planning Association*, 58:342–346.
Haggett, P. and R.J. Chorley (1972) *Network analysis in geography*. London: Arnold.
Hall, P. (1995) "A European perspective on the spatial links between land use, development and transport", in D. Banister, ed., *Transport and urban development*, London: Spon.
Jansson, J.O. (1984) *Transport system optimization and pricing*. New York: Wiley.
Jansson, J.O. (1993) "Government and transport infrastructure – pricing", in J. Polak and A. Heertje, eds., *European transport economics*. Oxford: Blackwell.
Jara-Diaz, S.R. (1986) "On the relations between users' benefits and the economic effects of transportation activities", *Journal of Regional Science*, 26:379–391.
Jara-Diaz, S.R. (2000) "Allocation and valuation of travel time savings", in K.J. Button and D.A. Hensher, eds., *Handbooks in transport 1. Transport modelling*. Oxford: Pergamon.
Jara-Diaz, S.R. and C.E. Cortes (1996) "On the calculation of scale economies form transport cost functions", *Journal of Transport Economics and Policy*, 30:157–170.
Kansky, K.J. (1963) "Structure of transportation networks", University of Chicago, Department of Geography Research Paper No. 84.
Keeler, T.E. (1983) *Railroads, freight and public policy*. Washington, DC: Brookings Institution.
Krugman, P. (1991) "Increasing returns and economic geography", *Journal of Political Economy*, 99:183–199.
Krugman, P. (1998) "Space: The final frontier", *Journal of Economic Perspectives*, 12:161–174.
Labbé, M., D. Peeters and J.-F. Thisse (1995) "Location on networks", in: M.O. Ball, T.L. Magnati, C.L. Monma and G.L. Nemhauser eds., *Handbooks in Operations Research and Management Science*, vol.. 8. *Network routing*. Amsterdam: Elsevier.
Mirrlees, J.A. (1972) "The optimum town", *Swedish Journal of Economics*, 74:114–135.
Newman, P.W.G. and J.R. Kenworthy (1989a) *Cities and automobile dependence: A sourcebook*. Aldershot: Gower.
Newman, P.W.G. and J.R. Kenworthy (1989b) "Gasoline consumption and cities: A comparison of US cities with a global survey", *Journal of the American Planning Association*, 55:24–37.
Peeters, D., J.-F. Thisse and I. Thomas (1998) "Transportation networks and the location of human activities", *Geographical Analysis*, 30:355–371.
Schaeffer, K.H. and E. Sclar (1975) *Access for all: Transportation and urban growth*. Harmondsworth: Penguin.
Solow, R.M. (1973) "Congestion costs and the use of land for streets", *Bell Journal of Economics and Management Science*, 4:602–618.
Standing Advisory Committee on Trunk Road Assessment (1994) *Trunk roads and the generation of traffic*. London: HMSO.
Standing Advisory Committee on Trunk Road Assessment (1999) *Transport and the Economy*. London: HMSO.

Venables, A. (1995) "Equilibrium locations of vertically linked industries", *International Economic Review*, 37:341–359.
Venables, A. (1999) "Road transport improvements and network congestion", *Journal of Transport Economics and Policy*, 33:319–328.
Venables, A. and M. Gasiorek (1999) *The welfare implications of transport improvements in the presence of market failure Part: 1*. London: Standing Advisory Committee on Trunk Road Assessment, Department of the Environment and the Regions.
Vickerman, R.W., K. Spiekermann and M. Wegener (1999) "Accessibility and economic development in Europe", *Regional Studies*, 33:1–15.
von Thünen, J.H. (1826) *Der Isolierte Staat in Beziehung auf Landwirtschaft und Nationalökonomie*. Hamburg.

Chapter 6

ECONOMICS OF TRANSPORT NETWORKS

KENNETH J. BUTTON
George Mason University, Fairfax, VA

1. The concept of networks

Transport is a network industry akin to such other sectors as telecommunications and energy distribution. As such it exhibits a set of economic characteristics that influence the way it is viewed by economists and the way it has been treated by regulators (Economides, 1996). These in part stem from the potential separation in the supply of the network infrastructure links (roads, rail track, etc.), the hubs and interchange points (stations, ports, airports, etc.), and the operations (cars, trucks, trains, ships, etc.) (Brooks and Button, 1995). But equally important are the interdependencies that exist between the costs and benefits of providing the various elements in any transportation network.

In some cases the networks are provided by a single mode, but more often transport is provided by a series of interconnected networks; the road network can be combined with the air transport network to offer door-to-door service. In some instance the networks are combined and provided by a single operator, but in others independent actors offer interactive networks of services. Transport networks also overlap with other complementary and substitute networks. Telecommunications can be a complement to transport (a Boeing aircraft involves about 10 000 calls before it can provide a commercial flight) but in other cases it may be a substitute and, for instance, a potential replacement for work travel (Button and Taylor, 2001). Transportation infrastructure can also service a variety of transport networks (e.g., roads cater for passenger and freight trips) but may also meet the needs of other networks (rail rights of way are often employed to carry fiber optic cables).

There are also important links between transportation networks and the spatial economy. The form and technology of a transportation network affects location and production decisions. The pioneering economic analysis of von Thünen and others of spatial economics explained concentric patterns of agriculture around towns largely in terms of transportation economics. The early 20th century

Handbook of Transport Systems and Traffic Control, Edited by K.J. Button and D.A. Hensher

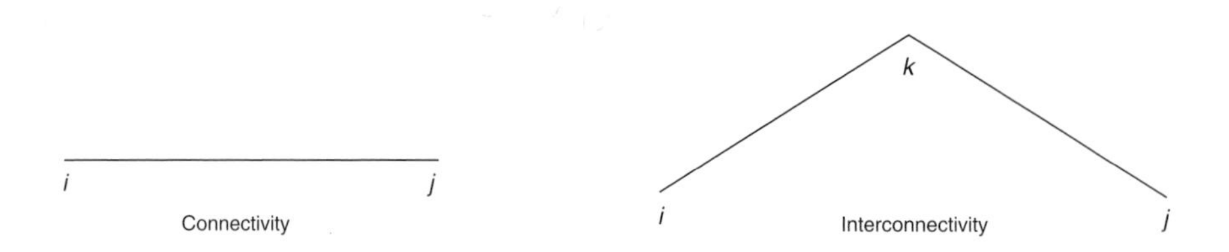

Figure 1. Connectivity and interconnectivity.

phenomenon of axial cities (i.e., with a star-shaped configuration) was due to the radial development of suburban rail and tramway networks.

Networks issues can be considered in several ways. The discussion here largely concerns networks involving interconnectivity. Connectivity occurs, as seen in Figure 1, when there is a connection between two points. Interconnectivity involves at least three nodes, as in the right portion of Figure 1. For it to be worthwhile to develop a network of any configuration rather than simply provide single-link services there must exist some form of network economy. These economies may be on the cost side but they may also reflect the possibility of earning higher revenues on the demand side. On the cost side there may simply be traditional economies of scale linked to having larger operations and the ability to spread fixed costs over a larger number of customers but there may also be additional network benefits associated with particular network configurations.

This contribution, whilst concerned with some of the key economic features of transportation networks, is not comprehensive. It does not cover, for example, the modeling issues that arise when seeking to forecast how any network would be used (see Chapters 2, 3, and 10 in Volume 1 of this series for papers on this topic), nor does it dwell on the important issue of how transportation networks affect activities such as land use development. There is also little discussion of overlapping and complementary networks to transportation, such as telecommunications, although this is an increasingly important topic (see Chapter 27 in Volume 1 of this series).

2. Standardization and networks

To make effective use of networks there is a need for users to be able to access the entire system and potentially to access competing or complementary networks. To adopt the jargon favored by the European Union (E.U.), this introduces notions such as interoperability (the ability of transportation systems to offer harmonized interfaces and an acceptable level of service, thus giving easy access to users) and interconnectivity (the physical linkages and co-ordination of transportation systems to facilitate the transfer of freight or passengers between links or networks).

To attain this level of integration across networks there is a need for some form of standardization. In the transportation context the classic issue was that of rail gauges and, more recently, of electrical power sources for rail engines. In the U.S.A., for example, there were seven different railway gauges in 1860, which impeded internal trade (although this proved useful to the South in the Civil War when the Confederate army could not use Southern track as it advanced!). The problem of different gauges was not unique to the U.S.A. and it is only comparatively recently that large parts of the Spanish rail network were brought into line with other E.U. countries.

Lack of technical standardization imposes costs and these can be increased if there is a lack of institutional co-ordination. The U.K. drives on the left by law, and Continental Europe on the right, and this imposes a cost over the E.U. transportation network. Borders (see Chapter 23) can impose institutional costs if there are different standards on either side or if legal barriers prevent operating units from one country from using the infrastructure network of another (e.g., as with trucking at the U.S./Mexico border).

There are, however, costs in standardization. It may mean adopting what transpires in a technologically dynamic world, a suboptimal standard. It may mean that potential economies of specialization are lost because a generic standard is adopted that is suboptimal for parts of the network. There is also the issue of incentives. If a standard is in place then the economic incentives are to optimize within that constraint. This reduces incentives that could improve the efficiency of a network by introducing new standards or removing them. Changes in standards are costly but can prove efficient; this was seen in Sweden when the country joined the rest of Continental Europe and switched to driving on the right.

The decision as to adopt a uniform standard across a transportation network essentially involves employing a cost–benefit calculation. This basically involves trading off the benefits of optimizing the use of an existing network and its associated technology against the longer-term gains of developing an optimal network and technology.

3. Economies of scale, scope, and density

Network suppliers may benefit from a range of cost features. Traditionally, economies of scale have been seen to reflect declining costs of production as, say, an airline's output increases, but these concepts have been supplemented by notions of economies of scope, networks, and density. Also, on the demand side, ideas of economies of market presence have come to the fore in terms of influencing the optimal scale of network activities. Econometric models have been able to accommodate these features and to allow parameter estimation through the use of flexible form models. The transcendental-logarithmic (translog) function is, for instance, now widely deployed because of its flexibility, although putting input prices on items such as capital still poses serious problems.

Strictly, economies of scope relate to falling costs of providing services as the range of services offered by a carrier increases, while economies of traffic density refer to falling costs as the amount carried increases between any given set of points served. In this sense economies of scale can be seen as a special, single product case of economies of scope. The technical distinction between economies of scale and scope can be seen by reference to the following equation, where C denotes cost and Q output. Economies of scope are assessed as

$$S = \{[C(Q^1) + C(Q^2)] - C(Q^1 + Q^2)\}/\{C(Q^1 + Q^2)\}, \tag{1}$$

where $C(Q^1)$ is the cost of producing Q^1 units of output 1 alone, $C(Q^2)$ is the cost of producing Q^2 of output 2 alone, and $C(Q^1 + Q^2)$ is the cost of producing Q^1 plus Q^2. Economies of scope exist if $S > 0$. There are said to be economies of scale if C/Q falls as Q expands.

These economies may materialize through natural growth in a network but frequently they involving combining existing networks. Market concentration through mergers is only likely, however, if there is adequate incentive for suppliers to combine. Take the simple network case discussed in Varian (1999). If it is assumed that the benefit v of a network is proportional to the number of users on that network (n) and, for simplicity, the constant of proportionality is taken as unity, then, following Metcalfe's law, the value of the network is n^2. A simple calculation shows that combining two networks (1 and 2) of size n_1 and n_2 yields

$$\Delta v_1 = n_1(n_1 + n_2) - n_1^2 = n_1 n_2, \tag{2}$$

$$\Delta v_2 = n_2(n_1 + n_2) - n_2^2 = n_1 n_2. \tag{3}$$

Each network gets equal value (v) from the interconnection. Those on the small network (2) each get considerable value from linking with the large number on the other network. The large number of people on the other network (1) each get smaller additional utility but there are a lot of them. This offers scope for reciprocation, with the networks having settlement-free access to the other network. The problem is that the large supplier may need to keep its market power to allow adequate price discrimination to recover costs and make an acceptable long-term return. In this case the larger concern may merge with the smaller and attain twice the value of interconnecting:

$$\Delta v_1 = n_1(n_1 + n_2)^2 - n_1^2 - n_2^2 = 2n_1 n_2. \tag{4}$$

Suppliers of network services may also benefit from additional revenues as the number of links in their system grows. These benefits come from economies of market presence. Essentially, a supplier of many interconnecting services can meet the needs of a large number of potential customers. For example, a bus company offering a service from A to B via C is effectively meeting the needs of three markets (A→B, A→C, C→B) and can reap revenues from each. If a direct service is offered then only A→B traffic generates revenue. There is a trade-off because indirect services over the longer route are slower, but the additional traffic to and from C allows overhead costs to be spread more widely and hence average fares or rates to be lower.

4. User benefits and costs

There are also user benefits that may or may not be exploited by suppliers of transport network services. Larger networks can offer the advantage of access to a much wider range of destinations and generally greater service frequency. This can be important in terms of making efficient use of operating capital. On any single link in a network, demand may be asymmetrical. The demand to carry coal from a mine to a steel mill, for instance, is seldom balanced by return loads of any kind. This is the "back-haul problem." The existence of a complex network allows scope for a variety of routings that have the potential for minimizing empty running on back-hauls. The recent developments in information technology and in scheduling algorithms have permitted this exploitation of networks to be refined, but even in the days of sailing ships the use of trade winds often involved triangular routings between Europe, the Americas, and Africa to ensure hold capacity was used to the maximum.

Additionally, any individual user helps ensure that this range of options is available to others. What this means in terms of an individual user of an air transport network, for example, is that by paying the full attributable cost of using the service, the user helps ensure that the service is also available for other potential users as well as benefiting from their support of the network.

The main reason for this is that the costs borne by air travelers extend beyond those of simply the fares being paid for the opportunities that larger networks may offer. In Figure 2 it is assumed that the costs C of providing networks rise with the size of the network, proxied by passenger numbers, but at a decreasing rate (i.e., some scale benefits are implied but these diminish with the size of the network). For the user there are increased benefits associated with larger networks (U) but the rate of increase declines with the size of the network. To obtain any passengers there is a minimum size of network (X_{min}) but after X_{max} the additional traffic the transport service supplier will gain from adding links is less than the additional costs, and further expansion will not be justified.

From the perspective of passengers, however, the optimum network is at X_m, where there is a maximum deviation between the financial costs they must pay to cover airline costs and the network benefits to be enjoyed. Passengers, however, will continue to seek flights up to X_{max} since their marginal benefit exceeds the costs they will be charged.

Another way of looking at this is to take the simple case of a positive externality, as seen in Figure 3. Here an individual's or company's demand (marginal benefit) for a transportation service is shown as MB, with the marginal costs involved being MC. For simplicity, it is assumed that there are no negative environmental externalities arising due to congestion. The journey made by the individual helps sustain the network for other potential users so that the marginal social benefit curve, where society is deemed in this club good context to embrace other travelers on the same network, is to the right of the MB curve at MB^*. It is assumed for simplicity that, in drawing this diagram the larger the number of trips taken by the individual, the greater the social benefit. If no allowance is made for this external effect then the fare is set at F and the volume of service used is V^*. Allowing for the externality pushes the optimum volume of service out to V^{**} and the fare up to F^*. There is no incentive for the traveler to move out to V^*, however, because he/she does not recognize these benefits. One policy solution is to subsidize fares down to F^{**}, which would achieve the optimal volume of traffic. Another approach would be to adopt a form of price differentiation that allows costs to be recovered by airlines at an output of V^* by extracting consumers' surplus from intramarginal users of the network.

In addition to the positive externalities that may be associated with larger networks, there are also issues of possible congestion that can arise on links and are often a major problem at hubs (see Chapter 7). There is an optimum level of congestion that that arises simply because there is interaction between those using elements of the network, and without this the facility would be underused in an

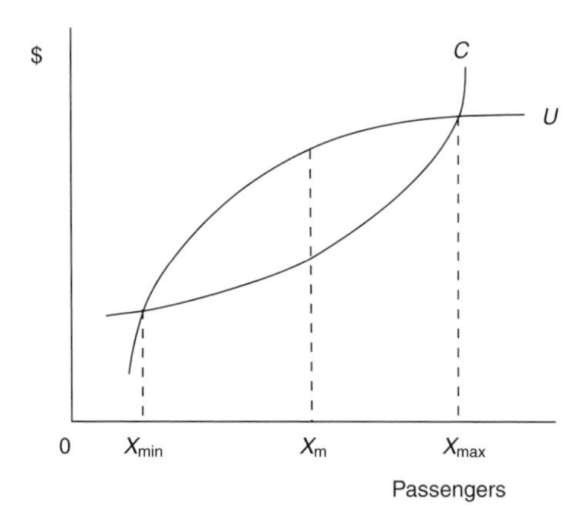

Figure 2. Optimal networks.

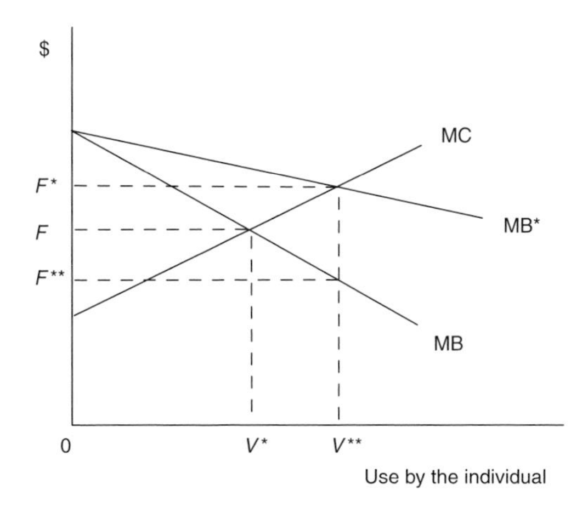

Figure 3. Positive network externalities.

economic sense. The bigger concern is that of excess congestion, when a facility is overused and the benefits that are being derived from it are not maximized. This normally stems from inappropriate pricing of the facilities involved.

The extent to which congestion is an issue depends to some degree on how much of it is already internalized within the fares paid by passengers and cargo shippers. Unlike externalities such as those associated with noise or air pollution, the inefficiency costs of congestion are borne by the users of network services and those that provide those services. In situations where there is only one carrier monopolizing an airport, for example, there is thus an incentive for that carrier to make efficient use of facilities and to allow for congestion costs when setting fares and establishing services. This is not the situation on road networks, where, without congestion charges, users bear the costs implicitly in terms of longer travel times and operating costs but have no incentive to modify their behavior to optimize them.

The issue of how equilibrium is attained across congested networks and the use made of individual links is a topic in its own right and is largely covered in Chapter 7 of this volume and Chapter 21 in Volume 1 of this series. Of particular importance are the generic rules developed by Wardrop (1952), notably his first principle: that at any equilibrium, depending on the degree information is perfect and on whether tolls are charged, the sum of either the actual or the expected average user cost (including time costs) plus the contingent fee should be equal for alternative routes or people would shift from one route to another; this sum should also be equal to the benefits enjoyed by the marginal road user.

5. Hub-and-spoke networks

Transportation networks take a variety of forms. Some of these have been imposed by institutional factors, but there are market forces that tend to favor particular configurations of links and hubs in the absence of regulation and without extreme geographical features. The radial network is perhaps the most common. In urban transport this is seen in the road and rail networks that serve to carry commuters into and out of cities. Historically, they often stem from the days when long-distance transportation (canals and railways) was more efficient than local distribution systems (walking and horses). The economies of scale inherent in focusing the trunk haul on a limited number of nodal points (ports and stations) led to cities growing around terminal and interchange points. Local distribution and collection then radiated out from the hub.

More recently, with the advent of new technologies, this hub-and-spoke configuration has been repeated in the context of road infrastructure, in maritime transportation, and, more recently, in air transport. Air transportation in most countries was, until recently, heavily regulated. The standard regulatory structure, which also applied to international services, was based on regulating entry and fares (or rates in the case of freight) on individual links (Button and Stough, 2000). The result was that a carrier would have a pattern of operations of the type seen in Figure 4. There would be no focal point, and interlining with the carrier's own services to provide an integrated network would be very limited. If someone wanted to travel from C to G, for example, the carrier could well offer a service that went from C to B to I to G with the plane stopping at each point *en route*. Often the core routes were established prior to regulation, with the carrier gaining additional ones piecemeal over time.

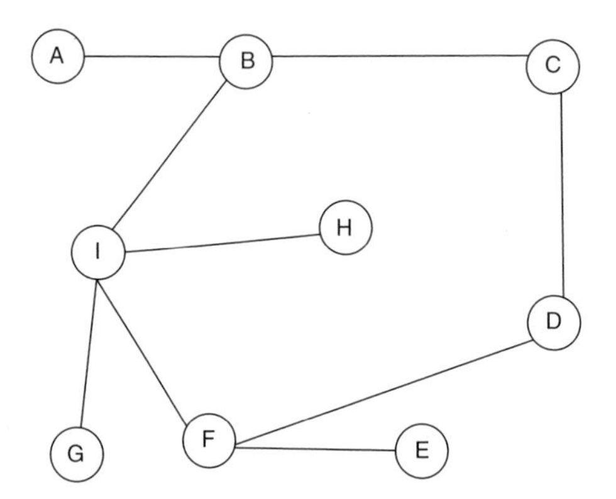

Figure 4. Prederegulation airline linear route structure.

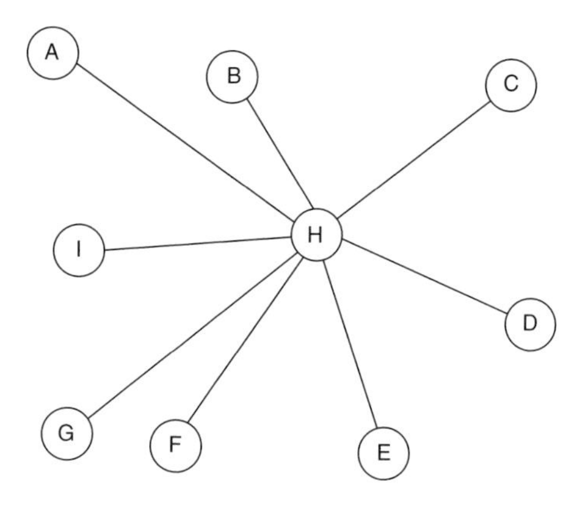

Figure 5. U.S. postderegulation hub-and-spoke airline route structure.

The relaxation of many of these regulations in most major air transportation markets has allowed more efficient networks to emerge. The outcome has differed according to the nature of the carriers involved. In particular, there have been important differences between the air cargo operators and predominantly passenger carriers. One reason is that passengers are sensitive to routings, while packages, being inanimate, have no such feelings and consignors generally have little preference regarding routes, provided time windows are met. Cargo carriers find it easier, therefore, to focus on megahubs and to channel considerable volumes of traffic through them at times passengers would see as antisocial. The growth of package services also means that good road accesses allowing intermodal operations with trucking play a role in the hubbing decision.

There were, from the passengers' perspective, some advantages to this structure in that numerically many direct services existed, although, set against this, frequency was often low and fares were high, because of the inability of airlines to reap the synergy of cost savings from a more market-oriented route structure. From a technical perspective, monopoly control over routes combined with the scale benefits of operating large aircraft led to a tendency to move to the use of wide-bodied aircraft as they became available, and this further reduced service frequency. The system also did not mean that there were no hub airports where consolidation of traffic could take place. There were major interchange nodes but the passengers often had to change between carriers as well as between aircraft. Schedules were not fully co-ordinated, and through ticketing could pose problems.

The deregulation of the U.S. domestic air transport market from 1977 showed how market incentives could lead to a very rapid adoption of hub-and-spoke operations across transportation networks (Figure 5). The U.S. airlines (with one

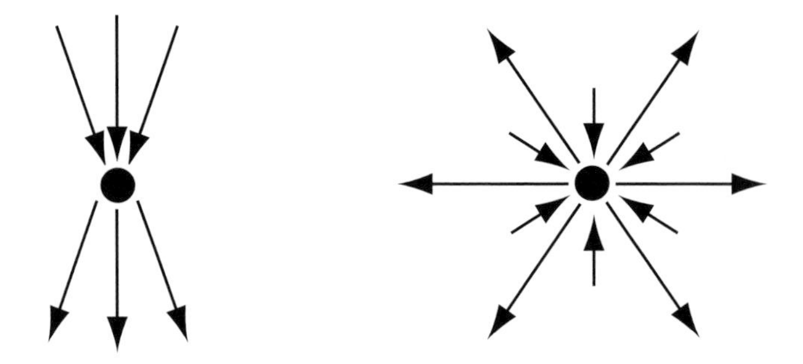

Figure 6. "Hourglass" (left) and "hinterland" (right) hubs.

or two notable exceptions such as Pan American) were quick to exploit the potential economies of scale, scope, and density that a freer market situation permitted. On the demand side there also came the gradual appreciation that network services provided economies of network presence that could add to the revenue flow. The situation is not unique to passenger transportation, and undertakings such as the post office in all countries have long exploited the cost advantages of hub-and-spoke network considerations.

In Figure 5, city H, which previously enjoyed only one air link, becomes the hub for an entire range of airline services. Why H? In most cases with passenger transportation an airport hub is located in a major city that generates significant "local" traffic as well as being the point of interchange for transferring, "flow" traffic. The airport must also have the physical capacity (runways, terminals, etc.) to handle the high volume of transit traffic. Its selection is also determined by location – a central position normally allowing for linkages to a larger number of spoke airports. The same types of argument can apply to seaports or rail interchanges.

With this form of hubbing structure, flights are funneled in banks into a number of large hubs where substantial numbers of passengers change aircraft to complete their journeys. These banks involve the co-ordinated arrival of a large number of flights in a short space of time and then an equally co-ordinated departure of flights within a narrow time window. Larger hubs may well have up to seven or more such banks a day. Travel time was be longer for many people but fares fell and the range of potential flight combinations available to any particular destination expanded considerably.

There have been a number of attempts to delineate different forms of airline hub system. Doganis and Dennis (1989), for example, separate the single hub model into "hourglass" and "hinterland" hubs. The former is represented in the left part of Figure 6, where the hub is operated with flights from one region to points broadly in the opposite direction. The hinterland hub on the right feeds short-haul connecting traffic to long haul-flights. The hourglass hub operation tends to use aircraft all of a similar size, whereas the hinterland hub has aircraft of mixed sizes. Oum and Tretheway (1990) develop structures that combine hub-and-spoke systems with linear systems to allow for the fact that some routes offer sufficient demand that a carrier will find it economical to provide some direct services over a hub.

From a market structure perspective, the creation of hubs has led away from competition along links in a network to competition between networks themselves. If one considers Figure 5 again and superimposes another air service hub-and-spoke network supplied by a second carrier focused, say, on city Z, which is not shown, where the second carrier offers services from Z to I, D, and E, then competition exists between I and D, I and E, and D and E, although there are no direct services between any of the pairs. Network competition may be between networks involving the same mode, but equally it may be between hub-and-spoke systems involving different modes (e.g., air and high-speed rail in Europe).

6. Public sector assessment of network investment

While many networks are provided by the private sector, others are often seen as a public sector responsibility. A number of reasons exist for this such as the view that some networks are public goods, it is impossible to regulate private monopoly networks, or that co-ordination of links would be suboptimal left in private hands. These arguments are not considered here but, rather, some aspects of public sector network provision are discussed.

Private networks are developed on commercial criteria whereas public authorities often used social appraisal techniques, and most notably cost–benefit analysis. This involves making network additions that generate a positive increase in consumer surplus. Most transport infrastructure forms a link in a much larger, interacting network and changes in any one link tend to affect demand on competitive and complementary links. Although this sort of complexity exists for virtually all forms of transport, the problem of assessing the overall effect on road transport of improving a single link has, because of the dominance of this form of transport in modern society, attracted most attention. The general principles are, however, more widely applicable

If there are two roads, one from X to Y and the other X to Z, where Y and Z are to some degree substitute destinations, then an improvement in route XY will affect three groups. It is assumed that all demand curves are linear and that the pre-investment traffic flows on XY and XZ are T_{xy} and $(T_{xz} + R)$ respectively. The three groups of users to consider are then:

(1) Existing users who remain on their original routes (T_{xy} and T_{xz}). These will gain consumers' surplus because those on route XY will now be using a higher-quality facility while those on XZ will benefit from reductions in demand for this route as some former users switch to the improved XY. If this latter traffic which has diverted from XZ to XY is denoted as R, then the total benefit to those remaining loyal to their initial routes is

$$T_{xy}(C_1 - C_2) + T_{xz}(D_1 - D_2), \tag{5}$$

where C_1, C_2, D_1, D_2 are the pre- and postinvestment costs by roads XY and XZ respectively.

(2) Generated traffic consisting of those not previously traveling (G_{xy} and G_{xz}). On average (with linear demand curves) each of these groups of new road users will benefit by half as much as existing, non-switching traffic. The total benefit of the investment to this group will thus be

$$0.5G_{xy}(C_1 - C_2) + 0.5G_{xz}(D_1 - D_2). \tag{6}$$

(3) Diverted traffic that switches from route XZ to route XY as a consequence of the investment (R). The switch, given the free choice situation open to travelers, must leave this group better off – they would not have switched otherwise – and the additional welfare they enjoy is equal to half half of the difference in benefit between the cost reductions on the two routes,

$$R[(D_1 - D_2) + 0.5\{(C_1 - C_2) - (D_1 - D_2)\}], \tag{7}$$

which may be reduced to

$$0.5R\{(C_1 - C_2) + (D_1 - D_2)\}. \tag{8}$$

The total benefit (TB) of the investment is the summation of these three elements, namely

$$\begin{aligned}\text{TB} = {} & T_{xy}(C_1 - C_2) + T_{xz}(D_1 - D_2) + 0.5G_{xy}(C_1 - C_2) + \\ & 0.5G_{xz}(D_1 - D_2) + 0.5R\{(C_1 - C_2) + (D_1 - D_2)\}.\end{aligned} \tag{9}$$

This equation offers a useful pragmatic basis for assessment. If we denote the *ex ante* and *ex post* traffic flows on XY as and $Q_{xy} = T_{xy}$ and $Q'_{xy} = (T_{xy} + G_{xy} + R)$, and *ex ante* and *ex post* flows on XZ as $Q_{xz} = (T_{xz} + R)$ and $Q'_{xz} = (T_{xz} + G_{xz})$, then substituting into eq. (9) gives

$$\text{TB} = 0.5(Q_{xz} + Q'_{xy})(C_1 - C_2) + 0.5(Q_{xz} + Q'_{xz})(D_1 - D_2) \tag{10}$$

or more generally the oft-cited "rule of half,"

$$\text{TB} = 0.5\sum_n (Q_n + Q'_n)(C_n - C'_n). \tag{11}$$

The rule of half can be applied to transport schemes that interact with other components of the transport system where demand curves are linear. It must, however, be used with a degree of circumspection when routes are complementary, where demand for the non-improved links may shift to the right, but the broad principle applies. One problem is that there are many possible sequences in which the price changes on routes XY and XZ could follow; each would yield a different level of aggregate social welfare. The general measure set out in eq. (11) assumes that the demand fluctuations are linear in their own prices and with respect to cross-price effects. How realistic this assumption is depends on circumstances, but the rule of half has proved a very powerful tool in road appraisal activities.

7. Regulatory issues

The widespread deregulation of network industries that can be traced back to the freeing of trucking in the U.K. under the 1968 Transport Act has aroused considerable interest not only in the economic theory of transportation and other networks but also in measuring their performance. In particular, because of the sudden nature of the changes (which reduced the "noise" created by parallel events) and the time that has elapsed, the U.S. transportation market has generated significant numbers of studies – Peltzman and Winston (2000) offer a set of recent U.S. papers on the subject, whilst Button et al. (1998) cover European topics. Overall, it can be concluded that in most cases deregulation has proved relatively successful. The aim here is not to go into detail about whether individual reforms have yielded an economic benefit, but rather to offer some insights into the reasons economic regulation was felt necessary in the first place and what has replaced it.

The economic regulation of transportation networks such as railways initially came about because of fears of monopoly exploitation, because of the perceived need to co-ordinate systems, and through a desire to ensure almost ubiquitous geographical coverage for political and social-integration reasons. This resulted in some countries such as the U.S.A. and, initially, the U.K., in regulating rates and establishing a set of carriage rules. In other countries, such as Canada, policy sought to inject competition by fostering the development of more than one physical network across the country. In France it was felt that the monopoly issue applied only to rail track, which was co-ordinated and provided nationally, and that competition was effective amongst different users of the track network.

Later, the advent of buses and trucking raised another issue regarding network features. Transportation network infrastructure is generally costly, inflexible, and durable but the services that it provides have an extremely low marginal cost because they are non-durable. The cost of carrying an extra passenger on a plane is virtually zero but, since the output vanishes the moment the plane's door is shut, the service is not durable. The cost curve associated with a service on a transportation network is steeply downward sloping, but the incentive is to try to pull in as much traffic as possible before the service vanishes. This leads to pressures to price down to marginal cost and, *ipso facto*, the tendency to make losses. The result is undersupply and market instability. This sort of problem is often described as a market lacking a "core" – see Telser (1994). Regulations to limit supply and/or control rates and fares on air and road networks where there was a separation of the network infrastructure from its use became widespread in the 1920s and 1930s for these economic reasons. Subsidies, either direct or cross, were also common to cover fixed cost deficits.

More recent policies have focused on stimulating competition. This may be "between" networks (as in the case of air transport), "for" networks (as in the case of the U.K. bus industry, where many services are put out to competitive tender for set periods of time), or "on" networks (where track has been a monopoly track network and is separated from competitors, who may make use of it, as with part of the Swedish railway network). In some instance, as with the U.K. passenger rail network, it may involve a combination of measures. The use of the physical infrastructure of many networks is, thus, now largely controlled through economic instruments. One notable exception is that involving the use by automobiles of road networks. With a very small number of exceptions such as toll roads and urban-area licensing schemes, the use of roads is controlled by physical restraints or through time costs (namely congestion). The econometric evidence is that such methods result in a highly inefficient use of road networks (Newberry, 1989).

References

Brooks, M. and K.J. Button (1995) "Separating track from operations: a look at international experiences", *International Journal of Transport Economics*, 22:235–260.

Button, K.J. and R. Stough (2000) *Air transport networks: Theory and policy implications*. Cheltenham: Edward Elgar.

Button, K.J. and S.Y. Taylor (2001) "Towards an economics of the internet and electronic commerce", in: S.D. Brunn and T.R. Leinbach eds., *The wired worlds of electronic commerce*. London: Wiley.

Button, K.J., P. Nijkamp and H. Primus, eds. (1998) *Transport networks in Europe: Concept, analysis and policies*. Cheltenham: Edward Elgar.

Doganis, R. and N.P.S. Dennis (1989) "Lessons in hubbing", *Airline Business*, March:42–45.

Economides, N. (1996) "The economics of networks", *International Journal of Industrial Organisation*, 14:673–699.

Hensher, D.A. and K.J. Button (2000) *Handbooks in transport 1. Transport modelling*. Oxford: Pergamon.

Newberry, D.M.G. (1989) "Cost recovery from optimally designed roads", *Econometrica*, 56:165–185

Oum, T.H. and M.W. Tretheway (1990) "Airline hub-and-spoke system", *Transportation Research Forum Proceedings*, 30:380–393.

Peltzman, S. and C. Winston, eds. (2000) *Deregulation of network industries: What is new?* Washington: Brookings Institution.

Telser, L.G. (1994) "The usefulness of core theory in economics", *Journal of Economic Perspectives*, 8:151–164.

Varian, H.R. (1999) "Market structure in the network age", in: *Understanding the Digital Economy, Conference*, Washington, DC.

Wardrop, J. (1952) "Some theoretical aspects of road traffic congestion", *Proceedings of the Institute of Civil Engineers*, 1:325–378.

Chapter 7

TRAFFIC CONGESTION AND CONGESTION PRICING

ROBIN LINDSEY*
University of Alberta, Edmonton

ERIK VERHOEF*
Free University of Amsterdam

1. Introduction

For several decades growth of traffic volumes has outstripped investments in road infrastructure. The result has been a relentless increase in traffic congestion. Congestion imposes various costs on travelers: reduced speeds and increased travel times, a decrease in travel time reliability, greater fuel consumption and vehicle wear, inconvenience from rescheduling trips or using alternative travel modes, and (in the longer run) the costs of relocating residences and jobs. The costs of increased travel times and fuel consumption alone are estimated to amount to hundreds of dollars per capita per year in the U.S.A. (Schrank and Lomax, 1999), and comparable values have been reported for Europe.

Traffic congestion is a consequence of the nature of supply and demand: capacity is time-consuming and costly to build and is fixed for long time periods, demand fluctuates over time, and transport services cannot be stored to smooth imbalances between capacity and demand. Various policies to curb traffic congestion have been adopted or proposed over the years. The traditional response is to expand capacity by building new roads or upgrading existing ones. A second method is to reduce demand by discouraging peak-period travel, limiting access to congested areas by using permit systems and parking restrictions, imposing bans on commercial vehicles during certain hours, and so on. A third approach is to improve the efficiency of the road system, so that the same demand can be accommodated at a lower cost. Retiming of traffic signals, metering access to highway entrance ramps, high-occupancy vehicle lanes, and advanced traveller information systems are examples of such measures.

*The authors would like to thank Richard Arnott, André de Palma, Claude Penchina, Stef Proost, and especially Ken Small for corrections and insightful comments on an earlier version of this paper. Any remaining errors, however, are the authors' responsibility alone. Erik Verhoef's research has been supported by a fellowship of the Royal Netherlands Academy of Arts and Sciences.

Handbook of Transport Systems and Traffic Control, Edited by K.J. Button and D.A. Hensher

This chapter is concerned with congestion pricing as a tool for alleviating traffic congestion. The insight for congestion pricing comes from the observation that people tend to make socially efficient choices when they are faced with all the social benefits and costs of their actions. As just noted, various demand management tools to accomplish this can be used. But congestion pricing is widely viewed by economists as the most efficient means because it employs the price mechanism, with all its advantages of clarity, universality, and efficiency. Pigou (1920) and Knight (1924) were the first to advocate it. But it was the late William Vickrey, who steadfastly promoted congestion pricing for some 40 years, who was arguably the most influential in making the case on both theoretical and practical grounds. In one of his early advocacy pieces, Vickrey (1963) identified the potential for road pricing to influence travelers' choice of route and travel mode, and its implications for land use. He also discussed alternative methods of automated toll collection. Another of his early proposals was to set parking fees in real time as a function of the occupancy rate. An overview of Vickrey's contributions to pricing of urban private and public transport is found in Arnott et al. (1994).

As Vickrey's work makes clear, true congestion pricing entails setting tolls that match the severity of congestion, which requires that tolls vary according to time, location, type of vehicle, and current circumstances (e.g., accidents or bad weather). Congestion pricing is common in other sectors of the economy – from telephone rates and air fares to hotels and public utilities. But despite the efforts of Vickrey and other economists, congestion pricing is still rarely used on roads. Tolls are not charged on most roads, and fuel taxes do not vary with traffic volumes. And costs of registration, licensing, and insurance do not even depend on distance traveled. Nevertheless, the number of applications and experiments in road pricing is slowly growing, spurred on by the combined impetus of worsening traffic conditions and advances in automatic vehicle identification technology. Descriptions of various road pricing schemes, including Singapore's pioneering toll system, Scandinavian toll rings, and Californian pay-lanes, are found in Gómez-Ibáñez and Small (1994) and Small and Gómez-Ibáñez (1998).

This chapter is organized along similar lines to the review of congestion modeling in Lindsey and Verhoef (2000). Section 2 starts by outlining the basic economic principles of congestion pricing in a simple static ("time-independent") setting with one road. Section 3 adds a time element by considering travelers' time-of-use decisions and time-varying tolls. Various complications are addressed in Section 4, including pricing in networks, heterogeneity of users, stochastic congestion, interactions of the transport sector with the rest of the economy, and tolling on private roads. Section 5 considers the implications of congestion pricing for optimal road capacity. Explanations for the long-standing social and political resistance to road pricing are offered in Section 6, and conclusions are drawn in Section 7. Due to space constraints, some topics related to congestion pricing are not covered in this review. There is no explicit treatment of freight transportation.

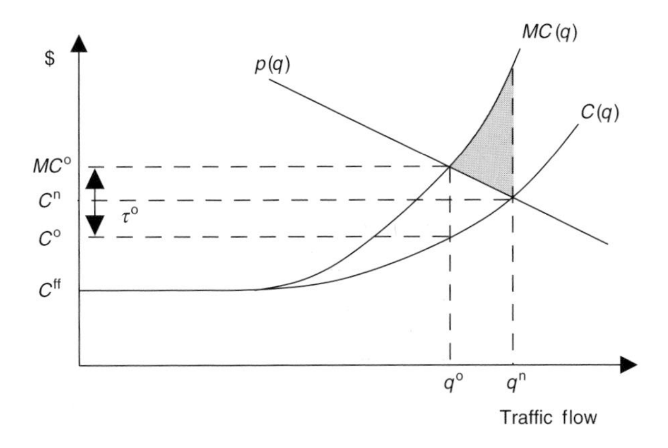

Figure 1. Optimal road pricing in a time-independent model.

Nothing is said about the implications of congestion pricing for urban structure or the location of new developments. And only passing mention in Section 4 is given to the potential effects of congestion pricing on traffic noise, pollution, and traffic accidents.

2. Congestion pricing in time-independent models

The basic principles of congestion pricing can be illustrated in the following simple setting. Consider one origin and one destination connected by a single road. Individuals make trips alone in identical vehicles. Traffic flows, speeds, and densities are uniform along the road and independent of time. Equilibrium in this setting is described in Figure 1, which is due to Walters (1961). The horizontal axis depicts traffic flow or volume: the rate at which trips are initiated and completed. The vertical axis depicts the price or "generalized cost" of a trip – which includes vehicle operating costs, the time costs of travel, and any toll. At low volumes vehicles can travel at the free-flow speed, and the trip cost curve, $C(q)$, is constant at the free-flow cost C^{ff}. At higher volumes congestion develops, speed falls, and $C(q)$ slopes upwards. (Figure 1 ignores the possibility of "hypercongestion" that would cause $C(q)$ to bend backwards on itself; see Lindsey and Verhoef, 2000.)

If flow is interpreted to be the quantity of trips "demanded" per unit of time, then a demand curve $p(q)$ can be added to Figure 1 to obtain a supply-demand diagram. The demand curve is assumed to slope downwards to reflect the fact that, as for most commodities, the number of trips people want to make decreases with the price. The unregulated "no-toll" equilibrium occurs at the intersection of $C(q)$ and $p(q)$, resulting in an equilibrium flow of q^{n} and an equilibrium price of C^{n}. Since "external benefits" of road use are not likely to be significant (benefits are normally either purely internal or pecuniary in nature), $p(q)$ specifies both the private and the marginal social benefit of travel. Total social benefits can thus be measured by the area under $p(q)$. Analogously, $C(q)$ measures the cost to the traveler of taking a trip. If external travel costs other than congestion, such as accidents and air pollution, are ignored, then $C(q)$ measures the average social cost of a trip. The total social cost of q trips is then $TC(q) = C(q)q$, and the marginal social cost of an additional trip is $MC(q) = \partial TC(q)/\partial q = C(q) + q\, \partial C(q)/\partial q$.

The social optimal is found in Figure 1 at the intersection of $MC(q)$ and $p(q)$, where the marginal willingness to pay for trips is MC^{o} and the number of trips, q^{o}, is less than in the unregulated equilibrium. The optimum can be supported as an equilibrium if travelers are forced to pay a total price of MC^{o}. Because the price of a trip is the sum of the individual's physical travel cost and the toll, the requisite toll is $\tau^{o} = MC(q^{o}) - C(q^{o}) = q^{o}\, \partial C(q^{o})/\partial q$, where $q^{o}\, \partial C(q^{o})/\partial q$ is the marginal congestion cost imposed by a traveler on others. This toll is known as a "Pigouvian" tax, after Pigou (1920).

The fact that a toll is required to support the social optimum reveals the fallacy that travelers fully pay for the congestion they cause through the time they personally lose. Analogously, a person squeezing through a crowd, a shopper queuing in line at a supermarket counter, or a person reading a popular library book after a long wait for it, imposes costs on others that he/she does not him/herself bear. Note, however, that at the optimum as drawn in Figure 1, some congestion remains since the generalized cost net of the toll, C^{o}, exceeds the free-flow cost, C^{ff}. Efficient tolling therefore does not necessarily dictate that congestion be eliminated.

The efficiency gain derived from the optimal toll can be expressed as the increase in social surplus, defined by the reduction in total costs minus the reduction in total benefits due to the decrease in traffic. This gain is measured by the shaded area in Figure 1. The toll revenue, $q^{o}\tau^{o}$, nets out because it is a transfer from road users to the government. Nevertheless, the transfer leaves road users worse off. The q^{o} users who continue to travel per unit time suffer a cost increase of $MC^{o} - C^{n}$. And the $q^{n} - q^{o}$ individuals who stop traveling suffer a loss of surplus that ranges from zero for the pretoll marginal user at q^{n}, to $MC^{o} - C^{n}$ for the new marginal user at q^{o}. These losses are the root of opposition to congestion pricing. As discussed in Section 6, ways are being sought of using toll revenues to make congestion pricing more politically palatable.

3. Congestion pricing in time-dependent models

Time-dependent models of congestion build on time-independent models by adding two elements: a specification of how travel demand depends on time, and a specification of how traffic flows evolve over time and space. To maintain focus on the time elements of congestion pricing, attention will be limited as in Section 2 to a single road joining one origin and one destination. Until near the end of Section 3 it will also be assumed that the total number of trips is fixed, i.e., price inelastic. But heterogeneity in the trip-timing preferences and time costs of travelers is allowed.

First consider the modeling of demand. In Vickrey's (1969) pioneering approach, an individual i is assumed to have a preferred time t_i^* to complete a trip, and to incur a *schedule delay cost* $D_i(t - t_i^*)$ if arriving at time t instead, where $D_i(0) = 0$ and $D_i(x) \geq 0$ for $x \neq 0$. Let $T(t)$ denote travel time or trip duration, α_i denote i's unit cost of travel time, and $\tau(t)$ denote the toll (if any) at time t. The cost incurred by i in arriving at time t is assumed to be linear in trip duration and additively separable:

$$C_i(t) = \alpha_i T(t) + D_i(t - t_i^*) + \tau(t). \tag{1}$$

Time-dependent models also specify how speed and flow evolve over time and space. Various modeling approaches have been developed; see Hall (1999) and Lindsey and Verhoef (2000). It is assumed here that traffic flow is governed by a form of flow congestion with no overtaking possible, consistent with what is assumed for the steady state in Figure 1. Details of how flow changes with time and location on the road are unnecessary for our purposes here and are therefore omitted. Vickrey's bottleneck model provides an alternative description of supply, in which congestion takes the form of queuing. The bottleneck model yields distinctive results in terms of congestion pricing that will be noted in due course.

This section focuses on first-best congestion pricing, leaving second-best pricing considerations to Section 4. The analysis proceeds by first characterizing the no-toll equilibrium (NTE), then the social optimum (SO), and finally identifying the effects of congestion pricing by comparing the NTE and SO.

The NTE is a Nash equilibrium in which each individual minimizes his/her trip cost defined by (1) with $\tau(t) = 0$, while taking the travel time choices of other travelers as given. Individual i's choice of t can be characterized using indifference curves by fixing the trip cost parametrically at C_i and solving (1) with $\tau(t) = 0$ for the implied travel time, denoted by $T_i(t, C_i)$:

$$T_i(t, C_i) = [C_i - D_i(t - t_i^*)]/\alpha_i. \tag{2}$$

$T_i(t, C_i)$ can be interpreted as a *congestion delay indifference curve*. The absolute slope of $T_i(t, C_i)$, $||D_i'(t - t_i^*)||/\alpha_i$, reflects i's marginal willingness to incur delay in order to arrive closer to t_i^*. The curve is steep if either the marginal schedule delay cost is high or the unit travel time cost is low. An individual with a steep curve can be said to have a high *congestion tolerance*.

Figure 2a depicts an NTE for a time span encompassing a morning peak period. The curve $T^{\mathrm{n}}(t)$ shows how travel time starts rising above the free-flow travel time (T^{ff}) at time t_0^{n}, grows smoothly to a maximum, and then falls back to free-flow

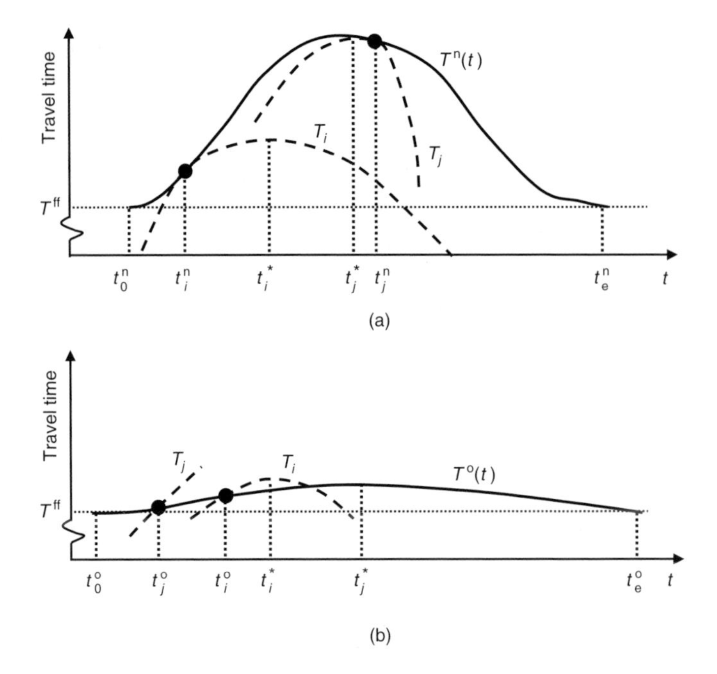

Figure 2. (a) No-toll equilibrium. (b) Social optimum.

conditions at t_e^n. Choice of trip time is shown for two individuals, *i* and *j*. Individual *i* is a highly paid professional with an early preferred work start time x, a very high value of time (α_i), and a strong but somewhat weaker preference for arriving on time – so that *i* has a relatively low congestion tolerance. Individual *i* chooses to arrive early at a time x where the travel time curve $T^n(t)$ is tangent to the dashed curve T_i: the lowest attainable of *i*'s congestion delay indifference curves. By choosing to arrive early, *i* gains a saving in travel time that more than compensates for the schedule delay. Individual *j* is a service-sector employee with an official work start time x near the peak, a relatively low value of time (α_j), and little aversion to arriving early, but a strong aversion to arriving late because of the risk of employer censure. Individual *j* chooses to arrive late at $t_j^n > t_j^*$ when travel time is declining. But because *j* has a high congestion tolerance for avoiding late arrival, *j* is unwilling to arrive late by very much.

In the NTE, the travel time curve $T^n(t)$ forms an upper envelope of all the travelers' equilibrium indifference curves, its slope at each point matching the congestion tolerance of the traveler who arrives then. For reference in Section 4, note that heterogeneity in preferred arrival times has a moderating effect on congestion: when some individuals want to arrive early and others late, there is less competition for road space than if everyone wants to arrive at the same time. Heterogeneity in congestion tolerance also has a moderating influence: individuals with a low tolerance travel at the beginning and at the end of the peak period contribute relatively little to the buildup of congestion.

Now consider the social optimum (SO). The SO is derived by choosing the time at which each individual arrives to minimize the sum of aggregate schedule delay costs and aggregate travel time costs. This is an optimal control problem that involves both a choice of the *rate* at which individuals arrive and a choice of the *order* in which they arrive. (The relationship between the arrival rate and the departure rate from the origin is determined by the flow congestion curve. Given no overtaking, the *order* of departure and the order of arrival coincide.) The optimal arrival rate is governed by a trade-off between schedule delay costs and travel time costs. A high arrival rate compresses the arrival period and reduces schedule delay costs, but boosts congestion and travel delay. Correspondingly, a low arrival rate increases schedule delay costs but dampens congestion. Let $(t_0^{\mathrm{o}}, t_{\mathrm{e}}^{\mathrm{o}})$ denote the arrival period. The arrival rate at t_0^{o} is restricted to maintain free-flow speed, since otherwise it would be better to let the first individual arrive a moment earlier to prevent congestion, with only a minimal increase in schedule delay. Similarly, the arrival rate at $t_{\mathrm{e}}^{\mathrm{o}}$ is restrained to free-flow conditions. In between t_0^{o} and $t_{\mathrm{e}}^{\mathrm{o}}$, speeds are below the free-flow level. But travel time costs are still reduced relative to the NTE. For this reason the order of arrival in the SO is governed primarily by the goal of minimizing schedule delay costs, which dictates that individuals with strong arrival time preferences arrive at, or near to, their preferred times.

Figure 2b shows an SO that might obtain with the same set of travelers as in the NTE of Figure 2a. (The indifference curves T_i and T_j in Figure 2b are explained later.) Arrivals occur over a longer time interval than in the NTE, and the travel time curve $T^{\mathrm{o}}(t)$, is lower and flatter than $T^{\mathrm{n}}(t)$. Individual *i* still arrives early, at t_i^{o}. But since *i* has strong arrival time preferences, *i* arrives close to t_i^*. Individual *j* on the other hand has only a weak aversion to arriving early. So *j* is scheduled to arrive early in the rush hour and *before i*, rather than after t_j^* as in the NTE. Thus, the SO not only involves changes in individual arrival times relative to the NTE, but can also feature changes in their arrival order.

The properties of the SO differ somewhat in the bottleneck queuing model, in which travel speed remains at free-flow speed for flows right up to the bottleneck's capacity. Flow is maintained at capacity throughout the travel period to maximize throughput while avoiding any queuing. Therefore there is no trade-off between schedule delay costs and travel time costs. Because the arrival rate matches bottleneck capacity in both the SO and the NTE, the duration of the travel period is the same, although arrivals can begin either earlier or later in the SO than in the NTE.

With descriptions of the NTE and SO to hand, it is now possible to consider congestion pricing. The question naturally arises of whether the SO can be decentralized by tolling. Arnott and Kraus (1998) show that it indeed can be as long as travelers cannot overtake each other, and provided the toll can be varied freely over time. This is a deep result that takes time to appreciate fully. It follows essentially from the fact that the only choice drivers are assumed to have is when to arrive, and the externalities associated with arrival time can be fully internalized through a time-varying toll.

For ease of reference, the optimal time-varying congestion toll will henceforth be called a fine toll. The *fine toll* incorporates both a static component analogous to the Pigouvian tax in Section 2, and a dynamic component; see Arnott and Kraus (1998, eq. 20). (Carey and Srinivasan (1993) derive an equivalent toll for a model with exogenous trip timing. Yang and Huang (1997) also derive the toll for a variant of the bottleneck model in which the bottleneck's capacity depends on the length of the queue behind it.) Because the fine toll depends not only on the flow congestion technology, but also on the joint frequency distribution in the population of preferred arrival times, values of time, and schedule delay costs, its time evolution can be quite complex. And the toll does not necessarily rise and fall in perfect synchrony with contemporaneous congestion (Carey and Srinivasan, 1993). But because free-flow conditions are maintained at t_0^o and t_e^o, the toll must be the same at these times, as well as before and after the peak period when there is no congestion. With price-inelastic demand any constant amount can be added to or subtracted from the fine toll throughout the day without upsetting the SO.

Reconsider the SO shown in Figure 2b. Individual i's congestion delay indifference curve T_i is steeper at t_i^o than the travel time curve $T^o(t)$. To induce i to arrive at t_i^o, rather than earlier or later, the fine toll must increase at the appropriate rate. Similarly, the toll must increase at the appropriate rate to induce individual j to arrive at t_j^o. Later in the travel period, the toll has to fall in order to return at t_e^o to the level at t_0^o.

How is the fine toll affected if demand is price-elastic? In the static model of Section 2 in which demand is elastic, the optimal congestion toll is positive in order to deter individuals from taking socially unwarranted trips. Thus, in the dynamic model too it would appear to be necessary to add a positive constant to the fine toll in order to restrict demand. Yet Arnott and Kraus (1998) show that this is not the case (see also Carey and Srinivasan, 1993). The fine toll supports not only an optimal time pattern of trips conditional on a given demand, but also the optimal set of users. (As discussed in Section 4, this is not true when the time variation of tolling is constrained.) To see this recall that there is no congestion at the beginning or end of the travel period, so that individuals who arrive then (or outside the peak period) do not create a congestion externality and therefore should pay no toll.

Consider now the efficiency gains from the fine toll. The toll brings about a reduction in travel time costs – which in the bottleneck model amounts to the full cost of queuing in the NTE. The toll has opposing effects on aggregate schedule delay costs: by spreading out the travel period it tends to boost costs, but by reordering individual arrivals according to strength of travel time preference it reduces costs. The net effect on schedule delay costs can go either way, so that the efficiency gains from tolling can be greater or smaller than the savings in travel time costs. (In the bottleneck model there is no spreading of the travel period, so that schedule delay costs either remain unchanged or fall, and efficiency gains equal or exceed the savings in travel time costs.)

Tolling of course affects the welfare of travelers. In the simple world of Section 2 tolling leaves travelers unequivocally worse off. The case is not as clear-cut in the dynamic model because of the efficiency gains derived from altering trip timing. Indeed, depending on the congestion technology and the joint frequency distribution in the population of preferred arrival times, values of time, and schedule delay costs, total private costs can rise or fall. On balance, individuals with high values of time (high α) stand to gain more (or to lose less). In Figure 2, for example, individual i, with the very high value of time, enjoys both a reduction in travel time and a decrease in schedule delay in return for paying the toll.

Several policy lessons can now be drawn. First, congestion pricing not only reduces travel times but also affects schedule delay costs. A cost–benefit analysis that considered only travel times could be biased either for or against a congestion pricing project, and might lead either to unwarranted acceptance or to unwarranted rejection of it. Second, the efficiency gains from congestion pricing are of the same order of magnitude as the toll revenue, and can even exceed it. By contrast, the efficiency gains from flat tolls computed using static models can be dwarfed by the toll transfers involved. This suggests that the economics of dynamic congestion pricing schemes are not as sensitive to the costs of infrastructure and operation as are the economics of tolling schemes in static models. Third, congestion pricing has welfare distributional effects on travelers that tend to favor those with high values of time. Because value of time is positively correlated with income, this is consistent with the conventional view that tolling is regressive. Finally, under the fine toll all individuals pay the full marginal social costs of their trips, regardless of their respective characteristics and of when they travel. Section 4, following, extends consideration to various real-world complications to examine the robustness of this auspicious result.

4. Second-best issues in congestion pricing

Sections 2 and 3 have outlined the principles of congestion pricing when tolls can be set to match the external costs generated by each traveler. Such pricing is called "first-best" congestion pricing because it supports a first-best optimum in which roads are used at maximum efficiency. Although useful as a theoretical benchmark, first-best pricing is increasingly recognized as of limited practical relevance. Attention has turned in the recent literature to more realistic types of "second-best" congestion pricing, in which various costs or constraints deter or prevent the setting of first-best tolls. Examples of second-best tolling include the use of toll cordons around cities instead of tolling each road in the network, the use of step tolls instead of smoothly time-varying tolls, tolling according to a fixed daily schedule rather than day-specific traffic conditions, etc. The rules for setting optimal second-best tolls are generally quite complicated because they must reflect all sorts of indirect effects, both good and bad. (For an overview of second-best pricing see Bohm, 1987.) This section will discuss a number of examples of second-best congestion pricing without attempting a general treatment.

4.1. Networks

The first-best rules for tolling a single road were identified in Sections 2 and 3. As Beckmann et al. (1956), Dafermos (1973), and Yang and Huang (1998) have shown for the static modeling framework, these rules continue to apply to each link of a road network provided every link is efficiently priced. But for several reasons it is quite unlikely that tolling will be implemented throughout a network. First, collecting tolls is costly. Conventional tollbooths involve substantial investments in space-intensive infrastructure, have high operating costs, and delay travelers when they stop to pay. Electronic tolling has much lower operating costs and imposes no delay, but it too requires investments in roadside infrastructure as well as a means of vehicle identification. Because of these collection costs it is generally not economic to toll every street, particularly in a large urban road network. A second constraint on road pricing is that most countries where pricing has been implemented require that toll-free alternatives to toll roads exist. And third, due to the expenses and political resistance to road pricing, pricing is likely to be implemented incrementally rather than all at once. The U.S.A., for example, now has a few demonstration projects that feature "value pricing," whereby one of several parallel highway lanes is tolled while the other lanes remain free. Thus, even under optimistic assessments about the future of road pricing, much of the road network is liable to remain untolled for a long time. This raises the question of how second-best tolls should be set on toll roads given unpriced congestion on untolled roads elsewhere in the network.

Lévy-Lambert (1968) and Marchand (1968) were the first to address this question, using a simple network featuring one toll road (call it road A) and one untolled road (road B) running in parallel between a common origin and destination. With no toll, B is overutilized. Excessive usage of B can be alleviated by reducing the toll on A below its first-best level in order to draw traffic off B and on to A. The optimal second-best toll is determined by balancing the gains from reducing usage of B against the costs of inducing excessive usage of A. Verhoef et al. (1996b) demonstrate that if route B is particularly congestion-prone, the optimal second-best toll on A can be negative. More generally, they show that the optimal toll depends on the relative free-flow travel times and capacities of the two routes, and on the price elasticity of travel demand. They also find that the welfare gains from second-best pricing are typically only a small fraction (e.g., 10%) of the gains from first-best pricing. Liu and McDonald (1998) corroborate these results using model parameters descriptive of one of the California road pricing demonstration projects (State Route 91 in Orange County).

These studies may underestimate the efficiency gains from second-best tolling because they ignore some of the ways in which pricing can alter driver behavior. Braid (1996) and de Palma and Lindsey (2000) allow for trip-timing adjustments by considering time-varying tolls in the Vickrey (1969) bottleneck model applied to the same two-parallel-routes network. Second-best tolling yields higher absolute efficiency gains than in the static model, as well as a greater fraction of the first-best efficiency gains. This is because the toll not only curbs excessive total usage, but also eliminates queuing on the tolled route.

Another way in which second-best tolling can enhance efficiency is through the sorting of drivers according to value of travel time. A pay-lane, for example, offers an expensive but quick trip that attracts users with high values of time, while the untolled lanes offer cheaper but slower service that caters to other travelers. Using a model with two groups of travelers with different values of time, Small and Yan (1999) find that the efficiency of pay-lanes relative to the first-best optimum is higher than in the equivalent model with no heterogeneity in value of time. Verhoef and Small (1999) obtain broadly similar results using a frequency distribution of value of time based on a Dutch survey of morning peak road users.

The two-parallel-routes network is just one of many network settings where second-best congestion pricing is relevant. A similar setting arises when travelers have a choice between driving and using public transit (Tabuchi, 1993). Public transit systems typically feature significant economies with respect to ridership that often outweigh any congestion externalities. The marginal social cost of a passenger trip is then below average cost, and first-best marginal-cost pricing results in a deficit. If transit is obliged to be self-financing for political or other reasons, then fares must be set at average cost and overpricing of transit results. The second-best toll on the road is then set above marginal cost too in order to boost transit ridership. If, alternatively, the self-financing constraint applies to transportation as a whole, then the road toll should again be set above marginal cost to cover part of the transit deficit and allow transit fares to be set closer to marginal cost.

Another instance of second-best pricing arises in the setting of public parking fees when roads are underpriced. Glazer and Niskanen (1992) consider a network in which users of public parking, users of private parking, and through traffic all drive on the same road link. Glazer and Niskanen derive the optimal combination of lump-sum parking fees, hourly parking fees, and capacity for public parking. They show that if an optimal road toll can also be set, then the optimal lump-sum parking fee is zero, and the optimal hourly fee equals the marginal cost of supplying parking space per hour. In the second-best solution in which no road toll is levied, the first-best optimal hourly parking rate still applies, whereas a positive lump-sum parking fee is levied, equal to a fraction of the first-best road toll. The lump-sum fee does not fully substitute for the road toll because, by suppressing trips by public-parking users, it exacerbates excessive travel by private-parking users and through traffic. Thus, analogously to the two-parallel-routes setting, the second-best congestion toll (i.e., the lump-sum part of the parking fee) is set below the first-best congestion toll, although the second-best tax rules differ between the two cases.

A general treatment of second-best congestion pricing in a network is found in Verhoef (1998), who derives optimal static tolls on any subset of links (including parking spaces) in an arbitrary network. The toll formulae, which are quite complicated, include terms reflecting marginal external costs on other links, and weights that depend on various demand and cost elasticities. Because the information required to use these formulae may be very costly to obtain, third-best pricing (i.e., setting "quasi"-first-best tolls, as if second-best distortions do not exist) or other rules of thumb may be worth considering in practice. Careful consideration, however, is in order in such cases: the use of more or less arbitrary tolls, on a few links of a network only, may well lead to a welfare reduction compared with the no-toll situation.

Closely related to the task of setting second-best tolls is the task of deciding which links in a network to toll, also considered in Verhoef (1998). Some locations where road pricing has been adopted or tested, including Singapore, Hong Kong, and Bergen, have well-defined geographical boundaries that simplify the decision. But in many areas this is not the case, and designers must contend with the problem that (as in the two-parallel-routes network) congestion gets displaced from tolled roads to untolled roads. The best candidates for road pricing appear to be freeways and major urban arterial roads because of their high traffic volumes, their role in providing rapid travel over longer distances, and the fact that they do not have close substitutes.

Evidence that congestion pricing may well produce adverse network effects is suggested by a recent study by May and Milne (2000). Using steady-state equilibrium simulations on a road network in Cambridge, U.K., they compare congestion pricing with three other road pricing schemes: cordon pricing, time-based pricing, and distance-based pricing. The tolls considered were not second-best optimal tolls as described above, but various exogenously determined toll levels instead. All four schemes were found to be prone to adverse boundary effects, including "rat-running" (where road users seek untolled routes). More discouragingly, a smaller percentage of travelers enjoyed travel time reductions with congestion pricing than with the other schemes. By inducing travelers to reroute to less congested roads, congestion pricing also had a tendency to increase travel distances – with potentially adverse environmental effects. These findings suggest that route-choice decisions deserve particular scrutiny in the design and evaluation of real-world congestion pricing projects. De Borger and Proost (2000) report on the relative efficiency of kilometer charges, fuel charges, parking charges, and public transport pricing for different cities and countries, using the multimodal TRENEN model. They find that parking charges combined with cordons can achieve efficiencies of more than 70% of the first-best ideal in some cases, while the potential of fuel pricing and public transport pricing is rather limited.

4.2. Heterogeneity of users

Both road vehicles and travelers vary in a number of characteristics. Vehicles differ in the road space they occupy, the visual obstruction they impose on drivers of other vehicles, their weight and acceleration capabilities, and the number of people they carry. Travelers differ in their values of time, trip-timing preferences, desired speed, and so on. Important questions in the practical design of congestion pricing schemes are whether first-best congestion pricing can still be implemented given these dimensions of heterogeneity, and if it cannot be, how second-best optimal tolls are determined. In addressing these questions it is useful to distinguish between tolling schemes that are constrained to be *anonymous*, i.e., independent of driver or vehicle type, and schemes that can impose *non-anonymous*, or type-specific, tolls.

Consider first heterogeneity in drivers' values of time and trip-timing preferences. As mentioned in Section 3, Arnott and Kraus (1998) have shown that first-best pricing remains possible using anonymous tolls with heterogeneous user groups provided the tolls can be varied freely over time. The optimality of anonymous tolling derives from the fact that the appropriate toll depends only on the marginal externality costs that drivers impose, and not on their individual preferences.

Optimal anonymous tolling may entail segregation of vehicle or driver types onto separate routes or traffic lanes. Using a static model, Verhoef and Small (1999) consider differentiation of tolls across parallel traffic lanes. The higher-priced lanes attract drivers with high values of time who are willing to pay for quicker trips, leaving other drivers to use the cheaper but slower lanes. For each lane separately an anonymous toll is still optimal, and efficient segregation of drivers is achieved without other forms of regulation. Nevertheless, the extra benefits turn out to be rather small, so that a second-best single toll applied to the entire highway does not impose much of a welfare loss.

The benefits from having drivers with different values of travel time use different routes must be weighed against the benefits from having drivers with different trip-timing preferences travel on the same roads at different times (see Section 3). Arnott et al. (1992) investigate this trade-off, using the bottleneck model with two driver groups and two routes when pricing is limited to time-invariant tolls. Spatial segregation of the groups onto separate routes turns out to be desirable when the groups have a similar congestion tolerance so that little is gained from segregating them temporally. Spatial segregation can be accomplished by imposing differential anonymous tolls on the two routes. A similar analysis could be done for time-varying tolls, which would produce a different set of conditions for optimal segregation.

Next, consider heterogeneity in travel speed – which may be due to differences in driver preferences or vehicle capabilities. Verhoef et al. (1999) investigate optimal pricing for two groups using a road on which overtaking is impossible. Traffic is assumed to be sufficiently light, that in the absence of desired-speed differences, there would be no congestion and no need for congestion tolls. Verhoef et al. show that the optimal toll for slow vehicles is always higher than the optimal toll for fast vehicles, and *decreases* with the fraction of slow vehicles in the total traffic. The toll decreases for two reasons: first, a slow vehicle delays fewer fast vehicles on average, and second, the average speed of fast vehicles declines asymptotically toward the speed of slow vehicles so that addition of a slow vehicle delays a given fast vehicle by a lesser amount.

Implementation of the optimal pricing scheme requires non-anonymous tolls because slow vehicles must pay more than fast vehicles. This is straightforward if vehicles are observationally distinguishable, such as cars and trucks. It may also be feasible for different groups of car drivers by measuring speeds and using automatic vehicle identification technology. But for practical or political reasons congestion tolls may be constrained to be equal. Verhoef et al. (1999) show that the optimal second-best toll is a weighted average of the congestion externalities created by each group. The formula for the toll is a special case of the general formula for second-best tolls derived in Verhoef (1998).

There are obvious advantages to segregating vehicles with different travel speeds onto separate lanes or routes. In practice this is done in rough-and-ready fashion on multilane roads by encouraging or requiring slow-moving vehicles to use the shoulder lanes, as well as by prohibiting slow-moving vehicles from certain routes. In principle, it could be accomplished by tolls. Tolls might also be based on other aspects of driving style that affect congestion and safety (e.g., non-use or misuse of headlights, failure to maintain steady speeds on mountain roads, lane weaving, parking on highway shoulders) using sophisticated electronic monitoring. Such monitoring would supplement or replace current methods using police surveillance, fines, and demerits for driving infractions, and adjustments of insurance premiums.

Heterogeneity in the physical characteristics of vehicles can be treated along similar lines to heterogeneity in drivers. Indeed, the principles of static congestion pricing with heterogeneous vehicle types were established by Dafermos (1973). Charging by vehicle type requires non-anonymous tolling. Currently this is done according to vehicle class (car, bus, truck), number of axles, number of trailers, laden weight, and axle weight. Traffic engineers often account for the greater contribution of trucks and buses to congestion by computing passenger car equivalents that depend on the type of road and terrain, and tolls could be set on this basis.

4.3. Step tolls

The discussion of dynamic congestion pricing in Section 3 focused exclusively on the optimal continuously time-varying (fine) toll. Fine tolling is certainly feasible using electronic toll collection technology. But there is conflicting evidence on whether drivers appreciate smoothly varying tolls. And most existing road pricing schemes employ either constant tolls, or step (piecewise constant) tolls – as is the case with Singapore's former area licensing scheme and Trondheim's toll ring.

Because step tolls increase or decrease in jumps, they do not rise or fall in perfect synchrony with congestion externalities and are therefore not fully efficient. Indeed, surges of traffic can occur just before increments and just after decrements, as were observed in Singapore and (to a lesser extent) in Trondheim. Arnott et al. (1993) show that in the bottleneck model with inelastic demand, a one-step toll provides one half or slightly more of the efficiency gains of the fine toll. Chu (1999) obtains broadly similar results in a more elaborate model featuring flow congestion and a traveler population with heterogeneous values of time and schedule delay costs. The efficiency of a step toll scheme naturally improves with the number of steps it embodies (Laih, 1994). But unlike with a fine toll, marginal-cost pricing requires that a fixed component be added to the step toll in order to support the optimal number of trips under elastic demand. Thus, the optimal step toll is positive at the beginning and end of the congested travel period.

As noted above, anonymous tolling is efficient with heterogeneous travelers when tolls can be varied freely over time. Since this is not the case for step tolls, non-anonymous tolling can in principle enhance the benefits of a step-toll congestion pricing scheme, although setting tolls according to values of time and time-of-use preferences is unlikely to be possible at a high level of precision.

4.4. Uncertainty and information provision

Up to this point it has been assumed that travelers know how long a trip will take and what cost they will incur, inclusive of toll. In reality travelers are usually faced with one or both of two types of uncertainty: idiosyncratic uncertainty and objective uncertainty. *Idiosyncratic uncertainty* exists when traffic conditions are predictable, but individual travelers do not know travel times precisely and form their own idiosyncratic perceptions. The standard approach to describing traveler behavior given idiosyncratic uncertainty is as a stochastic user equilibrium (Daganzo and Sheffi, 1977) in which drivers minimize their perceived travel costs. Smith et al. (1995) and Yang (1999) have shown that under reasonable assumptions standard static Pigouvian tolls based on actual travel costs remain optimal in the stochastic user equilibrium framework.

In contrast to idiosyncratic uncertainty, *objective uncertainty* exists when travel conditions vary unpredictably on account of traffic accidents, bad weather, roadworks, surges in demand due to special events or transit strikes, and so on. Various studies (e.g., Schrank and Lomax, 1999) have determined that a large fraction of time lost in congestion is attributable to these shocks. One way to model traveler behavior in the face of objective uncertainty parallels the treatment of idiosyncratic uncertainty in stochastic user equilibrium. In this approach, called stochastic network stochastic user equilibrium by Emmerink (1998, Chapter 4), drivers minimize their expected trip costs conditional on any information about traffic conditions available to them from radio or other sources.

Two approaches to congestion pricing with objective uncertainty are possible. One is to charge tolls that do not vary with traffic conditions. For example, the toll for trucks on Route A at 8 a.m. on weekdays would be the same every day regardless of how congested Route A actually was. The other approach, termed responsive pricing by Vickrey (1971), is to condition tolls on any information available about traffic conditions in order to match actual congestion externality costs as closely as possible. To implement responsive pricing effectively it is necessary to collect information about traffic conditions, calculate the appropriate tolls, and convey the updated information and tolls to drivers – all on an ongoing basis. While infeasible in the past, this goal is becoming practicable through the use of advanced traveler information systems (ATIS) that transmit information to travelers by phone or internet, or directly to vehicles equipped with on-board computers and Global Positioning Systems receivers. (For overviews of ATIS see Emmerink et al., 1994, and Emmerink and Nijkamp, 1999.)

Recent research on congestion pricing under uncertainty has focused on the use of either responsive or "non-responsive" pricing in conjunction with the implementation of ATIS. It is clear that there are technological synergies between ATIS and road pricing as far as their use of road infrastructure, centralized computing capability, and communications with drivers is concerned. It is not so evident whether there are also synergies in their benefits; that is, whether the benefits are superadditive or subadditive. In favor of superadditivity, it can be observed that in the presence of unpriced congestion, providing information to drivers can be welfare-reducing, as has been shown by Ben-Akiva et al. (1991) and Catoni and Pallottino (1991), *inter alios*. In favor of subadditivity, it can be argued that if one technology improves travel conditions, it reduces the scope for further gains from the other technology.

A few recent studies have investigated the additivity question. Verhoef et al. (1996a) use a static model featuring endogenous route choice and elastic demand. They find that non-responsive pricing and (perfect) information are approximately additive in their benefits, as well as complementary in the sense that under conditions when one instrument does not yield much benefit the other instrument does particularly well. El Sanhouri and Bernstein (1994) adopt a dynamic model with endogenous trip-timing decisions, and likewise find the benefits of non-responsive pricing and ATIS are approximately additive. Yang (1999) finds that additivity depends on how many drivers receive information, while de Palma and Lindsey (1998) show that information can be welfare-reducing unless it is supplemented with *responsive* pricing.

One consideration not addressed in these studies is the attitude of drivers toward tolling under uncertainty. It is unclear as yet from the limited formal research whether drivers are better off in terms of expected travel costs under non-responsive or responsive pricing. Some surveys have found that drivers dislike uncertainty about how much they will have to pay in tolls. Aversion to uncertainty about payment was one of the reasons for opposition to the congestion metering project planned for Cambridge, England, in which vehicles would have been charged on the basis of actual congestion experienced. However, drivers have been receptive to the recent adoption of responsive pricing on Interstate I-15 north of San Diego. Further research is clearly called for on the economics and politics of road pricing under uncertainty.

4.5. Interactions with other economic sectors

Economic transport models typically consider only demand and supply conditions pertaining to transport, and thus implicitly assume that the rest of the economy operates under first-best conditions. Although this assumption simplifies the modeling, it is usually well off the mark. In particular, distortionary taxes in labor and commodity markets are the norm, motivated by the need to raise government funds. This is recognized in the literature in environmental economics on the controversial "double-dividend hypothesis," which examines the interactions between environmental externality charges and distortionary taxes (e.g., Bovenberg and Goulder, 1996).

Following the environmental-economics lead, several recent studies have addressed second-best aspects of road pricing that arise from interactions with the rest of the economy. For a number of reasons a congestion tax on transport is likely to have a non-marginal impact on efficiency elsewhere in the economy. First, the extra taxes on transport as a consumption good reduce the real purchasing power of the wage, thereby aggravating the distortion from a preexisting tax on labor (the "tax-interaction effect"). Second, by affecting the costs of commuting, peak-period congestion tolls also more directly influence labor supply (the "complementarity effect"). Third, the revenues from congestion tolls can be used in various ways: to finance road capacity expansion or public transit services, to reduce labor taxes, to increase government spending on other services, and so on (the "recycling effect"). The benefits from these alternative expenditures may vary considerably with the direction and magnitude of the distortions involved. Fourth, all road users pay the tax but not all road users will share in the tax revenues (the "tax shifting effect"). This may lead to direct welfare effects if different groups have different welfare weights in the social welfare function, and may furthermore induce efficiency effects through interactions with the other efficiency effects mentioned.

Mayeres and Proost (1999) evaluate the efficiency effects of transportation charges in Belgium by computing the marginal welfare cost of public funds for a number of tax instruments using a general equilibrium model, for different degrees of social income inequality aversion. They find that a marginal increase in peak-period road transport prices yields the highest benefit when revenue is spent on road capacity expansion. The only negative benefit obtains for expenditure on public transport (which is already heavily subsidized), provided the degree of social inequality aversion is not too high, so that the benefits received by lower-income groups are not weighted too heavily.

Parry and Bento (1999) also find that the general equilibrium effects of road congestion pricing schemes are sensitive to the allocation of revenues, and may deviate considerably from partial-equilibrium estimates. In particular, they find that a lump-sum redistribution of congestion tax revenues can make the tax welfare-reducing because of its depressing effect on labor supply, whereas using the revenues to lower labor taxes doubles the overall gains. Their results thus demonstrate that incorporation of general equilibrium effects may greatly magnify the difference between welfare when revenues are, or are not, used to reduce preexisting distortions.

General equilibrium studies such as these provide at least two important lessons. One lesson is that partial equilibrium analyses of road congestion pricing can miss out on important indirect efficiency and welfare effects. Second-best optimal congestion taxes that take these effects into account can differ significantly from the first-best taxes derived using partial equilibrium methods. The second lesson is that it is dangerous to allocate revenues from congestion pricing solely to "buy" public acceptability. Poorly chosen allocations can have such adverse indirect effects that they outweigh the direct benefits of congestion pricing, leaving society worse off.

4.6. Other traffic-related externalities

Accidents, noise, local air pollution, and global warming are all negative side effects of road traffic. By altering the timing, location, speed, and volume of traffic levels, congestion pricing affects these other externalities. But these externalities cannot be properly internalized through pricing aimed solely at congestion relief. While congestion may contribute to extra fuel consumption and pollution emissions, it has proved difficult to establish a tight statistical relationship between them (Small and Gómez-Ibáñez, 1999). And by speeding up traffic, congestion pricing may reduce the frequency of accidents but increase their severity, although again statistical evidence is hard to come by. Thus, it is safe to say that the incidence of congestion and the incidence of other transport externalities are far from perfectly correlated. Additional policy instruments are required to address these other externalities, such as the gasoline tax, periodic vehicle emissions testing, road design, safety legislation, etc. These other instruments are unlikely to operate perfectly, however, so that the impact of congestion pricing on other externalities should in principle be accounted for.

4.7. Congestion pricing by private operators

Private toll roads have long existed in Europe and the Pacific Rim (Gómez-Ibáñez and Meyer, 1993), and privately operated pay-lanes are emerging in the U.S.A., and being considered for the Randstad area in The Netherlands. Private toll-road operators are typically interested in maximizing profit rather than social surplus, so that first-best pricing cannot be expected of them. The purpose of this section is to compare profit-maximizing and welfare-maximizing congestion tolls on the same facilities. To simplify, regulatory constraints on private tolls – which are often imposed in the form of price or rate-of-return caps – are ignored, as are other distortions from first-best conditions.

Consider first monopoly pricing. Various (static) monopoly settings have been considered in the literature, including a single-road monopolist competing with train service (Edelson, 1971), monopoly pricing of a single facility facing no competition (Mills, 1981; Mohring, 1985), and a monopolist operating two roads (Verhoef et al., 1996b). As long as all potential users incur the same disutility from congestion, the monopoly toll in each of these settings is found to be

$$\tau_i = N_i\,\partial C_i/\partial N_i - N_i\,\partial p/\partial N, \tag{3}$$

where τ_i is the toll on road i (possibly the only road), N_i is the usage of road i, $C_i(N_i)$ is the travel cost on road i, N is the total usage, and $p(N)$ is the inverse demand curve for travel. The first component of the toll in eq. (3) is the first-best congestion toll, and the second component is a markup that depends on demand but not travel costs. The monopolist therefore fully internalizes the congestion externality. The reason is that by reducing congestion costs, the monopolist can charge a higher toll without losing patronage. But, as is true of monopolists in general, the monopolist adds a markup that is larger the steeper the demand is. (Indeed, by rewriting eq. (3) as $\tau_i + N_i \, \partial p/\partial N = N_i \, \partial C_i/\partial N_i$, it can be seen that the monopolist equates marginal revenue, rather than price, to marginal social cost.) Thus, except in the limiting case of perfectly elastic demand, the monopolist sets a toll above the first-best toll and accommodates too little traffic. Indeed, the markup may be so high that welfare is actually reduced relative to no tolls (e.g., Verhoef and Small, 1999).

In most countries (and in California) private road monopolies are not allowed because of legislation requiring that free alternatives be available (Gómez-Ibáñez and Meyer, 1993). Accordingly, several studies have examined private-sector road pricing using the two-parallel-routes network (discussed under "Networks" above), where one of the two routes is private, and the other is either free-access or publicly operated. Using static models, Verhoef et al. (1996b) and Liu and McDonald (1998) show that the private operator will again set a toll higher than the second-best toll, and may end up reducing welfare relative to no toll. de Palma and Lindsey (2000) show that the welfare effects of private-sector pricing are more salubrious in a dynamic model with endogenous trip-timing decisions. As in the case of second-best pricing (see "Networks" above), this is because a private operator has an incentive to adopt time-varying tolls to reduce peak-period congestion.

Whether private toll roads can operate profitably depends on various factors, including infrastructure and operating costs, regulatory constraints on tolls, and competition from other roads and modes of transport (Nijkamp and Rienstra, 1995). Viton (1995) concludes that private toll roads can be profitable under a wide range of conditions even when competing against a free public road, particularly for urban areas in which high tolls can be charged during peak periods. In part, Viton's results are driven by his assumption that drivers have strong idiosyncratic preferences for roads, which limits their willingness to switch from one to the other. He also assumes that a private operator can price discriminate on the basis of automobile size, with the result that in his model the toll per mile for large cars is over triple the toll for small cars.

One consideration ignored in the discussion thus far is heterogeneity in drivers' congestion costs. As mentioned above under "Heterogeneity of users," welfare-maximizing tolls depend on the congestion costs incurred by all drivers, including inframarginal ones. By contrast, as both Edelson (1971) and Mills (1981) observe, (undifferentiated) profit-maximizing tolls depend only on the congestion costs incurred by marginal users. When inframarginal users are more averse to congestion on average than are marginal users, private tolls are biased downward, and in theory might even be lower than first-best tolls. Nevertheless, like a welfare maximizer, a private firm operating more than one facility has an incentive to provide differentiated quality service (Chander and Leruth, 1989) by charging different tolls on different roads or traffic lanes. Because firms are inclined to set high tolls that may nearly eliminate congestion, however, the scope for differentiation on the basis of speeds may be limited (Verhoef and Small, 1999) – although it is still possible when the routes differ in length.

5. Congestion pricing and investment

Building new roads was once almost a conditioned response to road congestion. But it is now severely constrained by environmental concerns, and by shortages of funds and space. Still, selective construction continues, and as roads wear out decisions must be made about rehabilitation and replacement. How much should be spent on roads thus remains an important question. The purpose of this section is to explore the dependence of optimal investment on how road usage is priced. (Another interesting question is the degree to which optimal investments are self-financing under congestion pricing; for summaries and references see Hau, 1998, Sections 3.7, 3.10, and 3.11, and Lindsey and Verhoef, 2000.)

The optimal capacity of a single road is easily characterized. Following Arnott et al. (1998) let K denote the capacity, N the number of trips taken, $C(N, K)$ the trip cost net of tolls, $G(N)$ the total gross benefits from trips, and $B(N, K) = G(N) - NC(N, K)$ the total net benefits. Then the marginal benefit of capacity expansion (gross of construction cost) is

$$\begin{aligned}\frac{\mathrm{d}B}{\mathrm{d}K} &= \frac{\partial B}{\partial K} + \frac{\partial B}{\partial N}\frac{\mathrm{d}N}{\mathrm{d}K} = \frac{\partial B}{\partial K} + \frac{\partial B}{\partial N}\frac{\mathrm{d}N}{\mathrm{d}p}\frac{\mathrm{d}p}{\mathrm{d}K} \\ &= -N\frac{\partial C}{\partial K} + \left[\left(\frac{\partial G}{\partial N} - C(N, K) - N\frac{\partial C}{\partial N}\right)\frac{\mathrm{d}N}{\mathrm{d}p}\frac{\mathrm{d}p}{\mathrm{d}K}\right].\end{aligned} \tag{4}$$

The first term on the right-hand side ($-N\,\partial C/\partial K$) is the direct benefit of the capacity expansion that would occur with no change in travel behavior. The second term, in square brackets, is the indirect benefit that derives from an increase in the number of trips due to the reduction in trip price caused by the capacity expansion. With marginal-cost pricing, the envelope theorem applies, and the indirect benefit is zero because the marginal social benefit of the induced increase ($\partial G/\partial N$) in traffic equals the marginal social cost ($C + N\,\partial C/\partial N$). But with underpricing (or no pricing) of road use, the marginal social cost exceeds the marginal social benefit. The indirect benefit is therefore negative and, in the limiting case of no tolling and perfectly elastic demand, completely offsets the direct benefit. To see this, consider a stationary traffic setting and suppose that capacity is increased. This shifts the travel cost curve ($C(q)$ in Figure 1) to the right. With no tolling, equilibrium is established at the new intersection of $C(q)$ with the demand curve. If the demand curve is horizontal, traffic volume increases until the trip cost is back to its previous level and the investment yields no benefit.

This reasoning might suggest that with underpriced (or unpriced) congestion, the optimal capacity is lower than in the first best because of the negative indirect effect. But because underpricing of congestion results in greater usage, the direct benefit of a capacity expansion is higher than in the first best. The net effect of these opposing forces has been investigated extensively in the literature; Arnott and Yan (2000) provide an insightful review and synthesis. One result is that if a toll is reduced slightly below the first-best level, optimal capacity rises because the increase in the positive direct effect dominates over the increase in the negative indirect effect. (In eq. (4), both terms of the direct effect, $-N\,\partial C/\partial K$, increase to first order in the capacity expansion, whereas the term in brackets of the indirect effect $(\partial G/\partial N - C(N, K) - N\,\partial C/\partial N)$ is zero at the first-best solution.)

Another comparison of potential practical importance is between the second-best optimal capacity and the capacity that would be chosen if the indirect effects of expansion associated with underpricing of congestion were ignored. Such practice is dubbed "naive cost–benefit analysis" (CBA). Arnott and Yan (2000) show that in the standard static model, naive CBA leads unambiguously to overestimation of the benefit from expansion and hence to excessive investment. Indeed, they show, using a plausible numerical example, that the capacity chosen using naive CBA can exceed the second-best optimal capacity by more than the second-best capacity exceeds the first-best capacity.

Naive CBA rules have also been investigated using dynamic models of morning peak travel by Small (1992a) and Henderson (1992). Small (1992a) adopts the bottleneck queuing model with linear schedule delay costs for arriving early or late, and fixed total demand. The planner who chooses capacity is assumed to err by treating the departure times of travelers as given. This introduces two opposing biases: by ignoring the fact that a capacity expansion induces travelers to reschedule trips toward the peak, the benefits from a reduction in schedule delay are missed, but the savings in travel (queuing) time are overestimated. Small finds that the second effect dominates if unit travel time costs are large relative to unit schedule delay costs. This is because the perceived travel time savings are then large, whereas the overlooked benefits from rescheduling are small.

Henderson's (1992) model differs from Small's in two respects. First, travel is subject to *flow* congestion rather than queuing, with travel time determined by the instantaneous arrival rate of vehicles at work. And second, the planner errs by treating the *arrival* times, rather than departure times, of travelers as given. Henderson concludes that the misguided planner overinvests in capacity regardless of the relative unit costs of travel time and schedule delay. The contrast with Small's conclusion highlights the sensitivity of the results to assumptions, and suggests that further research on naive CBA is warranted.

To conclude this section, it is instructive to see how the impact of congestion pricing on optimal capacity is affected by some of the complications considered in Section 4.

5.1. Networks

If marginal-cost pricing is adopted throughout a network then all the indirect effects of a capacity expansion net out, and the marginal benefit from expanding a single link is still given by the direct effect in eq. (4). In the absence of marginal-cost pricing of all links, however, a capacity investment can have outright perverse effects. An extreme case of this is the "Braess paradox" (Braess, 1968) that occurs when adding a new link to an untolled network increases total travel costs.

5.2. Step tolls

Eq. (4) remains valid for any pricing regime, including step tolls, as long as the appropriate trip cost function $C(N, K)$ is used. Arnott et al. (1993) show that in the bottleneck model with homogeneous users, a clear-cut ranking of optimal capacity according to the time sensitivity of the tolling regime obtains. When the price elasticity of demand is less than unity (as it typically is for peak-period travel), optimal capacity is lower the more refined the tolling regime. This is so because total trip costs are lower for any given capacity in a more refined tolling regime – so that the marginal direct benefit from expansion is lower, and in addition because with relatively inelastic demand the direct effect of expansion dominates over the indirect effect. The opposite ranking obtains if the price elasticity exceeds unity.

5.3. Congestion pricing by private operators

The profit-maximizing capacity of a private operator can be derived in an analogous way to the socially optimal capacity. Provided the operator chooses a profit-maximizing toll, the envelope theorem applies and the marginal benefit of capacity is given by the same formula as for the direct effect in eq. (4): $-N\,\partial C/\partial K$ (see Small, 1992a). But because the private operator sets a higher toll, usage of the road is less than under public management. Therefore, as long as $-N\,\partial C/\partial K$ is an increasing function of N, the private operator underinvests in capacity (De Vany, 1976). This is true whatever (common) form of tolling the private and public operator adopt, e.g., flat, single-step, multiple-step, or fine.

6. The social and political feasibility of congestion pricing

Despite its economic appeal, road congestion pricing appears to enjoy little support outside academia. The limited social and, consequently, political support for congestion pricing has caused many proposed schemes to be abandoned before implementation, or at least to be postponed, sometimes indefinitely. These include detailed plans for Hong Kong, London, the Randstad, and Stockholm (Small and Gómez-Ibáñez, 1998).

One important reason for opposition to congestion pricing was identified in Section 2: before redistribution of toll revenues, everybody except the taxman appears to be worse off. This result should be qualified on two counts. First, if congestion takes the form of pure queuing and users have identical values of time, a fine toll leaves users equally well off. Second, users with a high value of travel time may benefit from congestion pricing even before revenue recycling; see Richardson (1974), and supporting empirical evidence from a Dutch survey by Verhoef et al. (1997a). Still, because value of time is positively correlated with income, this implies that road pricing is regressive – which is unlikely to improve the social acceptability of congestion pricing.

Because tolling makes many users worse off, revenue allocation and its impacts on income distribution have been identified as key determinants of the acceptability of congestion pricing. Various allocation schemes have been proposed that leave all major user groups better off. Goodwin (1989) and Small (1992b) both propose schemes that allocate revenues in three ways. In Small's (1992b) scheme revenues are used to reimburse travelers as a group, to offset regressive taxes, and to fund new transportation services (particularly transit). Small finds that each of the six prototypical residents he considers benefits from the scheme. In the survey of Verhoef et al. (1997a) road users expressed the following preferences (in decreasing order) for the use of toll revenues: investment in new roads, reduction in vehicle ownership taxes, reduction in fuel taxes, investment in public transport, subsidies for public transport, investment in car pool facilities, general tax cuts, and expansion of other public expenditures. The survey suggests that the use of toll revenue to finance road infrastructure as a substitute for other funding sources is politically attractive. It is also transparent, and efficient from an economy-wide perspective according to the general equilibrium analysis of Mayeres and Proost (1999). Indeed, it is worth repeating that general equilibrium studies of road pricing strongly suggest that revenue allocation schemes designed solely to improve the public acceptability may induce welfare losses elsewhere in the economy, leading to efficiency losses that may even outweigh the initial improvements. A trade-off between the efficiency and acceptability impacts of revenue allocation schemes will generally exist, and should be given careful attention in their design.

Apart from the above considerations, which reflect a more or less rational attitude of road users towards congestion pricing, other reasons for opposing congestion pricing have been identified. From a review of public attitude studies Jones (1998) identifies the following reasons for opposition: (1) drivers find it difficult to accept the idea of being charged for something that they wish to avoid (congestion), and also feel that congestion is not their fault but rather something that is imposed on them by others; (2) road pricing is not needed, either because congestion is not bad enough or because other measures are superior; (3) pricing will not get people out of their cars; (4) the technology will not work; (5) privacy concerns; (6) diversion of traffic outside the charged area; (7) road pricing is just another form of taxation; and (8) perceived unfairness. Similar concerns were recently voiced in The Netherlands when the Dutch Automobile Association (ANWB) successfully launched a large public campaign to prevent the implementation of a full-scale congestion pricing scheme for the Randstad area. The Minister of Transport managed to save part of the original plan (as of this time of writing) only after radically downsizing it to a few toll points and pay-lanes, and after offering extra money for infrastructure investments to the cities affected.

U.S. experience also suggests that congestion pricing may have to begin at a modest level involving a few demonstration projects such as pay-lanes, and may very well end there too. As discussed in Section 4, however, pay-lanes may yield only a small fraction of the potential welfare gains from congestion pricing. Verhoef and Small (1999) suggest that instead of pay-lanes, highway pricing should involve "free-lanes" whereby all but one lane is tolled. This would have the merit of pricing a larger fraction of capacity while still offering, via the free lane, a lifeline service for travelers with low values of travel time.

Another way to improve the acceptability of road pricing is to reduce the amount of toll collected in the first place. Daganzo (1995) has proposed a combination of pricing and rationing whereby each traveler is prohibited from driving on some fraction of days. As a variant, Daganzo and Garcia (1998) suggest that a fraction of drivers be exempted from paying the toll on any given day. An analogous idea, proposed by Verhoef et al. (1997b), is to issue free of charge a large number of tradeable smart-card units for use with electronic road pricing. Yet another suggestion, by DeCorla-Souza (2000), is to convert some existing freeway lanes to toll lanes, and to use the toll revenues to provide drivers who use the remaining free lanes with credits for future trips on the toll lanes, for parking, or for trips by transit.

According to economists, congestion pricing is the best instrument for controlling congestion. Yet the bulk of empirical evidence suggests that, even with cleverly designed tolling mechanisms and toll revenue allocation schemes, congestion pricing is likely to remain difficult to implement. It is no coincidence that the only area-wide congestion pricing scheme currently in operation is in Singapore, where the culture allows the government to implement unpopular policies. Indeed, Frick et al. (1996) infer from the series of failed attempts to implement congestion pricing on the San Francisco Bay Bridge that both the public and officials must be convinced that *no* feasible alternative to congestion pricing exists before it will be accepted.

7. Conclusions

Given forecasts of continuing growth in road travel, and the reduced scope for expansion of road infrastructure, traffic congestion is not a problem that will go away soon. Recent advances in electronic vehicle identification and automated charging technologies have made congestion pricing a viable means of combating traffic congestion, rather than just an academic curiosity. This chapter reviewed the economic principles behind congestion pricing, which derive from the benefits of charging travelers for the externalities they create. Attention was paid to various complications that make simple textbook congestion pricing models of limited relevance, and dictate that congestion pricing schemes be studied from the perspective of the theory of the second best.

Despite the economic case for congestion pricing, it has attracted strong social and political opposition, and assorted legal and institutional constraints create further barriers to implementation. It thus seems safe to conclude with the prediction that coming decades will witness an increasing number of attempts to implement congestion pricing, many of which will fail in their early stages, but some of which will succeed, if only on a piecemeal basis. The design and evaluation of such schemes will probably require a deeper understanding of the direct and indirect impacts of congestion pricing than is available to date. One can hope, therefore, that this review will be obsolete by the time a second edition of this handbook is published.

References

Arnott, R. and M. Kraus (1998) "When are anonymous congestion charges consistent with marginal cost pricing?" *Journal of Public Economics*, 67:45–64.

Arnott, R. and A. Yan (2000) "The two-mode problem: Second-best pricing and capacity", Boston College: Working paper.

Arnott, R., K. Arrow, A.B. Atkinson and J.H. Dreze, eds. (1994) *Public economics: Selected papers by William Vickrey*. Cambridge: Cambridge University Press.

Arnott, R., A. de Palma and R. Lindsey (1992) "Route choice with heterogeneous drivers and group-specific congestion costs", *Regional Science and Urban Economics*, 22(1):71–102.

Arnott, R., A. de Palma and R. Lindsey (1993) "A structural model of peak-period congestion: A traffic bottleneck with elastic demand", *American Economic Review*, 83(1):161–179.

Arnott, R., A. de Palma and R. Lindsey (1998) "Recent developments in the bottleneck model", in: K.J. Button and E.T. Verhoef, eds., *Road pricing, traffic congestion and the environment: Issues of efficiency and Social feasibility*. Cheltenham: Edward Elgar.

Beckmann, M., C.B. McGuire and C.B. Winsten (1956) *Studies in the economics of transportation*. New Haven: Yale University Press.
Ben-Akiva, M., A. de Palma and I. Kaysi (1991) "Dynamic network models and driver information systems", *Transportation Research A*, 25(5):251–266.
Bohm, P. (1987) "Second best", in J. Eatwell, M. Milgate and P.K. Newman, eds., *The new Palgrave: A dictionary of economics*, Vol. 4. New York: Macmillan.
Bovenberg, A.L. and L.H. Goulder (1996) "Optimal environmental taxation in the presence of other taxes: A general equilibrium analysis", *American Economic Review: Papers and Proceedings*, 86:985–1000.
Braess, D. (1968) "Über ein Paradoxen des Verkehrsplanung", *Unternehmenforschung*, 12:258–268.
Braid, R.M. (1996) "Peak-load pricing of a transportation route with an unpriced substitute", *Journal of Urban Economics*, 40:179–197.
Carey, M. and A. Srinivasan (1993) "Externalities, average and marginal costs, and tolls on congested networks with time-varying flows", *Operations Research*, 41(1):217–231.
Catoni, S. and S. Pallottino (1991) "Traffic equilibrium paradoxes", *Transportation Science*, 25:240–244.
Chander, P. and L. Leruth (1989) "The optimal product mix for a monopolist in the presence of congestion effects", *International Journal of Industrial Organization*, 7:437–449.
Chu, X. (1999) "Alternative congestion pricing schedules", *Regional Science and Urban Economics*, 29:697–722.
Dafermos, S. (1973) "Toll patterns for multiclass-user transportation networks", *Transportation Science*, 7:211–223.
Daganzo, C.F. (1995) "A Pareto optimum congestion reduction scheme", *Transportation Research B*, 29(2):139–154.
Daganzo, C.F. and R.C. Garcia (2000) "A Pareto improving strategy for the time-dependent morning commute problem", *Transportation Science*, 34(3):303–311.
Daganzo, C.F. and Y. Sheffi (1977) "On stochastic models of traffic assignment", *Transportation Science*, 11:253–274.
De Borger, B. and S. Proost, eds. (2001) *Reforming transport pricing in the European Union*. Cheltenham: Edward Elgar.
DeCorla-Souza, P. (2000) "Making pricing of currently free highway lanes acceptable to the public", *Transportation Quarterly*, 54(3):1–20.
de Palma, A. and R. Lindsey (1998) "Information and usage of congestible facilities under different pricing regimes", *Canadian Journal of Economics*, 31(3):666–692.
de Palma, A. and R. Lindsey (2000) "Private toll roads: Competition under various ownership regimes", *Annals of Regional Science*, 34(1):13–35.
De Vany, A. (1976) "Uncertainty, waiting time, and capacity utilization: A stochastic theory of product quality", *Journal of Political Economy*, 84(3):523–541.
Edelson, N.E. (1971) "Congestion tolls under monopoly", *American Economic Review*, 61(5): 872–882.
El Sanhouri, I. and D. Bernstein (1994) "Integrating driver information and congestion pricing systems", *Transportation Research Record*, 1450:44–50.
Emmerink, R.H.M. (1998) *Information and pricing in road transportation*. Berlin: Springer.
Emmerink, R.H.M. and P. Nijkamp eds. (1999) *Behavioural and network impacts of driver information systems*. Aldershot: Ashgate.
Emmerink, R.H.M., P. Nijkamp, P. Rietveld and K.W. Axhausen (1994) "The Economics of motorist information systems revisited", *Transport Reviews*, 14(4):363-388.
Frick, K.T., S. Heminger and H. Dittmar (1996) "Bay Bridge congestion-pricing project: Lessons learned to date", *Transportation Research Record*, 1558:29-38.
Gómez-Ibáñez, J.A. and J.R. Meyer (1993) *Going private: The international experience with transport privatization*. Washington, DC: The Brookings Institution.
Gómez-Ibáñez, J.A. and K.A. Small (1994) "Road pricing for congestion management: A survey of international practice", in: *National Cooperative Highway Research Program, Synthesis of highway practice 210, TRB*. Washington, DC: National Academy Press.
Glazer, A. and E. Niskanen (1992) "Parking fees and congestion", *Regional Science and Urban Economics*, 22:123–132.

Goodwin, P.B. (1989) "The rule of three: A possible solution to the political problem of competing objectives for road pricing", *Traffic Engineering and Control*, 30(10):495–497.
Hall, W.R., ed. (1999) *Handbook of Transportation Science*. Norwell, MA: Kluwer Academic.
Hau, T.D. (1998) "Congestion pricing and road investment", in: K.J. Button and E.T. Verhoef, eds., *Road pricing, traffic congestion and the environment: Issues of efficiency and social feasibility*. Cheltenham: Edward Elgar.
Henderson, J.V. (1992) "Peak shifting and cost–benefit miscalculations", *Regional Science and Urban Economics*, 22:103–121.
Jones, P. (1998) "Urban road pricing: Public acceptability and barriers to implementation", in: K.J. Button and E.T. Verhoef, eds., *Road pricing, traffic congestion and the environment: Issues of efficiency and social feasibility*. Cheltenham: Edward Elgar.
Knight, F. (1924) "Some fallacies in the interpretation of social costs", *Quarterly Journal of Economics*, 38(4):582–606.
Laih, C.-H. (1994) "Queuing at a bottleneck with single- and multi-step tolls", *Transportation Research A*, 28(3):197–208.
Lévy-Lambert, H. (1968) "Tarification des services à qualité variable: application aux péages de circulation", *Econometrica*, 36(3-4):564–574.
Lindsey, R. and E.T. Verhoef (2000) "Congestion modelling", in: D.A. Hensher and K.J. Button, eds., *Handbooks in transport 1. Transport modelling*. Oxford: Pergamon.
Liu, L.N. and J.F. McDonald (1998) "Efficient congestion tolls in the presence of unpriced congestion: A peak and off-peak simulation model", *Journal of Urban Economics*, 44:352–366.
Marchand, M. (1968) "A note on optimal tolls in an imperfect environment", *Econometrica*, 36(3–4): 575–581.
May, A.D. and D.S. Milne (2000) "Effects of alternative road pricing systems on network performance", *Transportation Research A*, 34(6):407–436.
Mayeres, I. and S. Proost (1999) "Marginal tax reform, externalities and income distribution", Centre for Economic Studies, Catholic University Leuven, working paper.
Mills, D.E. (1981) "Ownership arrangements and congestion-prone facilities", *American Economic Review, Papers and Proceedings*, 71(3):493–502.
Mohring, H. (1985) "Profit maximization, cost minimization, and pricing for congestion-prone facilities", *Logistics and Transportation Review*, 21:27–36.
Nijkamp, P. and A.S. Rienstra (1995) "Private sector involvement in financing and operating transport infrastructure", *Annals of Regional Science*, 29:221–235.
Parry, I.W.H. and A.M. Bento (1999) "Revenue recycling and the welfare effects of congestion pricing", Resources for the Future, Washington, working paper.
Pigou, A.C. (1920) *Wealth and welfare*. London: Macmillan.
Richardson, H.W. (1974) "A note on the distributional effects of road pricing", *Journal of Transport Economics and Policy*, 8:82-85.
Schrank D, and T. Lomax (1999) "The 1999 annual mobility report information for urban America", Texas Transportation Institute, Texas A&M University System, College Station, Texas, http://mobility.tamu.edu.
Small, K.A. (1992a) "Urban transportation economics". *Fundamentals of pure and applied economics*. Chur: Harwood.
Small, K.A. (1992b) "Using the revenues from congestion pricing", *Transportation*, 19(4):359–381.
Small, K.A. and J.A. Gómez-Ibáñez (1998) "Road pricing for congestion management: The transition from theory to policy", in: K.J. Button and E.T. Verhoef, eds., *Road pricing, traffic congestion and the environment: Issues of efficiency and social feasibility*. Cheltenham: Edward Elgar.
Small, K.A. and J.A. Gómez-Ibáñez (1999) "Urban transportation", in: P. Cheshire and E.S. Mills, eds., *Handbook of regional and urban economics*. vol. 3. Amsterdam: North-Holland.
Small, K.A. and J. Yan (2001) "The value of 'value pricing' of roads. Second-best pricing and product differentiation", *Journal of Urban Economics*, 49:310–336.
Smith, T.E., E.A. Eriksson and P.O. Lindberg (1995) "Existence of optimal tolls under conditions of stochastic user-equilibria", in: B. Johansson and L.-G. Mattsson, eds., *Road pricing: Theory, empirical assessment and policy*. Boston: Kluwer Academic.
Tabuchi, T. (1993) "Bottleneck congestion and modal split", *Journal of Urban Economics*, 34:414–431.

Verhoef, E.T. (1998) "Second-best congestion pricing in general static transportation networks with elastic demands", Free University of Amsterdam, working paper.

Verhoef, E.T. and K.A. Small (1999) "Product differentiation on roads: Second-best congestion pricing with heterogeneity under public and private ownership", Tinbergen Institute, Amsterdam-Rotterdam, Discussion Paper TI 99-066/3.

Verhoef, E.T., R.H.M. Emmerink, P. Nijkamp and P. Rietveld (1996a) "Information provision, flat- and fine congestion tolling and the efficiency of road usage", *Regional Science and Urban Economics*, 26(5):505–530.

Verhoef, E.T., P. Nijkamp and P. Rietveld (1996b) "Second-best congestion pricing: The case of an untolled alternative", *Journal of Urban Economics*, 40(3):279–302.

Verhoef, E.T., P. Nijkamp and P. Rietveld (1997a) "The social feasibility of road pricing: A case study for the Randstad area", *Journal of Transport Economics and Policy*, 31(3):255–267.

Verhoef, E.T., P. Nijkamp and P. Rietveld (1997b) "Tradeable permits: Their potential in the regulation of road transport externalities", *Environment and Planning B: Planning and Design*, 24:527–548.

Verhoef, E.T., J. Rouwendal and P. Rietveld (1999) "Congestion caused by speed differences", *Journal of Urban Economics*, 45:533–556.

Vickrey, W.S. (1963) "Pricing in urban and suburban transport", *American Economic Review*, 53: 452–465.

Vickrey, W.S. (1969) "Congestion theory and transport investment", *American Economic Review (Papers and Proceedings)*, 59:251–260.

Vickrey, W.S. (1971) "Responsive pricing of public utility services", *Bell Journal of Economics and Management Science*, 2:337–346.

Viton, P.A. (1995) "Private roads", *Journal of Urban Economics*, 37(3):260–289.

Walters, A.A. (1961) "The theory and measurement of private and social cost of highway congestion", *Econometrica*, 29(4):676–697.

Yang, H. (1999) "Evaluating the benefits of a combined route guidance and road pricing system in a traffic network with recurrent congestion", *Transportation*, 20:299–321.

Yang, H. and H.-J. Huang (1997) "Analysis of time-varying pricing of a bottleneck with elastic demand using optimal control theory", *Transportation Research B*, 31(6):425–440.

Yang, H. and H.-J. Huang (1998) "Principle of marginal-cost pricing: How does it work in a general road network?", *Transportation Research A*, 32(1):45–54.

Chapter 8

MODAL DIVERSION

DAVID A. HENSHER
University of Sydney

1. Introduction

The reasons why passengers and freight operators choose one form of transport over another are many and varied. There is, however, a view that we can broadly classify the influences into those that are related to time, to cost, and to service quality (the latter being additional to time-related influences). The challenge for any investigation of influences on modal choice is to identify the attributes that describe each class of influence and to place weights on each attribute to establish its relative influence. In addition, it is recognized that not all attributes are easy to measure, and hence are often ignored in an analysis even though they continue to influence modal choice.

In addition to identifying attributes that are important, it is generally accepted that behavioral responses to attribute levels and their impact on modal diversion vary by market segment. Whilst the determination of suitable market segments remains an important research activity, there are a number of broadly agreed bases for segmentation of the population of travelers and commodity movements. For example, passenger trips are commonly stratified by trip purpose such as commuting and non-commuting to recognize that commuters have different degrees of behavioral responsiveness to travel times and costs than do non-commuters. The view is that a commuter typically is willing to pay a higher price to save travel time than a non-commuter for reasons related to the fixity of the arrival time and because commuting is linked to income-earning whereas non-commuting is more leisure-based. It is also assumed that high-value, low-volume, non-bulk commodities need to be delivered faster than low-value, high-volume, bulk commodities, and hence the weights attached to travel time and reliability should reflect this circumstance. Other influences on modal diversion include the socio-economic characteristics of passengers and the profile of the commodities being transported.

These introductory remarks suggest the flavor of the chapter. Our objective is to review the key elements of the methods used in practice to evaluate the role of

Handbook of Transport Systems and Traffic Control, Edited by K.J. Button and D.A. Hensher

modal attributes and market segment in the explanation of modal split and modal diversion. We focus on the discrete-choice modeling approach to studying influences on modal diversion (see Louviere et al. (2001) for a recent reference source on such methods). The chapter begins with an overview of the changing context for a study of modal diversion, followed by an overview of the methods used to establish the weights for each attribute and the types of policy outputs that analysts typically seek from such models (such as values of travel time savings, share elasticities, and market shares). Readers will benefit by cross-referencing to the many chapters in the *Handbook of Transport Modelling* in this series (Hensher and Button, 2000).

2. The changing context for studying modal diversion

Choosing one form of transport over another for a specific trip is a complex decision. What do we mean by "form of transport" and "a specific trip"? It has historically been assumed that most trips can be defined in terms of an outward trip by a single mode and a return trip by the same mode between a common origin and destination for a single trip purpose (e.g., commuting). Typically a leg of a journey (known as a one-way trip) is described by its access, line-haul (or main mode), and egress sections, each defined by a set of times and costs. The time components for passenger trips include walk time, wait time, in-vehicle time, parking search time, and reliability of arrival time; the cost components include fare or vehicle fuel costs, tolls, and parking costs. There is an extensive literature of empirical studies identifying the determinants of modal choice for such trips (see Hensher and Dalvi, 1978). For freight trips, the supply chain includes the time spent loading and unloading, line-haul time, time reliability, and transactions time (the latter including processing the delivery).

Trip-making is becoming more complex, often with multiple destinations and purposes, so that flexibility of the travel mode becomes more important. This makes rail and bus potentially less attractive and finding generic ways of getting car users to switch to public transport (and freight to switch to rail) increasingly unsuccessful. A continuing challenge for transport policy makers is to identify opportunities to reduce the dominance of the automobile in urban travel, especially during periods of high traffic congestion (Hensher, 1998a). There are many segments of car users, with varying degrees of dependence on the car, who need to be assessed differentially in the search for the barriers to using public transport. The identification of the boundaries of switching propensity will enable planners to concentrate more of the individual segments where the greatest prospects for modal switching might take place. This is known as individualized marketing of transport (Brog and Schadler, 1998).

Table 1
Home-based trip chains in a sample of Sydney journeys in 1991

Trip chain number	Trip chain	Mode of transport	Trip configuration (a)	Frequency
1	Simple work	Public transport	h–w(–w–)–h	39
2	Simple work	Car	h–w(–w–)–h	145
3	Complex work	Public transport	h–nw(–nw/w–)–w–(nw/w)–nw–h	32
4	Complex to work	Car	h–nw(–nw/w–)–w–h	16
5	Complex from work	Car	h–w(–nw/w–)–nw–h	59
6	Complex to and from work	Car	h–nw(–nw/w–)–w–(nw/w)–nw–h	33
7	Complex at work	Car	h–w(–nw/w–)–nw–(nw/w)–w–h	30
8	Simple non-work	Public transport	h–nw–h	41
9	Simple non-work	Car	h–nw–h	211
10	Complex non-work	Public transport	h–nw(–nw–)–h	16
11	Complex non-work	Car	h–nw(–nw–)–h	69

Note: (a) w = work, n = non-work, and h = home. The bracketed terms represent additional trips that may be present in the chain.

One recognized barrier to modal switching is trip chaining. Formally, a trip chain exists, in varying degrees of complexity, where one or more individuals undertake a trip that involves more than a single activity at one or more intermediate and final destinations. Strathman and Dueker (1995) distinguish between *simple* journeys involving a trip from home to a given destination and then returning home, and *complex* journeys, involving a sequence of more than two trips that begins and ends at home. Trip chain typology tends to distinguish trips involving work and non-work purposes. Passenger examples for home-based trip chains are shown in Table 1 (based on Strathman and Dueker 1995), with Sydney travel activity frequency used to illustrate the extent of trip complexity. The same logic applies to freight activity.

In addition to modal attributes, the nature and formation of trips and potential to switch modes tends to be influenced by the composition of the household. Strathman and Dueker (1995) and Oster (1997) investigated trip chain formations across gender, age, household structure, employment status, trip purpose, travel time, urban area, income, and mode. They found that different household structures impacted on the relationship between work and non-work travel. Specifically, it was suggested that with reductions in household size coupled with an expansion of multiple-worker households, the tendency to link non-work trips to the work commute would increase. This brief discussion of the structure of travel is important in preconditioning the context in which to identify the influences on modal diversion. Modal diversion models should incorporate all of these sources of influence.

3. Representing behavioral responsiveness

There are two useful ways of calculating the behavioral responsiveness of individuals and firms to changes in the levels of all influences on modal choice. These are:

(1) The use of an elasticity calculation, typically obtained from a modal choice model or a before-and-after monitoring study of the impact of a policy change;
(2) The calculation of changes in choice probabilities and shares using a modal-choice model.

To gain a good understanding of these methods we will use a series of examples.

3.1. Elasticities

An elasticity indicator in the context of modal choice defines the relationship between a percentage change in the level of an attribute (e.g., travel time) and the percentage change in the modal share (e.g., of train), all other influences held constant. A direct elasticity relates an attribute of a mode to the same mode; a cross-elasticity relates an attribute to a competing mode's share.

A comprehensive review of the concept of elasticity is provided in Handbook 1 of this series by Oum and Waters (2000). There is also an extensive literature on empirical elasticities, summarized by Oum et al. (1992) and Goodwin (1992). Typical direct elasticities for a number of applications are presented in Table 2. The range for many of the direct price elasticities is very large, and is a warning about carefully sourcing published elasticities. The differences, due to the particular context such as existing market share, attractiveness of competing modes, the discretionary vs. mandatory nature of travel, the quality of data, and the method of modeling, all contribute to explaining differences in the values. Nijkamp and Pepping (1998) have investigated sources of differences in public transport demand elasticities in The Netherlands, Finland, Norway, and the U.K., and conclude that country, number of competitive modes, and type of data collected have the strongest explanatory power on the magnitude of elasticities. Thus care should be taken in comparing elasticities from different countries even when the estimation methods are the same. Cultural differences do affect sensitivity to prices and service levels. Increasingly, site-specific travel demand studies are being undertaken to gain more confidence in how the local market will respond to changes in prices and other attributes influencing travel choice and demand.

The evidence on cross elasticities is somewhat limited. Typical passenger commuter studies suggest that the cross elasticities for rail and bus with respect to bus and rail fares are very similar, with an unweighted average value of 0.24 ± 0.06. The car-to-public-transport and public-transport-to-car cross elasticities, however, are quite different. The average cross elasticity of car demand with respect to bus

Table 2
Illustrative direct price elasticities

Setting	Low estimates	High estimates
Car use – U.S.A. short run	–0.23	–0.27
Car use – Australia short run	–0.08	–0.24
Car use – U.K. short run	–0.28	–0.28
Car use – U.S.A. long run	–0.28	–0.71
Car use – Australia long run	–0.22	–0.80
Car use – U.K. long run	–0.24	–0.71
Urban public transport – time series	–0.17	–1.32
Urban public transport – cross section	–0.05	–0.34
Urban public transport – before/after	–0.10	–0.70
Air passenger – leisure time series	–0.65	–1.95
Air passenger – business time series	–0.4	–0.67
Air passenger – leisure cross section	–1.52	–1.52
Air passenger – business cross section	–1.15	–1.15
Intercity rail – business time series	–0.67	–1.0
Intercity rail – non-business time series	–0.37	–1.54
Intercity rail – business cross section	–0.70	–0.70
Intercity rail – non–business cross section	–1.40	–1.40
Rail freight – all commodity classes	–0.09	–1.06
Rail freight – food products	–1.04	–2.58
Rail freight – iron and steel products	–1.20	–2.54
Rail freight – machinery	–0.16	–3.50
Truck freight – all commodity classes	–0.69	–1.34
Truck freight – machinery	–0.78	–1.23
Truck freight – food products	–0.52	–1.54

fares is 0.09 ± 0.07; and with respect to train fares it is 0.08 ± 0.03. These values are significantly higher for travel to central business district (CBD) destinations, where the propensity to use public transport is greater (i.e.. higher initial modal share). Glaister and Lewis (1978) have stated that the evidence on elasticities for the impact of public transport fares on car traffic for the off-peak are largely guesswork. Twenty-three years on, little appears to have changed.

3.2. *Elasticities in the context of passenger activities: An example of ticket choice*

In predicting the response of the market to specific fare classes and levels (e.g., weekly ticket), knowledge of how various market segments respond to both the choice of ticket type within a public transport mode and the choice between modes is important. Missing in many studies is a matrix of appropriate direct and cross fare elasticities that relate to specific *fare classes* within a choice set of fare class opportunities.

A popular method of obtaining such elasticities is the estimation of a discrete-choice model of the multinomial logit (MNL) form. It is well known, however, that the independently and identically distributed (IID) assumption of the random errors underlying the MNL model imposes a restriction on all cross elasticities, making them identical. Alternative choice models such as the heteroscedastic extreme-value (HEV) logit model relax the constant-variance assumption of the MNL model, allowing the cross-elasticities to be alternative-specific. Bhat (2000) in Handbook 1 reviews these methods.

An example set of direct and cross elasticities for commuters in Sydney is shown in Table 3, obtained from MNL and HEV models using mixtures of revealed-preference (RP) and stated-preference (SP) data (Hensher, 1998b). Each column provides one direct share elasticity and six cross share elasticities. A direct or cross *share* elasticity represents the relationship between a percentage change in fare level and a percentage change in the proportion of daily one-way trips by the particular mode and ticket type. For example, the column headed TS tells us that a 1% increase in the train single fare leads to a 0.218% reduction in the proportion of daily one-way trips by train on a single fare. In addition, this 1% single-fare increase leads to a 0.01% higher proportion of one-way trips on a train travel pass and a 0.001% increase in one-way trips on a train weekly ticket.

Transport studies employ typically two types of data to derive these elasticities, RP and SP data. The former refers to data measuring actual travel behavior and perceptions of attribute levels of existing modes. The latter refers to data that is derived from a hypothetical choice experiment in which attribute levels are predefined and individuals evaluate the alternatives and select a preferred alternative mode. This literature is vast and growing and is summarized in Hensher (1994), Louviere and Street (2000), and Louviere et al. (2001). The set of fare elasticities is based on the use of the SP parameter estimates for fare and cost, rescaled into the RP model (by the variances of the random errors for RP and SP data pertaining to the same mode and ticket type), which provides the choice probabilities and fare (or car cost) attribute levels. For comparison we report the direct and cross elasticities from the SP model and the MNL direct elasticities (noting that the cross elasticities for an MNL model are uninformative). The issue of rescaling of parameters when they are transferred between SP and RP alternatives is discussed in Section 3.4.

The results offer many implications. The differences in direct elasticities between the SP and RP choice sets reflect the different probabilities of choice. As is well known, although often ignored, studies which derive elasticities from stand-alone SP models tend to obtain exaggerated switching propensities, which arises from the accumulating evidence that respondents have a tendency to exaggerate their stated responses, no matter how well the choice experiment is designed. Since an elasticity calculation uses three inputs – a predicted choice probability, a taste weight (and a scale parameter in an HEV model), and an attribute level –

Table 3
Direct and cross share elasticities (a) for Sydney commuters

	TS	TW	TP	BS	BT	BP	Car
Train single (TS)	**–0.218** (–0.702) [–0.161, –0.517]	0.001 (0.289)	0.001 (0.149)	0.057 (0.012)	0.005 (0.015)	0.005 (0.009)	0.196 (0.194)
Train weekly (TW)	0.001 (0.213)	**–0.093** (–0.635) [–0.057, –0.313]	0.001 (0.358)	0.001 (0.025)	0.001 (0.024)	0.006 (0.019)	0.092 (0.229)
Train travel pass (TP)	0.001 (0.210)	0.001 (0.653)	**–0.196** (–1.23) [–0.111, –0.597]	0.001 (0.023)	0.012 (0.022)	0.001 (0.017)	0.335 (0.218)
Bus single (BS)	0.067 (0.023)	0.001 (0.053)	0.001 (0.031)	**–0.357** (–0.914) [–0.217, –0.418]	0.001 (0.248)	0.001 (0.286)	0.116 (0.096)
Bus travel ten (BT)	0.020 (0.020)	0.004 (0.037)	0.002 (0.023)	0.001 (0.206)	**–0.160** (–0.462) [–0.083, –0.268]	0.001 (0.163)	0.121 (0.090)
Bus travel pass (BP)	0.007 (0.025)	0.036 (0.063)	0.001 (0.034)	0.001 (0.395)	0.001 (0.290)	**–0.098** (–0.700) [–0.072, –0.293]	0.020 (0.103)
Car (C1)	0.053 (0.014)	0.042 (0.023)	0.003 (0.013)	0.066 (0.009)	0.016 (0.011)	0.003 (0.006)	**–0.197** (–0.138) [–0.130, –0.200]

Note: (a) Elasticities relate to the total ticket price, not price per one-way trip. SP direct and cross elasticities from the HEV model are given in parentheses. The MNL direct elasticities from the RP and SP choice sets are given in square brackets. The interpretation for a specific fare class is given under each column heading.

the appropriate probabilities must come from the RP model. The RP direct elasticities for public transport are lower than the SP equivalences; however, since the results are driven primarily by probability differences, some elasticities must be higher for the SP model. This is the case for the car mode; this is explained by the fact that the SP percentage choosing the car is less than the actual market share.

For direct elasticities, sensitivity within the commuter rail and bus markets decreases as we move from a single ticket through to multiple-trip tickets. This has interesting implications for a fares policy – increasing the price of a multiuse ticket offers higher revenue growth prospects for small losses of patronage than is the case for single tickets. The cross elasticities suggest that there is more movement between modes for a given fare class than between fare classes within modes. The strongest cross-mode substitution occurs between train and bus single tickets, although it is not symmetrical, with cross elasticities of 0.067 and 0.057 for train to bus and bus to train, respectively. The largest cross elasticity is 0.335, for the switch from car to train travel pass in the event of a price increase in car use. A travel pass per trip is the best value-for-money train fare, where the price per one-way trip is $1.28, compared with $1.64 for a train single and $2.46 for a travel ten ticket. All the cross elasticities associated with car operating costs are sizable compared with the other modal-switching contexts. Interestingly, changes in public transport fares across all ticket categories have less of an impact on car use than a change in car costs has on public transport use.

A comparison of the HEV and MNL revealed-preference elasticities shows a systematically lower set of direct-elasticity estimates for all alternatives in the MNL model; thus, on the one hand, we might conclude that an SP model tends to produce lower elasticities than its RP counterpart, where the SP choice probabilities are higher than the RP probabilities; and on the other hand MNL direct-elasticity estimates tend to be lower than their HEV counterparts in both RP and SP models. These results illustrate how important it is to use the best available data and modeling methods.

3.3. Elasticities in the context of freight activity

In the freight sector, the selection of an appropriate elasticity measure is complicated by the distinction between travel demand and commodity demand. Travel is a derived demand, and as such is dependent on the demand for commodities. That is, the demand for freight services is derived from the demand for commodities being transported. To appreciate this important difference, we set out the relationship between the elasticity of travel demand (as illustrated in Table 2) and the elasticity of commodity demand in the freight sector. We will develop two scenarios – the situation where there is only one modal supplier and the situation where there are competitive alternatives (Hensher and Brewer, 2000).

Situation A: One supplier (e.g., rail or air only)

Define the transport price elasticity of demand as ε_T = (% change in Q_D)/(% change in P_T), where P_T is the freight rate per unit of commodity shipped, and Q_D is the number of units of commodity shipped. The percentage change in the quantity demanded depends on the relative importance of the freight rate to the delivered price of the commodity (i.e., $P_T/(P_T + P_C) = \alpha$, where P_C is the free-on-board (f.o.b.) delivered price of the commodity, and the elasticity of demand for a commodity in its market is ε_D (i.e., % change in delivered price = α^* (% change in P_T)). If freight rates change by S%, the impact on the price of the commodity being shipped is αS in percentage terms. Since ε_D = (% change in Q_D)/(% change in P_C), which also equals (% change in Q_D)/αS, then the percentage change in quantity shipped = $\varepsilon_D \alpha S$. But given that ε_T (transport demand) = (% change in Q_D)/(% change in P_T), then $\varepsilon_T = \varepsilon_D \alpha$; α is typically less than 10% in practice (but clearly varies between situations).

Situation B: Existence of other suppliers (road vs. rail, road vs. air)

Define the elasticity of transport (rail) demand as $\varepsilon_{RT} = 1/\Gamma[\varepsilon_D \alpha - (1-\Gamma)^* \varepsilon_{cross}]$, where Γ is the rail market share, ε_D is the elasticity of commodity demand, and ε_{cross} is the cross elasticity of demand between rail and truck or rail and air. The total change in rail traffic due to rail rate decreases depends on (assuming no truck retaliation) the "all other things equal" conditions: (i) the increase in total traffic due to rail rate decrease and (ii) the amount of traffic shifted from truck to rail. To account for the elasticity of supply ε_S of a commodity, define $\varepsilon_D = \alpha\{\varepsilon_S \varepsilon_D / [\varepsilon_S - \varepsilon_D(1-\alpha)]\}$. The elasticity of demand for transport varies directly with the product of ε_D and ε_S, directly with α, and inversely with the sum of elasticities since, $\varepsilon_D \leq 0$ and $\varepsilon_\Sigma \geq 0$ Unfortunately, poor data in practice often inhibits the use of such useful formulae. We usually end up assuming (without realizing it) that ε_S approaches infinity (i.e., perfectly elastic supply) and that ε_T is independent of ε_D. If $\varepsilon_D > \varepsilon_T$ and $\varepsilon_S < \alpha$ then ε_S tends to reduce the elasticity of demand for transport. That is, existing estimates are upward biased.

3.4. The calculation of changes in modal shares using a joint mode and parking choice model

Although elasticities are very useful in assessing modal diversion under a specific attribute policy change, they impose the requirement that all other potential

Table 4
Models for joint mode and parking location choice

Attribute	Commuter market		Non-commuter market	
	Alternative	Parameter (t-value)	Alternative	Parameter (t-value)
Parking price ($ per hour)	rpcar, spoffsta, spoffstb	–1.5330 (–7.6)	rpcaron, rpcaroff, sponst, spoffst	–0.1626 (–1.71)
Line-haul in-vehicle cost ($)	all	–0.2182 (–5.3)	all	–0.1440 (–2.37)
Access cost ($)	rppt, sppt	–0.2182 (–5.3)		
Line-haul in-vehicle time (min)	all	–0.0214 (–4.2)	all	–0.0094 (–1.96)
Parking search time (min)	rpcar, spoffsta, spoffstb	–0.0448 (–3.38)	rpcaron, sponst,	–0.1443 (–2.75)
Parking search time (min)			rpcaroff, spoffst	–0.0716 (–3.19)
Egress time (min)	all	–0.0448 (–3.38)	all	–0.0094 (–1.91)
Access time (min)	rppt, sppt	–0.0424 (–2.71)	rppt, sppt	–0.0094 (–1.91)
Wait time (min)	rppt, sppt	–0.0512 (–2.55)	rppt, sppt	–0.0094 (–1.91)
Train-specific constant (1, 0)	rppt, sppt	2.0397 (12.12)	rppt,sppt	2.0988 (11.0)
Tripmaker pays parking (1, 0)	rpcar	–3.4745 (–2.80)		
Tripmaker pays parking (1, 0)	spoffsta, spoffstb	–1.3882 (–2.10)		
Shopping trip (1, 0)			rpcaron, sponst	–1.4194 (–4.86)
Shopping trip (1, 0)			rpcaroff, spoffst	–1.3447 (–6.12)
Education trip (1, 0)			rpcaron, sponst	–2.2356 (–3.35)
Education trip (1, 0)			rpcaroff, spoffst	–1.8384 (–3.61)
Alternative-specific constant	rpcar	3.2729 (2.68)	rpcaron	3.0119 (3.63)
Alternative-specific constant	spoffsta	1.9602 (2.54)	rpcaroff	1.4179 (3.80)
Alternative-specific constant	spoffstb	2.2639 (3.02)	sponst	0.9935 (1.64)
Alternative-specific constant	spout	–1.3079 (–1.70)	spoffst	1.3690 (2.77)
Alternative-specific constant			spout	–6.3119 (–2.45)
Inclusive value	RU1: rpcar, spoffsta, spoffstb, spout	0.5780 (5.47)	RU2: rpcaron, rpcaroff, sponst, spoffst, spout	0.4719 (2.83)
Log–likelihood at convergence		–817.35		–938.50
Adjusted pseudo-r^2		0.538		0.529
Sample size		964		1005

influences remain fixed at their base level. In contrast, with a model, it is possible to allow for more than one attribute to be changed. If there is a sophisticated feedback and equilibrium procedure in place it is also possible to manipulate one attribute and see how it affects the levels of other attributes. In this section we take an example of a choice study that focuses on the choice between public transport (bus, train) and a number of car options with respect to parking for commuting and non-commuting trips. Each model comprises a mixture of RP and SP choice sets to recognize that the necessary behavioral information on parking options cannot be identified from capturing actual modal behavior. There is no market data available on specific choices such as paid on-street parking because currently all on-street parking is free in the study area.

The choice set for commuters consists of two RP and SP modes (car and public transport). The RP car trips all entail off-street parking (commuters tend not to park on-street). For the SP choice set, the car alternatives are distinguished by parking location – close to the center of the CBD, on the boundary the CBD, and outside of the CBD. The two parking locations in the CBD are off-street parking, whereas the external parking is on-street and free. The RP and SP line-haul public transport is either train or bus.

In the specification of the joint mode and parking choice models we investigated a large number of potential influencing attributes of the modes, the parking location, and the trip maker. We also considered alternative nested logit (NL) structures to account for different distributional assumptions on the variances of the unobserved (or random error) components that contribute, together with the observed set of attribute influences, to the overall relative utility of each competing mode and parking location option. The final models are summarized in Table 4. The commuting model is a random utlity (RU1) specification of the NL model and the non-commuting model is an RU2 specification. The differences between these model forms are explained in Hensher and Greene (1999), but in essence they reflect alternative ways of handling the variances of the unobserved random-error components, by normalizing from either the top level (RU2) or the bottom level (RU1).

These variances are very important in determining the structure of the NL model since we partition the alternatives to reflect differences in these variances. Since each of the parameter estimates of the attributes is implicitly scaled by the inverse of the square root of the corresponding variance (arbitrarily normalized to 1.0 for all alternatives in an MNL model and hence ignored), it is important that we establish whether these scale parameters might vary across the alternatives. If they do, then the unity scale for all alternatives is wrong and will lead to incorrect inferences on the importance of each attribute. Another way of saying this is that failure to account for any differences in scale (i.e., imposing an MNL form) will confound scale parameter estimates with true attribute parameter estimates. Although earlier studies had assumed that the differences in scale were unique to differences in the data source (i.e., RP and SP), recent studies such as Hensher (1998b) have shown that, often, the differences in scale are between alternatives within and between data sources. We have found herein that the differences are between car and public transport and not between RP and SP. What this tells us is that the unobserved influences on mode and parking choice tend to be more similar within the car and within the public transport alternatives, regardless of whether the data is derived from real markets or the SP experiment. There is intuitive appeal to this interpretation.

The final models have a rich array of modal and parking attributes, all statistically significant and of the correct sign at the 95% level, except for parking cost per hour for non-commuters, which has a *t*-value of 1.71. The overall goodness of fit for a non-linear NL model is impressive, 0.538 and 0.529 for commuting and non-commuting, respectively. These overall fits are above what is normally found in mode choice models (typically in the 0.2–0.4 range).

A comparison of absolute parameter estimates in a non-linear discrete choice model is meaningless and so we look for suitable behavioral measures of performance. We have derived a set of behavioral values of travel time savings for the time components as one basis of determining the plausibility of the models. The results are summarized in Table 5 on a per-person-hour basis, as well as on the basis of a percentage of the average gross wage rate of the sample (based on a working year of 2000 hours, typically adopted by studies that convert time savings values to a percentage of the wage rate). The commuter and non-commuter values are not directly comparable, because of the different representations of the components of travel time. We can see for commuters that egress time, access time, search time, and wait time all have a mean value of travel time saving (VTTS) higher than line-haul in-vehicle time, as expected (see Hensher and Dalvi, 1978), and that the ratios of such values relative to line-haul and each other are in the range that other studies have found (typically 1.5–3.0). That is, the value of waiting-time savings for public transport is 2.39 times higher than line-haul in-vehicle time, access time is 1.98 times higher, and the composite of searching for parking and walking to the destination is 2.09 times' higher. The reasoning is that the marginal (dis)utility of a minute of in-vehicle line-haul time is less than that of other time components. Another way of putting this is that an individual is willing to pay more to save a minute waiting for a bus or train, or searching for parking and walking to the destination, than to save a minute traveling in public transport or a car. The mean VTTSs as a percentage of the average gross wage rate are also acceptable.

For non-commuting we have a more limited basis for comparative comment since there is relatively little literature on non-commuter values. Furthermore, we have had to aggregate all sources of travel time except searching for parking in order to derive a statistically significant time effect. This may say something about the relative homogeneity of travel time for non-commuters compared with commuters, although this needs further research. The relativities of the values of

Table 5
Behavioral values of travel time savings ($/person hour). Percentages in parentheses are relative to the average gross wage rate of sample

Travel time component	Commuter segment Alternatives	Commuter segment VTTS (a) (as % average wage rate)	Non-commuter segment Alternatives	Non-commuter segment VTTS (as % average wage rate)
Line-haul in-vehicle time	all	$5.90/h (28%)		
Egress and search time	Rpcar, spoffsta, spoffstb	$12.33/h (58%)		
Search time			rpcaron, sponst	$60/h (350%)
Search time			rpcaroff, spoffst	$29.9/h (173%)
Access time	rppt, sppt	$11.67/h (56%)		
Wait time	rppt, sppt	$14.09/h (67%)		
Total time excluding search time			all	$3.94/h (23%)

search time savings for car parking are very informative, suggesting that non-commuters are willing to pay about twice the amount to save a minute of search time looking for on-street parking than for off-street parking. This makes intuitive sense. Search time is relatively long in the study area, averaging 6.5 minutes for on-street parking and 5.8 minutes for off-street parking for individuals who actually park (and 6.6 and 8.4 minutes on-street and off-street respectively, for all parkers and potential non-parkers). This converts to a willingness to pay on average $1 and $0.50 to save a minute.

Assessing changes in market shares through sample enumeration

The joint mode and parking choice models provide the behavioral base for the specification of a set of utility expressions for each market segment that utilizes the information in both the RP and the SP utility equations. Since the scale differences have been identified between car and public transport rather than between the RP and SP data, it is necessary to scale the public transport utility expressions when they are combined with the car utility expressions in the finalization of the application equations.

The commuter application model comprises five utility expressions representing (1) off-street parking close in, (2) off-street parking further out in the central city area, (3) parking outside the city and walking in, (4) using the train, and (5) using the bus. The final empirical form is given below. The scaling parameter for bus and train is the inverse of the inclusive value parameter derived from the RU1 model form of nested logit.

(1) Uoffsta=1.9602-1.533*pcperh1-.2182*invhcst1-.02144*travelt1-1.3882*youpay1-.04483*egresst1-.04483*searcht1
(2) Uoffstb=2.2639-1.533*pcperh2-.2182*invhcst2-.02144*travelt2-1.3882*youpay2-.04483*egresst2-.04483*searcht2
(3) Uout=-1.30798-.2182*invhcst3-.02144*travelt3-1.3882*youpay3-.04483*egresst3
(4) Utrain=(1/0.578)*(2.03979-.042463*accesst4-.04483*egresst4-.02144*travelt4 -.051214*waitt4 -.2182*invhcst4-.2182*accessc4)
(5) Ubus=(1/0.578)*(-.042463*accesst5-.04483*egresst5-.02144*travelt5-.051214*waitt5 -.2182*invhcst5-.2182*accessc5)

The non-commuter application model also comprises five (different) utility expressions, representing (1) on-street parking in the central city area, (2) off-street parking further out in the central city area, (3) parking outside the city and walking in, (4) using the train, and (5) using the bus. The final empirical form is given below. The scaling parameter for bus and train is the inclusive value parameter derived from the RU2 model form of nested logit.

(1) Uonst=3.01194-.16268*pcperh1-.14395*invhcst1-.009455*egresst1-.009455*travelt1-.14431*searcht1-1.4194*shop1-2.23557*educ1
(2) Uoffst=1.41796-.16268*pcperh2-.14395*invhcst2-.009455*egresst2-.009455*travelt2-.0716228*searcht2-1.3447*shop21.8384*educ2
(3) Uout=-6.311948-.14395*invhcst3-.009455*egresst3-.009455*travelt3
(4) Utrain=0.4719*(2.09882-.143952*invhcst4-.09455*accesst4-.09455*egresst4-.09455*waitt4-.09455*travelt4)
(5) Ubus=0.4719*(.143952*invhcst5-.09455*accesst5-.09455*egresst5-.09455*waitt5-.09455*travelt5)

To illustrate the types of policy scenarios that can be evaluated, we have run a number of scenarios for each market segment, with the findings summarized in Table 6. A full sample enumeration procedure has been used, in which a sample of individual trip makers was selected to represent the distribution of travel times, costs, and socio-economic attributes for the population segment. This is preferred to the use of average levels for the entire population, which suffers severely from aggregation error.

The policy scenarios are all examples of opportunities to alter the modal split. There is clear evidence from these examples that a doubling of off-street parking prices has a significant influence on the modal share in favor of public transport, with the impact being greater in the commuter segment, especially for parking off-street at the boundary stations. The doubling of metered on-street parking in the non-commuter market from the base average of 31.8 cents per hour (and a

Table 6
Illustrative modal-share (MS) changes under various policy scenarios (all figures are absolute percentages)

(a) Commuter segment

	Double off-street parking prices in central area	Halve parking search time	Reduce walk time from public transport to 25% of current levels
MS off street close	7.51	7.51	7.51
Change MS	–5.18	0.097	–1.72
MS off-street boundary	29.3	29.3	29.3
Change MS	–10.9	0.99	–6.24
MS outside	2.68	2.68	2.68
Change MS	1.22	–0.125	–0.63
MS train	58.7	58.7	58.7
Change MS	14.5	–1.79	8.33
MS bus	1.72	1.72	1.72
Change MS	0.042	–0.052	–0.245

(b) Non-commuter segment

	Double on-street parking prices in central area	Double off-street parking prices in central area	Halve parking search time	Reduce walk time from public transport to 25% of current levels
MS on-street	64.0	64.0	64.0	64.0
Change MS	–2.68	2.5	7.2	–2.79
MS off-street	18.86	18.86	18.86	18.86
Change MS	1.51	–3.2	–2.0	–0.94
MS outside	0.049	0.049	0.049	0.049
Change MS	0.004	0.002	–0.013	–0.003
MS train	11.2	11.2	11.2	11.2
Change MS	0.77	0.46	–3.3	2.46
MS-bus	5.77	5.77	5.77	5.77
Change MS	0.40	0.24	–1.7	1.26

standard deviation of 72 cents) reduces the modal share from 64% to close to 60%, suggesting that metering will not have a significant impact on modal diversion but will clearly be a good revenue source. From this limited set of examples we can see that halving parking search time has a very small impact on the commuter modal share but a very noticeable impact in the non-commuter segment. In the latter setting we see a 7.2% absolute increase in the share for on-street parking and losses in shares across all other modes, with the greatest loss in train use. This is an interesting policy – clearly there is a benefit to public transport in "preserving" the high search time for on-street parking even though it is likely that there are significant user benefits from such a policy. Finally, improving accessibility from public transport by improving egress walk time to 25% of its current level has much to offer in terms of improving public transport modal share, especially the train share in the commuter market and the bus and train shares in the non-commuter market segment. In the latter segment, the public transport mode share increases by an absolute 3.72%, from 16.92% to 20.64%. These findings must be placed in the context of the base modal shares, which have been calibrated on the existing data sets.

4. Conclusions

The most popular methods available to investigate influences on modal choice and modal diversion have been presented in this chapter. Although much of the focus has been on the passenger market, the approaches are equally applicable to commodity movements. The state of practice has moved forward with great pace over the last ten years, especially with the inclusion of mixtures of data types as a way of studying behavioral responsiveness to attribute levels that are beyond those observed in actual markets and which are potential policy applications.

Any study of modal diversion is only as good as the data, the models, and the careful selection of the choice setting in which we expect trip makers to react. We have limited the discussion to modal diversion, holding other transport and non-transport choices fixed. It is possible, however, that policies geared towards influencing modal diversion may influence trip-timing, destination, and route choices in addition to or instead of modal split. The broadening of the behavioral-response options is an active research area, taking advantage of the contributions in data and model specification that have been made, in the main, through applied research focused on modal choice.

References

Bhat, C. (2000) "Discrete choice models", in: D.A. Hensher and K. Button eds. *Handbooks in transport 1. Transport modelling*. Oxford: Pergamon.

Brog, W. and M. Schadler (1998) "Marketing in public transport is an investment, not a cost", *Papers of the Australasian Transport Research Forum*, 22(Part 2):619–634.

Glaister, S. and D. Lewis (1978) "An integrated fares policy for transport in London", *Journal of Public Economics*, 9:341–355.

Goodwin, P.B. (1992) "A review of new demand elasticities with special reference to short and long run effects of price changes", *Journal of Transport Economics and Policy*, 26(2):155–169.

Hensher, D.A. (1994) "Stated preference analysis of travel choices: The state of practice", *Transportation*, 21(2):106–134.

Hensher, D.A. (1998a) "The balance between car and public transport use in urban areas: What can we do about it?", *Transport Policy*, 5(4):193–204.

Hensher, D.A. (1998b) "Establishing a fare elasticity regime for urban passenger transport", *Journal of Transport Economics and Policy*, 32(2):221–246.

Hensher, D.A. and A.M. Brewer (2000) *Transport: An economics and management perspective*. Oxford: Oxford University Press.

Hensher, D.A. and K. Button eds. (2000) *Handbooks in transport 1. Transport modelling*, Oxford: Pergamon Press.

Hensher, D.A. and M.Q. Dalvi eds. (1978) *The determinants of travel choices*. Farnborough: Saxon House Studies.

Hensher, D.A. and W.H. Greene (1999) "Specification and estimation of nested logit models", Institute of Transport Studies, The University of Sydney.

Louviere, J.J. and D. Street (2000) "Stated preference methods", in D.A. Hensher and K. Button, eds., *Handbooks in transport 1. Transport modelling*. Oxford: Pergamon.

Louviere, J.J., D.A. Hensher and J. Swait (2001) *Stated choice methods: Analysis and applications in marketing, transportation and environmental valuation*. Cambridge: Cambridge University Press.

Nijkamp, P. and G. Pepping (1998) "Meta-analysis for explaining the variance in public transport demand elasticities in Europe", *Journal of Transportation and Statistics*, 21(1):1–14.

Oster, C.V. (1997) "Second Role of the work trip – visiting non work destinations", *Transportation Research Record*, 728:79–81.

Oum, T. and W.G. Waters (2000) "Travel demand elasticities", in: D.A. Hensher and K. Button, eds., *Handbooks in transport 1. Transport modelling*. Oxford: Pergamon.

Oum, T., W.G. Waters and J. Yong (1992) "Concepts of price elasticities of transport demand and recent empirical evidence", *Journal of Transport Economics and Policy*, 26 (2):139–154.

Strathman, J.G. and K.J. Dueker (1995) "Understanding trip chaining", U.S. Department of Transportation, Special Reports on Trip and Vehicle Attributes, 1990 NPTS Report Series.

Chapter 9

INDUCED TRAVEL AND USER BENEFITS: CLARIFYING DEFINITIONS AND MEASUREMENT FOR URBAN ROAD INFRASTRUCTURE

PETER W. ABELSON
Macquarie University, Sydney

DAVID A. HENSHER
University of Sydney

1. Introduction

There is a popular and not entirely unjustified belief that the construction of new roads, especially in urban areas, creates more problems than it solves. It is observed that new roads often fill up rapidly and may do little or nothing to improve travel speeds in the long run. Moreover, the increase in motor vehicle traffic often worsens air quality, increases greenhouse gas emissions, increases transport noise, and reduces residential and pedestrian amenity.

On the other hand, most new roads do reduce travel times across the network for a period at least, with the length of time depending on the latent demand. Moreover, even if traffic speeds do not increase in the long run, there are benefits to traffic that was previously deterred by inadequate network capacity but which can now travel on the network. Any incremental environmental costs should be netted off such benefits.

In practice, however, road evaluations frequently ignore all effects (both negative and positive) of induced travel. Most urban road studies in many countries base their evaluation on traffic models that use fixed trip matrices, in which both total travel and its distribution are independent of the state of the network. This is often unsatisfactory, because after-the-event studies have found that induced travel may be as high as 20% of initial travel (Goodwin, 1996). However, many estimates of induced trips are crude, given the neglect of formal methods to quantify this effect. In any case, there are many different notions of what is, or is not, induced travel.

The objective of this chapter is therefore to describe how to evaluate and model induced traffic in the presence of new road infrastructure. The chapter starts by

Handbook of Transport Systems and Traffic Control, Edited by K.J. Button and D.A. Hensher

defining the various kinds of induced travel along with some empirical findings about induced travel. We then describe how induced travel should be evaluated and show how some common evaluation methods can produce erroneous results. The last substantive part of the chapter outlines traffic-modeling methods for producing the appropriate travel data.

2. The meaning of induced travel

In order to discuss induced travel, we must first have a unit of measure of travel. There are many possible units, including vehicle, freight (tonnes) or person trips, or vehicle, tonne, or person kilometers. There is no one correct unit. For most purposes, vehicle kilometers are the most useful measure because urban networks must accommodate vehicles. Thus, if people make longer trips as a result of an improved network, vehicle kilometers increase although vehicle trips may not. There is induced travel by the vehicle kilometer measure, but not by the vehicle or person trip measure. In this paper, induced travel is considered to occur if a road investment results in additional vehicle kilometers on either the network as a whole or on that part of it where the infrastructure is improved.

The U.K. Department of Transport (1993) identifies five main ways in which road improvements may result in induced travel.

(1) Trip generation: new trips are made that were not made previously on any transport mode.
(2) Trip redistribution: trips change destination. Faster travel speeds may encourage people to switch from a close destination to a more distant one that is more attractive. In some cases a network upgrade may make a closer destination more attractive than before.
(3) Changes in modal split: trips are now made by car instead of by some other mode, such as rail, air, coach, or bus (see Chapter 8).
(4) Route reassignment: traffic traveling between A and B switches routes, so that there are induced trips on the improved route although not necessarily in the network as a whole.
(5) Time of day: trips traveling between A and B switch trip times, so that there are induced trips at certain times (probably peak hours) on the improved route although not necessarily more trips on the network as a whole over the day.

There are significant differences between these scenarios. In the first three scenarios vehicle kilometers increase on both the improved link(s) and over the whole road network. Many writers identify induced demand only with these increases in vehicle kilometers. Indeed, induced travel is sometimes limited to generated traffic (scenario 1). On the other hand, in scenarios (4) and (5), traffic increases on certain routes or at certain times, but not over the network as a whole.

Just as there is no single correct definition of traffic, so there is no single correct definition of induced traffic. However, there are two reasons for generally preferring a wider definition, embracing all five scenarios, to a narrower one. First, it is often difficult to decompose total observed induced traffic on individual routes into the five separate components identified above. Second, as we argue below, the method of evaluating all forms of induced traffic is similar, regardless of the source of the induced traffic.

3. Evidence on induced travel

There have been many attempts to identify the extent of induced travel, for example Goodwin (1996), Luk and Chung (1997), Litman (1999), DeCorla-Souza and Cohen (1999), and Lee et al. (1999). We summarize below the main elasticity evidence for urban road projects that have some measure of portability to similar urban contexts. In order to do this, we have reviewed the available evidence and culled the many studies focusing on interurban projects (including high-speed rail), natural barrier crossings (such as bridges and tunnels), and public transport projects. We have also focused on the evidence relevant to cities at a similar stage in development in terms of population density and economic wealth.

Before citing the evidence, the limitations of much of the survey work should be noted. First, as we have noted, there are various definitions of induced travel. Some studies focus on travel over the network but effectively include trip generation and modal change. Other studies may include the impacts of trip redistribution. Still other studies that focus on network links or on peak hour travel may include scenario (4) or (5). Thus, a few studies include all forms of induced travel. Second, some studies are based principally on *estimates* of the effects of changes in travel time – they estimate induced travel from assumed travel time elasticities. Other studies draw on survey evidence. However, survey evidence has to be interpreted carefully to ensure that other growth factors are fully accounted for. Thus, care must be taken in generalizing or transferring the results.

The British evidence is a good start, with the most comprehensive summaries provided by Goodwin (1996) and Coombe (1996). Studies undertaken in Norwich, Belfast, and West London used elasticity techniques in which travel between individual origins and destinations is adjusted in inverse proportion to travel costs, subject to an assumed travel time elasticity value that is imposed on the data. The major conclusions are the following.

The number of trips crossing the study area cordon in West London increased by 1% based on an elasticity value with respect to travel time of –1.0 (a fixed travel time budget). Improvements in Belfast resulted in an increase in forecast travel of 2% based on an elasticity value with respect to travel time of –0.5. The completion of the inner ring road in Norwich increased the number of trips in the study area by 2.3%, whilst the outer ring road led to an increase in trips of 2.9% based on an elasticity with respect to travel time of –0.5.

This evidence suggests that induced traffic effects are modest even though the (predetermined) elasticity values applied were substantial, and the models covered congested urban areas. Because the estimated induced-traffic effects in these studies were based on aggregate models of travel behavior, the composition of the induced demand is unknown. However, the authors suggest that because these are mature urban contexts, the induced travel probably consisted mainly of trip redistribution and modal shifts, rather than generated traffic.

Goodwin (1996) reworked a number of U.K. studies and concluded that:

(1) In the period 1978–88, a 10% increase in motorway length led to about a 1% increase in motorway traffic (implied elasticity of 0.1).
(2) In a model study by the Centre for Economic and Business Research, a 7% increase in trunk road capacity in 2010 would result in 0.77% extra traffic (an implied elasticity of 0.11).
(3) A comparison of forecast and observed traffic flows on 151 improved roads indicates that observed flows average 10.4% higher than predicted. Goodwin suggests that this is consistent with an elasticity of 0.1 with respect to capacity.

Goodwin concludes that an average road improvement will bring about an additional 10% of base traffic in the short term and 20% in the long term. These estimates have been widely quoted throughout the literature since 1996. The caveat is that these estimates are averages, the projects are relatively small scale, and the input elasticities are based on other elasticities that have a wide range.

Litman (1999), drawing on Goodwin's review and a number of U.S. studies, proposed a schematic diagram (Figure 1) to summarize the range of elasticity estimates over a 20 year period under varying assumptions of the magnitude of unsatisfied (latent) demand. The more congested a road system is, the more traffic is likely to be generated by increased capacity, due to high levels of latent demand. In the congested conditions common in urban areas, within five years more than half of added capacity is typically filled by vehicle trips that would not otherwise occur. A careful reading of the sources used by Litman indicates that most of these studies included both generated and redistributed traffic. Small (1992), also cited by Litman, states that "...latent demand consists of people who, because of congestion, now chose an alternative route, mode, time of day, or home or workplace location, or who do not travel at all." Small was unable to separate out the generation effect. Many of the cited sources preselect an elasticity value without an apparent justification of its source. This is especially the case with

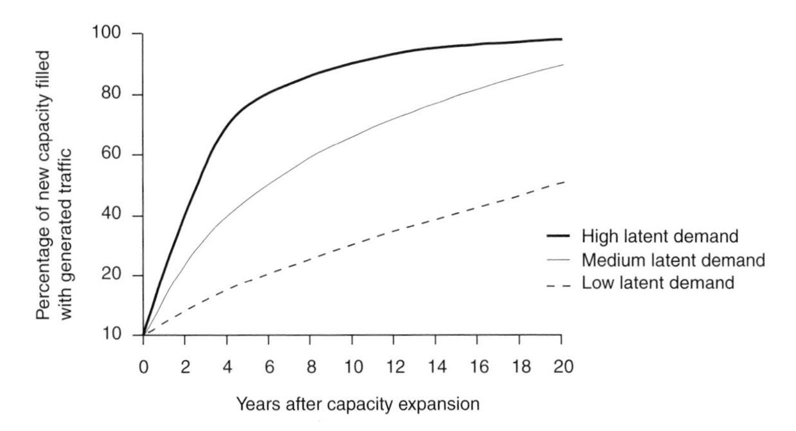

Figure 1. Elasticity of traffic volume with respect to road capacity. This illustrates traffic growth on a road after its capacity increases. About half of the added capacity is filled with new traffic within a decade of construction under normal circumstances.

estimates derived from modal-choice models (the most common source – see Chapter 8).

Hansen (1993) evaluated the traffic impacts over 20 years of additional lanes on 32 projects completed between 1970 and 1977 in urban counties in California. Again route diversion and latent demand are combined to produce interregional lane-mile elasticities, which are (surprisingly) almost identical at 0.32. An important finding is that intraregional (i.e. more local-area-specific) lane-mile elasticities are in the range 0.46–0.50. These elasticity values applied to an entire urban area generate more traffic on other parts of the system than is removed. That is, the complementary effect is often greater than the supplementary effect.

Finally, Luk and Chung (1999) attempt to unravel the sources of traffic growth in the southeastern corridor for metropolitan Melbourne in the period from 1975 to 1995. Induced demand here includes generated (latent) demand, mode shifts, and land use changes but excludes route switching. Luk and Chung conclude that most of the growth in traffic was natural growth of 1.7% per annum, though some growth was mode switching.

One major theme emerges from this review of the empirical literature, namely that we have very little substantive evidence on the contribution of each of the components of induced travel (as defined in Section 2).[*] The net increase in traffic appears to depend on the level of congestion and the spatial and temporal context

[*]On the basis of the literature review and cognizant of the broad interpretation of induced demand (excluding only route switching) the literature suggests a (conservative) range for the induced-demand elasticity with respect to generalized cost of 0.1 in the short run (up to 5 years) up to 0.3 in the long run (over 5 years).

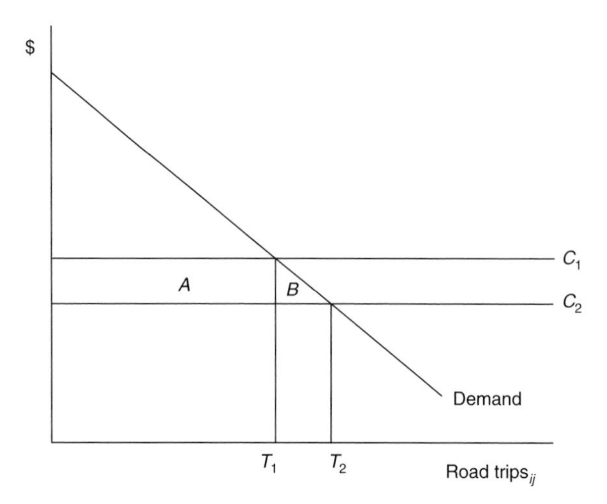

Figure 2. User benefits due to a road improvement.

of the empirical study as well as the (often) preassigned set of demand elasticities. It does appear, however, that trip generation (the narrower definition of induced traffic) is in the range from almost zero to 20%, with noticeable variations across the studies. Furthermore, there appear to be no known systematic influences to explain this variation. There is thus a very good argument for ensuring that all sources are accommodated in the methods used regardless of an ability to decompose total traffic into its constituent sources of inducement.

4. The valuation of benefits of induced travel

To examine the benefits of induced travel, we start with a simple case and then examine some complications. Figure 2 shows a demand curve for a road from i to j. After the road is improved, the generalized (behavioral) cost of travel,* inclusive of travel time and vehicle operating costs, falls from C_1 to C_2. Trips increase from T_1 to T_2. Induced trip-makers here include all people who use the improved road, inclusive of those who have changed route or mode, those who have changed destination, and those who make new trips as a result of the new road. Note also that generalized cost of travel is assumed here to equal resource costs. There are no taxes or externalities in this model. In this case, existing road users (T_1) gain cost savings given by area A, which equals $T_1 (C_1 - C_2)$.

*Note that for simplicity of illustration the cost curves are drawn as horizontals (with no congestion).

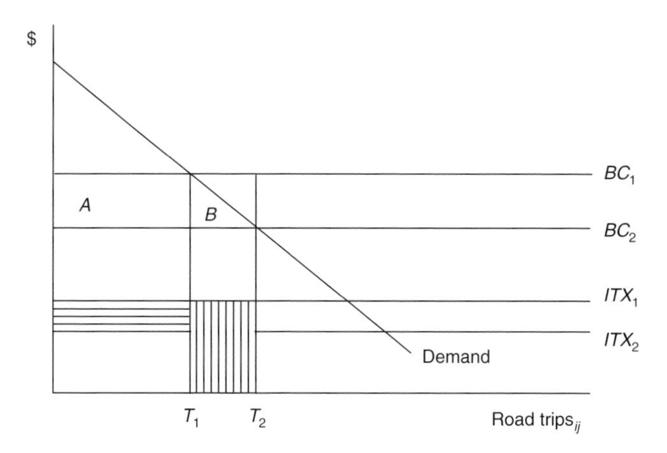

Figure 3. Road user benefits and indirect taxation.

Induced trip-makers $(T_2 - T_1)$ gain a consumer surplus (area B), which is the difference between what they are willing to pay for the trips and their actual costs. If the demand curve is linear, the consumer surplus equals $0.5(T_2 - T_1)(C_1 - C_2)$. This surplus* is a function of the change in trip costs. If total trip cost falls from say \$15 to \$10, some new trip-makers obtain a consumer surplus of nearly \$5. Others obtain a benefit of only a few cents. The average benefit is \$2.5 per new trip.

Algebraically, if the demand curve is linear, total user benefits (UB) are given by

$$UB = 0.5(T_1 + T_2)(C_1 - C_2) \tag{1}$$

When changes to the network are small, a linear demand curve is a reasonable approximation and eq. (1) a reasonable measure of user benefits. If changes to the network are large, as for example with a river crossing, the demand curve may be non-linear and the benefit of induced trips may be less than $0.5(T_2 - T_1)(C_1 - C_2)$. Now suppose that the generalized trip costs include indirect taxes, such as fuel taxes, that do not represent resource costs. Figure 3 shows the behavioral trip costs (BC) and the indirect-tax components before and after a road improvement. As shown, the road upgrade reduces the indirect tax paid by existing users (e.g., because it reduces fuel use). Now, the user benefits are as before (areas $A + B$). But the government loses some indirect tax (ITX) revenue associated with existing trips, $T_1(ITX_1 - ITX_2)$. On the other hand, it gains tax revenue from the induced trips, $ITX_2(T_2 - T_1)$. In order to estimate the total social gains from the road improvement, these losses and gains should be netted off user benefits. This is sometimes described as a "non-resource or taxation correction" (U.K. Department of Transport, 1996).

*This is referred to as the "rule of a half" in the evaluation literature.

However, a caveat should be noted. In so far as induced trip-makers purchase taxed fuel instead of some other taxed commodity, only the *net* increase in tax paid should count as a benefit to the government. When there is widespread indirect taxation, the government's gain in tax from the induced travel may be considered too small to be worth estimating.

The same arguments and equivalent results can be presented in a different way, but one that may be more familiar to some readers. Figure 4 shows the behavioral and the resource costs of travel, where the resource costs of travel include a trip-maker's valuation of the cost of his/her travel time savings. Assuming that trip-makers perceive vehicle operating costs reasonably accurately, behavioral costs are the sum of resource costs and indirect taxes. After a road improvement, behavioral costs fall from BC_1 to BC_2, and resource costs from RC_1 to RC_2. Trips again increase from T_1 to T_2.

In this case, the net benefit of the road improvement for existing users and government is the saving in resource costs. This is represented by area A, which equals $T_1(RC_1 - RC_2)$. Any savings in indirect tax are a transfer payment between road users and government. They can be ignored in the net result. On the other hand, the gross benefit of induced trips is the whole area between the demand curve and the RC_2 curve. This is the sum of the consumer surplus generated (area B) and the increase in tax revenue (the hatched area) that accrues to government, assuming that the alternative purchase would be untaxed.

Now we should recognize some other important assumptions that have been implicit in the above analysis. One is that that all individuals face a similar set of prices. This is not actually the case, because individual values of travel time savings

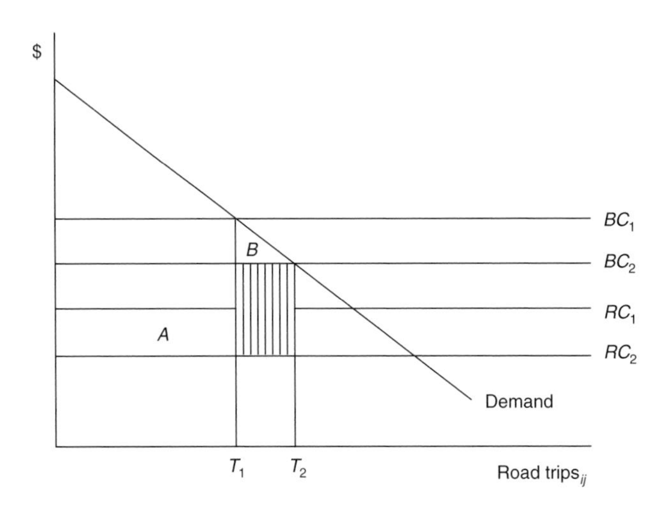

Figure 4. Benefits due to a road improvement (an alternative presentation).

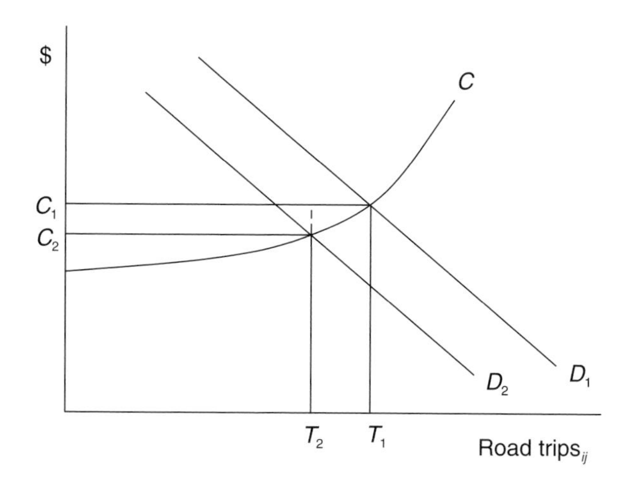

Figure 5. Reduced travel costs on substitute route.

vary. In principle, different generalized trip cost functions are required for each income group and trip purpose. However, this does not affect the evaluation principles described above.

A more important issue for evaluation is the implicit assumption that travel costs elsewhere on the network are unchanged. If the improved road results in significant traffic diversion, the demand for substitute routes will fall. In Figure 5, the demand curve for a substitute route shifts from D_1 to D_2 and trips on this substitute fall from T_1 to T_2. If this route suffered from congestion originally, the generalized cost of travel on this route will fall and users of it obtain a saving in trip costs, equal to $T_2(C_1 - C_2)$. These cost savings may in turn reduce the demand for trips on the improved route. The demand curve shown in Figure 2 (or 3 or 4) would shift to the left, thus reducing the consumer surplus from induced travel.

Now suppose that we introduce a negative externality of travel such as air pollution. Is induced travel now beneficial? Figure 6 shows a simplified situation. This shows four cost curves: two private marginal-cost curves (PMC_1 and PMC_2) and two social marginal-cost curves that include air pollution costs (SMC_1 and SMC_2), where subscripts 1 and 2 again represent the situation before and after a road improvement. Actual travel (T_1 and T_2) is determined by the intersection of the demand and *PMC* curves. The user benefits are areas $A + B$. However the benefit to society, inclusive of third-party effects, is quite different. The net social benefit for existing trips is represented by the vertically hatched area, which equals $T_1(SMC_1 - SMC_2)$. On the other hand, in this case induced trips create a social loss equal to the difference between the SMC_2 curve and the demand curve (the horizontally hatched area). This social cost can be avoided by appropriate pricing policy that ensures that trip-makers pay for environmental costs (which they may do when road taxes or prices are high enough). But in the absence of such policy, the net social benefit is the difference between the vertically hatched area and the horizontally hatched area.

Another important assumption of this simple model of benefits of induced travel is that prices equal marginal cost in other transport modes or sectors of the economy. If they do, there are no changes in consumer or producer surplus in these sectors as people switch out of these sectors. The assumption that, at the margin, people who switch to the new road are not sacrificing consumer surplus elsewhere is generally reasonable. However, prices may not equal marginal costs in some alternative transport modes, for example in the rail sector. If prices exceed marginal cost, there will be a loss of producer surplus as people switch modes. This loss may partially or wholly offset the gain in consumer surplus to induced travel on the road. On the other hand, if prices are below marginal cost, as sometimes occurs, there will be a gain in producer surplus as people switch modes.

Finally, brief mention should be made of the relationship between induced travel and land use changes. The general view of the economic literature (for example Mackie, 1996) is that when markets are competitive and prices reflect the marginal cost of production, the benefits of induced travel capture approximately the benefits of land use changes, for example the rationalizing of warehousing facilities. However, when there are economies of scale in production or when prices of products or factors do not reflect marginal resource costs, the benefits of induced travel do not capture all the benefits of changes in land uses. These changes in production may then have to be valued independently.

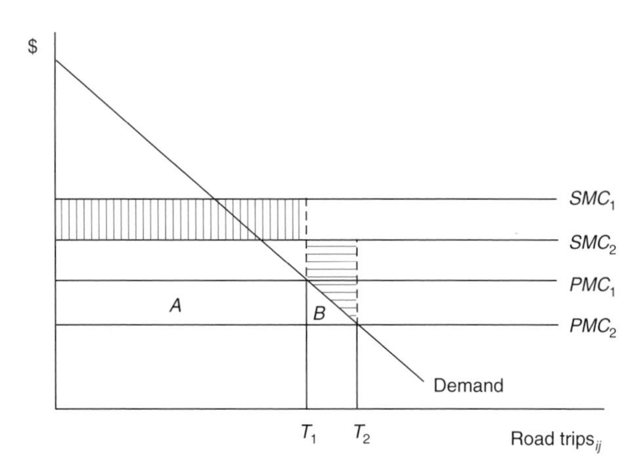

Figure 6. Allowing for environmental costs.

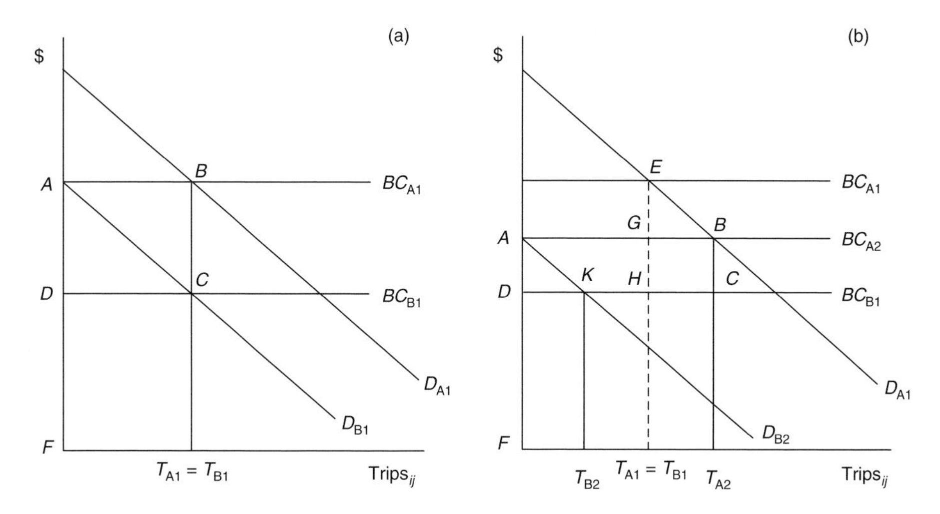

Figure 7. (a) Trips between i and j before route A is improved. (b) Trips between i and j after route A is improved.

In summary, the benefits of induced travel are the sum of the changes in the consumer surpluses of induced trip-makers, in indirect tax, in producer surpluses, and in environmental impacts.

5. Errors in the valuation of induced travel

Many road evaluation procedures use changes in resource costs as the measure of road user benefits. A strong feature of the evaluation of induced travel described above is that consumer surplus depends willingness to pay for travel – it is not measured by changes in resource costs. A simple example may illustrate some issues. We assume here that behavioral costs equal resource costs (taxes are ignored). Suppose that there are two routes (A) and (B) between two towns. A is an attractive, winding coastal route and trip costs are \$15. B is a direct inland route with trip costs of \$10. Initially, 80% of trip-makers use route B. Now it is proposed to improve the coastal route so that trip costs fall to \$12 and traffic is forecast to split equally between the two routes. A resource cost approach would penalize all who transfer to the coastal route by \$2 per trip. The willingness-to-pay approach would assign a benefit of \$1.50 (\$3.0 × 0.5) per diverted trip (assuming a linear demand curve). The latter is obviously a more accurate measure of the benefit of induced trips.

We now illustrate more generally the error of using resource costs to estimate road user benefits. In Figure 7a, the subject is now road trips from i to j. The demand curves D_{A1} and D_{B1} show the initial demand for trips from *i* to *j* by routes A and B, respectively. Curve D_{A1} is to the right of D_{B1} because A is a more attractive route than B. But route A is also the higher-(generalized)-cost route ($BC_{A1} > BC_{B1}$). To simplify the figure, trips are shown to split equally between each route, with the attraction of A exactly offsetting the higher cost (so that $T_{A1} = T_{B1}$). The total resource costs (assumed equal to behavioral costs) for these trips are given by area ABT_{A1}F for users of route A plus area DCT_{B1}F for users of route B.

Now suppose that route A is improved so that costs fall from BC_{A1} to BC_{A2}, as shown in Figure 7b. The demand curve D_{A1} is unchanged. But road users increase from T_{A1} (the initial position) to T_{A2}. On the other hand, the costs of using route B have not changed. However, demand for route B has fallen to D_{B2}, because the cost of the substitute route A has fallen. There are now T_{B2} users of route B, instead of T_{B1}. Total resource costs are now area ABT_{A2}F (for route A) plus area DKT_{B2}F (for route B). The total resource costs will generally be lower than before the road improvement because of the lower costs to existing users of route A.

But how are the benefits to road users who change to route A estimated? Using the willingness-to-pay approach, route changers would be *credited* with a benefit of area EBG in Figure 7b. Using the resource cost approach, they are *debited* with a cost of area GBCH. In general, when a relatively high-cost route is improved, using changes in resource costs as a measure of the benefits to induced traffic *understates* their benefits.

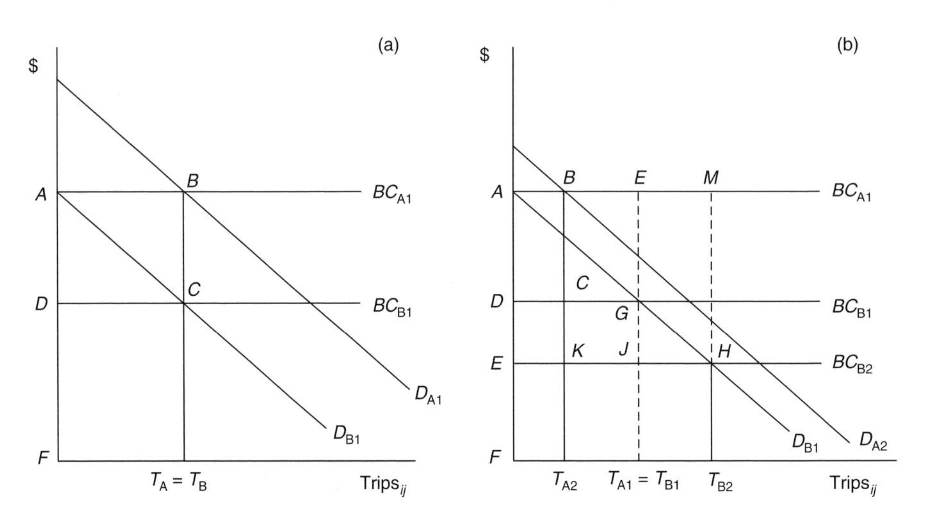

Figure 8. (b)Trips between *i* and *j* before route B is improved. (b) Trips between *i* and *j* after route B is improved.

Table 1
Resource cost, price, and willingness to pay in the example considered in the text

	Resource cost	Price	Willingness to pay
Rail mode	10	16	15
New road	8	8	18

What happens when a relatively low-cost route is improved? Figure 8a repeats the situation in Figure 7a. But now, as shown in Figure 8b, trip costs of route B fall from BC_{B1} to BC_{B2}. The demand curve D_{B1} does not shift, but trips on route B increase from T_{B1} to T_{B2}. On the other hand, the cost of using route A has not changed. However, the demand for route A has fallen to D_{A2} because the cost of the substitute route B has fallen. There are now T_{A2} users of route A, instead of T_{A1}. How are the benefits to road users who change to route B estimated? Using willingness-to-pay, route changers would be credited with a benefit of area GHJ in Figure 8b. Using resource cost savings, they are credited with the far larger area EMHJ. In general, when a relatively low-cost route is improved, savings in resource cost *overstate* the benefits to road changers.

5.1. Evaluating induced travel from modal changes

Now let us consider how induced travel as a result of modal changes should be evaluated. In particular, we want to know whether willingness-to-pay and resource-cost-saving methods of valuation produce similar results. We analyze this issue by way of two similar examples. Suppose that the marginal resource costs, prices, and willingness-to-pay values for a rail mode and a new road are as listed in Table 1. Resource costs, including the cost of travel time, are assumed to equal the price for road use in the absence of taxes.

In this example, the new road is more attractive than the rail mode – willingness to pay is higher for the road than for rail. What are the benefits when a trip-maker switches from rail to the new road? The resource cost saving per trip is \$10 – \$8 = \$2. The willingness to pay benefit is

User benefit (consumer surplus): \$18 – \$8 = \$10
Producer loss: \$16 – \$10 = \$6
Net benefit: \$4

Evidently, the willingness-to-pay method of evaluation produces a higher net benefit than the resource cost approach. In general, if the new road is preferred to the original mode, the use of the resource-cost-savings approach leads to an *underestimate* of the true benefits of the new road.

Now suppose that the new road is less attractive than the rail mode and that people are willing to pay only $10 for the road. The resource cost saving per road trip is still $2. However, the willingness-to-pay net benefit now falls to –$4. The willingness-to-pay benefit is

User benefit (consumer surplus): $10 – $8 = $2
Producer loss: $16 – $10 = $6
Net benefit: –$4

In general, when a new road is less attractive than the existing mode, the use of the resource-cost-savings approach leads to an *overestimate* of the true benefits of the new road.

It can easily be shown that when a trip-maker is indifferent between the two modes, the resource cost method of evaluation produces the same result as the willingness-to-pay method. However, whenever travel modes provide different levels of utility, as shown by willingness-to-pay values, the willingness-to-pay method (based on the sum of consumer and producer surpluses) provides the correct evaluation result.

6. Modeling induced travel

There are a number of methods typically used to estimate induced traffic. The most appealing starts with a definition of level of service based on either generalized cost or an index of expected maximum utility. Since the popular form of the induced-traffic model draws on a market share model of the multinomial logit form that is used to estimate a mode choice and a destination choice model, it is appropriate to set this model out and to show its limitations as the empirical source of all inputs into an induced-traffic model. It will provide some hints for an alternative specification that includes a generative model form. The multinomial logit (MNL) model takes the simple form in eq. (2) (see McFadden, 1981, Louviere et al., 2000):

$$P_i = \frac{\exp V_i}{\sum_{j=1}^{J} \exp V_j}, \tag{2}$$

where V_i is a utility expression for the ith alternative defining the role of attributes such as travel time and cost on the choice being studied (e.g., mode choice), and P_i is the probability of choosing alternative i from the set of $j = 1, \ldots, J$ alternatives. We can rearrange eq. (2) as follows:

$$\log P_i = V_i - V_J, \tag{3}$$

where

$$V_j = \log \sum_{j=1}^{J} \exp V_j.$$

V_J is the expected maximum utility (emu) index. It is essentially a probability-weighted generalized cost index that accounts for the role of all the alternatives in the choice process (including no travel) leading to a choice of travel alternative (e.g., mode or destination). If the MNL model is used as the basis of a distributive (or share) element and the emu is used as the measure of generalized cost, the proportional rate of change in *total* travel (∂T) with respect to emu must be less than unity (eq. 4):

$$(\partial T/\partial V_J)/T \leq 1. \tag{4}$$

Equation (4) can be modified by replacing V_J by the generalized cost (g) and transformed into a travel elasticity form (as shown in equation (5)):

$$(\partial T/\partial g)/(T/g) = \eta. \tag{5}$$

Thus the condition in eq. (4) implies, given eq. (5), that

$$\partial g/\partial V_J = (\partial T/\partial V_J)/(\partial T/\partial g) \leq g/\eta. \tag{6}$$

This formulation has some limitations that reduce its value when implemented in a context of a generative model that permits total travel to change (in contrast to changes in share conditional on a fixed trip matrix). In particular, eq. (6) is unlikely to hold over the entire range of applications, especially when dealing with short trips, where the differential must become very small. Thus rather than look to a share model framework to derive an appropriate elasticity-based formulation for a generative model, we need to seek out a form in which the elasticity declines as the travel increases (see Taplin et al., 1999). An appealing form (with ease of estimation) is given in eq. (7):

$$T = \alpha \exp(\beta g). \tag{7}$$

The derivative of trips with respect to generalized cost is given in eq. (8):

$$\partial T/\partial g = \alpha\beta \exp(\beta g) = \beta T. \tag{8}$$

The elasticity of traffic levels with respect to generalized cost is given by eq. (9):

$$\eta = \beta g. \tag{9}$$

Thus the challenge is to obtain an empirical estimate of β, not from a share (or distributive) model but from a total travel model. The relevant elasticity estimate for eq. (9) is an ordinary elasticity that accounts for both conditional traffic (i.e., the allocation of travel to a fixed number of total trips or a share model such as a discrete choice model of the MNL form) plus the additional traffic as a result of changes in the aggregate level of service. Since eq. (7) is a total traffic model, we can directly determine η (an ordinary travel elasticity) by identifying the estimate of β. The total traffic model eq. (7) can be transformed (using natural logarithms) into the following form for estimation:

$$\log T = \log \alpha + \beta g. \tag{10}$$

While this is a relatively simple equation, sourcing appropriate data is quite a challenge. To obtain an estimate of both α and β one would need a sample of data of trips undertaken at some appropriate spatial and temporal level. Since we expect to have a variation in the elasticity of traffic levels over time (i.e., as the number of years after capacity expansion is provided), it is appropriate to use a time series of cross section data. It is essential that a temporal dimension is allowed for, plus a spatial dimension to define the unit of analysis as a time-specific route (i.e., infrastructure). Such data would ideally be sourced from the traffic statistics measuring activity on specific new infrastructure (and any substitute infrastructure) at the time of opening and at regular intervals up to today.* Explanatory variables to consider in the full specification of eq. (10) might include

(1) the generalized cost of travel on each route at time period t,
(2) the catchment area population density, car ownership rate, and economic wealth (e.g., real per capita income) as measures of natural growth in traffic,
(3) the volume-to-capacity ratio on competing routes,
4) any natural-barrier dummy variables, and
(5) other (dummy) variables to describe qualitatively the specific infrastructures (e.g., freeway).

An additional variable that is very important in signalling the contribution of new infrastructure to the total road task is the proportional increase in capacity in the relevant catchment area (in contrast to a narrowly defined corridor). We would expect that additional capacity in a mature urban transport environment would induce far less traffic than would the equivalent capacity in an immature urban transport environment. Increased maturity tends to promote greater modal shifts and (especially route and time of day) redistribution than generated traffic (see DeCorla-Souza and Cohen, 1999).

*Since there is unlikely to be sufficient variation in the range of road projects providing data in a single metropolitan area one should compile such data for a number of metropolitan contexts. These data items would be collected for different road project contexts throughout as many metropolitan areas internationally as possible where broad comparability is meaningful.

7. Conclusions

This chapter has sought to clarify the meaning of induced travel in the context of determining the benefits to users of changes in the levels of service provided by urban road infrastructure. We have reviewed the theoretical literature as a way of highlighting the precision necessary in defining what is meant by induced travel. The empirical literature was used to illustrate the ambiguity in interpretation of what are the mixtures of behavioral responses that produce an overall level of traffic.

The importance of more reliable forecasts of traffic in response to new road infrastructure will remain a high agenda topic for both governments and potential investors (especially in a toll-road context). The framework presented is a useful one within which to establish a common understanding of the sources of all traffic so that the ongoing commitment to quality forecasting can be as unambiguous as possible.

References

Coombe, D. (1996) "Induced traffic: What do transportation models tell us?", *Transportation*, 23(1):83–101.

DeCorla-Souza, P., H. and Cohen (1999) "Estimating induced travel for evaluation of metropolitan highway expansion", *Transportation*, 26:249–262.

Goodwin, P.B. (1996) "Empirical evidence on induced traffic: A review and synthesis", *Transportation*, 23(1):35–54.

Hansen, M. (1993) "The air quality impacts of urban highway capacity expansion: Traffic generation and land use changes", University of California, Berkeley, Research Report UCB-ITS-RR-93-5.

Lee, D., L. Klein and G. Camus (1999) "Induced traffic and induced demand", *Transportation Research Record*, 1478:68–75.

Litman, T. (1999) *Generated traffic: Implications for transport planning*. Victoria, British Columbia: Victoria Transport Policy Institute.

Louviere, J.J., D.A. Hensher and J. Swait (2000) *Stated choice methods: Analysis and applications in marketing, transportation and environmental valuation*. Cambridge: Cambridge University Press.

Luk, J. and E. Chung (1997) "Induced demand and road investment – An initial appraisal", ARRB Transport Research Ltd, Vermont South, Victoria: Research Report ARR299.

Mackie, P.J. (1996) "Induced traffic and economic appraisal", *Transportation*, 23(1):103–119.

McFadden, D.L. (1981) "Econometric models of probabilistic choice", in: C.F. Manski and D.L. McFadden eds., *Structural analysis of discrete data*. Cambridge, MA: MIT Press.

Small, K. (1992) *Urban transportation economics*. Chur:Harwood Academic.

Taplin, J.H.E., D.A. Hensher and B. Smith (1999) "Imposing symmetry on a complete matrix of commuter travel elasticities", *Transportation Research*, 33B:215–232.

U.K. Department of the Environment, Transport and the Regions (1999) *A new deal for trunk roads in England: Guidance on the new approach to appraisal*. London: DETR.

U.K. Department of Transport (1993) *COBA manual revisions to 15/11/93*. London: Highways Economics and Traffic Appraisal Division, Department of Transport.

U.K. Department of Transport (1996) "Economic Concepts in COBA", in: *The COBA manual*, vol. 13, section 1. London: Department of Transport.

Chapter 10

COST–BENEFIT ANALYSIS IN TRANSPORT

PETER MACKIE and JOHN NELLTHORP
University of Leeds

1. Introduction

What is cost–benefit analysis? Is it useful, and in what contexts? What are the key components, what advances are taking place in its application in the transport sector, and what problems remain? These are the questions which we address in this chapter. There is no hope of being comprehensive, since complete books have been written on this subject (Pearce and Nash, 1981; Wohl and Hendrickson, 1984; Layard and Glaister, 1994, for example). Rather, the purpose of this chapter is to give an up-to-date view of the state of the art, referring to questions both of principle and of practice.

Appraisal is the generic term for the process of weighing up the impacts, positive and negative, of a project or policy action, so as to inform the decision-maker. Cost–benefit analysis is a specific branch of that genus. Its key characteristics are that the analysis is conducted from an overall social viewpoint rather than that of any particular agent, and that it uses monetary values, where feasible and valid, as the weights applying to the various impacts which are relevant to the decision. We begin by considering the need for project appraisal.

2. Why appraise projects?

Project appraisal is a necessary discipline because investment resources are limited and there are many potential opportunities for the use of these resources. Choices have to be made: transport appraisal can be seen as a value-for-money tool to inform these choices. Appraisal should therefore be thought of in a creative way, as an integral part of project development rather than merely an afterthought or ex post justification for what has been decided anyway.

Project appraisal is itself a sub-branch of a wider field of policy analysis. The literature in this field is massive, but from it two models of policy analysis can be briefly articulated. The first of these is the so-called rational/analytical

Handbook of Transport Systems and Traffic Control, Edited by K.J. Button and D.A. Hensher

model – see Box 1 – associated with Nobel Prize winner H.A. Simon (Simon, 1957). This is the model known to generations of engineers. It is a suitable framework for considering large one-off projects involving high sunk costs and significant risks. Within this model, there is plenty of room for debate – should the focus be on problem solving or should it be objectives-led? What should be the feedback loops, and how should these relate to the administrative processes through which the project passes as it proceeds along the project cycle?

A second model, less well known in the engineering disciplines, but very familiar to social scientists, is the "muddling through" model of Lindblom shown in Box 2 (Lindblom, 1959). This stresses different aspects of decision-making: the fact that decisions are often serial and incremental, seeking to address problems and improve on the status quo rather than finding some optimum, involving many parties with interests which need to be reconciled through bargaining and mutual adjustment. Whereas in the Simon model, analysis is seen as a central driver of decision-making, for Lindblom it is one element within an overall political process. As a description of the way transport policy evolves, Lindblom's model has much to commend it. Simon's remains a useful way of structuring the technical analysis which feeds the broader decision-making process.

Project appraisal is a comparative tool. It involves comparing alternative states of the world:

(1) A *do-something scenario* – in which the project is included in the transport network. There will be a separate do-something scenario for each alternative version of the project.

Box 1
The rational/analytical model (Simon)

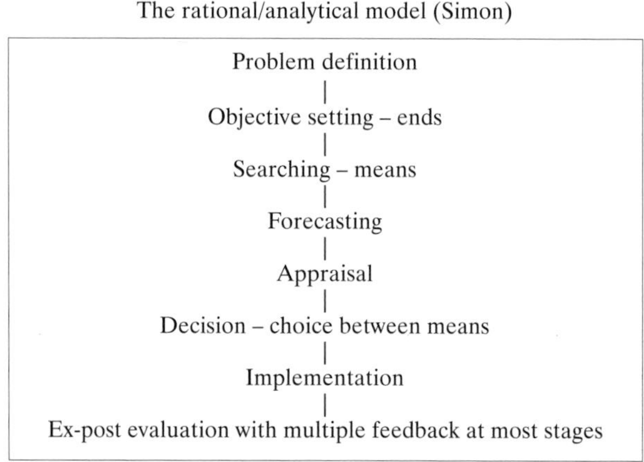

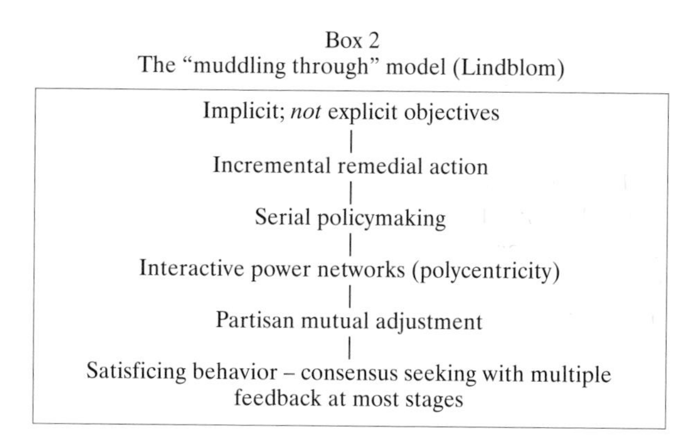

Box 2
The "muddling through" model (Lindblom)

(2) A realistic *do-minimum scenario* – assuming that the transport network remains as it would be if the project was not implemented. The do-minimum scenario is meant to include a realistic level of maintenance and a minimum amount of minor improvements where absolutely necessary, to avoid the transport network deteriorating – a pure do-nothing scenario would lead to unacceptable transport conditions, making it problematic as a base for appraisal.

The search process by which alternatives are selected for comparison is crucial; failure to do this properly, whether by accident or design, is an important source of appraisal bias (Mackie and Preston, 1998).

The format of the appraisal will depend upon the objectives of the organizations for whom the appraisal is undertaken. If the organization is in the private sector – say a logistics company – then the appraisal will be commercially driven, considering the costs and revenues of the scheme (e.g., a new depot, a new subnetwork, a new fleet of vehicles). If the project is analyzed on behalf of government, then a broader appraisal will be required, considering the relevant economic and social impacts of the scheme. For this social assessment, the analogue to commercial appraisal is known as cost–benefit analysis.

3. Principles of cost–benefit analysis

3.1. Overview

Some key features of cost–benefit analysis (CBA) are as follows. Cost–benefit analysis

(1) is undertaken from a social perspective, considering all relevant costs and benefits whoever they accrue to. There are practical issues of defining both what is relevant and who is included. For example, projects involving several countries and international traffic are problematic if appraised from the perspective of each country and its nationals separately.
(2) takes individuals' willingness-to-pay (WTP, see below) as the starting point for valuation. This is a key difference between CBA and most other forms of appraisal. Whereas the latter rely on "planners' values" or hold back from valuation altogether, CBA relies on individuals' values and preferences. Values for time, safety, and environmental impacts are derived from implicit or surrogate markets, for which revealed-preference and stated-preference methods are widely used (Louviere et al., (2000).
(3) may adjust from individual values to social values where appropriate. A key distinction is between behavioral (or pure willingness-to-pay) values used for modeling, and social values used for evaluation. Though controversial, it is commonly argued that willingness-to-pay values should be adjusted by some form of welfare or equity weights so as to correct for differences in WTP caused purely by income differences. A particularly simple, long-lasting, and effective example is the practice of the U.K. and other European governments of using standard values for non-working-time savings and safety benefits. This practice is currently under review in the U.K.
(4) involves the evaluation of costs and benefits against a consistent money baseline or numeraire. This means not only that prices and values must be consistent throughout the appraisal but also that all benefits and costs must be valued either at market prices (gross of indirect taxes) or at factor costs (net of indirect taxes). While the more usual approach has been to use the factor cost basis, the recent development of public–private partnership schemes tends to favor the use of market prices, and a switch to a market price basis is currently being implemented in the U.K. (Department of the Environment, Transport and the Regions, 2000a; Sugden, 1999).
(5) may involve bringing together disparate impacts, some of which are readily expressed in money terms whilst others are received in physical quantities. One of the major criticisms leveled at CBA is that it focuses the decision-maker on the monetized elements of the appraisal at the expense of the impacts which cannot be monetized. We return to this issue in Section 4.3.
(6) involves summing the benefits and costs over the life of the project, so that explicit procedures are required to value costs and benefits arising at different points in time, in which the social discount rate is a key parameter. In that context, problems of appraising very long-lived assets (e.g., transport infrastructure), irreversibility issues (habitat, biodiversity), and

cumulative effects (acid rain, global warming) all feature in transport sector applications. Risk analysis is a relevant tool which is probably underused in practice.

Figure 1 is generic – it outlines the CBA process in any typical transport application. For example, the specific application could be an appraisal of a road infrastructure improvement, a rail safety measure such as automatic train protection, or an urban transport plan.

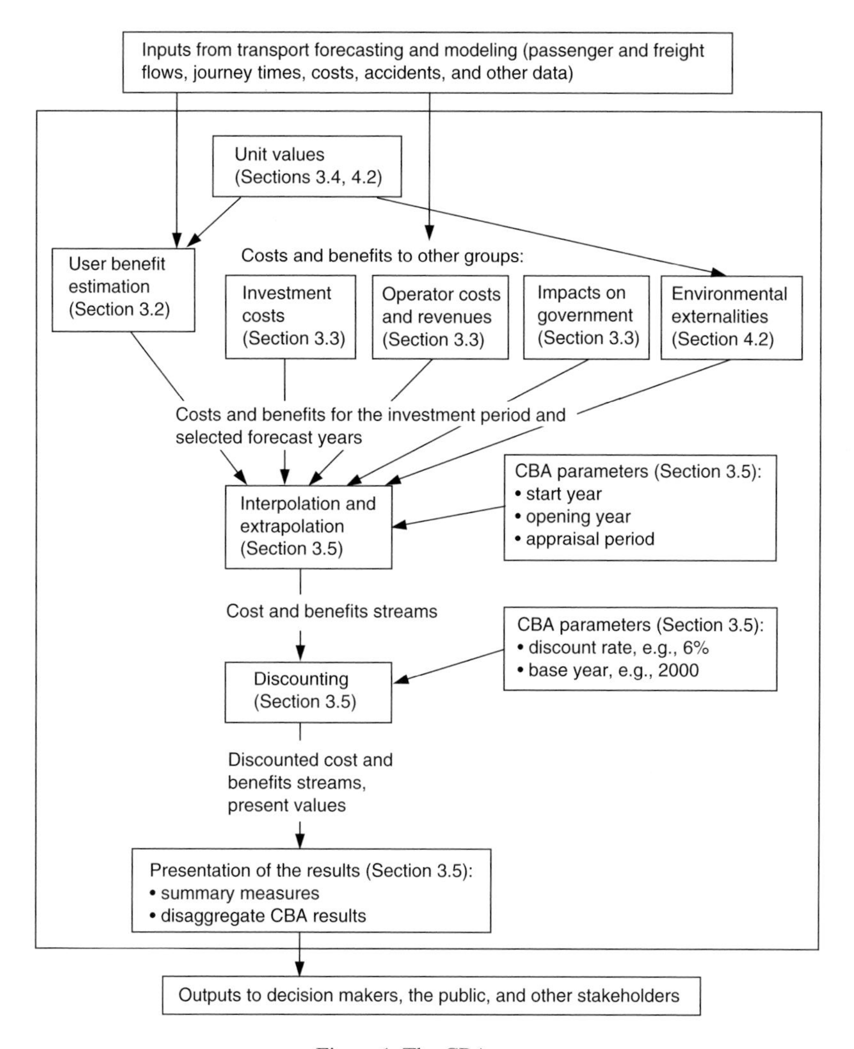

Figure 1. The CBA process.

Shown at the top of Figure 1 are the inputs to cost–benefit analysis from the previous step in the appraisal process – "forecasting." Alternative forecasting and modeling methods are covered in depth elsewhere in this handbook. Anything from complex network models to simple models of a single market may used to provide inputs to CBA. A key consideration is that the scope of the model is right to capture the main effects of the project.

The main box within Figure 1 includes the key components of CBA itself. These are:

(1) user benefits, principally time, safety, fares, and vehicle operating costs;
(2) investment costs;
(3) operator costs and revenues;
(4) impacts on government, especially taxation and subsidies;
(5) externalities, notably environmental impacts.

The analysis involves comparing the user benefits and revenue effects with the investment, operating, and other costs. Note the range of background data required in addition to the forecasting input – including unit values, investment cost estimates, and various other appraisal parameters.

Finally, the outputs of CBA are shown emerging at the foot of Figure 1. These are typically presented alongside other qualitative and quantitative information about the project, e.g., the project rationale, an environmental impact statement, and so on. The overall structure within which all this information is presented is commonly called the "appraisal framework." This information is provided to the "decision-maker(s)" in order to inform their decision (the next step in Box 1), but *not* in most cases to dictate it.

3.2. User benefit estimation

In many cases the motivation for a transport project will be to improve the conditions faced by transport users. How can the benefits or disbenefits of a project to users of the transport system be calculated from basic transport data? We begin by reviewing some definitions.

Definition of user benefit

Three fundamental concepts underlie the definition of user benefit in transport CBA: these are *willingness-to-pay*, *consumer surplus*, and *generalized cost*:

(1) *Willingness-to-pay* (WTP) is the maximum amount of money that an individual consumer would be willing to pay to make a particular trip from i to j by mode m. If the price of the trip is less than or equal to WTP, the

consumer is assumed to make the trip. If the price is in excess of WTP, the consumer will find an alternative, which may be not to travel to *j* at all. Different consumers have different levels of WTP for the same *ijm* trip opportunity. This enables us to construct the demand function for *ijm* trips, relating the price to the number of trips demanded. In general, as price increases, the number of trips demanded falls – in other words there is a downward-sloping demand curve.

(2) *consumer surplus* (CS) is a measure of the excess of willingness to pay over the cost of a trip. In the simplest terms, if a consumer is willing to pay $8 to travel by rail from *i* to *j* and if the cost of doing so is $5, that individual's consumer surplus is $3. What is of interest in cost–benefit analysis is the total consumer surplus across all individuals, and the extent to which this changes if a project of some kind is introduced:

$$\text{user benefit}_{ijm} = \Delta\text{CS}_{ijm} = \text{CS}^1_{ijm} - \text{CS}^0_{ijm},$$

where the superscript 1 indicates the do-something value and 0 indicates the do-minimum value. The total consumer surplus (CS^0) for a particular *i*, *j* and *m* in the do-minimum scenario is shown diagrammatically in Figure 2a. The user benefit, ΔCS_{ijm}, is shown by the shaded area in Figure 2b. Note that in economics, consumer surplus is related to the concept of *compensating variation* (CV), which is an ideal measure of user benefit in the sense that it is not influenced by income effects. However, if the income effects are close to zero, the consumer surplus and CV measures give the same result: this is in fact the conventional assumption in transport appraisal. Readers wishing to explore this point further are referred to Glaister (1981) or Jones (1977).

Operationalizing this in transport poses some practical problems. For most consumer goods the vertical axis in Figure 2a would simply indicate price. When it comes to transport, prices and money costs are only a proportion of the composite cost of travel, which in principle also incorporates the time spent by the individual, access times to public transport, discomfort, perceived safety risk and other elements. Price therefore needs to be replaced by *generalized cost* (Department of the Environment, Transport and the Regions, 2000a):

(3) *generalized cost* (GC) is an amount of money representing the overall cost and inconvenience *to the transport user* of traveling between a particular origin (*i*) and destination (*j*) by a particular mode (*m*). In principle it incorporates all aspects of this inconvenience, including "quality" factors like discomfort and unreliability. In practice, generalized cost is usually limited to time, user charges (e.g., fares/tolls), and vehicle operating costs (VOCs) for private vehicles:

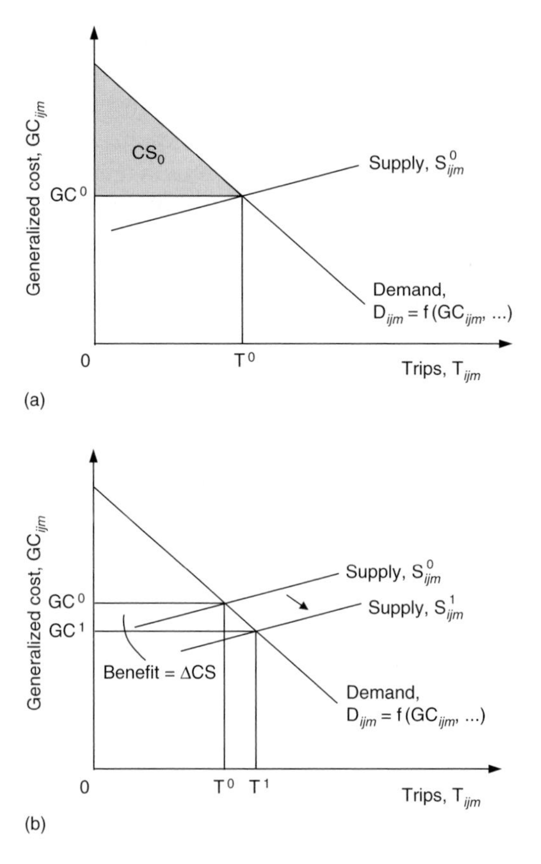

Figure 2. (a) Consumer surplus and user benefit. Consumer surplus in the do-minimum scenario. (b) Consumer surplus and user benefit. User benefit (= change in consumer surplus) in the do-something scenario compared with the do-minimum.

$$GC_{ijm} = \text{time cost}_{ijm} + \text{user charges}_{ijm} + \text{VOCs}_{ijm},$$

where "time cost" is defined as time in minutes × value of time in $/minute.

It is important to recognize that values of time vary between individuals and even for the same individual, depending on the trip purpose and possibly other factors (see below). There is no unique willingness to pay for travel time savings.

In Figure 2a, consumers' willingness to pay is represented by a downward-sloping demand curve. Transport supply conditions, i.e., the cost at which different quantities of trips can be provided by the system, are represented by an upward-sloping supply curve. The intersection of demand and supply determines equilibrium generalized cost. Consumer surplus is represented by the area beneath the demand curve and above the equilibrium generalized cost, area CS^0.

In Figure 2b, it is assumed that there is an improvement in supply conditions, due for example to an improvement in the road or rail infrastructure. The reduction in equilibrium generalized cost which results from this improvement leads to an increase in consumer surplus, which gives the user benefit equal to the area ΔCS. In an ideal application, but rarely in practice, ΔCS can be calculated as the integral of the demand function with respect to cost, between the do-minimum and do-something points,

$$\Delta CS = \int_{GC_{ijm}^{01}}^{GC_{ijm}^{1}} D_{ijm}(GC_{ijm})\, d\,GC_{ijm}.$$

Calculation of user benefits in practice

In practice, the cost–benefit analyst rarely possesses all the information embodied in Figure 2. Typically, the information available from the forecasting model will be in the form of matrices of trips, times, and costs on the network. The cost–benefit analyst therefore has basically two data points with which to work, for each *ijm* combination, i.e.,

(1) $(T_{ijm}^{0}, GC_{ijm}^{0})$ the "do-minimum" trips and costs, and
(2) $(T_{ijm}^{1}, G_{ijm}^{1})$ the "do-something" trips and costs.

It is conventional in transport appraisal to deal with this situation by making the assumption that the demand curve between (T^0, GC^0) and (T^1, GC^1) is approximately linear, and therefore that the user benefit can be approximated by the following function, known as the *rule of a half* (RoH) (see Chapter 6):

$$\Delta CS = \int_{GC_0}^{GC_1} D(GC)\, d\,GC \approx \text{rule of a half (RoH)} = \tfrac{1}{2}(GC_0 - GC_1)(T_0 - T_1).$$

In Figure 2b, the RoH is equal to the shaded area because the figure assumes the demand curve is a straight line. The RoH gives only an approximation to the true benefit – the more convex (or concave) the demand curve, and the larger the cost change, the less accurate the approximation will usually be. Given the multifarious sources of error in practical appraisals, the rule of a half is considered acceptable except in cases such as estuary or mountain crossings, where cost changes may be considered "large."

User benefits when demand shifts

Often, particularly in multimodal transport appraisal, we may wish to evaluate projects which result not only in supply shifts but in demand shifts. For example, where rail and car are substitutes, a project which improves rail services is likely to reduce the demand for car journeys – we need to know what the benefit for road users is, as well as the user benefit on rail. Conveniently, the RoH can be generalized to the case where any combination of demand and supply shifts occurs, subject to certain conditions on the demand function being met (see, e.g., Glaister, 1981, Section 2.7; Jones, 1977). It follows that whenever the effect of a project can be captured in the form of a change in generalized costs between particular origins and destinations, the rule of a half is a useful approximation to true user benefits.

New modes, closures, and new generators/attractors of trips

There are, however, certain special circumstances where the rule of a half breaks down entirely. These are

(1) introduction of completely new modes in the do-something scenario – e.g., high speed rail or urban light rapid transit – GC on that mode is not defined in the do-minimum scenario;
(2) closure of existing modes – i.e., GC is not defined in the do-something scenario;
(3) new attractors and generators of trips which do not feature in the *i–j* matrix – again the problem is one of costs being undefined, so the rule of a half cannot be calculated.

If any of these circumstances should arise, the analyst is thrown back on alternative measures of user benefit. Various approaches are discussed in Appendix D of the Common Appraisal Framework (MVA Consultancy et al., 1994). All of these basically attempt to sketch the demand curve to the left of the point which is known, so the consumer surplus can be calculated. This is becoming a more significant issue as appraisal becomes more multimodal in nature.

Total benefits and benefits by mode

Extending the user benefit measure from one *ijm* cell to a network and from passenger travel to freight is straightforward: user benefits calculated using the rule of a half for each of the components may be added together to give the total user benefit for the network as a whole.

As well as the total benefits, it is often of interest to decision-makers to know on which modes the benefits of a project arise. Strictly, there is no unique way of attributing user benefits between modes or indeed between *i–j* pairs, because it is not usually possible to identify an individual on the do-something network and trace back to find out which mode he/she used in the do-minimum. However,

Table 1
Potential user benefit items by mode

Mode	Travel time	User charges	VOCs
Public transport modes	Yes	Yes	No
Car	Yes	Yes	Yes

breaking down the total user benefit in proportion to the change in generalized cost on each mode is an intuitively appealing solution (e.g., Sugden, 1999; MVA Consultancy et al., 1994). This is effectively what happens when the rule of a half is applied at the mode, or the *i–j*, level. In the disaggregated CBA results, user benefits can then be presented split by mode, as well as by purpose (freight, working passenger, non-working passenger).

Components of generalized cost

The components of generalized cost will vary by mode. Public transport users (bus, coach, train, air, and ferry) will pay a money fare and give up time in order to travel to their destination. Car users and own-account freight users give up time, may be asked to pay an infrastructure access charge or toll, and pay for their own fuel and VOCs. Therefore there is a fundamental difference in the reported user benefits for users of different modes (see Table 1).

The total benefits can be broken down into components of generalized cost, again by applying the RoH separately to each component. Hence the formulae for time savings, VOC savings, and benefits from lower user charges are as follows:

$$\mathrm{RoH}_{\mathrm{time}} = \tfrac{1}{2}[(\mathrm{H}_0 - \mathrm{H}_1)\times \mathrm{VoT}](\mathrm{T}_0 + \mathrm{T}_1),$$

where H is the travel time per trip in hours, and VoT is the value of travel time in currency units per hour. Subscripts for *i*, *j*, *m* and for different trip purposes (which would carry different values of time (see below)) have been omitted for simplicity.

$$\mathrm{RoH}_{\mathrm{VOCs}} = \tfrac{1}{2}(\mathrm{VOC}_0 - \mathrm{VOC}_1)(\mathrm{T}_0 + \mathrm{T}_1),$$

where VOC is the vehicle operating cost in currency units per trip. Subscripts for *i*, *j*, *m* and for different vehicle types (which would incur different vehicle operating costs (see below)) have been omitted for simplicity.

$$\mathrm{RoH}_{\mathrm{user\ charges}} = \tfrac{1}{2}(\mathrm{U}_0 - \mathrm{U}_1)(\mathrm{T}_0 + \mathrm{T}_1),$$

where U is the user charge in currency units per trip. Subscripts for *i*, *j*, *m* have been omitted for simplicity.

Calculation of safety benefits

By convention, safety is treated differently from the other components of user benefit. Rather than being included as a component of generalized cost per trip, accidents and casualties are typically treated as random, occasional costs arising from the transport system, which can be evaluated by applying unit values per accident and per casualty, to forecast data on accident and casualty numbers by mode. The calculation is a simple multiplication of forecast accident numbers (by severity) with the costs of accidents (by severity) – the RoH formula is not used.

Software

Spreadsheet software and purpose-made computer code can both be very useful in carrying out all the above calculations, for the study area as a whole and for all the modes, trip purposes, benefit items and international/domestic traffic within it. Specialist software which can be applied to more than one type of transport model is beginning to emerge (Grant-Muller et al., 2001) to ease the process of carrying out repeated steps for many projects. Even so, care is still needed to ensure that the evaluation and the underlying forecasts are based on consistent assumptions, so that as fair a comparison as possible can be made between alternative uses of the available budget.

3.3. Revenues and costs

Although the user benefit analysis will often be the most testing part of the cost–benefit analysis, it needs to be undertaken alongside an analysis of revenues and costs. Only a few points of principle will be made here.

Producer surplus

Cost–benefit analysis is concerned not only with consumer surplus, but with total social surplus. This includes *producer surplus* (PS) as well:

$$\text{change in social surplus, } \Delta SS = \Delta CS + \Delta PS.$$

The greatest scope for changes in producer surplus arises from public transport projects, which can affect operators' fare revenues without having an equal and offsetting effect on operating costs. Producer surplus is defined simply as total revenue (TR) minus total costs (TC):

$$PS = TR - TC,$$

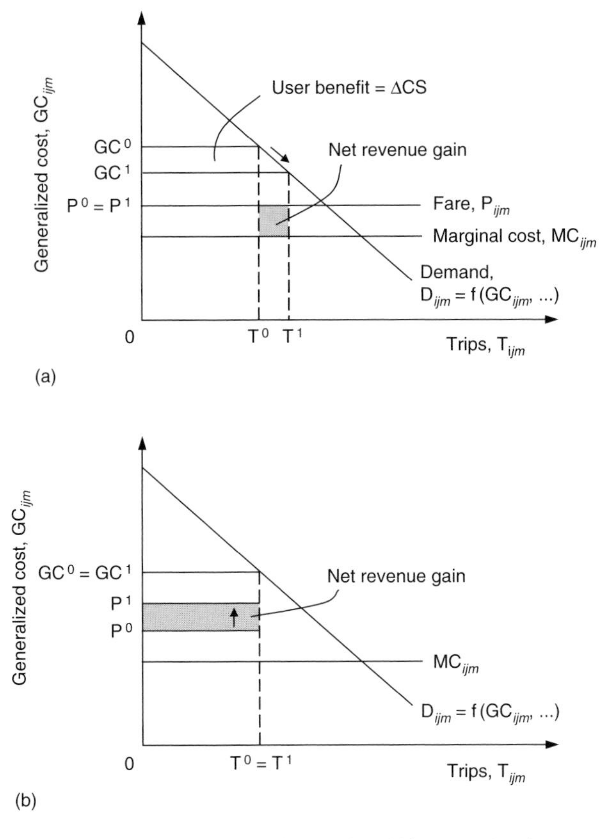

Figure 3. (a) User benefits and revenue effects of a rail journey time improvement. Price remains unchanged. (b) User benefits and revenue effects of a rail journey time improvement. "Pricing up" eliminates user benefits.

and therefore

$$\Delta PS = \Delta TR - \Delta TC.$$

For example, in the case shown in Figure 3a, where demand increases but marginal costs and fares (which are above marginal costs) stay constant, the analyst will need to calculate the change in producer surplus. This is the "net revenue gain" shown. Note, however, that there is an implicit assumption here that if the additional demand for this service is associated with reduced consumption of some other goods or services elsewhere in the economy, those goods and services are being priced at marginal cost, so that there is no offsetting or additional change in producer surplus elsewhere. This assumption is usually made, but it is worth making it explicit in the interests of transparency.

Revenue forecasts

It should not need saying (but does) that revenue forecasts depend on traffic forecasts, and both depend on pricing policy. Therefore it is essential in appraisal that the price policy assumptions on which the traffic and benefit estimates are based are consistent with those used for revenue forecasting. Consider a rail infrastructure project which speeds up the trains between A and B. Figure 3a shows the case in which the benefits are fully passed through to the travelers. Figure 3b shows the case in which there is complete "pricing up" by the railway operators. The size of the revenue and user benefit effects, as well as their distribution, depends upon the pricing policy. Although this seems obvious, in many practical situations, the appraisal may be undertaken before the details of the toll or price regime have been finalized, so that provisional assumptions made for the appraisal can turn out to be wide of the mark.

Operating costs

It also follows from Figure 3 that proper estimates are needed of the costs of operating both the infrastructure and the services, which are mode- and country-specific. They need to be estimated using a unit of account – market prices or factor cost – which is consistent with the rest of the appraisal. The main items will typically be

(1) the costs of infrastructure operation (e.g., signalling/traffic control);
(2) maintenance costs (cleaning, minor repairs, winter servicing);
(3) costs of renewals (road/rail reconstruction);
(4) changes in the vehicle operating costs of public transport services (some of these may have been calculated within the VOC model discussed in the previous section).

Taxation and government revenue effects

When a project leads to a shift in demand between private and public transport, the implications for government tax revenue may be significant because private transport is often relatively heavily taxed and public transport is often relatively lightly taxed. These changes in indirect-tax revenue to the government should ideally be shown in the cost–benefit outputs – see Table 5 for an illustration.

Such a project is also likely to involve reductions in operating costs for private transport and increases in operating costs for public transport. These changes in costs need to be treated consistently, so both private and public transport costs should be valued with any specific value added tax or duty taken out and shown separately. This is true whether the appraisal is being conducted on a factor cost or market price basis. Treatment of taxation in multimodal appraisal is an evolving area. The paper by Sugden (1999) was an important contribution and represents the published state of the art.

Investment costs

These costs will be derived from engineering design studies and estimates, which may themselves have been subjected to risk analysis and other formal assessments. However, for the purposes of the appraisal a number of adjustments may be required to allow for the following.

(1) *Mitigation measures*, since any such measures required in the light of the environmental impact assessment must be included in the cost, and the environmental impacts of the scheme assessed with the mitigation measures in place.
(2) *Disruption*, since delays to transport users during construction are a relevant scheme cost (or negative benefit) relative to the base case. In the case of public transport projects, bitter experience suggests a need to allow for the effect on traffic and revenues of any disruptions to existing service quality while new schemes are constructed.
(2) *The unit of account* – again it is important to be aware of the need to undertake the appraisal as a whole on a consistent basis, whether at market prices or at factor cost (i.e., net of indirect taxes), and to adjust the investment costs appropriately.

3.4. Values of time and safety

So far, nothing has been said about the techniques for valuing the benefits and costs of transport projects. Some of these – vehicle operating costs, investment costs, for example – will naturally be in money terms, though there is a need, as noted above, to ensure that valuation is consistent. Many of the user benefits, however, are not directly valued in money, and values need to be inferred from relevant market research studies. In principle these should cover all dimensions of consumers' willingness to pay, including comfort and reliability. In practice most work has focused on travel time and safety. That this presents a potential problem in multimodal appraisal where the comfort and reliability of modes differs is important to recognize. If it is possible to evaluate these effects in a robust way – for example using willingness-to-pay-based methods – for a specific project or strategy, then this will add to the usefulness of the appraisal results. However, this is not a standard part of most countries' official cost–benefit methodologies (e.g., Department of the Environment, Transport and the Regions, 2000a). Non-monetized or qualitative approaches to assessment tend to be preferred for these effects.

A basic distinction which is usually made when discussing values of travel time savings (henceforth "VOTs") is between

(1) travel in working time (or "on employer's business");
(2) travel in non-working time (usually defined to include all other travel purposes, such as shopping, commuting, education, personal business, and leisure);
(3) freight travel time.

We examine each of these separately below.

Working time

Working time is slightly easier to analyze, as there is clearly a market – the labor market – where working time is valued. There are two approaches to valuation in current use. The first is to adopt the gross wage, that is, the wage rate plus employee-related overheads, as a measure of the marginal product of labor. Each minute of time saved is assumed to be fully convertible to productive work. The gross wage per minute is therefore used as the value of savings in working time spent traveling.

The second approach, due to Hensher (1977), introduces some additional variables, since

(1) the ability to work whilst traveling varies between modes (e.g., it is relatively easy work whilst a passenger on a train, but relatively hard whilst driving a car);
(2) the ability to use productively any travel time saved also varies, depending upon the extent to which work tasks are divisible and flexible.

In this approach, the employer's valuation of travel time savings is given by

$$\text{VOT}_{\text{employer}} = \text{PVWT} \times [\%\text{W} - (\%\text{TW} \times \%\text{PTW})],$$

where PVWT is the productive value of an hour of work time, %W is the percentage of travel time savings returned to work, %TW is the percentage of travel time used for work, and %PTW is the relative productivity of work during travel.

Hensher goes on to add to this employer's value the employee's expressed WTP for travel time savings in work time. This method was tested in the Dutch, Swedish, and U.K. studies of the 1990s (Gunn and Rohr, 1996; Algers et al., 1995; Gunn et al., 1996) and found to give VOTs around 6% higher than the gross-wage approach, overall.

The working time values recommended by the U.K. government for use in cost–benefit analysis are shown in Table 2 – these still have a gross-wage-plus-overheads basis. Note the difference in magnitude between the working-time values and the non-working value, the ratio being around 4:1 overall.

Non-working time

For non-working time there is no market – for example, no payments are made for the time spent making the journey to work or to college. Therefore a WTP value, based on market research, is needed for use in appraisal. Both stated-preference (SP) and revealed-preference (RP) methods have been used to elicit values of time in various different travel choice situations – including route choice, mode choice, and others. For a review, see Wardman (1998).

Variations in the values of non-working time

There is at least some theoretical and empirical evidence to support the following propositions:

(1) values of time vary with disposable income (though possibly not proportionately);
(2) values of time vary additionally with socio-economic characteristics such as employment status, retired, etc.;
(3) values of time vary with the activity undertaken (walk and wait higher than in-vehicle; driving in congested conditions higher than free flow) and possibly with the mode (comfort, privacy, ability to carry out activities);
(4) values of time may vary by journey purpose, with commuting values if anything higher than for other non-work purposes;
(5) values of time may vary by size and sign of saving, but this is highly controversial;
(6) values of time may vary by journey length.

This range of variations places considerable stress on practical appraisal, since gathering the requisite data in all dimensions for the average project may not be cost-effective. Thus, as well as the arguments of principle, there are also pragmatic arguments for the use of standard values, as conventionally practiced in the U.K. (see Table 2). Other countries distinguish between modes (Netherlands), purposes (Norway, Sweden), journey distances (Norway, Sweden), or delay time (Sweden) but nevertheless maintain a high degree of averaging.

Freight travel time

Freight VOTs are an interesting case. Savings in *drivers' time* are usually the key component. Drivers' time savings on one trip may enable subsequent trips to be rescheduled and total labor costs to be reduced, although there is a literature in which this point has been debated (see, e.g., Mackie and Simon, 1986). Table 2 shows the drivers' time values recommended in the U.K., based on average gross wage rates.

Meanwhile *freight users*, that is, the businesses for whom the goods are being shipped, may also value savings in freight travel time if these enable them to adopt more efficient production processes. Studies suggest that these users' values tend to be low – $0.06 per minute according to Vägverket (1995). However, these values relate to expected journey time. It is possible that they hide a higher value for changes in freight journey time reliability. That is beyond the scope of most current cost–benefit analysis, and more research is needed on that topic.

Values for safety impacts

The valuation of accident savings has been transformed since the mid-1980s by the development, and widespread acceptance, of monetary values based on individuals' willingness to pay to avoid accidents. Jones-Lee (1992) sets out the theory; Persson and Ödegaard (1995) give some empirical results across 14 European countries. Table 3 illustrates this type of valuation using the UK appraisal values.

Previously, values per casualty (i.e., per person injured or killed) had been based largely on measures of lost output – that is the average reduction in GDP

Table 2
Appraisal values for travel time savings – an example: U.K. official values

Trip purpose	Value ($ per minute) (a)
Working time	
Road	
Car driver	0.44
Car passenger	0.36
Light goods vehicle driver	0.34
Other goods vehicle driver (b)	0.32
PSV driver (c)	0.33
PSV passenger	0.36
Rail	0.55
All modes	0.43
Non-working time	
Standard appraisal value	0.11

Source: Department of the Environment, Transport and the Regions (2000b, Table 2.1).
Notes: (a) All values have been converted to 1999 U.S. dollars. (b) Other goods vehicles include heavy goods vehicles. (c) PSVs are public service vehicles, principally buses. (d) Walking, waiting and, cycling in non-working time are given double this value.

Table 3
Values for safety improvements – an example: U.K. official values

Casualty severity	Value ($ per statistical casualty)
Fatal	1 760 000
Serious	198 000
Slight	15 300

Note: (a) All values have been converted to 1999 U.S. dollars.
Source: Department of the Environment Transport and the Regions (2000c, Table 1).

Table 4
Values for safety improvements – an example: U.K. official values

Item	Unit value ($)	Number per annum, 1999
Fatal accidents		3 326
Casualty value per fatal casualty	1 760 000	
Total cost per fatal accident	2 040 000	
Serious accidents		39 295
Casualty value per serious casualty	198 000	
Total cost per serious accident	238 000	
Slight accidents		191 633
Casualty value per slight casualty	15 300	
Total cost per slight accident	23 500	
Damage-only accidents		~4 000 000
Total cost per damage-only accident	2 100	

Source: Department of the Environment, Transport and the Regions (2000c, Table 3).

due to the injury or death of an individual member of the workforce. This was augmented in some cases by arbitrary allowances for "human costs" or "pain, grief and suffering." The implementation of willingness-to-pay-based values often led to a significant increase in safety values in appraisal. For example, the U.K. fatality value rose from approximately 0.75 to 1.5 million dollars at 1999 prices.

An important counterpart to the value per casualty is the remaining non-casualty-related part of accident costs. These costs include material damage, police and fire services, insurance administration, legal/court costs, and delays to other passengers and freight. For fatal and serious accidents these costs are orders of magnitude lower than the casualty costs. But for "slight" injury accidents and for damage-only accidents (of which there are many, compared with fatal accidents) these costs play a substantial role. Table 4 gives the U.K. accident values for 1999 as an illustration of the way unit values for safety are presented for use in CBA.

Of course, accidents happen on all transport modes. In principle, the casualty-related costs should be transferable, whilst the remaining costs are mode-specific. However, evidence from WTP surveys suggests that individuals value changes in accident risk on public transport modes more highly than for private car journeys. Whether this should be carried through into CBA, particularly multimodal CBA, remains a subject of debate.

3.5. *The process*

Returning to Figure 1, in order to carry out a cost–benefit analysis for a particular project, there is a process leading from the initial inputs to the final CBA outputs. To briefly outline the key steps, these are:

(1) *inputs* – modeling and forecasting inputs in terms of trip, cost, and time matrices (for origins i to destinations j by modes m); estimates of the investment costs, changes in infrastructure and service operating costs; unit values; estimates of the numbers of accidents and casualties and environmental impacts (chiefly noise and air pollution changes);
(2) *cost and benefit estimation* – using the rule of a half for user benefits and the simple do-minimum versus do-something comparison for other cost and benefit items, initially, these calculations will be for one or two years only;
(3) *interpolation and extrapolation* – requires growth rates for quantities *and* unit values, in order to obtain cost and benefit streams over the entire appraisal period;
(4) *discounting* – future costs and benefits are discounted in line with public sector conventions on discount rates;
(5) *summary measures* – giving an overall measure of the project's performance in cost–benefit terms.

The main summary measure of net social benefit is the net present value (NPV):

$$\mathrm{NPV} = \sum_{t=0}^{t=n} \frac{B_t - C_t - K_t}{1+r},$$

where B_t are the benefits in year t, C_t are the recurring costs in year t, are the investment costs in year t, r is the social discount rate, and n is the number of years within the appraisal period.

If the social discount rate reflects the social opportunity cost of capital, the rules are

(1) accept all projects with a positive NPV;

Table 5
Costs and benefits of a "quality bus partnership": present values over 30 years, in \$000 at 2000 prices, 6% discount rate

Incidence group	Cost		Benefit	
Bus users			Reduced journey times	933
			Improved reliability	650
			Better information	140
			Increased comfort	63
Bus operators	Capital costs	700	Tax Savings: ex car users^{T}	148
	Increased operating costs	392	Increased revenue	605
Rail users			Grants/subsidies^{T}	750
Rail operators	Reduced revenue	30	Reduced crowding	42
Private road users			Decongestion effects	120
City council	Capital costs	880	Capital grant^{T}	440
Other residents			Reduced noise	20
			Reduced air pollution	10
Central government	Reduced tax revenue	148		
	Capital grant	840		
	Operating subsidy	350		
Present value (exclusive of transfers)		2002		2263
NPV		261		

Note: ^{T}, transfer.

(2) where there are mutually exclusive options or alternatives, accept the project option with the highest NPV.

Frequently there are budget constraints; not all projects with positive NPVs can be undertaken, because of a shortage of capital. In that case, the benefit–cost ratio (BCR) comes into play. This is

$$\mathrm{BCR} = \sum_{t=0}^{t=n} \frac{B_t - C_t}{(1+r)^t} \Big/ \sum_{t=0}^{t=n} \frac{K_t}{(1+r)^t}.$$

Here the decision rules are

(1) considering the availability of capital and the returns on the pool of projects, define the marginal acceptable BCR, and accept or reject projects accordingly;

(2) when considering project options, accept capital outlays only providing the BCR on the incremental outlay exceeds the marginal acceptable BCR.

The decision-maker may wish to see not only the overall NPV or BCR, but also the distribution of gainers and losers. By calculating the results in a more disaggregated form using the methods described in this chapter, it is possible to produce this information (in particular, see "Components of generalized cost" above).

Presentation of the results is the final step. Table 5 gives an example of cost–benefit analysis outputs for a "quality bus partnership" project. The table shows the breakdown of costs and benefits for specific incidence groups as well as the aggregate social costs and benefits, which helps to identify gainers and losers. The results of the calculations described in this chapter are shown in the columns headed "Cost" and "Benefit," used here to indicate negative and positive effects, respectively. The investment costs for the project, which include better on-road infrastructure, bus priority measures, and better-quality vehicles, are shown as "capital costs" for the bus operators and the city council. These can be contrasted with the additional revenue expected by the operator and the "capital grant" funding from central government – the latter is a transfer (marked ^{T}) which appears as a cost to the group paying and a benefit to the group receiving the funds, and which therefore nets out to zero in the total calculation. The benefits to users of bus, whose service quality improves in several respects, are calculated using the rule of a half. As a result of mode switch to bus, there are benefits from reduced crowding on rail and decongestion on the road network, also calculated using the rule of a half. Given a reduction in car traffic and only a small net increase in bus traffic (using newer vehicles), there are some small net gains in environmental pollution. Finally, note that the table is explicit about the increase in operators' costs and revenues and also about the reduction in government tax revenue from the loss of fuel tax revenue on car use.

The net present value, summarizing the project's performance in cost–benefit terms, is shown in the bottom line of the table. At a 6% test discount rate, the project is worthwhile overall, subject to distributional concerns and any non-monetized impacts outside the cost–benefit analysis.

4. Challenges in cost–benefit analysis

CBA is addressing itself to a wider set of the socio-economic effects of transport projects. It will be apparent to many readers that environmental concerns and the role of transport in economic development are key issues on the policy agenda. These are discussed further in Sections 4.1 and 4.2 below. Then in Section 4.3 we review the relationship between CBA and framework approaches to appraisal.

4.1. Extending CBA to the environmental impacts of transport

The environmental impacts of transport are increasingly well understood. We now know that in addition to the more obvious local impacts such as traffic noise and poor air quality, there are a range of pollutants emitted by transport which contribute to acid rain at a regional level and to climate change worldwide. The challenge for cost–benefit analysis has been to find ways of bringing all of these within the cost–benefit framework in a way that does not impose problems of inconsistency and which reflects in an appropriate way the much greater *uncertainty* associated with many environmental impacts. In the following, we discuss some of the local, regional, and global environmental effects which have been brought into CBA.

Local and regional environmental impacts

Inclusion of *transport noise* in CBA is increasingly common in Europe The seminal study by Soguel (1994) was a hedonic analysis of property rents and the impact of road noise, whose results have been shown to be comparable with other contemporary studies, through meta-analysis. Transferability to other situations is aided by the finding that the value of a 1 dB(A) noise reduction does not vary to a statistically significant degree with the preexisting level of noise, so the same values may be applied to noise changes experienced by individuals in any initial setting (European Conference of Ministers of Transport, 1998). Furthermore, the expression of the values in units of dB(A) 16 hour Leq means that they are applicable to modes of transport producing intermittent noise (e.g., rail or air) as well as to the continuous noise of roads. Values generated by this approach are typically of the order of 30 euro per person per dB per annum at year 2000 prices (Grant-Muller et al., 2001). Value transfer between countries is often made using purchasing-power parity exchange rates, although for projects with localized effects it is probably preferable to carry out location-specific studies where resources allow.

For local and regional air pollution, part of the achievement of recent research has been to isolate more clearly which are the most important pollutants (in terms of their impact on people and the environment) and to clarify how the impact varies depending upon where the emissions occur. Amongst the key findings from this field of research are that

(1) particulates are the most significant local air pollutant;
(2) damage costs are much higher per unit mass of pollutant emitted in urban areas than for extra-urban areas (by a factor of up to 5 for smaller cities and as much as 50 for larger cities such as Paris or London), so in project appraisal it is essential to separate urban from extra-urban traffic before the valuation stage; and

(3) calculation of the total environmental damage due to transport needs to include the whole fuel cycle, including electricity generation facilities for electric vehicles and the processing of fossil fuels – the location of these facilities does not feature in a conventional transport model, so something more comprehensive may be needed in order to compare fairly the emissions of different transport modes.

As with noise, several meta-analyses (most recently European Conference of Ministers of Transport, 1998) have added to confidence in the range of values being identified and at least seven European nations have adopted money values for changes in air pollution, in their appraisal of transport infrastructure projects. However, the confidence intervals associated with these values are rather wide (Friedrich et al., 1998). For example, for particulates we can be only 70% confident that the urban value lies in the range 46 to 740 euros per kg. Given this level of uncertainty, sensitivity testing is highly desirable.

Global climate change

It is well known that the climate change impacts of CO_2 emissions are very uncertain, so in taking these into CBA, best practice is to flag clearly what the bounds of uncertainty are. Major studies reviewed by the European Conference of Ministers of Transport (1998) point to a central estimate of 50 euros per tonne of CO_2 emitted at current prices, with the convenient feature (from a CBA point of view) that it is unimportant whereabouts the emissions occur. The confidence limits are wide – according to Friedrich et al. (1998) there is 70% confidence that the value lies between 4.2 and 600 euros per tonne. Again, some form of robustness testing will be useful.

4.2. Wider economic impacts

The appropriate treatment within appraisal of the wider economic impacts of transport projects is another problem area. CBA methods tend to focus on the direct user benefits and costs together with the environmental impacts. The tacit assumption is that the primary transport benefit measure stands as an adequate proxy of the final impact on economic welfare which one would ideally like to measure. The pragmatic view has been that it is difficult enough to measure the transport impacts to an acceptable degree of accuracy. To measure the final economic benefits, involving as it does the use of a model which traces the impacts through from their initial to their final incidence on prices, wages, and rents, would be an order of magnitude more difficult and might not make much difference to the pattern of decisions. This was the argument used by the Leitch

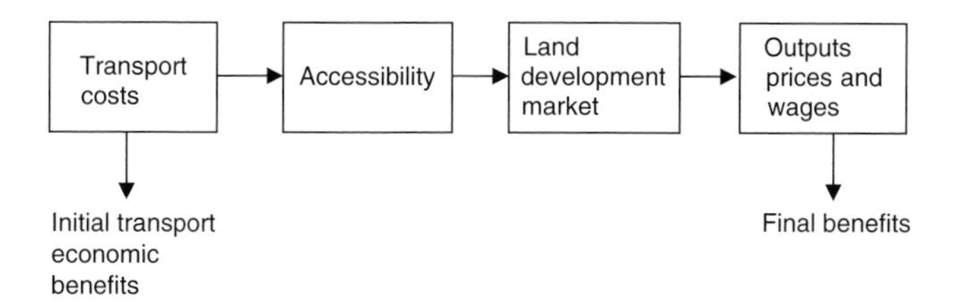

Figure 4. Wider economic impacts.

Committee in favor of focusing on the direct transport impacts and disregarding the wider economic impacts (Advisory Committee on Trunk Road Assessment, 1978).

The problem is that whereas this might seem broadly acceptable to analysts and other technocrats, it is much less so to the decision-making class who are the final customers of the appraisal process. Politicians want to know about the effect of public infrastructure on economic performance at the macro level and on the regeneration of particular places and regions. One consequence of this is that economic appraisal of transport projects has been less influential in practice in determining scheme prioritization and ranking than economists often like to believe. As a result the question of whether it is practical and desirable to extend CBA to cover the wider economic impacts as well as the direct transport and environmental impact has recently been revisited in the U.K. (Standing Advisory Committee in Trunk Road Assessment, 1999) and in the Netherlands (OEEI, 2000).

What exactly are these wider economic impacts? There are many processes and pathways but a generic description is given in Figure 4.

Changes in transport costs should be reflected in changes in accessibility, which in turn changes the pattern of demand for land. Finally, the level and pattern of prices and outputs is modified.

From this it follows that transport investments are expected to generate final economic benefits. When projects go in, things are expected to happen to distribution and manufacturing patterns, market areas served, and commuting catchment areas. From an appraisal point of view, the issue is not the presence or absence of such phenomena but the following.

(1) Whether and in what circumstances the absolute final economic impacts may exceed the initial transport impacts – so-called *additionality*.
(2) Where the incidence of the final impacts is more (or less) socially beneficial than the incidence of the transport impacts. Can transport projects generate benefits for target geographical areas or social groups? Even if there is zero additionality, are there *distributive effects*?

The case for additionality depends on the presence of imperfect competition in the goods market, the land market, or the labor market. The source of any additional benefits is to be found in divergences between prices and marginal social costs in relevant markets. In such cases, reductions in transport costs feed through to some extent into reductions in final goods prices and increases in output. Whereas under perfect competition, the price of this output reflects the marginal cost of production, in imperfect competition a correction is required to allow for the fact that price exceeds marginal cost. The size of this benefit depends on the size of the output effect, and the price : marginal-cost margin. A very useful framework for considering these issues at the conceptual level is the new economic geography, and the computable general equilibrium approach (Venables and Gasiorek, 1999). This approach stresses the imperfect-competition argument, considers forward and backward linkages between the transport sector and the economy, considers the effects of transport cost changes on changes in the location of economic activity (centripetal versus dispersion forces), and computes values for the ratio of final economic benefits to transport benefits. However, operationalizing this approach will require substantial effort and data. If it can be made operational, this approach is likely to be more relevant to evaluation of the road and transport program as a whole than to scheme evaluation.

When considering large schemes, or schemes such as bridges or tunnels which cross barriers to movement or schemes with specific development impacts, the likeliest approach to assessing these will be a qualitative or market research one. For a review of such approaches see ECOTEC (1999). A starting point is to consider the mechanisms which, if triggered, may lead to a net increase in output from the national perspective; relevant mechanisms are

(1) attracting footloose inward investment which would otherwise locate in another country;
(2) unlocking land for development in cases where land availability is an active constraint on development;
(3) promoting industrial reorganization and scale economies in production/distribution;
(4) reducing labor costs by increasing the quantity and/or quality of the available pool of labor to a center of employment;
(5) enabling increased output using otherwise unemployed resources, particularly labor, where market wages exceed opportunity costs.

There needs to be evidence of significant market imperfections. Our general view is that at the scheme level, an appropriate default position is that of zero additionality. Departures from the default should be based on researched evidence; in developed countries, significant additionality is, we believe, unusual. Normally, therefore it is more important to focus on measuring the direct transport and environmental impacts than on chasing additional wider economic impacts at scheme level. We recognize that this is not a welcome message to many politicians and planners.

Turning now to the distributive effects, transport infrastructure may well redistribute economic activity. Projects change the relative accessibility of different locations, and also the relative cost of centralized transport-intensive solutions versus decentralized production-intensive solutions. Changes in the transport system may well therefore affect the spatial distribution of activity, and this may be beneficial to society. Two logical steps are required in order to confirm this.

First, the impact on the distribution of economic activity needs to be forecast. It is essential to avoid jumping to conclusions. The fact that accessibility from a peripheral town to a regional center is improved is no guarantee that economic activity will migrate to the periphery. Low transport costs are a centralizing force, so if anything, improved transport facilities are more likely to cause migration to the regional center. This is the "two way road" effect. The demand and supply elasticity conditions required for a favorable result from the peripheral location's perspective are set out in Dodgson (1973). Perhaps the most important requirement is that the region should hold natural or cost advantages for a reasonable range of economic activities in which transport is a significant input. Primary products and tourism are two relevant examples where improved transport may permit exploitation of assets fixed in location.

If market conditions are favorable for a displacement of economic activity, then the second question is whether there is a net social advantage from the redistribution of economic activity. Is the gaining region a target area for economic regeneration? Is activity being displaced from more prosperous locations or from even more acute problem locations?

For major schemes, some form of land use/transport interaction model is needed in order to assess the transport–accessibility–economic-activity linkages in a coherent manner (see Simmonds and ME&P, 1999). For projects in regeneration areas, the impacts may best be considered alongside other non-transport measures within the context of the economic regeneration budget rather than the transport budget. Where the scheme impacts positively or negatively on target areas such as regeneration areas, the net social gain will usually be best estimated by considering the likely employment effects using some suitable shadow value for otherwise unemployed labor, and considering explicitly the issue of where that employment would otherwise be located.

4.3. CBA and frameworks

As noted earlier, one of the problems of CBA is how to deal with relevant impacts of the project which have not been valued in money terms. There is a spectrum of

Table 6
"Appraisal summary table": A6 Clapham, Bedford (GOER), 1996 scheme – 5 km D2 bypass, cost £30.9 million.

Problems
Poor safety and environment within Clapham (pop. 3200) where A6 carries up to 21 000 v.p.d. (7% HGV). 300 residential properties plus two schools front on to the road. Peak hour queuing occurs on length between village and northern outskirts of Bedford.

Other options
Two pelican crossings already provided in village. Large scale traffic calming scheme would have unacceptable noise, air pollution and severance effects. Other options considered include reduced standard single carriage way bypass on proposed line: an eastern bypass and improved rail services to new Bedford North station with park and ride. All have inferior benefits.

Criteria	Sub-criteria	Qualitative impacts	Quantitative measure	Assessment
Environmental impact CO_2 tonnes added: 0–2000	Noise	Properties within Clapham benefit from removal of through traffic	No. properties experiencing: increase in noise: 15 decrease in noise: 316	301 properties experience net decrease in noise
	Local air quality	Removal of through traffic by bypass will improve air quality within Clapham	No. properties experiencing: improved air quality: 400 worse air quality: 0	–409 PM_{10} –2549 NO_2
	Landscape	Bypass partially within local area of Great Landscape Value and would result in loss of pasture land		Slight decrease
	Biodiversity	County Wildlife Site affected		Slight decrease
	Heritage	No significant impacts		Neutral
	Water	Even with mitigation, there may still be: a significant risk of polluting a sensitive watercourse and an aquifer used for public water supply during both construction and operation; and an impact on flood risk as the scheme is within a flood plain and bridges a river		Moderate decrease

Safety		Bypass reduces pedestrian/ vehicle conflict in village and replaces a section of poor standard single carriageway	Accidents: 311 Deaths: 9 Serious: 94 Slight: 359	PVB £10.5 million 54% of PVC
Economy	Journey times and VOCs	Faster journey times on new bypass	Peak: 4.6 min Inter-peak: 3.5 min	PVB £27.6 million 141% of PVC
	Cost			PVC £19.6 million
	Reliability		Route stress: Before: 104% After: 38%	Slight Low relative to PVC
	Regeneration		Serves regeneration priority area?	No
Accessibility	Public transport	Will help to reduce peak journey times of existing local bus services		Slight increase
	Severance	Removes 80% of traffic from village		Moderate increase
	Pedestrians and others	Will improve accessibility for residents to local services		Moderate increase
Integration		Complements Bedford/Kempston package proposals and facilitates proposed Bedford North rail station with park and ride. Assists proposed local residential and commercial developments		Positive

Source: Department of the Environment, Transport and the Regions (1998).
COBA: PVB, £38.2 million; PVC, £19.6 million; NVP, £18.5 million; BCR, 1.95.

views about this. At one extreme, it can be argued that it is best to use monetary values, even based on sketchy evidence, throughout the appraisal, and so undertake comprehensive CBA. At the other, it is argued that the absence of values for many elements within the appraisal renders the problem of "horse and rabbit stew" so acute as to make CBA unworkable. Neither extreme is particularly helpful to good-quality decision taking.

Practical approaches to the problem tend to cluster around the framework, an early example of which was created by the Leitch Committee (Advisory Committee on Trunk Road Assessment 1978). This seeks to present data in monetary units where possible and defensible, and otherwise in physical units or in a qualitative way. The incidence of benefits and costs, at least in terms of community impact groups, is also shown. In the U.K. style of appraisal, adding up the monetized and non-monetized benefits and costs is left to the discretion and judgment of the decision-takers. Like many tools, therefore, appraisal is as good as the system within which it operates.

In recent years, the U.K. Government has sought to develop closer linkages between its transport policy objectives and the assessment of projects. This led the government to create its New Approach to Appraisal (NATA). One innovation within the New Approach is the summary presentation of the key scheme impacts on a single sheet of A4 which is published (Department of the Environment, Transport and the Regions, 1998). Table 6 shows such an appraisal summary table (AST). Features include a statement of problems, of other options considered, a list of scheme impacts mapped on to the government's five criteria, and qualitative, quantitative, and, where relevant, monetary measures of impacts. As with the earlier framework, this data provides input to the decision-making process. Work by the authors indicates that the decisions taken by ministers were related – to an encouraging extent – to the factors in the ASTs (Nellthorp and Mackie, 2000).

However, there are many weaknesses. Insufficient progress has been made with the derivation of money values for noise and pollution. Much more work is required in acceptable measurement and valuation of important impacts such as reliability and regeneration. Some elements, such as integration, seem politically rather than technically inspired, while in other cases, double counting is a risk. Nevertheless, the AST, within which CBA is a major component, has enhanced the contribution of appraisal to decision-taking in the U.K., and forms the basic reporting structure for the current wave of regional multimodal studies.

5. Conclusion

We conclude with one positive and one cautionary note. Serious analytical work to define and assess choices between projects, policies, and packages is in strong demand. Cost–benefit analysis, within a framework context, has an important contribution to make. But it is important to remember that any CBA can only be as good as the basic input data from the modeling and forecasting stages of the process. There is no substitute for a clear understanding of the base situation and a credible model with which to predict the impacts of the scheme or policy. Without that, any form of appraisal, including cost–benefit analysis, is built on shifting sands.

References

Advisory Committee on Trunk Road Assessment (1978) *Report of the UK Advisory Committee on Trunk Road Assessment*. London: HMSO.

Algers, S., J. Lindqvist Dillén and S. Widlert (1995) "The national Swedish value of time study", in: *Transportation planning methods: proceedings of Seminar E held at the PTRC European Transport Forum*. London: PTRC.

Department of the Environment, Transport and the Regions (1998) *Understanding the new approach to appraisal.* London: DETR.

Department of the Environment, Transport and the Regions (2000a) *Guidance on the methodology for mult-modal studies.* London: DETR.

Department of the Environment, Transport and the Regions (2000b) *Design manual for roads and bridges*, vol. 13. *Economic assessment of road schemes*, section 2, Highways Economics Note 2. London: The Stationery Office.

Department of the Environment, Transport and the Regions (2000c) *Highways Economics Note 1.* London: DETR.

Dodgson, J.S. (1973) "External effects and secondary benefits in road investment appraisal", *Journal of Transport Economics and Policy*, 7(2):169–185.

ECOTEC (1999) "Review of methodology for assessing the economic development impacts of new highway infrastructure", DETR., London, Report to SACTRA.

European Conference of Ministers of Transport (1998) *Efficient transport for Europe: Policies for internalisation of external costs.* Paris: OECD.

Friedrich, R., P. Bickel and W. Krewitt, eds. (1998) *External costs of transport*. Stuttgart: Institute of Energy Economics and the Rational Use of Energy (IER).

Glaister, S. (1981) *Fundamentals of transport economics.* Oxford: Blackwell.

Grant-Muller, S.M., P.J. Mackie, J. Nellthorp and A.D. Pearman (2001) "Economic appraisal of European transport projects – the state of the art revisited", *Transport Reviews*, 21(2):237–262.

Gunn, H.F. and C. Rohr (1996) "The 1985–1996 Dutch value of time studies", *PTRC International Conference on the Value of Time*. London: PTRC.

Gunn, H.F., M.A. Bradley and C. Rohr (1996) "The 1994 value of time study of road traffic in England", *PTRC International Conference on the Value of Time*. London: PTRC.

Hensher, D.A. (1977) *Value of business travel time.* Oxford: Pergamon.

Jones, I.S. (1977) *Urban transport appraisal.* London: Macmillan.

Jones-Lee, M.W. (1992) *The economics of safety and physical risk*. Oxford: Blackwell.

Layard, R. and S. Glaister (1994) *Cost–benefit analysis.* Cambridge: Cambridge University Press.

Lindblom, C.E. (1959) "The science of muddling through", *Public Administration Review*, 19(2):79–88.

Louviere, J., D.A. Hensher and J. Swait (2000*) Stated choice methods.* Cambridge: Cambridge University Press.

Mackie, P.J. and J.M. Preston (1998) "Twenty one sources of error and bias in transport project appraisal", *Transport Policy*, 5(1):1–7.

Mackie, P.J. and D. Simon (1986) "Do road projects benefit industry? A case study of the Humber Bridge", *Journal of Transport Economics and Policy*, 20(3):377–384.

MVA Consultancy, Oscar Faber TPA and Institute for Transport Studies, University of Leeds (1994) "Common appraisal framework for urban transport projects", Report to the PTE Group and Department of Transport. London: HMSO.

Nellthorp, J. and P.J. Mackie (2000) "The UK roads review: A hedonic model of decision-making", *Transport Policy*, 7(2):127–138.

OEEI (2000) *Appraisal of infrastructural projects: Guide for cost–benefit analysis.* The Hague: Central Planning Bureau, Netherlands Economic Institute.

Pearce, D.W. and C.A. Nash (1981) *The social appraisal of projects: A text in cost–benefit analysis.* London: Macmillan.

Persson, U. and K. Ödegaard (1995) "External cost estimates of road traffic accidents: An international comparison", *Journal of Transport Economics and Policy*, 29(4):291–304.

Simmonds, D. and ME&P (1999) *Review of land use/transport interaction models.* DETR, London, Report to SACTRA.

Simon, H.A. (1957) *Models of man.* Chichester: Wiley.

Soguel, N. (1994) *Évaluation monétaire des atteintes á l'environnement: Une étude hédoniste et contingente sur l'impact des transports.* Neuchâtel: Imprimerie de l'Évole SA.

Standing Advisory Committee on Trunk Road Assessment (1999) *Transport and the economy.* London: The Stationery Office.

Sugden, R. (1999) "Developing a consistent cost–benefit framework for multi-modal transport appraisal", University of East Anglia, Norwich, Report for DETR.

Vägverket [Swedish Road Directorate] (1995) *Reviderade värdninger 1998–2007* [in Swedish]. Stockholm: Swedish Road Directorate.

Venables, A.J. and M. Gasiorek (1999) "The welfare implications of transport improvements in the presence of market failure", DETR, London, Report to SACTRA.

Wardman, M.R. (1998) "The value of travel time: A review of British evidence", *Journal of Transport Economics and Policy*, 32(3):285–316.

Wohl, M. and C. Hendrickson (1984) *Transportation investment and pricing principles: An introduction for engineers, planners and economists*. Chichester: Wiley.

Chapter 11

TRANSPORT SUBSIDIES

WERNER ROTHENGATTER
University of Karlsruhe

1. Introduction

Subsidies are payments by the government to producers for which it receives no goods or services in return. In the system of national accounts they are treated as the counterpart of indirect taxes. Areas of subsidization exhibited in the national accounts are in the first instance agriculture, structural aid, exports, or new technologies. For example, the European Union's budget of 90 billion euros (= U.S. $85 billion) involves spending about 45% on agriculture and 35% on structural aid.

According to official statistics, the amount of subsidies which are given to the transport sector is comparatively low. For instance, in Germany only DM 1.4 billion from the total of DM 39 billion reported as federal subsidies in the GDP statistics of 1998 is given to the transport sector. This contradicts the observation that tremendously high flows of public money go to transport. For instance, in Germany about DM 40 billion are paid annually to the railway sector for various reasons: for old debt of the national railway company, for investment in the network, for contracting public services in regional transport, and for supporting regional or local investment in railway networks.

Furthermore, it is supposed that in many countries transport does not pay the cost of the infrastructure provided by the state. This statement depends, however, on the definition of costs and specific revenues from transport activities. There is an ongoing debate about which parts of transport-related taxation can be regarded as compensation for infrastructure costs, because usually the sum of transport-specific taxes is higher than expenditures for transport infrastructure.

Environmentalists often argue that road and air traffic are creating high social costs for society. As these costs are not completely internalized by taxation, they call this an indirect form of subsidization of these traffic modes. Estimations of external costs of transport show that the latter might be even higher than the infrastructure costs (INFRAS and IWW, 2000). Therefore the acceptance or non-acceptance of this argument will drastically change the ways in which of transport subsidies are viewed.

Handbook of Transport Systems and Traffic Control, Edited by K.J. Button and D.A. Hensher

These examples show that the statistical definition of subsidies may not be sufficient for the transport sector. GDP statistics embrace only a small part of the financial support given by the state to the transport sector. But there is no common agreement as to what the scope of an appropriate definition should be. Therefore Section 2 will be devoted to the pros and cons of a broader understanding of the subsidization phenomenon in the transport market. As a result the following classification is suggested:

(1) subsidies for the transport sector as defined in SNA statistics;*
(2) further direct and open financial support from the state to the transport sector;
(3) indirect and hidden subsidies.

Subsidies are usually initially justified for normative reasons but often continued after the normative reasons have disappeared. This coincides with evidence that in subsidized sectors there are close links between politicians and benefiting parties, and a common interest in stabilizing the state support arises over time. This "positive" explanation of subsidization gives rise to the analysis in Section 3 of the original goals of supportive actions of the state on the one hand and the incentives for rent-seeking parties on the other hand.

In Section 4 the particular market effects of the existing practices of subsidizing transport are illustrated, and counterproductive impacts with respect to sustainability goals are worked out. Examples will be given of some distortion effects in the airline, railway, and road haulage markets. This will be the baseline for a final conclusion in Section 5, which summarizes some recommendations to transport policy makers.

2. Typology of subsidization

2.1. Subsidies for the transport sector as defined in SNA statistics

The subsidies in SNA statistics are financial aid and tax reductions given to industries. Unfortunately, there is no uniform treatment in international statistics. For example, there are different definitions used by the Organisation for Economic Co-operation and Development, the World Trade Organization (WTO), and national statistical offices.

In general, national accounts list only a few items for transport subsidization. For instance, the German GDP statistics exhibit total subsidy payments from all public authorities (federal, state, E.U., ERP) amounting to DM 115 billion in 1998. The

*SNA: Standard System of National Accounts, applied in the countries of the OECD.

federal government paid a sum total of DM 39 billion in 1998 to industry sectors, of which DM 1.4 billion was allocated to transport. This sum is split up into

(1) financial support for sea shipping (reason: strengthening of competitive power);
(2) structural aid for inland waterway shipping (reasons: reduction and modernization of the fleet; support for small enterprises);
(3) tax reductions for journeys to work;
(4) tax exemption for ships and ports;
(5) tax exemptions for public transport (vehicle taxes; reduced ecological tax);
(6) tax reductions for private transport (e.g., reduced vehicle taxes for low-emission cars);
(7) tax exemption for air transport;
(8) tax exemption for diesel oil consumption of inland waterway shipping.

Although the list of benefiting parties in the transport sector is remarkably long, this is only the tip of the iceberg. This is because

(1) transport subsidies are classified differently and are included in public payments to other industrial sectors, and
(2) transport subsidies are not reported at all as state aid in the SNA statistics.

The first category, which is treated in this section, consists of

(1) support given to the shipbuilding industry;
(2) support given to the aircraft production industry;
(3) payment for research, development, and innovation in the transport sector;
(4) structural aids.

Subsidies for the shipbuilding industry were first given in the form of direct payments. For instance, the federal state and also the European Union have supported the East German shipbuilding industry with large amounts to make the industry strong enough for international competition. It is known that only part of this money has arrived in the right places and that the other part vanished through various practices of cash management in the industry when applying for investment aid.

Aircraft production is assumed to be a high-tech industrial sector producing strong spillover effects on other branches and thus fostering overall economic development. An outstanding example has been the development of the supersonic civil aircraft Concorde in France and in the U.K. Altogether 17 aircraft were produced until the late 1970s, of which 12 have been put into commercial operation by Air France (five aircraft) and British Airways (seven aircraft), beginning in 1977. The development costs have been completely carried by the states. For a long period, also, the cost of maintenance was covered by the public such that the air carriers only had to pay for short-term variable costs. Only in the year before the accident in Paris in July 2000 did financial performance improve, because higher prices could be charged and occupancy increased. Although Concorde was a financial flop from the commercial point of view, the production locations profited. The city of Toulouse in France, for instance, became a center of aircraft technology and today is producing the essential electronic components for Airbus Industries. The history of Airbus Industries is a history of subsidization for reasons of technology support, which will be presented in more detail in Section 4.3.

The MAGLEV experiments in Germany and in Japan are outstanding examples of the public support given to stimulate research and development in innovative technologies. In Germany, industry has received about DM 3 billion since the early 1970s. As the E.U. directive on financial aid prevents the federal government from significantly supporting technology development directly, the industry is calling most actively for building an initial MAGLEV reference track because in this case indirect and hidden subsidies can be given (see the following section).

Structural aid for regions and branches also contributes to support the transport sector. Payments allocated for this purpose are, in Germany about four times the payments specially dedicated to the transport sector (about DM 6 billion). The European Union has established a fund for regional structural development (EFRE) and a so-called cohesion fund, both taking about 35% of the E.U.'s budget. About one half of this money is spent on developing the transport networks in regions lagging behind economically (EFRE), and in countries at the periphery of the E.U. (cohesion funding for Ireland, Spain, Portugal, and Greece). In fact, the cohesion fund is spent primarily on transport projects.

If one were to reorganize the GDP statistics on subsidization in such a way that all payments officially classified as subsidies and dedicated to the transport sector were summarized, one would obtain a figure that was about 3–4 times the payments classified under the heading of "transport" (DM 5–6 billion instead of DM 1.4 billion in 1998 for the German case). The differences in national statistics might lead to different multipliers for other countries, but the basic insight that official statistics exhibit only a small part of the total subsidization given to the transport sector can be generalized.

2.2. *Further direct and open payments of the state to the transport sector*

The category of payments which is discussed in this section includes items which are reported in statistics but not defined as subsidies. Examples are

(1) investment in the public transport infrastructure;
(2) dedicated aid paid to the railway sector;
(3) payments to support regional and local public transport.

The reasons why such payments are not classified as subsidies* are that

(1) the payments can be regarded as compensation for public services produced by the transport companies (this holds for public transport);
(2) there might be public revenues correlated with the public payments, for instance income from transport-specific taxes or charges (this holds for road and air traffic).

Looking at the magnitude of such payments, one first notices that they are of a much larger dimensions than the classified subsidies. For instance, in Germany the federal expenditures for extending and upgrading the network of motorways and primaries add up to about DM 10 billion per year. For the railways the federal government planned to spend DM 8 billion in 2000. Because there are heavy complaints about insufficient investment in road and rail, the federal state will increase public spending by DM1 billion per year for highways and DM 2 billion per year for railways. This is only a part of the money that is allocated yearly to the transport sector. The railway system, in particular, consumes a big part of the public budget. This is because, in addition to the investment aid, the federal government pays about DM 13 billion per year for old debt (of which the federal railway company has been freed since the railway reform of 1994) and about DM 13 billion per year for supporting regional railway services in the states (*Länder*). Summing investment aid, old debt payments, and regional support gives a total of DM 36 billion for 2000, which is an annual payment of DM 450 per capita of the German population for the railways.

It would be too simple, however, to define all these payments to the railways as public subsidies. This is because a part of the investment aid has to be paid back through yearly depreciation payments and, furthermore, the railway company pays specific taxes (fuel tax, ecological tax). According to recent calculations (Deutsches Institut für Wirtschaftsforschung, 2000) the overall cost recovery of rail passenger transport is 103%, while it is only 16% for rail freight.† The amount of subsidization, following this calculation, is about DM 8.5 billion, based on the year

*Again, there are differences in the statistical classification of such payments. For instance the White Paper of the European Commission on Public Aid Payments includes payments for the railway and public transport in the subsidy definition, contrary to country statistics.

†To give some background information on these surprising figures: The income of the railway companies from contracted services for the public are treated as market revenues. Therefore the regional passenger traffic "pays" according to this definition. As there are no contracted services for the public with long distance traffic, in particular not with freight, the level of cost recovery is much lower for these segments. This shows that one has to be very careful when drawing conclusions from reported figures on infrastructure cost recovery.

1997. In the case of the road system a similar calculation leads to the result that the sum of vehicle and fuel taxes exceeds the yearly costs of road infrastructure in Germany by DM 23 billion and the degree of cost recovery is 148%.

Such figures have to be questioned, however, and there is a debate going on in the E.U. and in Germany as to which cost items are relevant to the infrastructure cost allocation and which kinds of taxation can be considered as revenue. For instance, a high-level commission on infrastructure finance in Germany has recommended that one should define only those parts of taxation as compensation for infrastructure costs that clearly have been dedicated to serve as financing instruments for the road infrastructure (Infrastruktur-Finanzierungskommission, 2000). In an appendix to the commission's report, Rothengatter (2000) argues that it would be misleading to define all revenues from transport-related taxes as compensation for infrastructure costs. The main reason is that the real costs of providing the road infrastructure are not only the fixed and variable costs of the network, but also the overhead costs for administration and planning, as well as the external costs of infrastructure provision and use. This argument will be elaborated further in the section on indirect and hidden subsidies.

2.3. *Indirect and hidden subsidies*

The cost items discussed in Section 2.2 are reported in the statistics but not addressed as subsidies. Therefore there is a problem in identifying the cost responsibility of the transport sector. The main problem is to figure out which part of the payments may be defined as subsidies. We now turn to the sort of support to the transport sector for which it is very difficult to identify clearly the cost responsibility because statistical reporting is incomplete. Such items are

(1) overhead costs for public administration and political insurance;
(2) external costs of transport infrastructure and infrastructure use.

Obviously the expenditures for investment costs and running costs for the transport infrastructure do not include all costs which the public sector puts in for the users. The list of overhead costs comprises, in the first instance,

(1) administration at the federal, state, and community level;
(2) planning activities by agencies supported by the public;
(3) agencies for operation and control of the network;
(4) agencies and public organizations for traffic control;
(5) police and fire brigade activities;
(6) law court activities unpaid by private parties;
(7) public agencies for road research and safety;
(8) public activities to safeguard energy supply (e.g., costs of the Gulf War in 1990/91).

It seems to be almost impossible to give a reliable estimate of the costs of these activities that are provided partly for the transport sector and paid for by the state. For The Netherlands, Bleijenberg et al. (1994) conducted a study on social costs of transport that gives a rough figure for the magnitude of public costs for these overhead services. The result was that the overhead costs for the road system were estimated as 51% of the road infrastructure costs. In the case of the railways one can argue that the overhead costs to the public are lower because the institutional arrangements are much clearer and most of these costs are allocated to the railway system through the infrastructure-costing exercise.

Estimating the costs of externalities of transport has been an issue for transport economists for decades. One recent study has been undertaken by INFRAS/IWW.* The elements of external costs listed in Table 1 were quantified.

The cost estimations have been based in the first instance on willingness-to-pay measures. Using this approach, the aggregate results shown in Figure 1 were obtained for the 17 European countries included. The sum of external costs of transport in Europe amounts to 530 euros in the year 1995, which is about 7.8% of GDP. The main contributions to this high cost level are from car traffic and trucking. Comparing this cost figure with the infrastructure costs, it turns out that the external costs are about five times as high. But one has to consider that the approaches used for the cost estimations are not completely compatible. Infrastructure costs are calculated on the basis of the lowest-economic value approach, taking the lower margin of the valuation range. Willingness-to-pay measures, which were used for the evaluation of externalities, reflect the higher margin of the valuation range. If there were actually a market for ecological goods (for instance a trade of certificates) one would expect substantially lower values for many of the externality items. IWW has suggested in follow-up studies that one should reduce the external-cost evaluations by 50% if they are to be compared with infrastructure cost calculations. Even under this assumption, the external costs come out to be about twice the infrastructure costs of transport.

In summary, there are clear indications that the transport sector is supported by state activities that are not reflected in statistics. Summarizing the overhead services which are provided by public institutions to the transport sector and the external costs, which are covered by society, the result is a multiplier of about 2.5 of the infrastructure costs in the case of the road transport sector. This would imply that in the German case the level of cost recovery of road transport would drop from 148% to less than 60%. The yearly deficit to society would amount to DM 60 billion, which could be called a real net subsidy. For the railways the multiplier expressing the overhead and external costs is much lower because a higher proportion of overhead costs is internalized and the external costs of

*INFRAS: a transport consulting company based in Zürich, Switzerland. IWW: Institute of Ecnomic Policy Research of the University of Karlsruhe, Germany.

Table 1
Overview of external costs that have been considered

Type of effect	Cost components	Leverage points and variability	Type of externality
Accidents	Additional costs of (1) medical care, (2) economic production losses, (3) suffering and grief	Depending on different factors (partly on vehicle-kilometers)	Partly external (part which is not covered by individual insurance), especially opportunity cost and suffering and grief
Noise	Damage (opportunity costs of land value) and human health	Depending on traffic volume and environmental performance	Fully external
Air pollution	Damage (opportunity costs) of (1) human health (2) material (3) biosphere	Depending on vkm, energy consumption and environmental performance	Fully external
Climate change	Damage (opportunity costs) of global warming	Depending on consumption of fossil fuels	Fully external
Nature and landscape, water and ground sealing	Additional cost to repair damages, compensation costs	Fixed costs (separation effects partly depending on traffic volumes)	Fully external
Separation in urban areas	Time losses of pedestrians	Depending on traffic volume	Fully external
Space scarcity in urban areas	Space compensation for bicycles	Depending on traffic volume	Fully external
Additional costs from up- and downstream processes (a)	Additional environmental costs (air pollution, climate change and risks)	Fixed costs (gray energy of infrastructure and rolling stock)	Fully external
Congestion (treated separately)	External additional time and operating costs	Depending on traffic amount (number of vehicles)	Separate issue (in relation to other costs): Average costs are internal to the users. Differences between marginal and average costs are external costs

Source: INFRAS and IWW (2000).
Note: (a) Upstream: process of producing vehicles, infrastructure, and energy. Downstream: process of disposing of vehicles after lifetime.

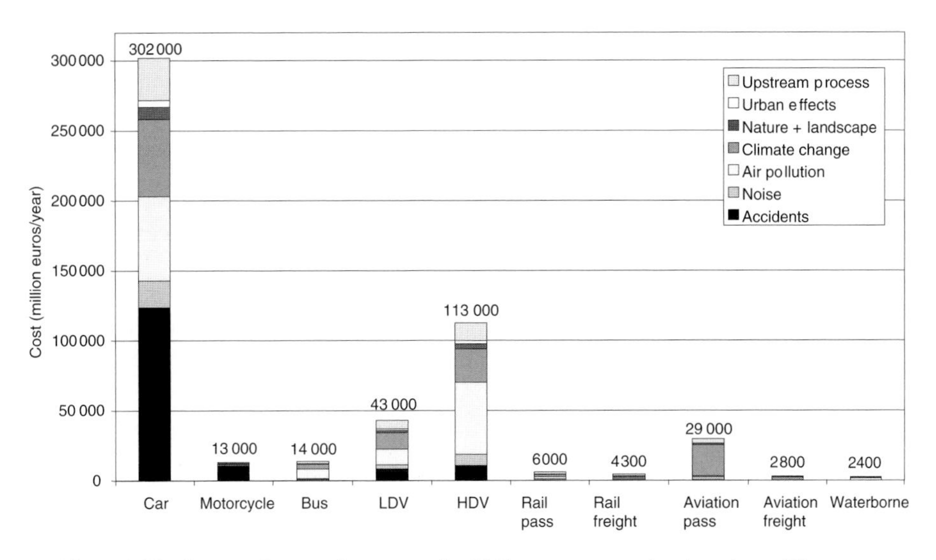

Figure 1. Total external costs of transport for 17 European countries, based on willingness-to-pay measures, for the year 1995.

railways are substantially lower compared with road transport. Nevertheless, applying a rough "guesstimate," one obtains a level of cost recovery below 50% and a real net subsidy of about DM 20 billion. Relating the total net subsidies to the performance figures in terms of passenger and ton kilometers, one obtains a higher subsidy per unit of traffic activity for rail compared with road. This can be interpreted as indicating that policy makers evaluate the non-market contribution of the railway sector to societal well-being more highly than has been done in our rough economic calculation.*

3. Normative reasons for subsidization in the transport sector

The reasons for correcting the market through subsidies can be summarized in two main arguments:

(1) correction for market failures, and
(2) changes in the distribution of income.

*Note that very low figures have been used for the external cost of transport. If the results of the INFRAS/IWW study were applied without the correction, the road sector would be allocated a higher share of subsidies than the rail sector.

Public finance literature has discussed at length the motivation for public interventions in the market processes. The classical instruments for implementing these interventions are establishing institutions, setting regulations, imposing taxes, and paying subsidies to adjust the decisions of private decision-makers in order to correct for market failures. The latter could occur through

(1) a public- or merit-good property of the traded goods or services,
(2) externalities of production and consumption,
(3) excessively high risk aversion of investors in the case of sunk costs, or
(4) natural monopolies and deficits occurring with pricing/investment strategies.

3.1. Public- or merit-good properties of traded goods and services

There are goods which cannot be allocated according to market principles, i.e., through price bids and offers, because nobody can be excluded from consumption (free-rider property). Beyond the pure public goods such as national defense, the legal system, or the provision of a basic framework for foreign contacts, there are a number of quasi-public goods or merit wants. While pure public goods undoubtedly call for public control for provision and distribution, the quasi-public merit goods are someway in between the public and the private sphere. The transport sector gives examples of goods and services of this type.

A standard example in the literature is lighthouses, which will not be supplied in sufficient quantity by private ship owners, because the latter calculate only their private benefits and not the benefits produced for others. A lighthous is a typical club good,[*] because it does not benefit the whole of society, but rather it benefits a set of users of the facility only. In principle, private solutions are possible if an arbiter – which is usually the state – provides an arrangement for forming partnerships and making the choices, such that the members of the club can determine the optimal layout of the common facility and the distribution of property rights, benefits, and costs. The provision of transport network infrastructure or communication networks is a similar case of club-good provision. Obviously the private users receive benefits from the common provision of essential facilities (compared with standing alone and building facilities separately) and they are willing to pay for this service (in this respect a club good differs from a pure public good, which is characterized by free riding).

[*]For the theory of club goods see Buchanan (1972) and the literature on co-operative games, summarized by Schotter and Schwödiauer (1980).

There is no reason for subsidizing club goods as such. But often club-good properties coincide with externalities or have merit want characteristics. In the case of merit wants, the state government assumes that there is a collective demand for a service, which exceeds the private demand at a given price. By lowering the price and paying the deficit through subsidies the state can increase private demand to a socially desired level. Education, health services, culture, and public transport are examples of this type of merit goods. In the case of public transport the provision of services under public conditions is supposed to attract more customers and guarantee a fair treatment of all user categories. It is seen as some type of public mobility guarantee regardless of the income of the users.

As the treatment of public transport varies in different countries, one can conclude that the social preferences for the merit-good property can differ widely. In many countries transport infrastructure is regarded as a public domain, in particular the provision of infrastructure for public transport is usually defined as a public task, not only with respect to the organization but also, in particular, with respect to finance.

3.2. Externalities of transport

Three properties characterize an externality (Rothengatter, 2000):

(1) involuntary interaction among agents who use a resource jointly, for which the property rights have not been defined;
(2) processing of the interaction outside the market, that is, without trading or bargaining, such that costs occurring or benefits generated are not allocated to the responsible party; and
(3) relevant market failure, which leads to reduced adaptive efficiency through false signaling.

In a Pigouvian scheme[*] negative externalities would be internalized through taxes and positive externalities through subsidies (Figure 2).

In Figure 2 the shaded areas depict the welfare loss ("deadweight loss") generated by the externality. This loss is removed by introducing optimal taxes AG or subsidies BF. In transport policy the subsidy given is often motivated by positive externalities or by differentials of negative externalities, i.e., a transport mode is subsidized because it produces lower external costs than others. The latter argument seems strange from the theoretical point of view, but is a usual instrument from the viewpoint of a policy-maker who faces a high resistance from lobby groups to Pigouvian taxation. In this case it is much easier to subsidize the

[*]This concept was developed by Arthur C. Pigou (1920) following the original ideas put forward by Alfred Marshall (1890).

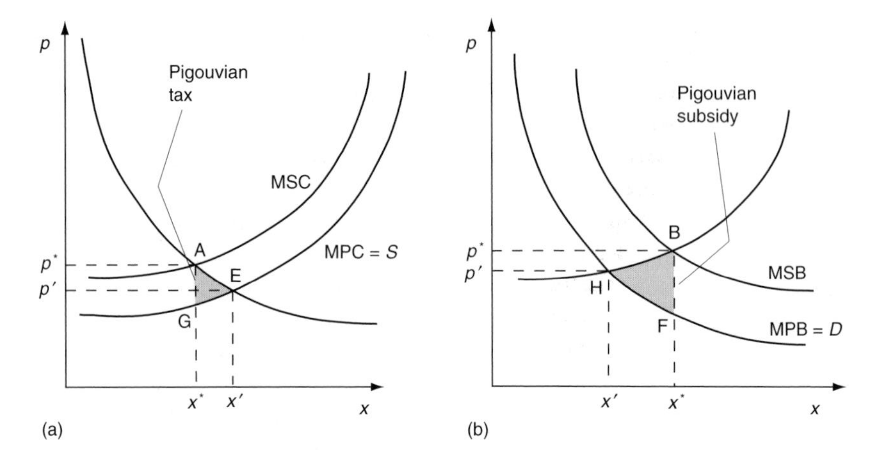

Figure 2. Pigouvian taxes and subsidies (MSC, marginal social costs; MPC, marginal private costs; MSB, marginal social benefits; MPB, marginal private benefits; D, demand; S, supply).

environmentally well-performing transport mode (e.g., the railways) rather than to penalize the party with poor environmental performance (e.g., the car or truck transport).

Externalities can be produced through the provision of the infrastructure and through infrastructure use. In the case of infrastructure provision the following effects can occur:

(1) Network effects through the interdependence among links. A provider of single facilities (bridge, tunnel, motorway section) in this case will not be able to capitalize all benefits produced by its activity.
(2) Growth effects from interactions between the transport sector and the spatial economy. Benefits can be induced which are realized outside the network segment in which investment has been made and outside the transport sector.[*]
(3) Benefits are produced for other state activities, such as national defense or social security.

External benefits of those types justify the fact that the state does not allocate the full infrastructure costs to the users. If parts of the infrastructure are provided by private investors, as in the case of large-scale projects promoted by BOT-type schemes,[†] the occurrence of external benefits will motivate the state to give grants to finance the project.

[*]Krugmann (1991), Venables (1999) and Venables and Gasiorek (1998) have analyzed such effects, which occur in particular if markets are imperfect.

[†]BOT: build–operate–transfer scheme for private investors.

In the past there has been an active debate on the external benefits generated by the use of the transport infrastructure. On one hand, some authors argue that positive externalities on a transport network can only be produced by the users, such that it is justified to increase the positive externalities. Willeke (1994), for instance, mentions that transport activities are closely integrated in developing consumption and production patterns and therefore participate in manifold ways in the evolution of the economy. In particular, in countries with a strong car-manufacturing industry, this argument is promoted intensively to give subsidization of road transport an economic rationale. However, these arguments are rejected by other authors, who stick to a rigid welfare-theory foundation and conclude that externalities from infrastructure use cannot occur. The effects denoted as positive externalities from transport actually stem from infrastructure provision. Better infrastructure provides option values for extended use, induced by lower time consumption and costs. These effects are included in cost–benefit analysis of infrastructure projects and internalized through the financial scheme. Accordingly, there is no reason for considering these effects a second time (Rothengatter, 1994).

The profound analysis of Coase (1960) on the nature of externalities has given rise to the question of whether every externality occurring in the market system should be internalized through Pigouvian instruments. According to Coase's theorem, the problems could be solved by a proper allocation of property rights, and negotiations or bargaining between the parties involved would be a step in the right direction. Only in the case of dilemma situations, high transaction costs, a high number of parties involved, unequal distribution of economic/political power, or imperfect information on the impact mechanism, would central action of the state be inevitable. From this theoretical basis two conclusions can be drawn:

(1) If positive externalities exist, one can expect strong incentives for private discovery and consideration, in particular if the state pays (as in the case of infrastructure).
(2) External costs of transport will not so easily be discovered and negotiated by the groups concerned, because decentralized solutions are less probable. But as an internalization scheme might imply high transaction costs, its design has to be optimized through cost–benefit analysis.

This result has to be kept in mind when it comes to state interventions to protect transport modes, which produce comparatively low external costs. Figure 3 shows the result of an external-cost calculation by INFRAS and IWW (2000), that clearly indicates that the average external costs of road and air traffic are much higher than the average costs of rail and inland waterways.

Interpreting the result of Figure 3 in the Pigouvian way, one would conclude that a huge transfer scheme is necessary to reduce the detrimental effects of transport. However, remembering the transaction cost argument of Coase,

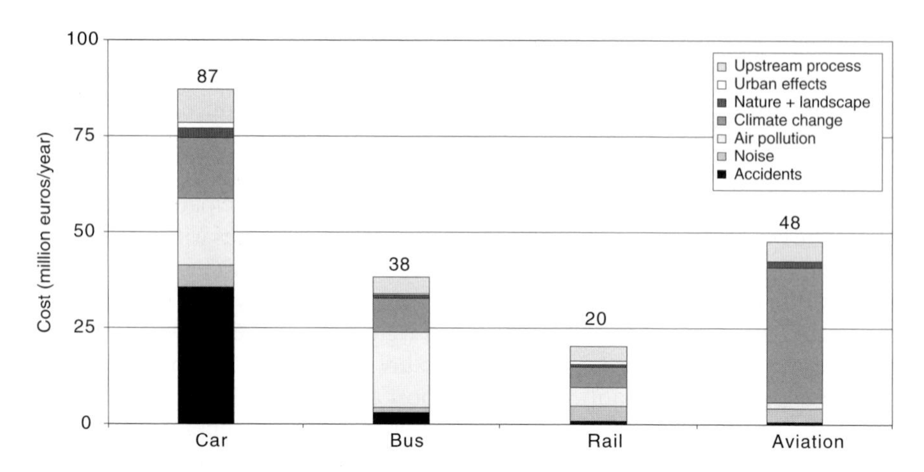

Figure 3. Average external transport costs of different modes for 17 European Countries (INFRAS and IWW, 2000).

one can find other schemes which achieve the same result with a much lower disturbance from public interventions in the market. Such a scheme, as has been developed in the context of the ASTRA project for the E.U. Commission (ASTRA, 2000), includes incentive-compatible standard setting for exhaust emissions and safety, differentiation of vehicle taxation, temporary subsidization of new technology, and local car bans for old-technology vehicles. As a scheme which includes multiple instruments, triggered so as to influence the right lever points, it is in general superior to a uniform policy, one should be careful when relying on pure Pigouvian taxation/subsidization policy. But the latter can be a very effective element of an overall policy mix.

3.3. Risk aversion of investors in the case of sunk costs

Large-scale projects and the introduction of new technology in the transport sector often face the barrier of a high risk of investment, which is hard to manage by private financial schemes. Flyvbjerg et al. (2001) have analyzed the economics of such projects in the transport sector and outlined the extent to which state participation is necessary to overcome barriers to private investors. For instance, in the Channel Tunnel project political risks were not properly allocated, while in the cases of the Scandinavian projects of the Great Belt and the Oeresund fixed links, the state has incorporated both the public and parts of the private risk. If projects are large and the lifetime and the amortization period are high it is often not easy to find a good balance between public and private risk taking.

When it comes to high-speed railway investment the French solution is to share the financial contributions among the railway company (SNCF; it usually insists on a minimum internal rate of return of 8%), the federal state, and the regions that benefit from the improved service. In Germany the investment costs have been shared, since the railway reform of 1994, by the federal government and the railway company (DB AG). The federal government is responsible for the initial finance and gives grants, and DB AG has to pay for the depreciation for the non-granted part of the investment. After some years it turned out that DB AG was not able to pay for substantial parts of the high-speed rail investment. This underlines the fact that creating a high-speed rail network, in accordance with the E.U. Commission's idea of trans-European networks is only possible if there is a sufficient public commitment to taking financial risk. This includes also the public institutions on the regional level according to the French example, because financial participation lowers the incentives to seek rents from the activities of the federal government and the federal railway company.

Finance of projects with high sunk cost is even more difficult if new technology is involved. The Japanese and German developments of new MAGLEV technologies provide examples of the problems that arise. Private industry is promoting the new technology (in Japan, MLU; in Germany, Transrapid) with arguments of innovation, industrial development, and competitive power in the world market. Industry tries to hedge the risk through implicit allocation to public companies or to the public budget. In Germany, the Transrapid project between Hamburg and Berlin has been promoted since the early 1990s by government and industry, on the basis of manipulated figures for costs and revenues. After a change in government and the board of DB AG, the project had to be stopped because both the government and DB AG insisted on a financial scheme which allocates the risks of cost overruns and of overestimation of passenger demand to private industry (the federal government limited its cost share for constructing the 290 km guideline to DM 6.1 billion). Now the federal government and industry are looking for suitable MAGLEV projects that are smaller and might be managed more easily. Two new projects have been preselected (Munich Main Station to Munich Airport; Dortmund to Düsseldorf). These projects are not financially viable, but it will be more transparent what the public contribution will be.

3.4. Natural monopolies and deficits occurring with pricing/investment strategies in these markets

As in many networks where there are economies of scale and of density, transport and communication are classic examples of natural monopolies. These industries are characterized by subadditive costs, i.e., at least partly decreasing average costs,

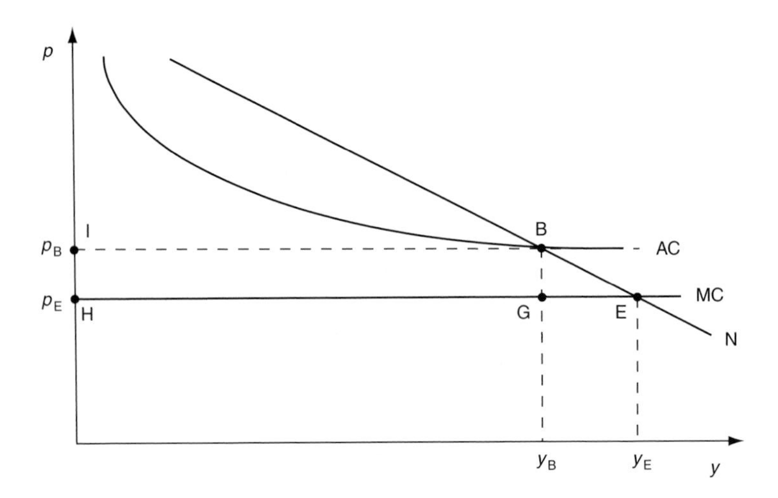

Figure 4. Marginal-cost pricing and the resulting deficit (AC, average costs; MC, marginal costs; N, demand curve; *P*, price; *Y*, quantity).

such that one supplier can produce the services at lower costs than a set of smaller suppliers. When it comes to optimal pricing for natural monopolies, there is one line of argument that marginal-cost pricing is the best strategy for the public because in this case the welfare position in the market considered is maximized.* In the *White Paper on Fair Pricing for the Use of Transport Infrastructure* (European Commission, 1998) the E.U. Commission takes up this theoretical argument and recommends a common pricing scheme based on the social marginal costs relevant to the member countries. The general problem of such a scheme is illustrated by Figure 4.

The marginal-cost pricing strategy would lead to the equilibrium (p_E, y_E), and the deficit resulting would be IBGH. Charging average costs, however, would reduce demand to y_B and incur a deadweight loss of BGE. The E.U. Commission has suggested that one should charge marginal congestion costs and external marginal costs additionally to cover the financial deficit. However, one can show that the income from charging marginal congestion costs would not be sufficient and, overall, external costs would not be recovered through such a scheme. A deficit would be left to society. In consequence, pricing the transport infrastructure according to the marginal-costing scheme would end in a direct or hidden subsidy.

*This can be shown as maximizing the social surplus consisting of the consumer's and the producer's surpluses.

3.5. *Redistribution of income*

Many types of transport subsidies are motivated at least partly by the idea of benefiting lower-income classes and economically backward regions. Often the journey to work is in itself regarded a reason for supporting travelers. Therefore the state has to provide people who do not own a car with basic facilities to move from residential areas to the places of employment. This gives a reason for providing rural areas with public transport, although public support might be excessively high and beyond economic rationale. For instance, rail transport service in Germany is even provided in regions where patronage is less than 1000 people per day. The rules for providing public transport in such regions allow such services to be offered, even if they are much more costly than carrying people by bus.

When it comes to infrastructure allocation, the public goals of balancing the endowments of regions with public resources and fostering the balance of regional development have to be considered. In most countries the issues of spatial development and supporting lagging regions are treated outside the cost–benefit calculus. In Germany, checking for the impacts on the regional economy and on employment is a part of the cost–benefit exercise done for federal transport infrastructure planning. Furthermore, there is an implicit rule, accepted by the states and the federal government, stating that the proportion of funds for transport infrastructure allocated to the states remains unchanged over time. This rule again causes hidden subsidization in the transport sector.

4. Positive reasons for subsidization in the transport sector

4.1. *Normative versus positive economics*

While normative economics starts from the basic value judgment that policy-makers try to maximize social welfare and that all public interventions in the market are governed by this issue, positive economics looks at the world as it is. George Stigler (1971) was among the first economists to systematically analyze the internal motivation structure of political bodies. The basic insight is that the failure of politicians to carry out what would seem to be in the public interest is not just the consequence of the greed or malevolence of politicians but may be the inevitable consequence of the working of political institutions in democratic societies.

Policy-makers might not wish to come into power to realize programs, but they make programs to come into power. Policymaking is a career like any other career and the players in this field try to maximize their own private benefits, in a way comparable to managers in private enterprises. They need playing fields to compete with others, networks of supporters and reporters, and, finally, the votes of the voters. In this sphere it is not important where a political actor is really moving with respect to defined goals; what is more important is what the voters believe he/she is doing. In this world, public and regulated markets serve partly as playing fields and give political actors the opportunity to present themselves.

Stigler's work gives an understanding of the phenomenon that regulated markets continue to be regulated and public governance persists even when the reasons have vanished. One of the main reasons is "captivity," which can be interpreted in three ways:

(1) Policy makers and managers in a regulated or public market are sitting in the same boat and are interested in protecting themselves against bad fortune. The best way to do that is to tell the public success stories, even though this might contradict reality.
(2) A position as manager in a public or regulated company might be a reserve career of a policy-maker. Therefore he/she will try to act supportively and helpfully.
(3) The network of public/regulated enterprises provides close interfaces with the network of private businesses, which is important, at least as a financial platform for a political career.

4.2. Interest groups and stakeholders in the transport sector

The behavior of the players in political games is characterized by their actual decision situations, i.e., it can be interpreted by means of stimulus–response schemes. It is governed by routines resulting from cognitive processes working through learning mechanisms, i.e., good experiences increasing and bad experiences decreasing the probability of choosing a routine. Normative thinking in terms of maximizing social welfare is only necessary to sell a policy to the voters. In the following we briefly present the major interest groups in the transport sector and their predominant behavior in the "market for subsidies" (see Grüne, 1996; Ursprung, 1991).

The government and the ruling parties

This group comprises two types of actors: the protagonists of better transport, including the responsible ministry, and other bodies, depending on the political subject. For instance, in the case of a new transport technology, the ministry of economics might be in favor of subsidizing technology development (e.g., MAGLEV or aircraft development), while in the case of market regulation policy this ministry the promoters. Typically the protagonist part of the government will have to put through subsidization policy with the support of the benefiting parties, which are, in the case of technology support, the industries providing the technical supply.

The second wing of the government comprises non-protagonists that might be disadvantaged insofar as money is withdrawn from purposes which they favor (e.g.. education, social affairs). In many cases the ministry of finance has a very strong position in trying to reduce the public budget and to control subsidy payments. This leads to the typical consequence that the public protagonists and the industrial lobby both try to move the policy action forward through brightening the figures on costs and benefits. This holds in particular for large-scale projects and new transport technologies. The experience in Germany, that the final investment costs are twice as high and the demand comes out as only a fraction of the originally estimated figure, is largely the result of the rules of the political game.

The rest of the public organizations

The parties in opposition will attack government policy in relation to the responsiveness of the voters. This means that it is not so much the political failure in itself which is blamed, rather, its effects on particular groups. Examples of social injustice are very effective, such that social movements against unfair government policies are readily supported by the opposition and parts of the press. Examples are the movements against high fuel prices in the U.K., in France, and in Germany. In France the social problems associated with high gasoline prices have been solved by tax reductions for some groups, and in Germany the kilometer rates at which income taxes of commuters are reduced have been increased, including users of cars, of public transport, and of non-motorized transport.

It is a striking phenomenon that poorly justified transport subsidies, as such, are not in the center of the attacks of opposition parties. The reasons for this are twofold. First, in the case of large-scale projects or major industrial subsidization, the parties presently in opposition might have been the ruling parties when the policy action started. Then it will not be wise to attack that policy. Secondly political parties hope that supportive behavior will be rewarded by the industry in the form of assistance in forthcoming electoral campaigns.

Industrial-lobby and motorist clubs

The transport lobby and motorist clubs form a powerful group in favor of investment of any kind in the transport sector, as well as of special rules for the transport market. National and international associations of vehicle manufacturers and of transport haulage companies have been established which engage in rent-seeking for their clients. This means that any kind of disbenefit is heavily attacked and any kind of financial aid is welcomed. Usually these groups know how to use the instrument of the press to foster public support for their claims. Motorist clubs can form very strong opinion-leading organizations in the interest of consumers. In Germany, for instance, the General German Motorist Club (ADAC) has about 17 million members, and its monthly magazine is the most widely read journal. It addresses directly 17 million voters and their family members, so that it has a strong influence on policy makers. The tax relief given to commuters mentioned in the previous section is a result of an aggressive campaign by the ADAC against ecological taxation, which in the case of transport is linked to the fuel tax.

The influence of these lobby groups is dependent on the relative power of the automobile industry. In Denmark and in Switzerland, where there is no automobile manufacturing, the government is much less dependent on the appraisal of these rent-seeking actors.

Public opinion

We do not discuss the influence of further groups, such as trade unions, religious communities, or local and green movements, but politicians of course try to learn about potential feedback from all of the voters. This is done by personal intuition, by information from the local party organizations, and by opinion polls. There is hardly an important political decision taken without testing public opinion. As the way in which people are questioned is decisive in determining the result, the organizations performing the opinion polls might be seen as invisible power groups that exert the highest influence on political decision-makers.

Again, if people are questioned about their good and bad impressions of governmental transport policy, they will rarely put subsidization at the top of the priority list of political wants. Instead they will put fairness issues, which rank very high in the case of pricing policy, and wishes for better transport quality through investment in networks and rolling stock at the top. Environmental issues and the sustainability problems of the transport sector seem to have ambiguous influences. There are times when people feel that not enough is being done for the environment and call for public action. This holds in particular if the perception of the problem is high (El Niño, earthquakes in Japan, reports on the ozone hole, reports on climate change). But after a period of public discussion, the sustainability issues are crowded out by day-to-day problems, such as high energy prices. This makes it difficult to treat the problems of environmental costs through Pigouvian taxes. The only policy which is widely accepted by the stakeholding groups is the enforcement of environmental standards for vehicle propulsion technology and its implementation by means of tax relief and subsidies.

4.3. Example: The case of aircraft subsidies

Out of the great many cases of subsidization in the transport sector, one example will be considered in more detail, because it reveals all normative and positive arguments for subsidization and the consequences for fair competition in the international market for transport technology suppliers. Since the establishment of Airbus Industries in the early 1970s, the U.S. government has complained about illegal subsidies. Meanwhile Airbus Industries has won 40% of the market for civil aeronautics and is now attacking the last monopoly of the U.S. civil aircraft industry, which is the Boeing 747 megaliner. The new Airbus 380 wide-body aircraft will offer up to 620 seats and will threaten Boeing's position worldwide because the A 380 is designed in particular for the air traffic between the big international hubs.

The U.S. government has argued that Airbus Industries receives much higher subsidies than are allowed in the context of the WTO agreement of 1994. Actually, the governments of the U.K., France, Spain, and Germany have put considerable startup investments into the Airbus aircraft developments. The development of the A 380 will cost about U.S. $10.7 billion, with support of 33% by the states mentioned. Furthermore, the regions in which the production plants are located support the industry with additional aid. The city of Hamburg, for instance, plans to build an artificial peninsula to prepare construction ground for the large production hall.

The European Union is arguing on the basis of the Large Aircraft Agreement, which was signed by the E.U. and the U.S.A. in 1992 and allows subsidization of up to 33% by both parties. From this standpoint the Europeans do not see any violation of agreements on fair competition. Furthermore, they add that most of the subsidies are paid back. Since 1992 Airbus Industries has paid back about U.S. $3 billion, and also the support for the A 380, amounting to U.S. $3.5 billion will be paid back – conditional on the market success of the aircraft. Looking at the U.S. practice of supporting the aircraft industry, the E.U. mentions that the U.S. government has been giving indirect subsidies for a long time through military development payments. The Boeing 747, for instance, is a by-product of a military transporter, of which the development was fully paid by the state. Other public funds are allocated to the civil aviation industry through aerospace research activities. (It is estimated that Boeing has received about U.S. $400 million in the past years from NASA.) Last but not least, the U.S. government is subsidizing so-called dual-use technologies by some U.S. $2 billion per year. One can suppose that a part of this sum is flowing to the aircraft industry.

Remembering the discussions of the normative and positive reasons for subsidization, one can argue that:

(1) There is a common standpoint of the E.U. and the U.S. government that aircraft production is a key industry for national development and international competition.
(2) The industries argue that they cannot take the risk of new technological development, and that this is a case for state action.
(3) There has not been any problem, neither in the U.S. nor in the E.U., with public agreement to the subsidy strategies. There was even no resistance from the Green parties, although the problems with sustainability issues will grow if the airline industry is able to attract more passengers in the future.
(4) As a result a sector is subsidized, but it is hard to discover the normative reasons for doing so from a neutral position. Air traffic is the fastest-developing transport sector, and the customers' willingness to pay is sufficient to finance all investments needed.
(5) These subsidies add to the indirect financial support through tax exemptions (VAT, fuel tax). The international air transport associations are strong enough to prevent governments from agreeing on international rules to remove subsidies.

5. Conclusions

Transport is one of the most heavily subsidized sectors. This holds for old and declining transport industries, such as the railways in unprofitable market segments, as well as for fast-growing sectors like road haulage or the airline industry. In many cases the original reasons for subsidization were normatively well founded. But, surprisingly enough, the subsidies continued after the normative reasons for them had vanished. The explanation for this phenomenon is that in the real world the players in the transport market, the politicians, the industrialists, and the customers, all try to maximize their individual profits and benefits, regardless of the consequences for social welfare. From the viewpoint of positive economics it is almost impossible to cut subsidies which were introduced when they were justified normatively. This is because powerful coalitions of politicians, industrialists, associations, and clubs will form to protect the subsidized institutions against any political change. It seems that there are only two time periods in political life in which reductions of transport subsidies are possible:

(1) a financial crisis in the public budget, which forces the government to drastically reduce expenditure;
(2) a rising perception of the risk to sustainable development, if the environmentally risky modes of transport continue to be subsidized, directly and indirectly.

Regarding the individual fortunes of politicians – e.g., in New Zealand – who have tried such a political change in the past, one can conclude that a new generation of policy-makers supported by a new generation of voters will be needed to break with the traditional rules.

References

ASTRA (2000) Assessment of transport strategies. Prepared by IWW, Karlsruhe, TRT, Milan, ME&P, Cambridge, and CEBR, London for the European Commission. Directorate General VII. Fourth Research Framework. Project No. ST-97-SC.1049. Karlsruhe.

Bleijenberg, A.N., W.J. van den Berg and G. de Wit (1994) *The social costs of traffic – a literature review*. Delft: CE.

Buchanan, J.M. (1972) "Toward analysis of closed behavioral systems", in: *Theory of public choice*. Ann Arbor, MI: University of Michigan Press.

Coase, R.H. (1960) "The problem of social cost", *Journal of Law and Economics*, 3:1–44.

Deutsches Institut für Wirtschaftsforschung (2000) *Wegekosten und Wegekostnedeckung des Strassen- und Schienenverkehrs in Deutschland im Jahre 1997* [Transport infrastructure costs and their recovery in the road and rail sector in Germany]. Berlin: DIW

European Commission (1997) Fifth white paper on financial aid in the European Union. Brussels: European Commission.

European Commission (1998) White paper on fair payment for the use of transport infrastructure. Brussels: European Commission.

Flyvbjerg, B., N. Bruzelius and W. Rothengatter (2001) *Megaprojects and risk. Making decisions in an uncertain world*. Cambridge: Cambridge University Press.

Grüne, M. (1996) *Subventionen in der Demokratie*. Munich.

INFRAS and IWW (2000) *External costs of transport: accident, envirnmental and congestion costs of tansport in Western Europe*. Zurich: INFRAS.

Infrastruktur-Finanzierungskommission (2000) "Bericht der Infrastruktur Finanzierungskommission and den Bundesminister für Verkehr, Bau- und Wohnungswesen" [Report of the Commission on Transport Infrastructure Finance to the Minister of Transport, Construction and Housing]. Infrastruktur-Finanzierungskommission, Berlin.

Krugman, P. (1991) "Increasing returns and economic geography", *Journal of Political Economy*, 99:183–199.

Marshall, A. (18990) *Principles of economics*. London: Macmillan.

Pigou, A. (1920) *The economics of welfare*. London: Macmillan.

Rothengatter, W. (1994) "Do external benefits compensate for external costs?", *Transportation Research* 28A:321–328.

Rothengatter, W. (2000) "External effects of transport", in: J.B. Polak and A. Heertje eds., *Analytical transport economics. An international perspective*. Cheltenham: Edward Elgar.

Schotter, A. and G. Schwodiauer (1980) "Ecnomics and the theory of games", *Journal of Economic Literature*, 18:479–527.

Stigler, G. (1971) "The theory of economic regulation", *Bell Journal of Economics and Management Science*, 2:3–19.

Ursprung, H.W. (1991) "Economic policies and political competition", in: A.L. Hillman, ed., *Markets and politicians: Politicized economic choice*. Boston: Kluwer.

Venables, A.J. (1999) "Road transport improvements and network congestion," *Journal of Transport Economics and Policy*, 33:319–328.

Venables, A.J. and M. Gasiorek (1998) "The welfare implications of transport improvements in the presence of market failure", Report prepared for the Standing Advisory Committee on Trunk Road Assessment (SACTRA), London.

Willeke, R. (1994) "Benefits of different transport modes. Germany", in: European Conference of Ministers of Transport, Round Table 93, Paris.

Chapter 12

TRANSPORTATION DEMAND MANAGEMENT

JONATHAN L. GIFFORD and ODD J. STALEBRINK
George Mason University, Fairfax, VA

1. Introduction

Transportation demand management (TDM) is an umbrella term used to describe a variety of actions aimed at reducing or modifying the demand for transportation facilities and services. The term derives from a contradistinction with actions that respond to congestion by increasing the supply of transportation facilities or capacity (i.e., supply management), for example, the expansion of highways or airports to accommodate peak traffic. TDM does include some supply measures, however, when they focus on alternatives to the mode in question. Hence, increasing the supply of transit or other non-highway capacity is generally seen as a form of TDM, and in some cases expanding intercity rail capacity is seen as a form of TDM for airports.

TDM has gained attention since the 1970s primarily as a result of significant increases in travel that have not been accompanied by increases in infrastructure capacity. Demand for transportation facilities and services typically grows out of economic expansion and prosperity. However, commensurate expansions in infrastructure capacity have become more difficult. The cost of infrastructure expansions is often borne by public authorities that are unable or unwilling to commit the requisite financial resources. The construction of new facilities or the expansion of existing facilities also often brings with it "externalities" such as noise, air pollution, and incursions into open space or already built-up communities. TDM has become increasingly appealing as supply responses to congestion have become more difficult. Support for TDM also has a normative component, insofar as some groups believe that increasing mobility, at least in some modes of transport, is incompatible with environmental sustainability.

2. TDM measures

TDM encompasses a wide range of measures.

Handbook of Transport Systems and Traffic Control, Edited by K.J. Button and D.A. Hensher

2.1. Congestion pricing

Congestion pricing (also known in North America as value pricing) utilizes tolls and other fees to raise the price of travel at peak hours in order to induce a shift in the time of day or mode of travel (see Chapter 7). For example, authorities could raise airport fees at peak hours to encourage airlines to schedule fewer flights at peak times, or raise peak-hour road tolls or transit to encourage off-peak travel.

Congestion pricing has gained widespread support in the academic and policy communities as the most efficient mechanism to accomplish two goals. First, market prices would in the short term efficiently allocate the use of scarce resources such as highway and airport capacity. Second, market prices could provide a mechanism for determining the economically efficient quantity of transportation facilities and services. That is, if willingness to pay in the short term is high enough to expand capacity, then revenues resulting from congestion pricing could be used to do so.

In practice, congestion pricing has only successfully been implemented in a few places, and many pilot projects have either failed short of implementation, or not continued beyond the pilot stage. Perhaps the most successful implementation has been in Singapore in 1975, where single-occupant vehicles are required to pay U.S. $1.50 to enter the central business district during the morning peak hour. The program has been extremely successful in reducing automobile traffic into the urban core, although a small side industry has sprung up of professional passengers who agree to ride into the city for a small fee and then return to the periphery, thereby allowing the driver to avoid the toll charge.

A second successful implementation, in 1991, has been in Trondheim, Norway's third largest city with a population of 140 000, where a cordon-based fee was assessed to vehicles entering the city between 6 a.m. and 6 p.m. Tolls vary from U.S. $0.62 to U.S. $1.56, with the highest tolls between 6 a.m. and 10 a.m. Significant in this implementation was that in order to reduce the impact of multiple tollbooths and toll lanes in an already built-up community, the road ministry launched a ambitious campaign to persuade vehicle owners to obtain electronic vehicle tags that would allow collection of the fee in normal travel lanes. Trondheim is also significant in that it uses the revenues from the fee to support expanded public transit. The Trondheim system has also raised some concerns about safety because cars may stop suddenly or pull to the side of the road near toll stations in order to wait until after 6 p.m. to pass. Norway has implemented similar system in two other cities, Oslo and Bergen.

A third noteworthy application of congestion pricing has been along the Riverside Freeway (State Route 91) near Los Angeles. In this instance, a 10 mile toll road was constructed in the median of a toll-free expressway (i.e., a freeway) and dynamically priced so that the toll road lanes would always be congestion free. Hence, tolls are quite low at off-peak times (U.S. $0.75), but quite high at peak travel periods (U.S. $4.25). While tolls adhere to a fixed schedule, that schedule can be adjusted over time to ensure congestion-free conditions. Interestingly, the franchisees of the toll road recently sued successfully to stop the government from adding lanes to the adjacent toll-free highway. More than 86 000 have signed up to use the toll road (it requires the acquisition of an electronic tag and establishment of an account), and more than 25 000 vehicles use the road daily.

Opposition to congestion pricing grows out of concerns about its impacts on equity, that is, that it may lead to a "two class" transportation system, one for the rich and the other for the common man. Some have labeled them "Lexus lanes." While experience on the Riverside Freeway indicates that many of middle and lower income find it attractive to use the toll lanes, a perception of congestion pricing as elitist persists. A second concern about congestion pricing is what to do with the revenues. As indicated, pricing could be used as a way to measure users' willingness to pay and to adjust supplies accordingly. In practice, however, revenues are often siphoned into the government's general fund, or used to support alternative means of transport. Some advocates of road pricing wish to set prices not on an economic basis but rather as high as necessary to achieve a desired level of travel.

2.2. Electronic road pricing

Integrally related to congestion pricing is electronic road pricing (ERP). ERP uses dedicated short-range communication (DSRC) technology to collect tolls without requiring a vehicle to stop at a tollbooth. The most widespread ERP approach involves the placement of an electronic transponder or tag on a vehicle and the establishment by the vehicle owner of an account with the road authority. Each time the vehicle passes a tag reader, the reader reads and confirms the status of the tag. The tag readers may be installed either at tollbooth-like barriers, or on gantries over normal travel lanes. There are two types of toll systems, open and closed. Open systems assess tolls each time a vehicle passes a reader. Closed systems write a point of entry on the tag when the vehicle enters a toll facility and assess a fee based on that point of entry when the vehicle exits.

An alternative to DSRC has been proposed that would mount a transponder on the vehicle that would determine its location using the satellite Global Positioning System and assess tolls based on the vehicle's location.

Of course, ERP reduces the costs of collecting tolls enormously compared with traditional tollbooths. Savings include the labor cost of toll collectors. But significant savings also result from the elimination or significant reduction in the size and scale of toll plazas, as well as the congestion, delay, and pollution they often induce.

ERP does raise some privacy concerns, since road authorities can determine from tag-reading records the location of a specific vehicle at a specific time and place. This has alarmed some in the civil liberties community, especially if such records become available to law enforcement authorities or private investigators. Measures to ensure the privacy of such data include the destruction of data after it is no longer needed to reconcile tag accounts, or if such data is to be retained for planning and analysis purposes, the destruction of any information that would associate an individual vehicle with a particular tag transaction.

2.3. *Ridesharing*

Ridesharing (also known as carpooling and vanpooling) programs match travelers with proximate origins and destinations in order to reduce travel in single-occupant vehicles. Particularly popular in the U.S.A., ridesharing programs sometimes organize shared rides to originate in park-and-ride lots or other locations, and may encourage ridesharing by providing preferentially located or priced parking.

2.4. *Transit improvements*

A fourth category of TDM is the management of demand by improving the supply of alternative modes of travel. Many cities in the U.S.A. have developed light rail systems in an attempt to increase transit ridership and foster higher-density patterns of urban development.

TDM also includes a wide range of other measures, the objective of which is to reduce peak period travel demand:

(1) Park-and-ride lots provide parking for travelers to leave their private vehicles in order to switch modes to either transit or shared rides.
(2) High-occupancy vehicle (HOV) lanes are highway lanes that are reserved for the use of vehicles containing two or more travelers (the minimum varies among facilities). Lanes are sometimes divided from general-purpose travel lanes by physical barriers, or sometimes denoted with “diamonds” or other markings.
(3) High-occupancy toll (HOT) lanes are HOV lanes that also allow access for a price to single-occupant vehicles.
(4) Employer commute option programs utilize on-site or dedicated transportation co-ordinators to work with employees to encourage ride-sharing and other TDM measures.

(5) Telecommuting measures allow employees to work at home or at remote locations some or all of the time instead of commuting to a distant office.
(6) Alternative work schedules allow employees to work schedules that differ from standard 40 hour per week, fixed schedules (e.g., 9 to 5, or 8:30 to 5, five days per week). Such alternatives include working four 10 hour days with three-day weekends, nine 9 hour days with alternating three-day weekends, etc.
(7) "Cashing out" free parking provides an option to employees who have access to free parking to "cash out" their access to free parking in exchange for an employer-subsidized transit pass. Recent changes in U.S. law allow employers to treat such programs as a tax-deductible employee benefit up to $65/month.
(8) "Family friendly" transportation services are those that attempt to accommodate the needs of commuters with children by, for example, providing child care, grocery, pharmacy, and other services at places of employment, near transit stations, or park-and-ride lots.

This list is by no means exhaustive, but it suggests the range of measures available for a TDM program.

3. TDM objectives

The objectives of TDM programs fall in the following general categories (Victoria Transport Policy Institute, 2000):

(1) *Reduction in or modification of travel behavior.* One of the most common objectives for TDM is a reduction in vehicle travel, or other modification of travel behavior, such as a change of mode or time of travel. Serious peak-period highway congestion is common in many urban areas and many view TDM as an alternative to expanding highways and the attendant disruptions to communities and damage to the environment. Similarly, peak-period airport congestion is widespread, and TDM offers an alternative to costly and disruptive airport expansions.
(2) *Reduction of environmental damages.* Both the construction and the operation of transportation facilities carry environmental impacts. Construction often disrupts natural settings or existing communities. Motorized vehicle operations contribute to a number of adverse environmental effects, such as local air pollution (CO, ozone, particulates and toxics), global air pollution (climate change and ozone-depleting gases), and noise and water pollution. By influencing travelers to rideshare or travel less, TDM strategies may lead to reductions in these adverse effects.

(3) *Increased land use efficiency.* The location and design of transportation facilities and the types of transportation services available are one of many determinants of the use of land, especially in urban areas. Airports attract the development of associated land uses such as hotels and business parks catering to industries that are frequent users of the air transport system. Freeways or motorways accommodate automobiles and trucks (lorries) and give rise to businesses that service and utilize them, such as gas stations and truck stops. Moreover, transportation facilities are an integral component of "urban design" – that is, the combination of sidewalks, streets, highways, bike paths, houses, retail establishments, and offices that together define the character of a particular place.

Some urban designs (or urban development patterns) have raised concerns, especially among environmentalists, because they depend heavily on the use of private automobiles, they are very difficult to serve with traditional public transit, and they are often inhospitable to pedestrians and bicycles. Residential land development may occur at the outlying edge of urban areas and take the form of "subdivisions" that consist largely of detached, single-family houses arrayed along curvilinear streets, culs-de-sac, and courts. Such developments, often called "urban sprawl," are very conducive to access by private automobile but extremely difficult to serve with traditional public transit. It is difficult to design efficient bus routes given the pattern of streets and culs-de-sac, and walking distances to collector streets or trunk highways, where efficient transit is more feasible, as too far for easy pedestrian use.

Similarly, the urban design of business centers, termed "edge cities" by some, utilizes large purpose-built or speculative office buildings surrounded by large parking lots and parking structures. Again, servicing such business centers with traditional transit often entails long walking distances from bus stops to building entrances, sometimes across boulevards of six or eight lanes of heavy traffic. Such edge cities have grown rapidly in the U.S. since the 1980s.

One objective of some TDM programs is to achieve urban designs that are less reliant on single-occupant autos and more amenable to public transit, bicycles, and pedestrians. TDM strategies may be used to encourage higher density, mixed land uses, and multimodal travel, and thus reduce low-density, auto-dependent development. The ability to achieve urban design objectives depends heavily on how land use decisions are governed. The U.S. vests land use regulation largely at the local level, so that a large metropolitan area may have dozens or more independent jurisdictions governing land use in their jurisdiction, which hinders the development and implementation of region-wide comprehensive plans. Other countries, such as the U.K., give much less power to local entities, requiring all development to adhere to a specific plan.

Table 1
Examples of TDM initiatives

Program/objectives	Strategy	Measures/technologies
Improve mobility between points A and B, during peak periods	Reduce travel Stimulate alternative routes and modes Promote ridesharing	Peak-period road pricing (congestion or "value" pricing) Traveler information technologies Remote parking lots
Reduce environmental impacts from travel in area X	Stimulate ridesharing Improve non-motorized travel conditions Increased vehicle use fees	High-occupancy vehicle (HOV) lanes Bike lanes Transit Public financial telecommuting incentives
Reduce number of deaths and injuries on road X	Reduce speed Improve road conditions	Increased police visibility Speed cameras Traffic calming
Corporation-driven TDM programs	Financial subsidies Regulation	On-site employee transportation co-ordination Alternative work schedules Telecommuting

(4) *Enhanced travel safety.* TDM can enhance travel safety by, for example, reducing the number of travelers. Reduced travel may reduce the incidence of traffic crashes, thereby contributing to reductions in fatalities, injuries, vehicle damages, and emergency calls.

4. TDM programs

A TDM program typically employs one or more strategies, which in turn utilize one or more measures, as illustrated in Table 1 (Winters, 2000).

5. The rationale for TDM

The standard rationale for TDM is that traditional supply-oriented transportation approaches lead to two market failures. First, road space is provided at a low out-of-pocket cost to road users and may therefore lead to its overconsumption (i.e., a "club good" problem) and, hence, congestion. A second market failure relates to the impacts of road use on the environment and community, principally through tailpipe emissions but also through noise and public safety. In order to arrive at the "optimal" amount of road usage in the short run, the apparent price paid by road users should reflect these technological "externalities." Thus, TDM's fundamental objective is, in some sense, to redress deficiencies in the pricing of transportation facilities and services.

This rationale is subject to two caveats. First, a challenge to the establishment of "efficient" prices is that while the supply of transportation capacity may be fixed in the short term, it is not fixed in the long term. Full-cost pricing can efficiently allocate access to a fixed supply in the short term. But in the absence of reliable techniques for calculating these costs, or market mechanisms for expanding (or contracting) supply in response to users' willingness to pay, an efficient level of transportation supply and demand will remain elusive. A second caveat is that "efficient prices" are theoretical constructs that fail to account properly for transaction costs. As a result, efforts to establish efficient prices run the risk of being suboptimal at best and arbitrary at worst.

6. TDM effectiveness

There have been a number of efforts to evaluate the effectiveness of various TDM programs, in the U.S.A. and elsewhere. Orski (1990) summarized these evaluation efforts with the following six points:

(1) Little effort has been devoted to evaluating the effectiveness of demand management in a balanced and dispassionate way.
(2) Uniform comparisons are rendered difficult because there is no agreement on how to assess TDM effectiveness.
(3) Although some successes are evident, there is no evidence that they can be replicated widely enough to influence traffic congestion or air quality on a regional or subregional level.
(4) There is a tendency to ignore the principle of latent demand.
(5) Effective travel demand management is not cost-free.
(6) Congestion pricing, despite its theoretical appeal, faces many hurdles.

A decade hence, these points are still apt.

Despite these weaknesses, however, TDM still attracts considerable interest, and TDM programs are common. As these programs begin to generate results their benefits (or lack thereof) should become clearer. At this time, data are primarily available for assessing the short-term responses of various programs. A case in point is the HOT lane program on the State Route 91 express lanes in Orange County, California, where a toll facility has been built in the median of an existing freeway that offers reduced fees to vehicles carrying multiple occupants. A study has monitored the road conditions and traveler options before and after the implementation of this program. Preliminary results indicate that many travelers discriminate in judging when time savings justify paying the toll (about

45% of the peak-period travelers were reported to use the facility once a week or less) (ITE Taskforce on High-Occupancy-Tolls (HOT-Lanes), 1998). As these programs mature, analysis of their long-term effects will become available.

It is noteworthy that the emergence of intelligent transportation system (ITS) technologies may provide significant opportunities for enhancing the effectiveness of TDM in practice. Five areas in particular have been identified: pretrip planning, parking management, congestion pricing, telecommuting, and transit service enhancements (Orski, 1995).

7. Historical perspective

TDM in the U.S.A. is rooted in two different concepts: transportation system management (TSM) and ridesharing. TSM refers to a major U.S. policy initiative issued jointly by the Federal Highway Administration (FHWA) and the Urban Mass Transportation Administration (UMTA), in response to the 1973 Arab oil embargo. The objective of TSM was to provide transportation planners with " a short-term, emergency response planning tool with a much shorter turnaround time than traditional methods of urban transportation planning, which focus on long run capital costs, business cycles, infrastructure investments, and related highway (and transit) construction schedules" (Ferguson, 1999)

TSM never reached widespread adoption, however, due to institutional inertia and a lack of appropriate implementation mechanisms (Gakenheimer and Meyer, 1979). It was replaced at the end of the 1970s by an intensified interest in "ridesharing," which encompasses various efforts to co-ordinate and pool travelers across available transportation modes. While ridesharing is a type of TSM strategy, it may be viewed as an individual concept in that it, in contrast to alternative strategies at the time, was deemed feasible both from a technical and an institutional standpoint. However, the lack of an appropriate implementation mechanism prohibited many ridesharing strategies from gaining widespread acceptance (Ferguson, 1999). TDM followed as an outgrowth of these two concepts during the 1980s. It was triggered by a boost in traffic congestion and air pollution problems.

8. Current issues

While TDM has the potential to bring about a more efficient use of existing transportation systems, it also faces several challenges. While transportation planners and policy experts may be intent on managing demand, not surprisingly, many travelers resist having their demand managed by their government. The determination of whether a particular trip is worthwhile is a complex calculation, and the entry of government into this calculation can give rise to conflicts in values. Of course, the decision to accommodate increasing demand by expanding infrastructure raises value questions, too. And how such questions are resolved differs widely depending on a country's traditions and expectations about the role of government and the freedom of the individual.

In the U.S.A., perhaps the most important challenge to TDM concerns the ability of implementing entities to provide concrete evidence of its effectiveness. This challenge has arisen as a result of federal requirements for TDM and resulting evaluation needs within recent transportation and clean air legislation. This legislation has led to an increased need for developing appropriate criteria for performance measures (Schreffler, 1994). Due to the difficulties of calculating reliable measurements of the value held by most transportation systems and the complex behavior of transportation system users, this issue poses a significant challenge for the future of TDM.

An additional challenge derives from TDM's dependency on new information-facilitating technology, such as intelligent transportation systems (ITS) and geographical information systems (GIS). These technologies often exhibit a much shorter life cycle than is being used for traditional transportation planning, which often are based on forecasts of 20 years. Consequently, a significant challenge for TDM will be to develop methods through which these technologies can be mainstreamed into conventional transportation planning processes.

References

Ferguson, E. (1999) "The evolution of travel demand management", *Transportation Quarterly*, 53:57–78.

Gakenheimer, R. and M. Meyer. (1979) "Urban transportation planning in transition: The sources and prospects of TSM", *Journal of the American Planning Association*, 45:28–35.

ITE Taskforce on High-Occupancy-Tolls (HOT-Lanes) (1998) "High occupancy tolls (HOT-lanes) and value pricing: A preliminary assessment", *ITE Journal*, June:30–40.

Orski, K. (1990) "Can management of transportation demand help solve our growing traffic congestion and pollution problems?", *Transportation Quarterly*, 44(4):483–498.

Orski, K.(1995) "Thinking small: Applying its technologies to TDM", *ITE Journal*, 65(12):57–60.

Schreffler, E. (1994) "Travel demand management evaluation: Current practice and emerging issues", in: Transportation Research Board, *Transportation Research Circular*, vol. 433. Washington, DC: Transportation Research Board.

Victoria Transport Policy Institute (2000) *Online TDM encyclopedia*, http://www.vtpi.org. Victoria, BC: Victoria Transport Policy Institute. Accessed 23 June 2000.

Winters, P., ed. (2000) *Transportation demand management, Transportation Research Board. Transportation in the new millennium*, vol. 7. Washington, DC: Transportation Research Board.

Chapter 13

INFRASTRUCTURE CAPACITY

T.R. LAKSHMANAN and WILLIAM P. ANDERSON
Boston University, MA

1. Introduction and overview

The capacity of a transport infrastructure facility signals its ability to accommodate a flow of people or vehicles. Its measurement is sensitive to a variety of analytical assumptions about the *context of flow*, and thus not amenable to an unambiguous definition. A transportation facility's capacity (design or practical) is typically defined from an engineering perspective in a contextual fashion, with reference to a level of service or quality of flow, that will provide satisfactory traffic operations. Occasionally, capacity has been viewed, from an *economic* perspective, as a threshold notion, as the minimum traffic volume for which a facility, say a road project investment, is justifiable.

The recent growing interest in the analytical and policy issues relating to infrastructure capacity reflects a convergence of several factors. First, the rapid growth in transport demand and the resultant congestion and delays on motorways, in airports, and in airspace is currently a prominent issue in policymaking and in the media. Apart from time losses, traffic congestion and incidents lead to severe reliability problems. Whereas the direct costs of delays are significant – estimated for 1995 at about a billion (U.S.) dollars in a small country such as The Netherlands, and currently in the U.S.A. considerably higher than the $16 billion estimated in 1989 – the indirect costs (including follow-up costs because of arriving early or late at the destination, and prevention costs of trying to be on time) may be much higher (Small and Gómez-Ibáñez, 1999; Bovy, 1998).

The traditional solution for addressing the congestion problem, namely infrastructure capacity expansion, has been handicapped in recent times by several developments in affluent industrialized countries. Such developments hampering infrastructure capacity expansion include: national economic pressures in many countries and the consequent reduced infrastructure expenditures; resistance to infrastructure expansion deriving from environmental concerns and local communities in the path of infrastructure rights-of-way; and the growing recognition that traffic growth induced by capacity expansion soon swamps the increase in capacity.

Handbook of Transport Systems and Traffic Control, Edited by K.J. Button and D.A. Hensher

Consequently, there is increasing interest among analysts and policy-makers in a second class of complementary solutions that augment the capacity of *existing* transport networks. Such policies, which attempt to squeeze more capacity from existing transport networks and thereby bridge the infrastructure capacity gaps, fall into two classes: those that reduce the capacity shortfall by increasing the supply of infrastructure capacity and those that reduce the demand for infrastructure capacity.

On the supply side, a number of operational or physical changes and information provision measures serve to effectively increase current infrastructure capacity. These measures include (a) a variety of dynamic traffic management measures (TMM) – e.g., ramp metering, dynamic lane allocation, arterial signal timing, and creative design of entries, exits, and weaving sections, and (b) a range of information technology measures which raise capacity – radio-dispatched taxi cabs, automatic vehicle location and computer-controlled marshaling yards, advanced air traffic control, the provision of traveler information, automatic vehicle spacing, etc.

The demand side measures address the infrastructure capacity shortfall by manipulating demand – generally directed at a reduction of demand and the modification of its temporal and spatial usage patterns. Such measures fall into two broad classes: those that are shorter-term measures, such as transport demand management (TDM) and pricing policies, and those that may work in the long run, such as regulatory approaches, land use controls, and enhancement of public transit.

This chapter aims at describing and clarifying the concept of infrastructure capacity as it has been used in different countries and modes in order to support infrastructural planning, design, and operational activities. Section 2 surveys these concepts and the extensive literature on methods that have been developed to estimate road capacity – both physical and "economic" – in the context of a level-of-service notion. It notes the practical advances in traffic design and operations generated by these capacity assessment methods (largely derived from the U.S. Highway Capacity Manual) as well as the pressures for road investments such capacity estimation methods have created. Section 3 describes the emerging variety of policies and measures that expand the capacity of different modes of existing infrastructure. It also surveys the broad range of policies and measures which manipulate the demand for transport capacity. Finally, some analytical and policy inferences drawn from this brief survey of the concept, measurement, and augmentation of infrastructure capacity conclude the chapter.

2. Infrastructure capacity: Definition and measurement

Infrastructure capacity represents the maximum number of vehicles, persons, or freight which have a reasonable expectation of passing through a given section of a transportation right-of-way or terminal during a given time period under prevailing facility, traffic, and control conditions. This notion of capacity as a *physical flow* (of people or vehicles) developed by the Transportation Research Board (TRB) (1998) for roadways is broadly applicable to the various transport modal links and nodes.* The prevailing conditions of flow are those imposed by a number of contextual factors. Such factors that govern infrastructure facility capacity levels include (a) geometrical attributes of the right-of-way, such as curvatures, steepness and length of grades, sight distances, etc.; (b) type of right-of-way, such as type of road or port type; (c) composition of traffic, such as the vehicle size mix in highway lanes or aircraft type mix at airports or vessel mix at ports; (d) traffic control technology and regulation regime, as illustrated by the differences between the prevailing central control model or the emerging collaborative decision-making (CDM) model – made possible by the Global Positioning System (GPS) – of air space control; and (e) other factors such as climatic conditions, alleviation of incidents, etc.

There is a vast literature concerned with the elaboration and refinement of measures of physical capacity of infrastructure facilities under such different traffic, facility, and control conditions. This work, stimulated by the planning, design, and operational activities associated with expansion of infrastructure facilities, has largely been in the area of highway capacity analysis.

As noted below, road capacity is viewed as the highest possible traffic volume that can be conveyed by a particular road, specifically as the maximum of a volume–density or volume–speed curve (Figures 1a and 1b). The resulting flow definition of physical capacity has been in the context of the quality of flow or level of service. The methods developed in this analytical tradition for measuring and estimating road capacity and for deriving road design standards for a satisfactory level of service have been pioneered in the U.S.A. and have been influential in engineering practice in Europe and elsewhere. Recent economic pressures and lowered infrastructure expenditures in these latter OECD countries have drawn attention to the above road design standards derived from U.S. practice – for example, the use of the 30th hour peak traffic flow as a design standard leads to high investments, while the use of a 200th hour does not yield an unacceptable penalty on traffic operations.

It is in this context that the French threshold approach of "economic capacity" as a choice criterion in road investments in rural areas has emerged (Government of France, Ministry of Transport, Office of Highways, 1980). As framed in a cost–benefit analysis for a particular road project, economic capacity is defined as "the smallest of all traffic volumes which need to be attained to justify the road

*This notion of highway capacity was introduced in the widely used *The Highway Capacity Manual* (HCM) in 1950 by the U.S. Bureau of Public Roads. HCM has subsequently appeared as a second edition (1965) and a third edition (1985), whose third update by TRB arrived in December 1997.

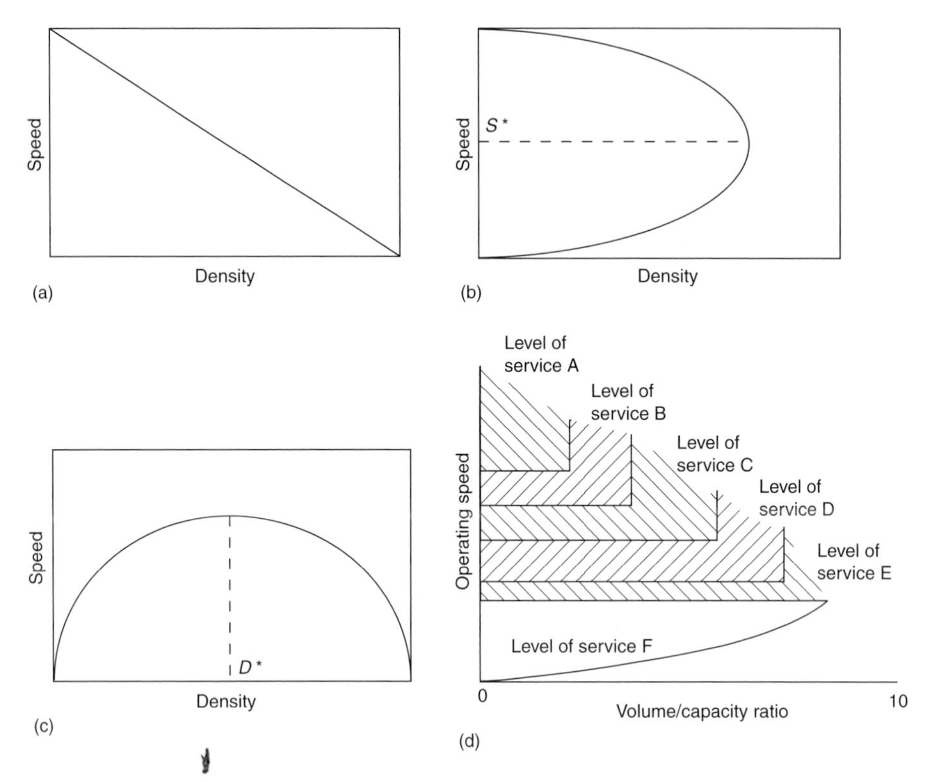

Figure 1. (a) Assumed relationships between speed and density. (b) Assumed relationships between speed and flow. (c) Assumed relationships between flow and density. (d) System by which roadway segments are rated (source: OECD, 1983).

project by cost–benefit analysis" – specifically signifying economic capacity in terms of average annual daily traffic volume (OECD, 1983). The time stream of benefits of a road project (monetized travel time savings, gains in safety, maintenance, operating expenses, driving comfort, toll rates, and fuel taxes in the context of induced traffic) is compared with investment cost. The economic or threshold capacity of a road improvement, in this calculus, is the level of traffic for which the short-run return of the road improvement equals the marginal rate of return.

Thus the notion of economic capacity of a road – rooted in a particular situation in an existing road network – is different from the concept of physical capacity commonly used in highway capacity analysis (see below). It is possible that in a congested high-density area the road project may not be justifiable in spite of high benefits because of even higher costs, while a low-traffic rural road improvement may be justified because of low right-of-way costs. While "economic capacity" offers a valuable decision criterion, the concept of physical capacity is the appropriate one for the discussions of infrastructure capacity in this chapter. What follows is an overview of the definition and analysis of physical capacity of infrastructure facilities by major mode of transport.

2.1. Highway capacity analysis

Highway capacity analysis is concerned with the ability of a particular road to accommodate an anticipated flow of people or vehicles. Capacity analysis is conducted to meet a variety of needs in transportation design and management, ranging from performance assessment of existing highway system elements to the planning for system extensions. In the U.S.A., methods for highway capacity analysis are set forth in the *Highway capacity manual*, first published in 1950 by the Bureau of Roads. This manual is now published and periodically updated by the Transportation Research Board of the National Research Council (Transportation Research Board, 1998).

The two critical concepts employed in capacity analysis are capacity and level of service (LOS). Capacity measures indicate the maximum number of vehicles or people that can pass over a highway element during some time interval. They do not, however, provide any information about congestion. It is possible that the absolute maximum flow could be achieved only at a level of congestion that would be unacceptable to most travelers. In assessing the ability of existing or planned highway infrastructure to satisfy current or anticipated demand, therefore, it is necessary to supplement measures of capacity with LOS measures, which incorporate information on congestion.

The standard definition of highway capacity is the "maximum hourly rate at which persons or vehicles can traverse a point or uniform section of a lane or roadway during a given time period under prevailing roadway, traffic, and control conditions" (Transportation Research Board, 1998). It is thus a flow measured in vehicles or people per hour. While this definition seems straightforward, there are a number of important considerations regarding its application and interpretation. First, capacity refers to a point or uniform section, and not to an entire facility. For example, sections of a limited-access road may be heterogeneous in terms of lane width, shoulder width, median, grade, etc. and may therefore have different capacities. In this case the segment with the lowest capacity will define the capacity of the road. Second, the capacity is a rate of flow rather than a volume of flow. Given that variations in flow can be expected over a one-hour period, capacity generally refers to the hourly rate that can be achieved during some peak period of 15 minutes or less. Finally, capacity is not so much an absolute maximum but rather one that can reasonably be expected to recur. So it is not impossible that flow rates above the capacity will be observed occasionally.

The methods of capacity analysis for uninterrupted-flow and interrupted-flow road facilities are significantly different. Uninterrupted flow refers to segments of multilane highway with access limited to widely spaced, unsignalized interchanges. Interrupted flow refers to the presence of signalized intersections or intersections with stop signs.

For uninterrupted flow, the basic approach is to define a maximum flow (capacity) under an ideal set of conditions and then apply adjustment factors for deviations from that ideal.

For freeways, for example, the ideal conditions include:

(1) minimum lane width of 12 feet (3.6 m),
(2) minimum right (shoulder) and left (median) lateral clearances of 6 feet (1.8 m) and 2 feet (0.6 m),
(3) only passenger cars in the traffic stream,
(4) ten or more total lanes,
(5) interchanges spaced at 2 miles (3.2 km) or more,
(6) level terrain, and
(7) all drivers familiar with the facility.

Under ideal conditions the highway capacity is about 2400 passenger car equivalents per hour per lane (pcphpl) (Transportation Research Board, 1998). This number is adjusted downward for deviations from ideal conditions such as narrow lanes, closely spaced interchanges and a high proportion of trucks.

Capacity analysis for interrupted-flow facilities is a more complex task, owing both to the dynamics of flow through intersections and to the multiplicity of intersection types. The maximum flow passing through a signalized intersection depends, naturally, on the proportion of time in an hour during which the signal light is green. It also depends on the dynamics of movement for vehicles as the light changes from red to green. The first n vehicles will experience some delay in movement, known collectively as the start-up lost time. All other vehicles pass through the intersection at a saturation flow rate. Thus the capacity of the intersection is equal to the saturation flow rate times the amount of green time in an hour, with an adjustment downward for start-up lost time. Since the start-up lost time occurs every time the signals change, the capacity is a decreasing function of signal cycle frequency. Capacity calculations are complicated by adjustments for geometry, traffic conditions, the proportion of vehicles turning, and other factors.

Intersections with stop signs rather than signal lights present equally complex problems for capacity analysis. In such cases, the time it takes a typical vehicle to traverse an intersection depends on the volume and distribution of the conflicting traffic flow. Delay depends both on what the typical driver considers an acceptable gap in the conflicting flow and on how frequently such a gap presents itself.

In many cases, it may not be enough to see whether the capacity of a road element is sufficient to accommodate the observed or anticipated demand levels, because operation at or near capacity may imply high levels of congestion. We can illustrate this by means of a set of assumed relationships among speed (measured as an average in miles per hour), density (passenger cars per lane mile), and flow (measured in passenger cars per hour), shown in Figures 1a, 1b, and 1c (May, 1990). Speed is a declining function of density (Figure 1a). Flow increases with density up to a point (D^*) where the increasing number of vehicles is offset by their declining speed as the road becomes congested. Thus flow is maximized at some speed (S^*) (Transportation Research Board, 1998).* This implies that capacity is reached only after speeds have been significantly reduced by congestion. The question is whether this level of congestion is tolerable.

The LOS is a measure of the quality of the flow, which may be described in terms of the speed of travel, room to maneuver, number of flow interruptions, etc. Given the multidimensional nature of the LOS, a system by which roadway segments are rated with the letters A through F is used, with A representing a nearly vacant roadway and F representing stop-and-go conditions under extreme congestion (see Figure 1d). The key variable used for LOS rating on uninterrupted facilities is the density of passenger cars per lane mile. Thus, within some range, capacity increases under conditions of declining LOS. The highest flow frequently occurs under LOS E, but since the quality of flow is deemed unacceptable at E, an alternative definition of capacity for planning purposes is the flow service rate, defined as the flow that can be achieved while maintaining a desired LOS.

Different criteria are needed to assign various interrupted-flow facilities to LOS categories. For example, in the case of signalized intersections LOS is determined on the basis of control delay, which includes deceleration time, queue move up time, stop time, and acceleration time. Intersections where the control delay for the typical vehicle is less than 10 seconds are assigned to LOS A. Inferior levels are assigned according to control delay intervals, ranging from 10 to 20 seconds for LOS B to more than 80 seconds for LOS F.

Applications of capacity analysis fall into three basic categories – operational analysis, design analysis, and planning analysis – depending on the type of information that is needed. Operational analysis seeks to predict the LOS on an existing or planned highway segment for which all geometric and traffic characteristics are known. Design analysis determines the physical characteristics (especially number of lanes) required for a new road segment that must serve a predefined traffic flow. Planning analysis is similar to design analysis, but it works

*These relationships can be analytically derived and they are roughly supported by a number of empirical studies, including Banks (1989) and Hall and Hall (1990).

with less detailed information – generally the average daily travel rather than a schedule of flows – to determine highway capacity requirements (Transportation Research Board, 1998). Planning analysis generally comes earlier in the highway development process.

Since demand conditions vary over time, capacity requirements have to be assessed at a specific hour. In metropolitan areas this is normally the weekday morning peak hour. Outside cities, there may be significant day-to-day and seasonal variation. For example, a rural area with tourist amenities may experience extraordinary demands on Friday or Sunday evening of a holiday weekend. Since it is not economical to design to a level of capacity that is needed only for a few hours per year, a design hour must be chosen upon which to base design analysis. The design hour is normally chosen between the 30th and 100th busiest hour of the year, with the 30th hour most typically chosen for rural highways.*

Highway capacity analysis is one of the most highly developed areas of transportation research. Despite this, there are areas where existing methods of analysis fall short. One such area is in defining the interactions among highway sections. The definition of capacity given above does not allow for any impacts from downstream traffic conditions. Methods for incorporating downstream and upstream impacts, permitting the analysis of capacity on a highway system, are still relatively approximate compared with other methods of capacity analysis (Transportation Research Board, 1998).

A second shortcoming is the failure of conventional methods to recognize the bidirectional causality between capacity and demand. Demand conditions are normally taken as exogenous, despite research that indicates that provision of new capacity may stimulate growth in demand (Downs, 1992; Standing Advisory Committee on Trunk Road Assessment, 1994). This phenomenon, known as induced travel, complicates the problems of design and planning analysis. In essence, the presence of induced traffic implies that in determining the number of lanes needed to support demand, some adjustment must be made to reflect the amount of new demand that will be stimulated by the expanded highway.

2.2. *Transit capacity analysis*

As with highways, transit capacity analysis must be concerned both with crude measures of capacity and with measures of LOS. The most common

*Choice of design hour is influenced by various factors. For example, the difference between the highest-flow hour and the 100th hour is generally much greater in rural areas, especially in recreational areas. Hence the 30th hour, rather than the 100th hour, is chosen. Also, a higher-volume design hour may be chosen to achieve a higher minimum LOS.

measure of capacity is person capacity, defined as the number of people that can reliably be moved past a certain point on a transit line during a time interval (normally one hour) without unreasonable delay, hazard, or discomfort (Transportation Research Board, 1998). Person capacity is sometimes called line capacity to indicate that it refers to a specific line rather than an entire system.

Person capacity depends on the capacity of vehicles and their headways. Headways in turn depend on safety considerations and the number of vehicles per interval of time that can take on and discharge passengers in a typical station. It is usually station capacity (maximum rate of flow through stations) rather than "way" capacity (maximum flow along right-of-way sections) that constrains system performance.

Transit systems vary considerably in terms of speed. Thus, a bus system with many vehicles moving at slow speeds may have the same person capacity as a rail system with fewer vehicles moving at higher speeds. Since the latter situation is preferable from the passengers' perspective, productive capacity – defined as the product of passenger capacity and average speed – is often used as an alternative measure. While the productive-capacity measure does not have a clear intuitive meaning, it provides a way of weighting trips by speed. High-speed trips contribute more to productive capacity than do slow trips.

Transit LOS depends on passengers per vehicle and vehicles per hour. The first reduces LOS due to crowding, while the latter increases LOS by limiting wait time. Holding frequency of service constant, therefore, capacity is inversely proportional to LOS. The "crush load" value of persons per vehicle, which is associated with LOS F, is therefore not used for planning purposes, even though it may provide the maximum capacity. Thus for most systems, "capacity" does not represent a desirable operating condition.

Capacity in transit is highly sensitive to the size of vehicle and to the acceptable LOS. For example, estimates of person capacity for rail rapid transit, assuming 50 foot (15 m) cars and 2 minute headways (30 cars per hour), vary from 9000 for six-car trains and LOS B to 30 000 for ten-car trains and LOS D. Capacities over 100 000 are in theory attainable for the largest possible trains at LOS F (Transportation Research Board, 1998).

For comparitive purposes, Table 1 shows ranges for person capacity and productive capacity for a variety of highway and transit options. Note that while the line capacities of road transit options exceed private auto lane capacities, the two options are more or less equivalent in terms of productive capacity. The table indicates the superiority of rapid transit rail options over all other options in terms of both person capacity and productive capacity. This capacity advantage only translates into an efficiency advantage if relatively high transit load factors are achieved.

Table 1
Capacity and productive capacity for highway and transit modes

	Person capacity	Person productive capacity
Private auto (lane)		
On street	720–1050	20 000–50 000
On freeway	1800–2600	60 000–120 000
On-street transit		
Bus	2400–12 000	20 000–90 000
Street car	4000–20 000	30,000–150,000
Semirapid transit (a)		
Bus	4000–12 000	75 000–200 000
Light rail	6000–20 000	120 000–260 000
Rapid transit		
Rail rapid transit	10 000–72 000	400 000–1 800 000
Regional (commuter) rail	8000–60 000	500 000–2 000 000

2.3. *Air travel capacity analysis*

Air travel is the fastest growing of the major transportation modes. Not surprisingly, the need to provide adequate capacity to accommodate this growth has been a major challenge for transportation planners. Two types of capacity come into play in air travel, capacity along flight routes and capacity at airports.

In order to maintain control and safety in a national or international air transportation system, all major commercial flights are assigned to specific flight routes. The capacity of these routes depends on the minimum separation that is allowed between planes by air traffic controllers. Especially in those areas around airports where flight routes converge, this capacity may place important limitations on the flow of airplanes. In the U.S.A., these limitations became problematic as early as the late 1950s, when the expanding fleet of commercial jets created growing demand for movement along high-altitude flight routes. Expanding capacity along flight routes has been achieved primarily through the introduction of technologies that allowed controllers to reduce separation distances. First, successive generations of ground-based radar and, more recently, the introduction of global positioning systems have made it possible to track the position of airplanes more precisely, thereby reducing the margin of error and thus permitting shorter separations (Federal Aviation Administration, 1999).

The capacity of flight routes is still an issue, however. For example, the current trend by regional air carriers to retire propeller-driven planes that use low-altitude flight routes and replace them with regional jets is placing pressure on the capacity of high-altitude flight routes.

The capacity of airports places the most important constraints on the overall air transportation system. Capacity of an individual airport is defined as the number of operations (takeoffs and landings) that can safely take place in a given period of time. In the U.S.A., Hatsfield Atlanta International Airport and Chicago O'Hare Airport both handle about 180 operations per hour between 7:00 a.m. and 10:00 p.m., while most of the remaining top 25 airports handle in a range from 75 to 125 operations per hour (Federal Aviation Authority, 1999). These are average values, however, and actual capacities are probably somewhat higher.

Person capacity for airports is equal to the capacity times the number of passengers per airplane. The average size of airplanes has increased, allowing person capacity to grow faster than capacity. Also, in recent years the airlines have been able to reduce the proportion of vacant seats, thus accommodating rapid growth in enplanements despite much slower growth in capacity.

As is true for highways and transit, there may be a significant trade-off between attaining capacity flow and LOS. Air travel delays, defined as the difference between actual flight time and unimpeded time, are caused primarily by weather problems, but a significant share of delays are due to excessive airport volume.

The case of air travel illustrates that assessments of capacity on individual links (flight routes) and nodes (airports) does not necessarily describe the state of capacity in the entire system. Clearly there is little benefit in improving one component of capacity (for example, reducing separation between airplanes to increase route capacities) if other components of the system (airports) are unable to support the expanded flow. Also, individual capacity shortfalls tend to rebound throughout the system. For example, if a hub airport is unable to handle its scheduled flow, other airports will be unable to use their full capacity as incoming flights are delayed or cancelled. Thus the system capacity may be severely reduced by small reductions in the capacities of one or more individual airports.

3. Infrastructure capacity expansion: Alternative approaches

A common feature of the transport experience in most countries in recent decades has been the explosive growth of demand, which appears to swamp infrastructure capacity, in turn leading to congestion in significant parts of the surface and aviation networks. Congestion is increasingly severe in OECD countries in dense urban road networks, where cycles of road capacity expansion followed by induced traffic and congestion appear to repeat. Car ownership and use patterns are rising (with the historical gaps in car ownership and road use intensity between the U.S.A. and other OECD countries narrowing in the last decade) faster than the

Table 2
Indicators of Road Network and Use, Selected Countries, 1992

Country	Road network extent (a) (1000 km)	Car ownership (per 1000 people)	Vehicle km car use per capita	Public transport (% of total urban trips)
France	805	420	5824	12
(West) Germany	–	492	6228	11
Sweden	131	419	6713	11
Spain	318	335	2003	–
U.K.	–	375	6000	14
Japan	1099	313	3108	–
Canada	–	486	8746	14
U.S.A.	6233	600	9728	3

Source: Pucher and Lefévre, 1996.
Note: (a) For year 1987.

buildup in road networks. Urban road network mileage in the U.S.A. rose only 31% from 1980 to 1995 as compared with a 74% increase in urban car use in the same period (Small and Gómez-Ibáñez, 1999).

Table 2 provides some indicators of road infrastructure and use in selected OECD countries for a recent year. In large urban areas of low-income countries characterized by an explosive rise in motorization rates, traffic congestion is far more severe, not only because of low stocks of road capital and limited resources for new investment, but also because of organizational factors that reduce road capacity such as the poorly developed traffic management regimes and the wide mix of motorized (and even non-motorized) modes in the traffic stream.

Broad approaches available to address this gap between available infrastructure capacity and the travel demands placed on these facilities are: generation of new road capacity, augmenting the effective capacity of existing road capacity, and manipulating demand for road transport, either by reducing it or by altering its temporal and spatial patterns. Each of these will be considered in turn.

3.1. New infrastructure

The first approach to address the growing gap between infrastructure capacity and demand and thus reduce congestion is investment in new road capacity, a process that took place in the remarkable era of road building during the 1950s and 1960s in the U.S.A. and Europe. Investment in new roads has dropped off sharply in the last two decades, failing to keep pace with the growth of auto and truck traffic. Highway capital grew in the U.S.A. at an annual rate of 6.2% during 1952–59, steadily declined till 1979 when it did not grow at all, and resumed in 1982 at a 1.2% growth rate (Nadiri and Mamuneas, 1996). The European pattern is similar, with real annual investment falling since 1975 to less than 1% of GDP (Banister, 1993).

As noted earlier, a confluence of financial constraints and political factors has sharply slowed the pace of new road construction. First, on the cost side, construction costs in high-density urban sections are steeply climbing. Second, in the U.S.A. on the revenue side, the tax source of highway revenue has effectively declined – the fuel tax effective real rate having declined from 3.56 cents/veh-mile (1995 prices) in 1960 to 1.67 cents/veh-mile in 1995 (Small and Gómez-Ibáñez, 1999). Third, concerns with car exhaust pollutants, urban air quality, and potential contributions of car usage to greenhouse gases are powerful brakes on road improvement projects, requiring elaborate and time-intensive environmental assessments; finally, opposition from neighborhood groups who are in the right-of-way and the relevant and necessary transaction and mitigation costs can raise road project costs.

As the economic and political resources for road expansion have thus declined, the role of new infrastructure capacity in relieving congestion has declined. Private toll roads can offer some relief in specific situations; in some cases, new roads are being financed by fees or levies charged to developers. Yet there is a significant and growing gap between infrastructure capacity and demand. Hence an increasing interest in the set of complementary solutions described next that generate more capacity from existing transport networks.

3.2. *More effective capacity from current infrastructure*

The capacity of a current infrastructure facility can be augmented in two broad ways:

(1) use of physical or operational changes in the system by which the transport management is carried out; such measures are denoted as TSM; and
(2) use of the complementary information technology to increase the capacity and functionality of a transport right-of-way or terminal.

TSM measures

A variety of dynamic traffic management measures have been introduced in many countries to augment road capacity, relieve congestion, and improve safety. A system of lane allocation signs and speed control is used to regulate lane use and traffic speeds in peak periods. Ramp metering is used at several on-ramps in order to regulate the access to a limited-access highway. Other measures include one-way streets, controlled left lanes, arterial signal timing, and novel designs for entry, exit, and weaving. All such measures squeeze more capacity out of an existing facility. A recent report by the Mitre Corporation on the very basic forms of traffic management measures notes travel time reductions of 20% to 48% for freeway management systems including ramp metering, and from 8% to 15% for advanced management systems (Mitre Corporation, 1996).

The use of special lanes for high-occupancy vehicles (HOVs) such as car pools, vans, and buses is another transport system management measure with growing application in the U.S.A. To the degree that the travel time reductions in HOV lanes induce more people into car pools and transit vehicles, HOV lanes can also be thought of as a demand management measure, discussed in the next section. HOV lanes, according to Mohring (1979) and Small (1983), generate significant benefits – about half of those possible from fully optimal pricing, at least under favorable circumstances.

Intelligent transportation systems (ITS)

The increasing role played by information technologies in transportation equipment, infrastructure, and operations derives from the ability of these technologies to provide vital information, enhancing responsiveness and efficiency, and often makes possible other transportation innovations. For example, as noted earlier, current integrated navigational and air traffic control systems permit pilots in individual planes to exchange information with control centers, other aircraft, and infrastructural elements, and to process and act upon that knowledge in real time, thereby operating with shorter separations with greater safety, and expanding the capacity of the airspace system.

Applications of ITS to road transport range from technologies that provide additional information to drivers, helping them avoid congestion, to automated systems that take control of vehicles away from drivers in order to optimize highway flow. All of these technologies have two things in common: they make intensive use of information technology and they seek to expand the flow of vehicles that can be accommodated by existing highway lanes.

The simplest highway applications of ITS are those that come under the heading of advanced traveler information systems (ATIS). Their purpose is to collect information on traffic conditions and distribute it to drivers throughout the system so that they can alter their routes to avoid traffic delays. This does not augment capacity per se, but rather transfers traffic to roads with excess capacity. Traffic conditions can be collected by pavement loop detectors or GPS systems installed on a sample of "probe cars" (Bowcott, 1993). More advanced ATIS not only provide information on travel conditions, but also propose alternative routes to individual drivers.

Advanced travel management systems (ATMS) take ITS a step further by using the information that is collected to exert some control over the traffic stream – thus ATMS apply information technology to the implementation of TSM measures. An example is a ramp metering system that adjusts the rate at which cars are allowed to enter the traffic stream based on real-time information about traffic conditions. Again, this does not increase capacity in the conventional sense, but rather redistributes demand. It may, however, make it possible to increase the volume accommodated by a particular facility on a daily basis. ATMS can theoretically be extended to road pricing systems whereby tolls are adjusted continuously to reflect congestion levels.

The most far-reaching application of ITS is the automated highway system (AHS), whereby the control systems of each car are completely taken over by the automated system as it enters the highway on-ramp, and relinquished to the driver only after the car enters an off-ramp. Besides making the driver's work easier, the main benefit of the ATS is that it packs cars into platoons with short following distances, thereby increasing the capacity of the highway by a factor of up to three. This requires a complex array of information and control equipment both for the roadway and for the vehicle, but can expand capacity more economically than by the addition of more lanes. Proof-of-technical-feasibility tests have already been conducted in the U.S.A.

ITS is now a major component of strategies to address highway capacity shortfalls in the U.S., Europe, and Japan. Some of the anticipated benefits of ITS, however, have been called into question. For example, it is not necessarily true that the significant benefits accruing to the small number of drivers who are currently participating in ATIS applications will accrue to all drivers as the systems become more widespread. (That is, the value of the information may in part be due to its exclusive nature.) The AHS has been the subject of particular criticism. The benefits of such systems have been shown to decline quickly with deviation from ideal operational conditions (DelCastillo et al., 1995) and it is unclear how the volumes of traffic on and off such a highway can be absorbed by local and arterial roads (Ewing, 1998).

ITS also apply to public transportation. For example, advanced electronic fare collection systems help speed vehicles through stations and bus stops, where a large proportion of delays normally occur. Also electronic tracking systems can be used to monitor vehicles, improving flow by helping to maintain optimal headways and quickly identifying incidents that cause delays (California Advanced Public Transportation System, 1994).

3.3. Manipulating infrastructure demand

A third approach to address the infrastructure capacity–demand imbalance is to lower the level of demand, modify its temporal pattern, or shift the spatial occurrence of demand. Some of these measures have impacts in the short run, while others operate in the long run. The former are exemplified by transport demand management (TDM) instruments, and pricing policies that modify the economic incentives governing road travel. Land use regulations and promotion of public transport illustrate the long-run instruments.

Transport demand management

These measures aim to influence the level, timing, and location of travel and are a heterogeneous lot. Some of them are directed at congestion management, while others modify the economic incentives confronting the traveler. The latter group is treated separately in the next section.

Metropolitan planners have used TDMs in their efforts to alleviate congestion. In the U.S.A., TDMs were a key component of travel and emission reductions envisaged in the Clean Air Act and the Intermodal Surface Transportation Efficiency Act (ISTEA), and its successor, the T-21. A number of these measures and some pricing policies are listed in Table 3. TDMs are used to encourage changes in mode choice (shifting trips from auto to transit), and to promote, in the work day or work week, scheduling by commuters. The HOV, noted earlier as a TSM measure, can also be viewed as a TDM in view of its capability to shift trips from single-occupancy to high-occupancy vehicles and raise effective infrastructure capacity.

Pricing measures

By modifying the costs of travel and associated costs, pricing measures can reduce the demand for trips of many types, and by promoting more flexible spatial and temporal patterns, can distribute the use of road capacity over time and space. Under these circumstances, individuals may utilize public transport or car pools when possible, thereby alleviating to a degree the capacity–demand imbalance.

Parking charges and pricing of congested roads are two such measures. Shoup (1994) notes that in the U.S.A. parking is free for 99% of all auto trips, and demonstrates from his case studies that charging realistic market rates for parking can reduce the proportion of commuters who drive alone by an average of 25%. Shoup's proposal of "cashing out" employer-provided free parking is handicapped by the fact that U.S. income tax laws tax cash subsidies but not the free parking provided by employers.

Congestion pricing, while popular in the economic literature, does not receive much public support. Experience in Singapore (charge for entering the central area during peak hours) since 1975 suggests that congestion pricing does lower congestion. However, major plans for congestion pricing in London, Cambridge, and Hong Kong have been abandoned for a variety of reasons. Road pricing experiments elsewhere (Oslo, Stockholm, etc.) have focused on revenue raising as well as on congestion relief. Recent interest in this measure derives from a public

Table 3
Transportation demand management (TDM) Measures and their effectiveness

TDM	Implementation mechanism	Examples	Vehicle mile travel (vmt) reduction (%): Range of study estimates	Trip reduction (%): Range of study estimates
Emissions/vmt tax	Economic incentive	Charged on gasoline or through differentiated registration fee	0.2–0.6	0.1–0.9
Buy-back of older cars	Economic incentive	Purchase of gross polluters above market value for scrapping	NA	NA
Area-wide ridesharing	Information and education	Commuter-matching programs and databases	0.1–2.0	0.5–1.1
Parking pricing (work)	Economic incentive	Charging full parking costs, parking "cash-back" program	0.5–4.0	0.4–4.0
Congestion pricing	Economic incentive	Charge for entering congested zone, congestion-based road pricing	0.2–5.7	0.4–4.2
Park-and-ride lots	Public facilities improvement	Peripheral lots at transit stops or as car and van pool staging areas	0.4 –0.5	0.0
Parking pricing (non-work)	Economic incentive	Charging full parking costs, parking "cash-back" program	3.1–4.2	3.9–5.4
Telecommuting	Information and education	Employees work at home or at "satellite" offices	0.0–3.4	0.0–2.8
Compressed work week	Information and education	Employees work more hours on fewer days	0.0–0.6	0.0–0.5

Source: Bureau of Transportation Statistics (1996).

sector need for resources for infrastructure expansion, and the advent of new IT that permits toll collection without tollbooths.

Table 3 summarizes the characteristics and effectiveness of a set of TDMs and pricing policies. Some TDMs (including employee trip reduction, and bicycle/pedestrian facilities) have relatively little potential for reducing vehicle miles traveled (vmt) and were not cost-effective. Economic incentives (congestion pricing, parking pricing) ranked high in reduction of vmt and cost-effectiveness. While taxes, tolls, and charges have low system-wide costs, they pose high direct costs to some individuals and thus energize political resistance to the measure. It is important to note that these TDM and pricing instruments have been evaluated individually in Table 3. Their potential larger joint effects in reducing vehicle miles traveled, when they are combined in practice, are yet to be assessed.

3.4. Promoting public transit

The idea is to enhance the quality of service in transit so as to attract travelers out of the cars during peak hours and thereby enlarge effective road capacity. The recent expansion and improvement of service quality of commuter rail, light rail, and other transit *has steadied* the aggregate level of transit ridership in the U.S.A. – halting the precipitous decline of transit ridership that started in the 1960s. By not dropping the 12 million daily transit trips on the roads, the current transit system has been helpful to vehicle flows on the road. Its potential beyond that as a major source of relieving congestion and augmenting effective road capacity is limited in the U.S.A., in view of the dispersed urban settlement patterns and the existing economic and political incentives for continuing sprawl.

3.5. Land use policies and regulations

Since the spatial pattern of activities and land uses determines the demand for urban travel, policies, and regulations, one way to influence travel demand is via changing land use patterns. A variety of post-World War trends and public policies in the U.S.A. – rising affluence, the Interstate Highway program, housing mortgage and tax subsidies, municipal mercantilism, and the "Tiebout" effect in metropolitan location decisions, etc. – have combined to promote a highly dispersed land use pattern that disadvantages transit, and is best served by autos. The economic and political incentives for this "sprawl" development continue. It is not surprising that the conclusion drawn by most analysts is that the anemic land use controls being discussed will be ineffective in affecting the demand for road capacity significantly.

4. Conclusion

The physical capacity of an element of an infrastructure system is measured as the number of vehicles or people that can pass by a given location per unit of time. For planning purposes, simple measures of capacity must be augmented by information on the level of service to ensure that observed or anticipated demand is served with acceptable levels of comfort and speed. Capacity analysis is critical to solving the fundamental problem of choosing levels of capacity provision to meet efficiently the continuously growing demand for transportation services.

Capacity is an issue of increasing concern to planners and policy-makers as all modes of transportation are typically experiencing high levels of congestion. Methods of analysis have been developed to determine the amount of a particular type of infrastructure that is needed to serve projected demands. The traditional approach of building more infrastructure to address capacity shortfalls, however, is currently being challenged due to the phenomenon of induced travel and the difficulty of siting facilities such as highways and airports that have negative local and environmental impacts.

The focus of capacity analysis is shifting toward ways of increasing the capacity of systems within existing physical dimensions by improving the efficiency of flow through them. Novel methods of system management that direct traffic flows into more nearly optimal patterns constitute one approach. Another complementary approach is the application of information technologies to help vehicle operators make better decisions, to allow more centralized control of geographically scattered vehicles, and possibly even to automate traffic flow. Road pricing seeks to optimize traffic conditions by creating an efficient market for access to road infrastructure. Finally, land use control and other demand-side measures seek to redress the imbalance between capacity and demand by contracting the latter rather expanding the former.

References

Banks, J.H. (1989) "Freeway speed–flow–concentration relationship: More evidence and interpretation", *Transportation Research Record*, 1225:53–60.

Banister, D. (1993) "Investing in transportation infrastructure", in: B. Banister and J. Berechman, eds., *Transport in a unified Europe*. Amsterdam: North-Holland.

Bovy, P.H.L. (1998) "Research questions on motorway traffic flow in The Netherlands", in: P.H.L. Bovy, ed., *Motorway traffic flow analysis*. Delft: Delft University Press.

Bowcott, S. (1993) *The Advance project*. New York: Parsons DeLeuw.

Bureau of Transportation Statistics (1996) *U.S. Department of Transportation Statistical Annual Report*. Washington, DC: U.S. Department of Transportation.

California Advanced Public Transportation System (1994) Technical Assistance Brief #4.

DelCastillo, J.M., D.J. Lovell and C.F. Daganzo (1995) "Technical and economic viability of automated highway systems: Preliminary analysis", *Transportation Research Record*, 1588:130–136.

Downs, A. (1992) *Stuck in traffic*. Washington, DC: Brookings Institution.

Ewing, G. (1998) "The impacts of technology on vehicle emissions and congestion reductions", *Energy Studies Review*, 8(3):279–284.

Federal Aviation Administration (1999) *1999 Aviation capacity enhancement plan*. Washington, DC: U.S. Department of Transportation.

Government of France, Ministry of Transport, Office of Highways (1980) "Instructions on the methods for evaluating the economic effects of highways in rural areas".

Hall, F.L. and L.M. Hall (1990) "Capacity and speed–flow analysis of the QEW in Ontario", *Transportation Research Record*, 1287:108–118.

May, A.D. (1990) *Traffic flow characteristics*. Englewood Cliffs: Prentice-Hall.

Mitre Corporation (1996) "Intelligent transportation infrastructure benefits: Expected and experienced", prepared for the Federal Highway Administration.

Mohring, H. (1979) "The benefits of reserved bus lanes, mass transit subsidies, and marginal cost pricing in alleviating traffic congestion", in: P. Mieszkowski and M. Straszheim, eds., *Current Issues in Urban Economics*. Baltimore: The Johns Hopkins University Press.

Nadiri, M.I. and T.P. Mamuneas (1996) "Contributions of highway infrastructure to industry and aggregate productivity growth", report prepared for the Federal Highway Administration Office of Policy Development.

OECD (1983) *Traffic capacity of major routes*. Paris: Organisation for Economic Co-operation and Development.

Pucher, J. and C. Lefevre (1996) *The urban transport crisis in Europe and North America*. Houndsmills: Macmillan

Shoup, D. (1994) "Cashing out employer-paid parking: A precedent for congestion pricing?", *Transportation Research Board Special Report*, 242:152–199.

Small, K.A. (1983) "Bus priority and congestion pricing in urban expressways", in: T.E. Keeler, ed., *Research in transportation economics*. Greenwich, CT: JAI.

Small, K.A. and JA. Gómez-Ibáñez (1999) "Urban transportation", in: P. Cheshire and E. Mills, eds., *Handbook of regional and urban economics*, Vol. 3. Oxford: Elsevier.

Standing Advisory Committee on Trunk Road Assessment (1994) *Trunk roads and the generation of traffic*. London: HMSO.

Transportation Research Board (1998) "Highway capacity manual", National Research Council, Washington, DC, Special Report 290, 3rd edn.

Vuchic, V.R. (1992) "Urban passenger transportation modes", in: G.E. Grey and L.A. Hoel, eds., *Public transportation*, 2nd edn. Englewood Cliffs, NJ: Prentice Hall.

229-40

Chapter 14

TRANSPORT SAFETY

IAN SAVAGE
Northwestern University, Evanston, IL

R41

(05)

1. Introduction

Transport has always been associated with the risk of death, injury, and the destruction of property. From the earliest days people have been thrown from horses and fallen out of canoes. The advent of mass transport simply turned private, localized grief into public spectacles in the form of shipwrecks, train crashes, and aviation disasters. While it is tempting to think that the risks are much higher in our technological world where we speed across the ground at 100 mph or more or defy gravity in an aluminum tube some five miles above the surface, you would be wrong. Travel on the rudimentary roads of two hundred years ago was very hazardous, as was venturing out on the ocean without proper navigational aids, or traveling on river steamboats that routinely exploded and sank.

The improvements continue. The risks in all modes of transport have fallen by at least a half since the 1950s. Yet the absolute level of harm is still very high. Even in a technologically advanced economy such as the U.S.A., one in 6000 of the population dies each year due to transport crashes (note that safety professionals prefer the word "crash" to "accident" because the latter suggests that occurrence is due to pure fate and cannot be influenced by human decisions). The annual death toll of about 44 000 represents half of all accidental fatal injuries when one includes workplace injuries but excludes homicides and suicides.

2. Measures of safety

Enumerating the harm caused by crashes is difficult. While one can be fairly certain about the number of fatalities, there is underreporting of non-fatal injuries and damage to property. In addition, while research has allowed reasonably accurate enumeration of the monetary costs of lost productivity and out-of-pocket expenses, the valuation of the associated pain and suffering is controversial. After correcting for undercounting, Miller et al. (1991) estimate that the direct costs associated with

Handbook of Transport Systems and Traffic Control, Edited by K.J. Button and D.A. Hensher

highway crashes in the U. S. were equivalent to 2.6% of gross national product. If a value is attached to pain and suffering and lowered quality of life for those injured, the total costs are much higher and equivalent to a staggering 6.6% of GNP.

Reasonably reliable cross-modal comparisons can only be made on the basis of fatalities. Table 1 shows typical annual modal fatalities in the 1990s for the U.S.A., a country of about 260 million people. The averaging gives a better representation for modes where there are infrequent high-fatality events such as major train wrecks and aircraft crashes. Analytically it is useful to divide the fatalities into three groups. The first we will term "private user" modes such as walking, cycling, driving, general aviation, and recreational boating, where the user is often the operator of the vehicle or is a friend or relative of the operator. The second is "commercial transport," where passengers and freight shippers contract with corporations. The third are those situations where private modes and commercial operators collide with each other, such as when trucks collide with cars, or trains collide with cars at grade crossings.

Only 15% of the fatalities involve commercial carriers. The remainder occur when private pilots, mariners, and highway users are involved in single-vessel/aircraft/vehicle crashes or collide with other private users. Furthermore, most of the risk in commercial transport does not fall on those directly involved in production and consumption of these services. The majority of the victims are road users and pedestrians involved in collisions with trucks, and trespassers and grade crossing users who suffer collisions with trains. In the global scheme of things, the number of commercial passenger fatalities is quite small. The number of true bystanders who get killed each year is also relatively small. However, one should not underestimate the public outrage associated with crashes that lead to the release of explosive or toxic substances, or oil spills that defile places of natural beauty.

Table 1
Average annual fatalities in the U.S.A., 1990–98

	Crashes solely involving private users	Collisions with commercial carriers	Commercial transport passengers	Commercial transport employees	Bystanders to commercial crashes
Cars/motorcycles	29 650	4500	–	–	–
Pedestrians	6 100	1120	–	–	–
Trucks	–	–	–	650	
Aviation	720	–	80	15	4
Railroads	–	–	10	45	1
Bus/subway	–	–	20	10	0
Maritime	820	–	0	170	0
Pipeline	–	–	–	15	5
Total	37 290	5620	110	905	10

Table 2
Modal passenger and employee risk

	Passenger fatalities per billion passenger miles 1990–98	Annual employee fatalities per 1000 employees 1998
Motor vehicle driving	9.58	–
Maritime	–	0.28
Trucking	–	0.21 (includes warehousing)
Railroads	0.79	0.07
Aviation	0.18 (commercial)	0.06
Bus	0.15	0.18 (transit and taxi)

Of course, Table 1 says nothing about risk, as exposure to crashes varies widely across modes. Table 2 shows driving and commercial passenger fatalities relative to passenger miles, and employee fatalities relative to total employees. Bus and commercial aviation have the best passenger safety records, at about one fatality for every five billion passenger miles. Riding the train is four times as risky, and driving is twelve times more risky than taking the train. However, this comparison is somewhat misleading. While passengers in commercial transport are victimized in a somewhat random fashion, driving risks are heavily dependent on the characteristics of the driver. A disproportionate number of auto crashes involve young male drivers and people who have been drinking. The fatality risk for sober middle-aged drivers is about 75% of the average driving risk.

Employment in transport is relatively risky as it entails working outdoors with heavy, moving equipment, often in hostile weather conditions and far from immediate medical attention. The maritime, trucking, and warehousing industries have the highest levels of employee risk. These risks are substantial, with employees facing more risk than they would if they were working in construction (0.14 per 1000 employees) and comparable to the amount of risk for miners (0.23 per 1000). The rate shown for buses is defined in government publications as workers in "local transport," and is unusually high due to the elevated risk of homicide of taxi drivers. In contrast, employees in the railroad and aviation industries face less risk, although the risks are still twice those in manufacturing (0.03 per 1000). It is not surprising that much of the public concern for safety has originated from organized labor.

3. The transport safety "problem"

While the level of harm is substantial, one cannot tell from risk data whether transport safety is "a problem." Ever since the dawn of civilization, humans have valued the ability to travel and to ship their goods, and have been prepared to endure the inherent risks. To the economist, if a person knows of, and can evaluate, both the benefits and the risks, and still decides to travel then there is no inherent problem. Even in the labor market, risks are not unreasonable if workers are aware of the risks of different jobs, and decide of their own free will to work in relatively risky occupations because of their own tastes and because riskier occupations pay premium wages.

Risk preferences can change over time. Innovative technology such as navigation systems and new materials has allowed firms to provide more safety at a lower or comparable cost than was possible a generation ago. The same is true for highway and automotive design. In addition, as a country becomes more wealthy, its citizens demand that all types of risks – transport, medical, food quality – are mitigated to ensure longer life expectancy. Consequently, risks that were acceptable at mid century would be regarded as unacceptable today. Transport crash risks are much higher in developing countries because society has different priorities for the use of its scarce resources, such as providing for basic education and health care. Consequently increasing wealth and product innovation mean that one should expect that there will always be pressure to reduce transport risks.

To summarize the discussion, if users of the transport system are fully aware of the risks, irrespective of their magnitude, and voluntarily accept them then there is not a "transport safety problem." Consequently society intervenes only minimally in the decisions made by private aviators and recreational boaters. It is assumed that these individuals are aware of the risks, and that in most cases they themselves are the only victims. There is only public concern when private planes crash on the houses of innocent third parties or when large amounts of public money are used to conduct search and rescue operations at sea. Likewise, society is less motivated to intervene in risks incurred during employment. Workers are generally assumed to have the expertise to appraise the risks they face. This is especially true in industries such as transport, where the threats are primarily due to observable mechanical rather than unobservable toxic and health-related sources, and where unions are highly motivated to investigate and report on job risks to their membership.

Society will be motivated to intervene when the market breaks down. Over the past 40 years academics in economics, law, and psychology have developed models to formalize the various possible market failures and examine their consequences and how they can be mitigated (see Savage, 1999, for a full set of basic references). There are six possible causes of market failure. The first occurs when people are not knowledgeable about the risks that they face. This is more likely in commercial transport, where passengers consume relatively rarely and do not have the technical expertise to understand the mechanics of the risks or have the knowledge and access to understand crash data. Shippers of freight, while removed from the actual operations, are more knowledgeable because they are typically repeat purchasers who deal with carriers on a daily basis to settle claims for minor loss and damage to their products. Because the losses are material rather than human, shippers can afford to take a rather analytical approach to assessing risks. Private users are generally better informed. It is immediately clear to drivers when they are driving while drunk or tired, or driving at higher speeds than the conditions might suggest are prudent. Most private drivers, pilots, and mariners are also aware of the dangers of various weather conditions, and the prudent action they should take. It must be pointed out, though, that private users do have to choose between purchasing various types and brands of vehicles, whose safety properties may not be apparent.

The second possible failure is that even fully informed people may make poor choices due to cognitive processes. People have a tendency to overestimate the possibility of low-probability events and events that kill multiple people at any one time (Lichtenstein et al., 1978). They are also particularly fearful of life-threatening events where they have no control over the outcome. Therefore aviation risks cause a disproportionate amount of fear compared with auto driving, where many drivers feel that they have the skill to mitigate or avoid hazardous situations. Working in the opposite direction is the possibility that some people underestimate risk partly because the consequences are too horrendous to contemplate, and partly because of a feeling of invincibility that bad outcomes "will not happen to me." This is particularly the case for young male drivers, where there is the additional factor that risky behavior, such as driving fast, powerful cars, may actually be a positive attribute of transport. Even for more mainstream drivers, there is evidence that risks are underappreciated. There is considerable evidence that most drivers believe that they are more skillful and safer than the average driver! Because most auto trips are completed without a crash (the U.S. average is one crash per ten years) there is daily reinforcement of drivers' beliefs. It is not an understatement to suggest that much of society's intervention in the safety market is to protect people from themselves rather than from avaricious carriers and third parties.

For commercial carriers the likelihood that most, or all, of your customers will be imperfectly informed allows the possibility of a third market failure. Safety has the characteristic that the costs of preventing crashes are incurred "up front," in the form of investment in equipment and staff training, whereas the "benefits" of a reduced number of crashes occur at unpredictable points in the future. Carriers who are very myopic in their preferences for short-term profits can reduce expenditures on crash prevention yet continue to masquerade as having high safety and charge a premium price. As crashes occur rarely, even for careless carriers, it may take some time before consumers become aware and either shun the carrier or demand a lower price. The victims of this "cheating" can be either the carrier's customers and/or society if the consequences of a crash affect bystanders. Inexperienced new entrants may also be myopic if they are well aware of the costs of crash prevention but unaware of the likelihood of crashes and how their decisions affect this probability. It is not surprising that public policy has focused on ensuring that new entrants are well qualified, and policing existing carriers to detect slipping safety.

The fourth possible failure occurs when crashes impose costs on innocent bystanders ("externalities"). Examples include oil spills, the release of explosive or toxic materials, planes crashing on peoples' houses, and innocent pedestrians hit by vehicles mounting the curb. Work by psychologists suggests that non-participants value the risks that they face much more highly that those who benefit from the risk either as users or as employees. Therefore, although the law has long recognized the right for these victims to claim compensation, the public outcry tends to be much in excess of the harm caused. One only needs to look at the consequences of the wrecks of the oil tankers *Exxon Valdez* and *Torrey Canyon* or the railcar explosion at Mississauga, Canada to appreciate this point. Indeed it is possible to argue that most of the safety concern about freight transport is solely due to the fear of externalities. While it is a legitimate issue between the carrier, its employees, and the affected shippers when ships sink on the open sea and freight trains wreck on the private right-of-way, the public only becomes concerned when bystanders are affected.

The fifth possible failure is associated with collisions between private users, and when private users have crashes with commercial carriers. These collisions, which represent more than 60% of annual fatalities, are called "bilateral crashes" because the actions of both parties influence the probability of occurrence. There is a complicated literature in law and economics that discusses the socially optimal actions of both parties, and the role of legal mechanisms to compensate victims, penalize perpetrators, and generally give both parties the correct incentives to take appropriate "care" (see Shavell, 1987, for an excellent summary). It is not an exaggeration to say that much of the development of accident law and insurance has occurred because the automobile allowed individuals unprecedented opportunities to interact with each other in a potentially harmful way.

The sixth and final possible failure is concerned with the amount of competition. As will be discussed later, individual drivers, passengers, and shippers may have varying tastes for the amount of safety that they desire. Individuals will prefer to purchase a transport service with safety characteristics close to their own tastes. In some markets, such as trucking, there are thousands of carriers of all ilks and shippers can find a service which matches their taste. However, in some other markets there is less competition and such a matching of tastes is less likely. The problem is compounded where a common safety-critical infrastructure, such as air traffic control services or airport runways, is used by all carriers or where a common highway infrastructure is used by all drivers. In these situations a common level of safety may be provided which may not be to the taste of those desiring very high or very low levels of safety.

Table 3
The magnitude (a) of the six market failures by mode

	Imperfect information	Cognitive failure	Carrier myopia	Externalities	Bilateral crashes	Imperfect competition
Private driving	*	***	NA	*	***	*
Private aviation and boating	Few failures					
Commercial passenger	***	***	***	*	**	**
Road freight	*	None	***	***	***	*
Maritime freight	*	None	***	**	*	None
Rail freight	*	None	**	**	***	***
Pipelines	*	None	**	***	None	***

Note: (a) *, limited failures; **, some failures; ***, substantial failures.

The applicability and magnitude of the six market failures vary significantly by mode. I have attempted to provide a summary in Table 3 using a star rating. It is only by recognizing where the market failures occur that there can be intelligent public policy prescription. Policy responses need to be tailored to the root causes of the problem in each mode. The remainder of the chapter is devoted to a description and evaluation of a century or more of public intervention in the transport safety market. Because the root causes of the safety problem are so different, we divide our discussion into two parts: private automobile driving, and commercial transport.

4. Public policy regarding private automobile driving

Traditionally public policy has been directed at the minority of drivers who do not appreciate the risks of driving and do not conduct themselves in a prudent manner (see Evans, 1991, for a comprehensive discussion of driver behavior). These drivers can be divided into four categories. The first are young drivers, especially males. Their crash rates are much higher than would be explained purely by inexperience behind the wheel. Of course, this age group also exhibits risky behavior in other ways such as their disproportionate involvement in criminal activity. There is a large literature discussing the effectiveness of new-driver education and testing and possible options of restricting the types of vehicles and the times of day at which young people can drive (the journal *Accident Analysis and Prevention* is a good source for this and all other aspects of highway safety).

The second group is older drivers. There is evidence that driver performance starts to deteriorate at the age of 50 and gets markedly worse after 65. However, societal risks are moderated as older people drive less and elect to drive at times of day and in places which are generally safer. Nevertheless this is a growing area of concern as the "baby boomers" move into retirement and medical advances prolong life expectancy. Therefore, we can expect to see calls for additional research regarding the appropriate response by licensing authorities.

The third group comprises people who drive while under the influence of alcohol. Some proportion of these people have a chemical dependency. However, the majority are more social drinkers who are heavily influenced by societal attitudes that condone and even encourage drinking as a desirable leisure pursuit. Because alcohol plays a large part in the social activities of young people, it worsens their already poor driver behavior. Since the 1960s most countries have attacked the problem by establishing laws that specify the maximum permissible blood alcohol content. In some countries it is legal to randomly test motorists, while in others police officers can only do so with cause (see Zaal, 1994, for a review of the law enforcement literature). While these programs have been very successful, Evans (1991) comments that a significant effect has been the ongoing change in social attitudes that now eschew excessive public drunkenness. If in the next 30 years attitudes toward alcohol change in a similar way to the change in attitudes toward tobacco in the past 30 years, then much of the traffic safety problem will go away.

The final group consists of drivers who fit into none of the above categories yet seem to be more accident-prone (or exhibit "differential crash involvement") than other drivers. This is a very controversial subject, especially when elevated driving risk is linked to other risky or deviant social behavior. Nevertheless a considerable amount of police time is spent in identifying these drivers by means of enforcing speed limits or catching drivers who do not obey signals at intersections (Zaal, 1994). Presumably the fines and possible loss of driving privileges act as a deterrent to some drivers, as does the insurance industry, which devotes considerable energies to assessing premiums that reward responsible drivers and penalize poor ones (Dionne, 1992).

Influencing driver behavior is just one part of the issue. Haddon (1972) introduced the useful concept that traffic safety can be characterized by a 3 × 3 matrix. The categories on one axis are the driver, the vehicle, and the highway. The other axis is composed of actions before a crash ("crash avoidance"), the crash phase, and the postcrash phase. All of the discussion so far has concentrated on the top left-hand cell of this matrix. Starting in the 1960s, there was a conscious move to attack traffic safety via other cells in this matrix. For example, a significant reduction in fatality risk has come from improved medical response in the postcrash phase. Highway design is clearly important both in promoting crash avoidance and in mitigating the harm when a crash occurs (Lamm et al., 1999). A controversial aspect of highway design has been the assignment of speed limits, which in the U.S.A. were reduced during the energy crisis of the 1970s and only liberalized in the past decade. The subsequent discussion of the relationship between risk and speed has led to a quite acrimonious debate between some economists and highway and automotive engineers.

Academic interest has been generated by regulations that imposed automotive solutions that sought to promote crash avoidance and increased survivability in the event of a crash. An example of the first is center-mounted rear brake lights at eye level. These are designed to mitigate the risks of driving too close to the vehicle in front by providing for a quicker response if the vehicle in front brakes. Examples of the second are requirements for collapsible fenders, strengthened passenger cabins, seat belts, and air bags. All of this equipment adds to the cost of new vehicles. Seat belts also require the driver to devote time to fasten them, and police resources must be deployed to enforce seat belt laws. Consequently, this presented the opportunity for economists to enter the debate to discuss whether the benefits were worth the costs (Loeb et al., 1994; Crandall et al., 1986; Blomquist, 1988). Not all changes have promoted safety. Requirements that automakers increase fuel efficiency have led to reductions in automotive size and weight which have increased the risk to occupants. Even the market-driven reversion to heavier sports utility vehicles (SUVs) is not without safety consequences, as these vehicles have higher centers of gravity and are more susceptible to roll-over crashes.

Economic evaluation is complicated and controversial because it involves assigning monetary values to the number of lives saved and injuries avoided, the extra time involved in buckling seat belts, and, in the case of changes in speed limits, longer or shorter journey times. In addition, it is very difficult to discern the effect of one particular policy measure as other contributory factors are usually changing at the same time. One particular problem is known as "risk compensation" by economists and as "human behavior feedback" by automotive engineers and human factors analysts. For example, it is well known that in the event of a crash a seat belt will reduce fatality risks by at least 40%, primarily because of reduced risk from being ejected from the vehicle. However, the national fatality risk reduction, even assuming everyone wore their belts, may be much less because drivers may partially compensate for the increased safety by, for example, driving faster or talking on a mobile phone. The magnitude of this effect provokes heated debate between automotive engineers and economists (see Evans, 1991).

5. Public policy regarding commercial transport

Before one can analyze public policy for commercial transport, it is important to gain an understanding of the marketplace for safety. Customers make consumption decisions regarding mode and carrier to patronize based on a whole array of considerations including price, speed, frequency, comfort, and safety. Safety is valued by consumers yet costly to provide. It is likely therefore that consumers will choose a lesser amount of safety than is technically possible because it saves them time and/or money, and be quite happy with this choice. If this was not the case, one would assume that consumers would demand five pilots in every cockpit and that trains operate at a maximum speed of 40 miles per hour.

Not all customers value safety equally. In freight transport, it is reasonable to suppose that shippers of expensive, delicate goods will be more prepared to pay a premium to obtain high-quality transport than would, say, shippers of inexpensive, bulk materials. While it is more controversial, it is quite reasonable to assume that passengers have a variety of safety tastes, albeit those demanding high safety may be wealthier individuals who can afford a premium price. A basic result in economics is that a "vertically differentiated" market will emerge with some carriers offering high quality at a premium price, and others offering lesser quality at lower prices (Shaked and Sutton, 1982). If consumers are knowledgeable about the array of services on offer and freely choose among them, an optimal situation occurs. Such a situation occurs in trucking and bulk maritime transport. To many lay observers the fact that some carriers can be shown to have worse safety records than their peers is typically taken as evidence of a market failure. On the contrary, when customers are fully informed, this should be taken as a sign that the market works.

The situation in passenger markets is a little different. Passengers are typically not as well informed and dispassionate about their choices as professional shipping managers. If they cannot recognize which carriers are offering premium quality, there is little incentive for carriers to differentiate their product. It is not surprising that most mainstream airlines, for example, offer roughly identical safety. However, even in these markets there exist low-price and low-quality choices, such as charter carriers, that some consumers decide to patronize. Of course this may, in itself, cause a problem in that some impecunious passengers would voluntarily select carriers that the more mainstream society would regard as posing unreasonable risk. Society may, and does, act to ban such carriers and consequently restricts the mobility of some people.

So how much safety should a carrier provide? For a monopolist or an infrastructure supplier, the answer is quite clear. The marginal revenue from providing a higher-quality service should be equated with the marginal cost of doing so (see the excellent empirical paper by Evans and Morrison, 1997). In a vertically differentiated market, the answer is less clear in that a carrier has to decide how to position itself in the market vis-à-vis other carriers. This is an area that has not been well researched. The somewhat controversial suggestion that carriers make conscious choices on the inherent safety of their production processes and the layers of "defenses" that they build into their systems to protect against errors has gained prominence in the past decade. The work of James

Reason and others has popularized the concept of "organizational accidents," which is totally in keeping with economists' thinking. Reason's co-authored book on aviation safety (Maurino et al., 1995) is required reading for those interested in the safety of all modes.

As described earlier, when some or all customers are imperfectly informed, there is the opportunity for some carriers to act in a myopic fashion. Some established firms can "cheat," and inappropriate decisions can be made by inexperienced new carriers. The net result is that some passengers end up consuming a service of a lower quality than they expect, and pay with their lives, and third parties and bystanders suffer from crashes involving myopic freight carriers. There is plenty of anecdotal evidence of this behavior in all modes. A number of academic studies have been conducted which primarily focus on whether new entrants or firms close to bankruptcy have elevated crash rates. It is reasonable to suggest that the fear of myopic behavior is behind most of the public policy action on safety, and specific instances of myopic behavior are usually the catalysts for tightened regulations and/or tougher enforcement action.

Of course, not all carriers will act in this way. Some will be constrained by morality, others will take a long-term perspective and not wish to lose future custom if their reputation is sullied. In addition society has instituted a number of institutional processes which act as behavioral modifiers. Injured parties can bring legal suits, and insurance companies who write policies to protect against legal settlements have some incentive to monitor and constrain the behavior of their policyholders (Shavell, 1987). However, these processes are not panaceas. In most markets there is a need for additional intervention. One option is to attack the root cause of the problem by providing information to make consumers better informed. This logical policy tool has traditionally been underutilized. But this may change. The revolution in information technology has decreased the cost of providing timely information on the safety performance of carriers directly to consumers. Of course, there are many unresolved questions such as which data are the most informative, how to interpret data on rare events, and the thorny issue that for cheaters, data on past performance are not an accurate predictor of the future.

A more common policy is to directly regulate the conduct of carriers. Safety regulation is both highly visible and controversial. High-profile disasters frequently lead to calls that "something should be done." Consequently there is a plethora of regulations in every mode that define acceptable safety standards, provide for inspections to certify the competence of new entrants, and check for continued compliance. Traditionally the designated standards have been expressed in terms of the quality and quantity of staff and equipment, which are easy to inspect, it is easy to and confirm compliance with standards. An alternative approach is "performance standards," which designate minimum acceptable crash rates. These have several advantages. First, carriers are allowed to use entrepreneurial skill to produce the desired level of safety at minimum cost. Second, new technology can be introduced quite quickly, unhindered by the need to change the regulations. Third, there is less susceptibility to politicking by avaricious parties who wish to use regulation of safety inputs to preserve old working practices, exclude new entrants, or promote the use of their own specific safety-related product. There are moves toward performance standards and this will provide a fruitful area of research for the coming years (see Savage, 1998, for a discussion with regard to the freight railroad industry). A further issue is that economic theory suggests that in effective law enforcement there is a trade-off between the frequency and nature of inspections and the scale of penalties assessed (Polinsky and Shavell, 2000). There are research opportunities to discover whether regulatory agencies currently have the balance correct. Because there are many unanswered questions, safety policy analysis and prescription remain an exciting multidisciplinary field of research.

References

Blomquist, G.C. (1988) *Regulation of motor vehicle and highway safety*. Boston: Kluwer.
Crandall, R.W., H.K. Gruenspecht, T.E. Keeler and L.B. Lave (1986) *Regulating the automobile*. Washington, DC: Brookings Institution Press.
Dionne, G., ed. (1992) *Contributions to insurance economics*. Boston: Kluwer.
Evans, A.W. and A.D. Morrison (1997) "Incorporating accident risk and disruption into economic models of public transport", *Journal of Transport Economics and Policy*, 31:117–146.
Evans, L. (1991) *Traffic safety and the driver*. New York: Van Nostrand Reinhold.
Haddon, W., Jr. (1972) "A logical framework for categorizing highway safety phenomena and activity", *Journal of Trauma*, 12:193–207.
Lamm, R., B. Psarianos and T. Mailaender (1999) *Highway design and traffic safety handbook*. New York: McGraw-Hill.
Lichtenstein, S., P. Slovic, B. Fisthhoff, M. Layman and B. Coombs (1978) "Judged frequency of lethal events", *Journal of Experimental Psychology: Human Learning and Memory*, 4:551–578.
Loeb, P.D., W.K. Talley and T.J. Zlatoper (1994) *Causes and deterrents of transportation accidents: An analysis by mode*. Westport, CT: Quorum Books.
Maurino, D.E., J. Reason, N. Johnson and R.B. Lee (1995) *Beyond aviation human factors: Safety in high technology systems*. Aldershot: Ashgate.
Miller, T.R., J. Viner, S. Rossman, N. Pindus, W. Gillert, J. Douglass, A. Dillingham and G. Blomquist (1991) "The costs of highway crashes", U.S. Government Printing Office, Washington, DC, Report DOT-FHWA-RD-91-055.
Polinsky, A.M. and S. Shavell (2000) "The economic theory of public enforcement of law", *Journal of Economic Literature*, 38:45–76.
Savage, I. (1998) *The economics of railroad safety*. Boston: Kluwer.
Savage, I. (1999) "The economics of commercial transportation safety", in: J. Gómez-Ibáñez, W.B. Tye and C. Winston, eds., *Essays in transportation economics and policy: A handbook in honor of John R. Meyer*. Washington, DC: Brookings Institution Press.
Shaked, A. and J. Sutton (1982) "Relaxing price competition through product differentiation", *Review of Economic Studies*, 49:3–13.
Shavell, S. (1987) *Economic analysis of accident law*. Cambridge, MA: Harvard University Press.
Zaal, D. (1994) *Traffic law enforcement: A review of the literature*. Canberra: Australian Government Publishing Service.

Chapter 15

ENVIRONMENTAL PROTECTION

EMILE QUINET
Ecole Nationale des Ponts et Chaussées, Paris

DANIEL SPERLING
University of California Davis

1. Introduction

Environmental protection is a major concern of modern societies, and transport is a major source of environmental damage. So it should not be surprising that environmental protection has played a central role in transportation policy and decisions, and that considerable resources have been devoted to mitigating adverse environmental impacts.

Environmental protection can be analyzed from different points of view, including technical, legal, ecological, and political. Here we focus secondarily on engineering and primarily on economic perspectives, with acknowledgment of their shortcomings. The engineer seeks technical solutions and designs technical devices to abate environmrntal damage; the economist seeks the best use of societal resources and identifies actions that yield the highest return for investments and other initiatives.

In market economies, prices are the primary mechanism for allocating resources. The difficulty, however, is that environmental impacts are largely outside the marketplace. When I drive a car, I make noise; I disturb my neighbors, but I do not pay for that damage, and my neighbors are not compensated in any way. If I was obligated to pay compensation – what economists call the polluter-pays principle – I would make less noise. Unfortunately there is no market for that "good" – or, more accurately, "bad"; there is no natural mechanism to ensure a proper balance between the demand for silence and the supply of silence. The same applies to air and water pollution, climate change, esthetics, loss of wetlands, loss of biodiversity, and most other forms of environmental damage.

This absence of markets for environmental damage is a clear case for public action; public authorities must devise mechanisms to respond to this market failure. Economics is a powerful tool, but because environmental effects are

Handbook of Transport Systems and Traffic Control, Edited by K.J. Button and D.A. Hensher

outside the marketplace, public action must be manifested in the creation of secondary markets or as administrative actions by government. Economists argue that these actions should be guided by economic principles. But, contrary to natural market mechanisms, which are relatively non-manipulable, devices created by public authorities are subject to discretion and can be more easily influenced by the strategic action of the private interests at stake.

While economics provides a powerful framework for guiding and designing actions to reduce environmental damage, the shortcomings noted above have led to other approaches playing a more dominating role in recent decades.

Government first actively intervened on behalf of environmental protection in the 1960s. First, regulatory programs were adopted to reduce air pollution, followed by fuel use regulations in the U.S.A. and Japan in the 1970s. The regulatory system that evolved was a creation of lawyers and engineers whose disciplinary paradigm was one of right and wrong. It was founded on highly specific rules of conduct and design, resulting in an approach that has come to be known as "command and control." While the various regulatory activities affecting vehicles and fuels are not strictly command and control – they contain some flexibility – the overriding framework continues to be one of directives that restrict the behavior of vehicle and fuel suppliers.

Most rules were premised on technological solutions and followed engineering principles of "best available technology." Examples include rules specifying the amount of pollution a car is allowed to emit, attributes of fuels (use of oxygenates, lead, etc.), allowable types of engine technologies (e.g., bans on two-stroke engines), required inspections of vehicle emissions, and bans on car use in city centers.

As part of a larger debate over the role and effectiveness of government intervention, conventional wisdom in both the political and the professional arenas is gradually shifting from an embrace of command and control to greater use of market instruments and greater emphasis on flexibility. Examples include use of tradable credits (for vehicle emissions), fuel taxes, vehicle taxes, tolls for use of high-occupancy vehicle lanes, and voluntary agreements with industry to reduce CO_2.

The perception exists that command and control rules are appropriate and effective when externalities are large and little effort has been expended. But after decades of efforts to reduce emissions, the marginal costs for further reductions can be large and the marginal benefits relatively small. It is widely accepted that the most cost-effective approach is to grant more flexibility to industry, though much of the environmental community remains skeptical. They fear that flexibility will result in industry finding loopholes to avoid action, and that market instruments will create environmental "hot spots" (often where residents are poor and less politically influential) and will not be as effective at changing behavior as regulations and rules.

These general considerations will be developed and woven into the following sections. In Section 2, we categorize environmental damage; Section 3 provides a review of economic tools to cope with environmental damage and criteria for when and how to use them; in Section 4, we describe how these tools are applied to general transport policy, infrastructure design, and vehicle design and use; and in the concluding section, we provide a general assessment, stressing shortcomings, successes, and future prospects for environmental protection.

2. Categories of environmental damage

The transport sector encompasses a wide variety of activities and facilities, from infrastructure building to vehicle use. Environmental damage includes the following:

(1) Esthetic effects resulting from new infrastructure and vehicle use. These effects are pervasive and highly subjective. They are usually viewed as negative intrusions, but in some cases can be positive, as when users gain access to previously inaccessible scenery.
(2) Habitat and community fragmentation, for animals, plants, and humans. Animal and plant migration is inhibited by the building of road and other infrastructure, exacerbating the difficulty of responding to environmental threats and changing climate.
(3) Soil and underground water pollution from construction and use of road and fuels infrastructure (leakage from fuel tanks and pipes, water runoff from roads, particles).
(4) Water pollution from tanker collisions, spills, and discharges, including such dramatic accidents as Exxon Valdez or Amoco Cadiz.
(5) Noise and vibration caused by all vehicles in the water, on land, and in the air. Road vehicles are most troublesome in dense urban areas, recreational (off-road) vehicles in pristine areas, and airplanes near airports.
(6) Air pollution, primarily from vehicle exhausts, but also from upstream production of energy and manufacture of vehicles and facilities. The principal concern is with road traffic, because of the large volume of pollution emitted and because it is in close proximity to people. About half of the air pollutants emitted in urban areas are from vehicles. Air pollution damages human health, as well as vegetation and buildings.
(7) Climate change caused by large emission of greenhouse gases, especially CO_2. About 20–30% of greenhouse gases in OECD countries are from transportation, largely from fossil fuel combustion in cars and trucks. Climate change effects have been small so far, but the continuing buildup of these gases is creating a growing threat.

In this chapter, we limit our focus to air pollution, noise, landscape and barrier effects, and climate change. These environmental impacts attract the greatest attention of decision-makers, and have received considerable study.

3. Economic tools to cope with environmental damage

A central issue is the extent to which public intervention to protect the environment is appropriate, and if so, what type of intervention is most effective. A celebrated theorem (Coase, 1960) tells that the use of natural market procedures, realized as bargaining, leads to an efficient result under some conditions:

(1) no transaction costs,
(2) perfect information, and
(3) all parties at stake are involved in the bargaining.

It is clear that these conditions are rarely satisfied in transport activities. Usually transaction costs are high, due to the large number of parties involved – the many people exposed to environmental risk and the many vehicle operators causing them. Perfect information is not available to all; those exposed are often not aware of the extent of damage until much later, and vehicle operators have even less information. And many parties at stake often do not participate in the bargaining, as with taxpayer financing of infrastructure. For these reasons, there is a strong case in favor of public intervention for environmental protection. This public intervention can follow various channels, which, following a classification close to Button (1999), can be characterized as follows:

(1) Direct public management. This approach is widely used for many types of modal and intermodal terminals (airports, buses, rail) and guideways, but there is a growing movement toward selective privatization.
(2) Regulation of actions by individuals and organizations, such as limiting emissions from cars and noise from airplanes (often referred to as "command and control").
(3) Economic instruments. These include mechanisms that alter price signals, such as fuel taxes, congestion fees, and payments for scrapping old polluting cars; and mechanisms that create property rights and new markets for environment goods, such as marketable credits for emissions, so that the Coase theorem could operate, and bargaining leads to an efficient result.

Policymakers and researchers carry on a continuing debate as to which tools to use for which problems in which situations. First-order criteria include the following:

(1) transaction costs (cost in money and time for information to be used in implementation, management, and control);
(2) accuracy and side-effects (e.g., an increase of petrol tax is quite an inaccurate way to master air pollution and has side-effects on car size);
(3) equity and distributional effects (taxing old cars and subsidizing accelerated scrappage are equivalent in terms of pollution abatement, but the distributional consequences of these two measures are quite different);
(4) collective efficiency (the economic optimum is not suppression of environment damage, but abatement to the point where the marginal cost of abatement is equal to the sum of the marginal willingness to pay of the people who suffer from it).

There are very few general findings about the advantages and drawbacks of the tools of public intervention vis-à-vis these criteria, and of those, virtually all are abstract and broadly theoretical. Certainly, economic instruments are more efficient than regulation under conditions of perfect information and without transaction costs; they allow for better decentralization of decisions and give more incentive to foster efficient technologies. And there are also some results relying on the choice between price and quantities regulation on the degree of uncertainty of costs and willingness to pay.

But economic instruments have more clear and distinct effects on redistribution and equity than regulations. Taxes for instance directly hurt the user's wallet,[*] while traffic regulations just impose time delay, and emissions regulations provide all drivers with a technologically improved car.

Adoption of any particular instrument implies a variety of trade-offs that are difficult to measure and foresee. Is it more important to reduce environmental damage at least cost to society or in a way that is deemed equitable? Even the definition of equity is elusive and difficult to measure; does it mean that costs are borne by those who are more affluent, or by those residing in the respective region, or by those more willing to pay to reduce the damage?

Equally problematic is the difficulty of designing and adopting instruments that elicit the desired effect without causing secondary effects. For instance, how should one reduce air pollution from vehicles? Some efficient devices such as catalytic converters increase energy consumption and greenhouse gases a small amount, and noise abatement devices increase the weight of the vehicle.

Consider also that vehicles emit a variety of pollutants. Some pollutants are of more concern in some locations (carbon monoxide near intersections), at sometimes of day (ozone when the sun is shining), and when people are outdoors. Moreover, emissions are highly sensitive to how a person drives (hard

[*] It can be shown that in many cases, road pricing has a negative direct effect on the user's surplus, and that the collective positive effect is due to the increase of government taxes.

accelerations cause 100–1000 times more pollution than simple cruising). An economist might argue that the ideal instrument in this case of air pollution is a tax that registers how much of each pollutant is emitted, exactly when, and the number of people likely to be exposed (perhaps using GIS and satellite data) – and that the fee is immediately communicated to the driver on the vehicle's instrument panel so that the driver has the information to alter his/her behavior. For this market instrument to operate efficiently, someone must calculate the value (or damage) of each pollutant at each location at each time of day; the technology must exist that accurately measures each pollutant on a real-time basis; and some means of actually billing each driver (or vehicle owner) must exist. Some day, this idealized market instrument may be possible due to the availability of new information and measuring technologies, and better understanding of pollutant damage. For now, this is no more than a dream.

Another difficulty in creating taxes and regulations that meet economic efficiency objectives is the uncertain environmental consequences of transport. Indeed, there is not even agreement on the health effects of pollutants and their importance relative to other social goals. For instance, consider the case of diesel engines. Diesel engines are about 10–20% more efficient than gasoline engines, and therefore emit that much less CO_2. They also emit less CO and hydrocarbons than gasoline engines, but higher levels of NO_x and particulate matter. In Europe, this trade-off is considered acceptable, even desirable. In the U.S.A., however, and increasingly in Japan, diesels are much less popular and regulators are taking a far more aggressive stance against them. The different regulations and policies are due in part to uncertainty about the health effects of particulate matter. It is now widely agreed that particles are the most threatening air pollutant to human health, but it is uncertain how much comes from diesel engines, whether the chief culprits are the very fine particles (with diameters in the nanometer range), and whether these fine particles are and will be reduced by new diesel emission control technology.

It is clear from this quick review that environmental protection is a matter of political decision – concerning trade-offs between equity and efficiency and various other social goals, and judgments about uncertainty. Researchers are becoming more sophisticated about quantifying damages and costs, and this information is finding its way into the policymaking process, but continuing uncertainties and conflicts mean that many decisions about environmental protection will continue to be based primarily on political factors. Private interests and lobbies use uncertainty to push governmental decisions toward their objectives. The story of environmental protection can be read as a struggle between pressure groups. But in a larger sense, the debate can be characterized as one of competing philosophies, values, and paradigms.

Another perspective, which we do not pursue further but which is fundamental to understanding the debate over environmental protection, is one put forth by ecologists. They believe that humanity's anthropocentric approach is inappropriate. We need to see humans as just one species within a larger ecological system, and that doing so would entail much more emphasis on preserving a sustainable balance between humans, plants, and animals. This perspective seems to be gaining more support over time.

Institutional mechanisms have evolved to ensure that externalities and public goods are taken into account – they include various governmental entities as well as non-governmental organizations (NGOs) that advocate the interests of those without direct economic interests, and even the interests of other species.

As there is no clear solution to any particular problem and a variety of forces and interests are at work, it should not be surprising that the equilibrium may be quite different from one country to another. It also explains why debates over environmental protection are often so disjointed; more than with most public decisions, they are the outcome of many partial struggles rather than the result of a comprehensive and coherent process.

4. The targets of environmental protection

We categorize the transport sector into three areas of activities:

(1) general transport policy,
(2) infrastructure design, and
(3) vehicle design and use

4.1. General transport policy

The transport sector is a complex mix of subsystems, often referred to as modes (buses, rail, airplanes, cars, etc.), all influenced to varying degrees by governmental entities. A major share of government resources and attention are devoted to transport because transport activities have major social costs and benefits. Governments at all levels constantly struggle over levels of support for different modes – for instance, to invest in more rail lines or more roads, or subsidize more buses or build more parking. Central to these policy decisions are widely held beliefs that road transport, especially cars, is a grave environmental threat, that more resources need to be devoted to collective transport, and that infrastructure pricing should be used to divert passenger and freight traffic away from cars and trucks.

This policy is explicit in many European countries; it is for instance the goal of pricing reforms supported by the European Union (European Commission, 1995,

1996). In urban areas, especially during the past few decades, the underlying premise of urban transport policy in virtually all cities of the world, including the U.S.A., has been to enhance transit. In Europe it is the result of a long process begun about 30 years ago, premised on the belief that travelers could be readily diverted from cars to collective transport. Considerable public investment has gone into transit in Europe and the U.S.A.; in both cases, though, the mode share of transit has continued to drop. Policy-makers in Europe, but not the U.S.A., are now shifting their emphasis from transit enhancement to suppression of car use. They are using various devices to decrease urban space dedicated to cars, to the benefit of pedestrians, bicycles or buses.

Support for rail and collective transport is explicitly justified by environmental arguments for reduced noise and air pollution. But social-cost studies show that measurable environmental damage is not sufficiently high to justify such policies (Greene et al., 1997). It can be said that strong public support for environmental protection is often a statement of distaste for the visual intrusion of cars and roads, delays caused by congestion, and a concern for what is often seen as a deterioration of community. This point is illustrated by the high cost of changes in the layout of infrastructure resulting from public hearings and the small benefits, especially in the case of noise abatement. In any case, pro-transit and pro-rail policies have had mixed success. Almost everywhere, transit has been losing market share to cars. U.S. cities, except for large city centers, are largely resigned to the marginalization of transit. European cities, which are older and denser, are making a stronger effort to slow or even stop the shift to cars. For interurban travel the dynamic is more complex. Great Britain, for instance, has in the past placed a moratorium on new road construction, and lies well behind other European countries such as France, where motorway construction continues at roughly the same pace as ten or twenty years ago, while TGV expansion is slowing after peaking around 1990.

A similar story can be told for freight transport, with rail continuing to lose market share to trucks almost everywhere. This shift has slowed in those cases, especially in the U.S.A., where rail operators have been most aggressive at partnering with trucking companies in using containers and encouraging intermodal shipments.

Transport pricing initiatives have also proceeded quite unevenly. Road pricing initiatives failed in The Netherlands, for instance, while urban collective transport prices have in general increased in recent years. Of course, it must be recognized that in Europe the price of collective transport was very low, resulting in large public subsidies, and that petrol taxes are high compared with the U.S.A. No ecotaxes have been adopted, except in Sweden. Several attempts have been made to bring road tolls nearer to social costs, but these efforts are justified by congestion cost externalities, not environmental costs, and have mainly consisted of differentiating tolls by time of day.

4.2. Infrastructure design

Infrastructure facilities can be designed and built to mitigate environmental damage, especially esthetics, soil and water pollution, noise, and barrier effects.

Landscape preservation relies on various devices and practices, including more esthetic designs of facilities, tunnels to avoid ecologically sensitive and esthetically attractive landscapes, and planting of attractive and native plants. It is common to devote 10% of the total cost of an infrastructure project to landscape preservation, and even more when historic buildings, scenic sights, or fragile ecosystems are nearby.

Barrier effects are reduced by building bridges and tunnels, specially designed for particular kinds of animals. The same is done for human beings, especially in urban areas, to maintain connection and community in neighborhoods. It is not possible, however, to eliminate the barrier effects caused by infrastructure. In the case of human pedestrians, the construction of tunnels and bridges often creates uneasiness, requires more travel time, and disrupts daily life.

Noise abatement is an important task that can be performed in part through infrastructure design. The three important means are soil movements, noise barriers, and tunneling; this last solution is the most expensive and is normally used only for dramatic situations of high noise levels involving large numbers of people.

Infrastructure investments are routinely subjected to cost–benefit analysis. It would seem normal that design decisions involving environmental protection would also be subjected to such criteria. This is rarely the case. First, cost–benefit analysis is not able, in the current state of the art, to cope with soil pollution, barrier effects, and landscape impacts. In the latter case, it is possible to value the effects through stated-preference methods but the results are often not reliable and are highly sensitive to the local situation: it not possible to draw general rules nor rely on findings from similar situations.

The unique monetizable effect is noise, and the practice of incorporating noise valuations in cost–benefit analysis is now becoming widespread. Unfortunately it suffers several drawbacks. First, as noise is a very local phenomenon that varies widely according to topography, it is difficult to measure it precisely in the early stage of the decision process of new infrastructure. Noise is better assessed during the last stages of the process, when the final track is fixed, the execution schemes are completed, and bargaining is in progress with the local bodies (neighbor associations, local political authorities, etc.). But it often happens, at this stage, that the bargaining power of these local bodies is tremendous. They are able to delay or postpone construction, they can use political threats, and they can inflict huge costs on the infrastructure providers. For these reasons it appears that ex post noise cost determinations, derived from observations of past decisions, are much higher than the cost of noise derived from willingness-to-pay methods – the ratio can be one hundred to one in the case of suburban highways, rail tracks, and new airports. (It must be acknowledged that the explicit argument about noise often covers implicit arguments about landscape, esthetics, visual intrusion, or barrier effect, as indicated earlier.)

4.3. Vehicle design and use

The transport sector contributes about half the urban air pollution and about 20–30% of greenhouse gases in most OECD countries. Most of the pollution and the majority of greenhouse gases are from road vehicles. This general statement camouflages many variations, however. Some modes have a larger effect in some countries; for instance, in the U.S.A., air transport has a much greater effect than in most other countries, and transit a much smaller effect. The technology mix can be very different, with most European cities having a much higher proportion of diesel engines than the U.S.A. or Japan. And the relative magnitude of emissions varies greatly and they have very different geographical reach. For instance, cars contribute as much as 90% of total carbon monoxide emissions in a city, but the health effects reach only about 50 m from the pollution source. On the other hand, greenhouse gas emissions from vehicles have a global reach.

In general, the quantity of air pollution emitted from a vehicle has been reduced dramatically in the U.S.A. and Japan during the past few decades, by about 75–95%, depending on the pollutant and taking into account actual in-use emissions over the life of the vehicle (Pickrell, 1999), and overall air pollution has also been slowly improving. In Europe, progress has been slower but is catching up, with strict new rules in place. Indeed, in all OECD countries, rules and institutions and enforcement mechanisms are in place to reduce vehicle emissions significantly below current levels. Air quality improvement is a major success story in the U.S.A. and Japan, and likely to be a big success story in Europe as well.

The situation with greenhouse gases is more problematic. Emissions of these gases, mostly carbon dioxide, have been increasing by about 1–2% per year in most countries, and somewhat faster in transport, despite the series of international protocols and agreements to reduce greenhouse gases. A voluntary agreement between the European Union and European automakers to reduce CO_2 emissions by 25% (per veh-km) between 1995 and 2008, and similarly stringent fuel economy standards in Japan suggest that past trends could be countered in those areas. In the U.S., proposals to tighten existing fuel economy standards continue to gain attention but standards have been frozen for cars since the late 1980s and for light trucks since the early 1990s.

When governments first intervened in the 1960s and 1970s to reduce emissions (and energy use in the U.S.A. and Japan), the initiatives contained a balance of technology and behavioral strategies. Over time, though, air quality regulators have gradually turned more to technical solutions – mostly through promulgation of increasingly stringent performance standards – and less to behavioral strategies aimed at reduced driving and speeds. In the U.S.A., "travel demand management" strategies first adopted in the 1970s to reduce car travel (for environmental and infrastructure efficiency reasons) now receive little attention; and 55 mph speed limits, imposed in 1975 on all U.S. motorways to reduce energy use, have now been relaxed (in some states rural speed limits have been completely abandoned).

Government regulation of vehicle emissions is highly formalized and structured. The paper flow between automakers and regulators contains minutiae on the design and conduct of emission tests and advises them of changes in reporting requirements, test procedures, and control technology. In most cases, a uniform standard is specified that each and every car, bus, and truck must meet. Initially, emission standards were established for three pollutants: carbon monoxide, hydrocarbons, and nitrogen oxides (plus standards were established for lead content of fuel), and later standards were developed for particulate matter and other toxic chemicals. Compliance is verified by running a sample of vehicles through a standard driving cycle.

Standards have been tightened intermittently, based on subjective judgments of the severity of ambient pollution problems and of what manufacturers were capable of achieving at reasonable cost. The strength of this uniform-standard approach is simplicity and apparent ease of enforcement.

Government intervention to reduce fuel consumption has been somewhat more flexible and less strict. No standards exist for heavy- or medium-duty trucks and buses. In the U.S.A. since 1978, light-duty vehicles supplied by each manufacturer have been required to meet an average standard. Compliance is measured by average fuel consumption across all vehicles sold in a particular year (divided into four separate groups of cars, light trucks, and imported and domestically produced vehicles). Japan has a long-standing set of fuel consumption rules that were recently tightened and redesigned to require roughly 23% improvement between 1995 and 2010 for each category size of gasoline-powered cars (and somewhat less for light trucks and diesel vehicles). Europe has never adopted fuel economy standards, but has relied on high fuel taxes to restrain fuel consumption. In 1998, however, automakers under pressure from the European Union, agreed to reduce CO_2 emissions of light-duty vehicles by 25% (per vehicle) by 2008. Various monitoring protocols were adopted but the program is voluntary and no rules or enforcement mechanisms were put in place

5. Conclusion

As humans gain wealth, they seek access to more goods, services, and activities, and seek to do so with greater safety, security, comfort, and convenience. Growing affluence translates into growing vehicle use (until or unless fundamental changes are made in the transport sector). More vehicles mean more resource consumption and greater stress on environments. The two fundamental responses are to alter the vehicles so that they have less impact on the environment, and to reduce vehicle usage. In the U.S.A., the greatest emphasis has been on rendering vehicles more benign; in most other OECD countries, at least equal attention has been given to reducing vehicle usage.

The greatest successes have come in air pollution, eliminating lead in most OECD countries, and greatly reducing vehicle emissions virtually everywhere. The result has been substantial improvements in air quality in most of the U.S.A. and Japan, and good prospects for doing so in Europe. The rules and institutions are in place, often with strong public support, to maintain these initiatives. Even so, given the expanding car stock, these air pollution successes will continue only if existing initiatives continue to be strengthened.

A similarly positive story can be told for noise pollution, though in this case government played a relatively minor role. The strongest force for change was vehicle owners, who demand and are willing to pay for quieter vehicles. Nevertheless, in dense settlements the problem of noise continues to be serious.

Improvements can also be noted in esthetics, landscaping, and barrier effects. With growing affluence comes more attention to esthetics, and a shift from pure functionalism to greater attention to beauty. Roads and other infrastructure built recently are far more attractive than those built decades ago, and landscaping along guideways is far more attractive and ecologically sound. Greater efforts are also being made, especially in Europe, to mitigate the fragmentation effects of infrastructure, with animal tunnels and bridges provided at key locations.

The greatest threat posed by expanding transportation systems appears to be climate change. Other impacts – air pollution, esthetics, noise – are local and reversible within fairly short time frames. In contrast, the effects of increased emissions of greenhouse gases are large and nearly irreversible. As more knowledge is gained, nations of the world are demonstrating greater resolve to reduce greenhouse gas emissions.

In a broader sense, transport continues to be at the center of debates over the future of our cities. The desire for more access and space leads to urban sprawl, as housing lots and office parks expand in size and number. What is the role of the auto in urban growth, and is it positive or negative? These questions are not easily answered, because values and preferences vary greatly, and because the process of urban development is complex, involving many forces and interests. Nevertheless, many public advocacy groups, as well as many researchers, indict the auto for not only consuming large amounts of resources and degrading the physical environment, but also for its role in loss of an urban esthetic, loss of community, and marginalization of the poor, old, and disabled. Many believe the only solution is to sharply curtail the use of vehicles.

These debates and these efforts to reduce the adverse effects of vehicles will continue. But what to do and how? The answer is not so clear, not even for the most well-defined and best-studied problems, such as air pollution. How much more effort should be devoted to reducing air pollution? How much effort should be devoted to each pollutant? How important is air quality relative to other goals such as reduction of greenhouse gas emissions?

The difficulty is insufficient knowledge of scientific phenomena, insufficient understanding of human values and preferences, uncertainties about technology development, fragmented policy and regulatory approaches, poor analysis and poor use of analytical tools, and undue influence by well-endowed interest groups. For broader issues such as climate change and urban community, the challenges are especially daunting. In the end, there are no unique and distinct answers. Clearly, though, decisions can be much better informed and environmental impacts can be reduced at less cost.

Consider again the case of air pollution. While carmakers have succeeded, in response to regulatory requirements, in dramatically reducing emissions of some pollutants, new medical research suggests that recently targeted pollutants, such as carbon monoxide and reactive hydrocarbons, may be less damaging than ultrafine particulate matter, which has received less attention. Have regulatory efforts been misguided? Have resources been allocated inefficiently? We shall not know until better scientific knowledge is available, better sharing of this knowledge takes place across scientific disciplines, and better measurement and analytical tools are developed. What we do know is that better integration of governmental initiatives would lead to more coherent and desirable policies – if only we knew how to design those instruments to be efficient, effective, and equitable, to have low transaction costs, and to be politically implementable. Similar stories can be told for noise and other environmental effects.

Looking toward the future, the climate change problem is probably most important. It inspires several questions. Will we be able to cope and adapt if the most pessimistic prognoses are realized? It is quite clear that measures now under review are inadequate to cope with the scale of the problem. A positive view is that humans have always discovered new tools and technologies to overcome problems which hitherto seemed unavoidable. But new solutions often lead to unintended and unanticipated consequences. Prudence is in order.

In the end, though, it is useful to remember that extraordinary progress has been made along all dimensions of policy, practice, and science, with knowledge, political commitment, and sophistication growing everywhere. The bad news is that many resources have been misallocated, many problems have worsened, and new challenges need to be confronted. Given the huge progress of recent decades, it is not inappropriate to believe that the current positive trajectory can continue. But it will require continued political and scientific commitment.

References

Button, K. (1999) "Environmental externalities and transport policy", in: Y. Hayashi, K. Button and P. Nijkamp, eds., *The environment and transport*, Cheltenham: Edward Elgar.

Coase, R.H. (1960) "The problem of social cost", *Journal of Law and Economics*, 3:1–44.

European Commission (1995) "Towards fair and efficient pricing in transport", Green Paper, COM 691.

European Commission (1996) "A strategy for revitalizing the community's railways", White Paper, COM 421.

Greene, D., D.W. Jones, and M.A. Delucchi (1997) *The full costs and benefits of transportation: contributions to theory, method, and measurement*. New York: Springer.

Pickrell, D. (1999) "Cars and clean air: A reappraisal", *Transportation Research*, 33A:527–547.

Chapter 16

TRANSPORTATION SYSTEMS AND ECONOMIC DEVELOPMENT

KINGSLEY HAYNES and KENNETH J. BUTTON
George Mason University, Fairfax, VA

1. Introduction

Intuitively there would seem to be a link between the transport system and the way and the pace at which economic and social development progresses. However, while the importance of transport in economic growth and development has never been seriously questioned, its exact role and influence have been subjected to periodic reappraisals. A major underlying problem is that our understanding of what causes economic development is poor and hence the role of any particular sector in the process is murky.

The exact importance of transportation infrastructure as an element in the economic development process has long been disputed. Much seems to depend upon the degree to which supply considerations are thought important. The demand-side Keynesian approach indicates that causality runs from economic exploitation to income and on to infrastructure generation. In contrast, neoclassical economics is supply-driven and transport and other infrastructure are generally seen as important elements in the production function. Much of this recent work follows the neoclassical mode in looking at the links between infrastructure provision and economic development through some form of aggregate production function analysis. It has sought to see how well the aggregate production function, and its individual elements, explains economic performance.

This chapter looks at some of the analysis that has been conducted on the role of transportation in the economic development process. This analysis is based upon a variety of different approaches that have ranged from the historical to the use of mathematical modeling. It also looks at development at various geographical levels spanning the national to the very local.

2. The overview arguments

Traditionally, it was argued that transportation exerted a strong positive influence on economic development and that increased production could be directly related

Handbook of Transport Systems and Traffic Control, Edited by K.J. Button and D.A. Hensher

to its improvement. In the U.K. context Baxter (1866) argued that "Railways have been a most powerful agent in the progress of commerce, in improving the conditions of the working classes, and in developing the agricultural and mineral resources of the country." Later Lord Lugard (1922) wrote, "the material development of Africa may be summed up in the one word – transport." Perhaps the strongest advocate of the positive role of transportation is Rostow (1960), who in accounting for economic growth maintains that "The introduction of railroads has historically been the most powerful single indicator to take-offs. It was decisive in the U.S.A., France, Germany, Canada and Russia." A broader-brush approach is adopted by Andersson and Stromquist (1988), who claim that all the major transitions in the European economic systems were accompanied (or initiated) by major changes in communications infrastructure.

Positive linkages between transportation provision and economic development can be divided between the direct transportation input and indirect, including multiplier, effects. Good transportation offers low distribution costs that have permitted wider markets to be served and the exploitation of large-scale production in an extensive range of activities. Further, most undeveloped countries are, for a variety of geographical, economic, and historic reasons, dependent upon international trade and an expansion of this trade is an essential prerequisite for their economic growth. In these circumstances the provision of efficient port facilities and other facilities will assist development. The indirect effects stem from the employment created in the construction of transportation infrastructure and the jobs associated with these services. Further, there may be multiplier effects stemming from the substantial inputs of iron, timber, coal, etc. required to construct a modern transportation system and that may be supplied by indigenous heavy industries.

This causal view of transportation and economic development has become less credible in recent years. The econometric work of Fogel (1964) in the U.S.A. offers evidence that American growth in the 19th century would have been quite possible without the advent of the railways – waterways supplying a comprehensive transportation system at comparable costs. The view that the railways were the motive force behind U.S. economic development has given way to a weaker position, namely that good transportation per se permits economic expansion.

Economic development is now generally seen as a complex process with transportation permitting the exploitation of the natural resources and talents of a country. It is necessary but not sufficient for development. Transportation can release working capital from one area that can be used more productively as fixed capital elsewhere, although a necessary prior condition is the existence of suitable productive opportunities in potential markets. Public infrastructure in this sense should be set in the context of the availability of private capital; many parts of the world, for example, would not benefit from more transportation infrastructure because of the lack of private resources. From a slightly different perspective, improved transportation can help overcome bottlenecks in production. A difficulty, of course, if this is true, is that the bottleneck may be some distance from the region and superficially appear unconnected with it.

While there are these arguments ascribing a positive role to transportation in development, albeit in different ways, there are also views that an excessive amount of scarce resources can get devoted to transportation improvements. As with any scarce input it is possible to define an optimal provision. It has been argued that there are economic forces that tend to lead to an excess of transportation provision (especially high-cost infrastructure) at the expense of more efficient and productive projects. In particular, the lumpiness of transportation capital, together with its longevity and associated externalities, makes it particularly difficult to estimate future costs and benefits. Consequently, decisions to devote resources to transportation are not easily reversible or readily corrected.

The political acceptability of transportation is highlighted by Hirschman (1958), who feels that the sector attracts resources quite simply because it is difficult for mistakes (of an economic nature) to be proved even after major projects have been completed. Also development planners tend to be mainly concerned with allocating public investment funds and it is natural that they should claim transport, communications, energy, drainage, and the like as being of fundamental importance.

3. Transportation infrastructure and development

Recent interest in the possible links between transportation infrastructure provision and economic development can be traced to a body of empirical work that began to emerge in the late 1980s. The initial high elasticities of output associated with public infrastructure investments, including transport, found by Aschauer (1989) at that time, have subsequently been refined downwards as data and estimation methodologies have improved.

The usual approach in this type of analysis is to take a production function of the standard form

$$Y = f(L, K, J), \tag{1}$$

with Y indicating output, L labor, K the stock of private capital, and J the stock of public capital (which is largely transportation infrastructure). The relevant parameters are then estimated – often employing a Cobb–Douglas specification. The models were used to estimate the output elasticities of public infrastructure investments. Many of the early studies, especially from the U.S.A., at the national and state levels, provided statistically significant and apparently robust evidence that well-designed and -operated infrastructure can expand the economic

Table 1
Estimated output elasticities of public infrastructure investments

Author	Aggregation	Output elasticity of public capital
Aschauer (1990)	National	0.39
Holtz-Eakin (1992)	National	0.39
Munnell (1990)	National	0.34
Costa et al. (1987)	States	0.20
Eiser (1991)	States	0.17
Munnell (1990)	State	0.15
Mera (1973)	Regions	0.20
Duffy-Deno and Eberts (1991)	Urban areas	0.08
Eberts and Fogarty (1987)	Urban areas	0.03

Source: Button (1998), where the full sources of references are cited.

productivity of an area. Aschauer, looking at data covering the period 1949 to 1985, concluded that a 1% increase in the public capital stock could raise total factor productivity by 0.39%. Munnell (1990) tabulates output elasticities of public capital derived from U.S. studies in the 0.03 to 0.39 range although with a preponderance of results toward the upper end of the range. Similar positive findings are recorded by Biehl (1991) regarding the European Union.

As more studies were conducted, however, the robustness of these results began to be questioned. A summary of some of the main studies is set out in Table 1. Button and Rietveld (1999) conducted a meta-analysis of the parameters derived in studies of this type. Their findings indicate some sensitivity to the method of estimation used, the country of origin of the data, and the time period when the analysis was conducted, but even allowing for this, the ultimate findings lack robustness.

More recently, these types of study have been subjected to a variety of criticisms.

(1) First, while econometric studies may throw up positive correlation between economic performance and the state of infrastructure, the direction of causation is not immediately clear. Wealthier areas may simply have more resources for infrastructure provision. The efforts at testing for causality are, as yet, minimal.
(2) Second, the term "infrastructure" is a flexible one with no agreed definition, and simply taking official accountancy data may disguise important measurement, qualitative, and definition factors. In U.S. work, for instance, while Aschauer and Munnell used a four-category "core" subset of the nine U.S. Bureau of Labor Statistics infrastructure categories, others have employed the full set.

(3) Third, the way in which infrastructure is managed and priced may be as important as the provision of infrastructure per se. In terms of policy, therefore, account must be taken of the short-term levels of utilization, maintenance, and so on in addition to the stock of, and investment in, infrastructure.
(4) Fourth, even within the very vague notion we have regarding what constitutes infrastructure, there are numerous sectors and elements. While the vast majority of these macro studies look at public infrastructure as a whole, a small number have focused on particular elements, such as Keeler and Ying's (1988) work on the productivity of the U.S. highway system. From a policy perspective it is, therefore, important to isolate the roles of, say, transport, energy, and softer infrastructure such as law, education, business services, and defense in influencing macroeconomic performance.
(5) There are similar issues on the output side, and Nadiria and Manuneas (1996) have pointed to closer links between infrastructure investment and particular industrial sectors.
(6) Sixth, as more studies emerge they are producing much wider ranges of results; as Morrison (1993) puts it, "A clear consensus about the impacts of infrastructure investment has as yet been elusive, at least partly because different methodologies generate varying results and implications." In a study by Sturm and de Haan (1995), deploying U.S. and Netherlands data, for instance, these authors point to the fact that the data series in most studies looking at the economic effects of public capital are neither stationary nor co-integrated and, thus, conclusions that public capital has a positive effect on private sector productivity are not well founded. Equally, Jorgenson (1991) has questioned the basic premise underlying the use of a production function approach. This sensitivity may, of course, go beyond simple matters of technique if there is, in fact, no underlying relationship.
(7) The studies that have been completed frequently indicate rates of return on public capital investments – Aschauer's (1989) calculations give returns of 38% to 56% – levels outside of the range of any a priori expectation or of those found in individual microproject appraisals.

4. Transportation and regional economic development

It is not only at the national level that transportation has been seen as a potentially important development stimulus. The interregional spread of economic activity within a country is of major concern to national governments. Geographical variations in unemployment, income, migration, and industrial structure are of importance because they both result in spatial inequalities in welfare and, in many cases, reduce the overall performance of the national economy. The policies, which have varied both in intensity and in form over time, have generally concentrated on giving direct financial assistance to industry and on improving the mobility of labor. In addition there have long been attempts to improve the economic infrastructure of the less prosperous areas, with specific emphasis being placed on providing better transportation facilities. Critics of this practice point to potentially marginal or even counterproductive effects, especially where infrastructure is already relatively comprehensive. It is now accepted by many developed countries that transportation policy motivated by regional policy objectives must be pursued with circumspection and that, in many cases, improved transportation facilities may prove counterproductive to the development of some areas.

A simple hypothetical example illustrates the difficulty (Sharp, 1980). We have two regions, A and B, producing a single, homogeneous commodity. The centers of the regions (see Figure 1) are M miles apart and the commodity can be transported over the area at a constant money cost per mile of t per ton. The markets served by the regions differ because it costs C_A to produce a ton of the commodity in region A and C_B to produce a ton in region B. Consequently, and assuming no production centers exist between the regions, a distribution boundary can be drawn (shown by the dashed line in the figure) which is m_A miles from the center of A and m_B miles from the center of B (where $m_A + m_B = M$). The boundary is determined by the relative production costs of the regions and the costs of transportation (i.e., $C_A + tm_A = C_B + tm_B$). Basic manipulation of the algebra gives the formula

$$m_A = 0.5\left(M + \frac{C_B - C_A}{t}\right). \tag{2}$$

Therefore, if production is relatively cheaper in region A then m_A will increase if infrastructure reduces the cost of transport. Thus if A is a depressed area then transportation improvements could assist in expanding its potential market and, therefore, generate more income and employment, but region A must be a low-cost producer for this to be automatically true. If region B is the depressed one, then quite clearly investment in improved transportation will only worsen the

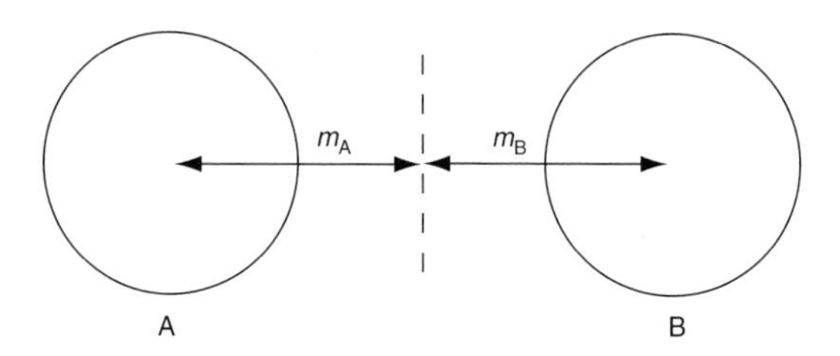

Figure 1. Market areas served by regions A and B.

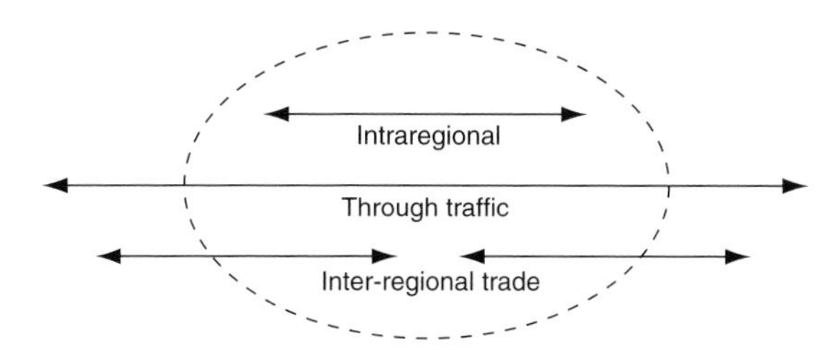

Figure 2. Influence of transportation infrastructure on a region.

regional problem by contracting the market area served by the region. Indeed, at the extreme (where $C_B - C_A > Mt$), region B may be forced from the market entirely by the expansion of the low-cost region's market area.

Of course, the model is a considerable simplification. Regions do not normally, for instance, specialize exclusively in the production of a single commodity, but produce a range of goods. Thus a transportation improvement, while damaging certain industries, may increase the competitiveness of others. The final effect of the improved transportation facility will then depend upon relative production costs between regions and the importance of transportation vis-à-vis production costs in the overall cost functions for the various commodities. Further, costs of production may vary with output and thus (following the "infant industry argument") it may be beneficial to reduce transportation costs if the government's regional policy also involves using grants and subsidies for encouraging the establishment of decreasing-cost industry in a depressed area. Supplementary measures of this kind may be necessary if the depressed area is sparsely populated and, to be successful, its industry needs to penetrate the markets of other, more populous regions to benefit from scale economies. It should be noted, however, that in these circumstances transportation improvements *must* be accompanied by other regional aids if the natural gravitation of decreasing-cost industries to centers of population is to be counteracted. Additionally, transportation costs tend not to increase linearly with distance, because of discontinuities and fixed-cost elements in the overall cost function. Consequently, the influence of any transportation infrastructure improvement is much more difficult to predict than the simple analysis implies.

Intellectually, the difficulty with the work linking transportation to economic development at the more spatially disaggregate level is also that transportation capital is diverse by nature, impacts on economies in a variety of ways, and is the subject of decision-making that can be efficient or not. Some of these problems can be seen by taking the example of transportation patterns seen in Figure 2. While it is true that the act of transportation investment itself may have primary multiplier implications combined with some secondary effects in terms of longer-term highway or rail track maintenance, if the facility principally serves transit traffic there is unlikely to be a great deal of value added. Equally, if it serves trade flows into and from the region then the implications for the area's local GDP will depend, as argued earlier, on the region's comparative and competitive advantages. There is no reason to suppose that automatically these would be positive.

Given the diversity of regional economies and the diversity of their related transportation systems, generalizations are difficult. Some are relatively closed with internal circulation dominating their transportation region. On the other hand, some are much more open by nature, with specific transportation channels that dominate their import and export activities. Others are transit areas, and accepting that it is unlikely that they will all have the same initial endowment of infrastructure, it is difficult to see why automatically one would anticipate that expanding the public capital base would lead to improved economic performance. This is particularly true if the opportunity costs of investments are not fully considered.

Despite these issues, as with recent debates at the national level on expanding transportation infrastructure, at the micro level a link between transportation provision and the economic development of local economies has become an oft-cited reason for the provision of new transportation facilities and improved infrastructure. Equally, though, the empirical evidence supporting such a link is not altogether solid. Table 2 provides details of some of the previous studies that have been undertaken of specific transportation infrastructure developments in the U.K. and the U.S.A. A variety of techniques have been used in these studies and there is always the question of the suitability of the design of any individual scheme to meet the desired objectives. Despite this there is certainly no indication that transportation infrastructure in itself has a major impact on local economic development. A number of possible explanations for this have been forthcoming.

From a narrow empirical perspective it is sometimes claimed that given the relatively small proportion of industrial costs which are expended on transportation in many countries – e.g., about 5.7% of total operating costs in the U.K. – it is unlikely that a strong link between economic performance and transportation provision will always be found to exist. There may, however, be quite significant differences in transportation costs by industry and by location even when, on average, they constitute only a relatively small part of costs. Also these types of statistics could be relating to the wrong question and the more important issue may be the importance of costs in profit determination. There are also the non-monetary aspects of transportation to consider – speed, reliability, and the like – which are increasingly important in sectors where just-in-time production management has been adopted. Finally, reliable interurban transport, good international transportation links, and high-quality local transportation are often found in empirical studies to be necessary to attract the type of labor required by high-technology industries.

Table 2
Micro studies of the development effects of transportation in industrialized countries

Author	Geographical scale	Infrastructure	Conclusions
Botham (1980)	28 zones (U.K.)	Changing nature of highway	Small centralizing effect on employment
Briggs (1981)	Non-metropolitan counties (U.S.A.)	Provision of highways	Presence of interstate highways is no guarantee of county development
Cleary and Thomas (1973)	Regional level (U.K.)	New estuarial crossing	Little relocation but changes in firm's operations
Dodgson (1974)	Zones in north (U.K.)	New motorway	Small effect on employment
Eagle et al. (1987)	87 counties (U.S.A.)	New highway expenditure	No increase in employment
Evers et al. (1987)	Regional level (Netherlands)	High-speed rail	Some effect on employment
Forrest et al. (1987)	Metropolitan areas (U.S.)	Light rapid transit	Property blight – good for urban renewal
Judge (1983)	Regional level (U.K.)	New motorway	Small economic impact
Langley (1981)	Highway corridor (U.S.A.)	Highway	Devalued property in area
Mackie et al. (1986)	Regional level (U.K.)	New estuarial crossing	Small overall effect
Mills (1981)	Metropolitan areas (U.S.A.)	Interstate highways	No significant effect on location patterns
Moon (1986)	Metropolitan areas (U.S.A.)	Highway interchanges	Existence of interchange villages
Pickett (1984)	Local districts (U.K.)	Light rapid transit	Properties close to the line benefit
Stephandes (1990)	87 counties (U.S.A.)	New highway expenditure	Could affect employment – depends on county's economy
Stephandes et al. (1986)	87 counties (U.S.A.)	New highway expenditure	Some positive association with employment
Watterson (1986)	Metropolitan areas (U.K.)	Light rapid transit	Modest growth in land use
Wilson et al. (1982)	Regional level (U.S.A.)	Existing highways	Transport affects location decisions but not development

Source: Button et al. (1995), where the full sources of references are cited.

Most of the empirical studies cited also tend to use revealed-preference approaches in their analysis and are typified by regression models relating location choices to a set of explanatory variables. This approach, while shedding useful insights, can, because of specification problems and interaction effects, sometimes prove a rather blunt instrument. While alternative methods also have their limitations, implicitly actually asking firms through a stated-preference approach about factors that have influenced their location choices and expansion plans does provide a viable alternative approach, especially when the sample is large enough to limit potential biases in responses.

5. Transportation provision in less developed countries

It is often argued that links between transportation systems and economic development for developing countries are different from those for industrialized nations. In particular, there is more likely to be a transportation infrastructure "shortage" in developing nations. As a result, transportation investment forms a major component of the capital formation of less developed countries, and expenditure on transportation is usually the largest single item in the national budget. Up to 40% of public expenditure is devoted to transportation infrastructure investment, with substantial supplements coming from outside international agencies such as the World Bank or in direct assistance from individual countries. At one level it is important to know whether this is, in aggregate terms, the most practical and efficient method of assisting the poor countries of the world, while at another level it is necessary to be able to assess the development impact of individual transportation schemes.

Broadly, transportation is seen by such agencies to have four functions in assisting economic development. First, it is an input into the production process permitting goods and people to be transferred between and within production and consumption centers. Because much of this movement is between rural and urban areas it permits the extension of the money economy into the agricultural sector. Secondly, transportation improvements can shift production possibility functions by altering factor costs and, especially, it reduces the levels of inventory tied up in the production process. Third, mobility is increased, permitting factors of production, especially labor, to be transferred to places where they may be employed most productively. The fourth factor is that transportation increases the welfare of individuals, by extending the range of social facilities to them, and also provides superior public goods such as greater social cohesion and increased national defense.

Making efficient use of transportation investment funds, however, poses problems. At the microeconomic level techniques of project appraisal have been developed that permit a more scientific assessment of the costs and benefits of individual transportation projects to be conducted. Many of the techniques of investment appraisal employed in the developed parts of the world are applicable in Third World conditions but local situations often require changes of emphasis. This is not surprising, considering these techniques were devised to look at transportation systems based almost entirely upon mechanical modes, while porterage and water transportation still account for the greatest proportion of goods movement in many less developed countries. The basic data are also often not so readily available or reliable in the Third World as in developed countries, thus limiting the precision of any analysis. Nevertheless, the development of investment appraisal techniques by agencies such as the World Bank permits consistent analysis across investment alternatives both within the transportation sector and between the transportation sector and other areas of economic activity. Such techniques emphasize the importance of estimating appropriate shadow prices for both inputs into transportation and the benefits derived from it. In particular, the shortage of foreign exchange suffered by many less developed countries is highlighted, while it is recognized that higher levels of underemployment and unemployment require adjustments to the wage costs of labor.

At the macroeconomic level, while it may be argued that ideally one should expand transportation provision to balance developments elsewhere in the economy, this is not always possible. The balanced-growth approach maintains that if transportation services are inadequate, then bottlenecks in the economy will curtail the growth process. If, however, the services are excessive this is wasteful in the sense that idle resources could be earning a positive return elsewhere in the economy and can become demoralizing if the anticipated demand for transportation does not materialize relatively quickly. Hirschman (1958) takes a somewhat different view, arguing that the relationship between economic development and the provision of social overhead capital, such as transport, is less flexible than members of the balanced-growth school believe.

In Figure 3 the horizontal axis shows the provision and cost of social overhead capital (SOC), which is normally provided by the public sector and embraces transportation as a major component. The vertical axis measures the total cost of direct productive activities (DPA), which are normally undertaken on purely commercial criteria. The balanced-growth approach assumes that DPA output and SOC activities should grow together (i.e., along the growth path represented by the ray from the origin), passing through the various curves from a to c representing successively higher amounts of DPA/SOC output. Hirschman, however, argues that less developed countries are in practice not in a position to follow such a path – partly because of the lack of the necessary expertise to ensure the balance is maintained and partly because of inherent indivisibilities in the social overhead capital schemes available. Consequently, growth is inevitably unbalanced and may follow one of two possible courses; one based upon excess

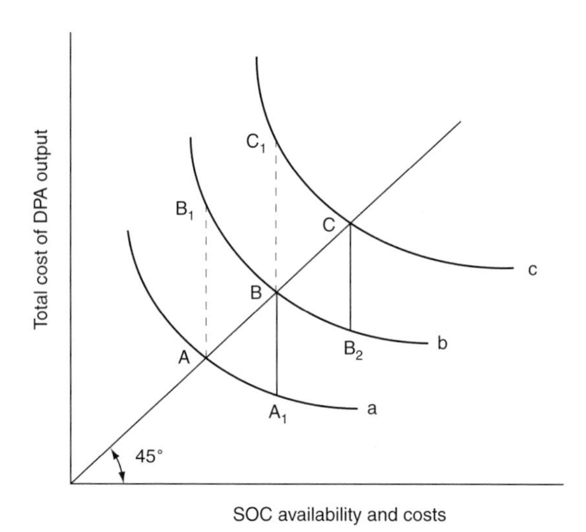

Figure 3. Balanced and unbalanced growth of social overhead capital (SOC) and direct productive activities (DPA).

capacity of SOC (i.e., path $A \rightarrow A_1 \rightarrow B \rightarrow B_2 \rightarrow C$), the other upon a shortage of SOC (i.e., path $A \rightarrow B_1 \rightarrow B \rightarrow C_1 \rightarrow C$). If a strategy of excess SOC capacity is preferred it is hoped that this will permit DPA to become less expensive and encourage investment in that sector. Alternatively, with the second approach, DPA expansion occurs first and DPA costs will rise substantially. As a consequence considerable economies will be realized through the construction of more extensive SOC facilities. The actual effectiveness of the alternatives depends upon the strength of the profit motive in the DPA sector, and the responsiveness of the public authority in the SOC sector to public demand.

The type of transportation provision most suited to developing economies is often of as much importance as the aggregate level of provision. Many developing countries tend to spend scarce development funds on prestige projects, especially international air transport, to demonstrate visually their capacity to emulate the performance of more developed nations. More critical is the way in which funds are spent on internal transportation provision and, in particular, whether there are advantages in concentrating limited capital resources in either the road or the rail modes.

The appropriateness of different modes often depends upon the geographic–demographic nature of the country. Most less developed countries may be categorized as one of the following (Fromm, 1965):

(1) densely populated tropical lands;
(2) tropical land with low population density;

(3) mountainous, temperate lands with a low overall density of population but a concentration on a coastal plain or altiplano;
(4) thinly populated desert areas with population concentrated along irrigated channels.

The appropriateness of different transportation modes for facilitating economic development changes according to the type of country under consideration, thinly populated, tropical lands having different transportation problems from highly urbanized countries with high population densities.

While the railways were important in the development of 19th century economies and characterized colonial development in many countries, the emphasis in recent years has switched to the provision of adequate road infrastructure. This is particularly true in areas where a skeleton of roads already exists and resources can be devoted to improving and extending an established, if rudimentary, network. This approach may be especially fruitful if it links isolated agricultural communities both with each other and with the more advanced areas of the economy. Millard (1959) argues that, unlike developed countries, in Third World nations "the benefits from road construction are almost entirely in the form of new development from traffic which the new road will generate." The effect is not purely on immediate output but can stimulate a propensity for further development. Wilson (1966) supported road development in Third World countries for this reason, arguing that "Investment options might usefully be analyzed in terms not only of their direct economic pay-off but also in terms of their influence on attitude" and that "The educational and other spill-over effects of road transportation appear to be greater than those of other modes of transport. This is especially significant at low levels of development." Having said this, there is a danger if integrated planning is not pursued that while improved road facilities may stimulate the agricultural economy, the new links between rural and industrial–urban areas could lead to increased polarization in the spatial economy with an enhanced geographical, as well as sectional, dualism resulting.

6. Conclusions

The links between transportation systems and economic development are much disputed and the empirical evidence does not provide conclusive evidence on the nature of the links. The topic is, however, an important one for public policy-makers interested in ensuring that national resources are used most efficiently. The nature of the links influences both the speed of economic development and the way in which it is spread through the various components of the economy. Since in some cases it appears that improving transportation access may actually adversely affect economic growth in some areas, this would indicate that caution should be exercised in too readily advocating transportation expenditure as a panacea for all regions' economic ills.

References

Andersson, A. and U. Stromquist (1988) "The emerging C-society", in: D.F. Batten and R. Thords, eds., *Transportation for the future*. Berlin: Springer-Verlag

Aschauer, D.A. (1989) "Is public expenditure productive?", *Journal of Monetary Economics*, 23:177–200.

Baxter, R.D. (1866) "Railway extension and its results", *Journal of the Statistical Society*, 24:549–595.

Brehl, D. (1991) "The role of infrastructure in regional development", in: R.W. Vickerman, eds., *Infrastructure and Regional Development*. London: Pion.

Button, K.J. (1998) "Infrastructure investment, endogenous growth and economic convergence", *Annals of Regional Science*, 32:145–163.

Button, K.J. and P. Rietveld (1999) "A meta-analysis of the impact of infrastructure policy on regional development", in: H. Kohno, P. Nijkamp and J. Poot, eds., *Regional cohesion and competition in the process of globalisation*, vol. 2. *General trends and policies*. Berlin: Springer-Verlag.

Button, K.J., S. Leitham, R.W. McQuaid and J.D. Nelson (1995) "Transport and industrial and commercial location", *Annals of Regional Science*, 29:189–206..

Fogel, R. W. (1964) *Railroads and American economic growth, essays in econometric history*. Baltimore: Johns Hopkins University Press.

Fromm, G. (1965) "Introduction: An approach to investment decisions", in: G. Fromm ed., *Transport investment and economic development.* Washington, DC: Brookings Institution.

Hirschman, A.O. (1958) *The strategy of economic development*. New Haven: Yale University Press.

Jorgenson, D. (1991) "Fragile statistical foundations: the macroeconomics of public infrastructure investment", comment on paper by Hulten and Schwab, American Enterprise Institute Conference on Infrastructure Needs and Policy Options for the 1990s, Washington, DC.

Keeler, T.E. and J.S. Ying (1988) "Measuring benefits of a large public investment: The case of the US federal-aid highway system", *Journal of Public Economics*, 36:69–85.

Lugard, F.D. (1922) *The dual mandate in British tropical Africa*. Edinburgh: Blackwoods.

Millard, R.S. (1959) "Road development in the overseas territories", *Journal of the Royal Society of Arts*, 107:270–291.

Morrison, C.J. (1993) "Macroeconomic relationships between public spending on infrastructure and private sector productivity in the United States", in: J.M. Mintz. and R.S. Preston eds., *Infrastructure and Competiveness*. Ottawa: John Deutsch Institute for the Study of Economic Policy.

Munnell, A.H. (1990) "Is there a shortfall in public capital investment?", *New England Economic Review*, Sept.–Oct.:11–32.

Nadiri, I. and T. Mamuneas (1996) *Contribution of highway capital to industry and national productivity growth*. Washington, DC: Federal Highway Administration Office of Policy Development.

Rostow, W.W. (1960) *The stages of economic growth*. Cambridge: Cambridge University Press.

Sharp, C.H. (1980) "Transport and regional development with special reference to Britain", *Transport Policy and Decision Making*, 1:1–11.

Sturm, J.E. and J. de Haan (1995) "Is public expenditure really productive? New evidence for the USA and the Netherlands", *Economic Modelling*, 12:60–72.

Wilson, G.W. (1966) "Towards a theory of transport and development", in: G.W. Wilson, B.R. Bergmann, L.V. Hirsch and M.S. Klein, eds., *The impact of highway investment on development*. Washington, DC: Brookings Institute.

Chapter 17

BUS SERVICES: DEREGULATION AND PRIVATIZATION (LESSONS FOR THIRD WORLD CITIES)

ALAN ARMSTRONG-WRIGHT
Independent Transport Consultant, London

1. Introduction

There is a growing tendency towards deregulation and privatization around the world (Armstrong-Wright and Thiriez, 1997). In parallel there has been greater participation of private operators in the supply of bus services. Sometimes this has been the result of deliberate government policy, for example, as in the case of the U.K. Here the aim was to introduce market forces, improve supply, and reduce the heavy subsidies being paid for publicly owned services. In developing countries local authorities have, in addition, been faced with very rapid growth in demand that has overwhelmed their publicly owned services. Few have resources to allow subsidies to go on growing in pace with increasing demand. As a result, publicly owned services, in terms of quality and quantity, have seriously declined. The vacuum created has been readily filled by a wide variety of privately owned transport modes from autorickshaws to double-deck buses. This situation has evolved rather than being the result of formal government decrees to deregulate and privatize.

This chapter examines the good and bad experiences of economic deregulation and privatization of urban bus services in several countries, which represent a variety of approaches. It goes on to consider how developing countries might benefit from these experiences and avoid some of the pitfalls.

2. The U.K.: Outside London

Bus deregulation was initiated in the U.K. at the intercity level (routes of over 30 miles) under the 1980 Transport Act and for local services (outside London) in 1986. The objective was to allow market forces to cut costs and improve efficiency. General subsidies were banned. The system included commercial and non-commercial bus services (see Bayliss, 1997).

Handbook of Transport Systems and Traffic Control, Edited by K.J. Button and D.A. Hensher

For commercial bus services operators simply register the details of their proposed services and any subsequent changes, 42 days ahead of commencement. Provided the operator and the vehicles meet safety and environmental standards the registration cannot be refused. Different operators can operate over the same route and can compete. There is no restriction on the fares charged. They may be compensated for taking part in concessionary travel schemes.

Non-commercial bus services are those services that have not been taken up by commercial operators but are considered by local authorities to be necessary for social or economic reasons. Operators are invited to submit competitive tenders for the amount of payment they require to operate the services. The routes, schedules, and fares are specified by the local authority.

The results of deregulation across the country have been very mixed. Generally it has been very successful in reducing costs and subsidies, and in increasing the supply and frequency of services. However, fares have increased above inflation and the use of bus services has declined, due partly to the trend towards the car culture (House of Commons Transport Committee, 1995).

The tendency for too many operators to concentrate on the same high-demand routes has caused severe congestion in some city centers. Also, experience shows that, given the opportunity, some large and powerful operators are able to squeeze out smaller competitors by severe price undercutting. This problem has been addressed by requiring operators who deliberately undercut and squeeze out other operators to continue with the same low fares for a year. However, the process for dealing with malpractices is cumbersome and has taken several years to effect corrections. To improve the situation the authorities are to be given greater and swifter powers to deal with malpractices.

Detailed consideration has been given by the U.K. authorities to franchising (on the lines of the London contracting system below) as an alternative to deregulation. On balance they have decided to continue with the deregulation process. The objections to franchises are that they place on local authorities a very considerable administrative burden. Also, local authorities suffer from a serious lack of the necessary experienced technical staff to oversee franchises. Other objections concern the difficulties in canceling or modifying franchises, which may become necessary in the event of poor performance or changing circumstances. They are also likely to inhibit sustained competition.

In an endeavor to improve bus services and reduce dependency on the use of private cars, a few local authorities have introduced what are termed "quality bus partnerships" (QBPs). These are agreements entered into by operators and local authorities on a voluntary basis. The agreements vary from place to place but generally are on the understanding that

(1) *Operators* are prepared to invest in new, higher-quality, and more accessible buses, staff training, and better ticketing;

(2) *Local authorities* will invest in traffic management schemes which give priority to buses, or in better bus stations, shelters, and other passenger facilities and information.

While QBPs cover less than 10% of bus networks in the U.K., the improvements have generally been very significant, with patronage increased by as much as 40% in some cases. As a result, the government is considering placing the agreements on a statutory basis. Local authorities will be given the opportunity to introduce statutory QBPs if they wish. However, many are expected to continue with a variety of voluntary QBPs. Under the statutory QBPs operators not meeting quality standards could be excluded from use of the bus facilities provided by the local authority. Also government is considering introducing legislation to allow local authorities to enter into "quality contracts" with operators giving them exclusive right to certain routes. However, the exercise of these powers will require the approval of the transport ministry and will apply when it is the only way in which local authorities will be able to meet the policies set out in their bus strategies. They will also be able to introduce congestion charging, whereby drivers pay for entering congested areas. These various measures are expected to encourage greater use of public transport and decrease use of private cars (Department of the Environment, Transport and the Regions, 1999).

3. London

For many years bus services in London were mainly operated by publicly owned companies controlled by London Regional Transport (LRT). These were all sold in 1993 (London Buses, 1993; Comptroller and Auditor General, 1995). The privatization process used was designed to stimulate highly competitive bids and to avoid monopoly ownership arising. To avoid asset stripping, the process provided for LRT to receive part of any profit from the resale of properties. LRT assisted the management and staff of their subsidiary companies to take a stake in the privatized companies.

All the services are now provided on the basis of contracts let by LRT to the private operators. LRT prescribes the fares and services. There is no direct on-road competition for passengers, but there is an element of short-term competition as a result of the contracts being let by competitive tenders. Originally the contracts mostly were "gross cost contracts" in which LRT pays the operator a fixed rate for operating the route and takes all the revenue. The main drawback to this is the strict supervision required and the lack of incentive for the operator to increase revenue or improve quality of service.

To overcome the drawbacks of gross cost contracts, "net cost contracts" have been introduced, in which LRT pays the operator a settlement figure representing the difference between the operator's projected costs and revenue. The operator keeps all on-bus revenue and is allocated a share of off-bus revenue (mainly from Travelcards, bus passes and concessionary fares) (London Transport, 1999).

While net cost contracts have encouraged operators to increase patronage, and hence revenue, in some cases the pressure to cut costs has led to a lack of investment in buses and a decline in the quality of services. This runs contrary to government policy, which is to enhance the quality of public transport and to provide an attractive alternative to the use of private cars. As a result London Transport plans to introduce "quality incentive contracts" for London's bus services during 2000.

Quality incentive contracts, as the name suggests, will provide operators with the incentive to provide greater focus on quality. They will receive increased payments for positive quality achievements as well as for the volume of services provided. However, payments will be reduced when standards set out in the contract are not reached. Also, to reduce the complexity and administrative cost associated with net cost contracts, London Transport may revert to fare revenue accruing to itself, rather than the operators.

The effect of contracting and privatization has been a very substantial increase in revenue and reduction in subsidies (down 70%), and improved service quality. However, fares have increased substantially. There has also been a small increase in patronage, party due to the extensive integrated ticketing system which covers all public transport throughout London (Enoch, 1996).

The success of the privatization exercise in London can be attributed to the strength of LRT: the sales process was particularly complicated and time-consuming. Also, the execution and management of contracts for the supply of bus services has called for very high standards of supervision and integrity.

4. Accra, Ghana

For many years public transport services in Accra have been dominated by small private buses, and minibuses and shared taxis. Up until recently, the publicly owned Omnibus Services Authority (OSA) also ran buses in Accra. However, its Accra services have been withdrawn in the face of rising losses, overwhelming operating difficulties, and stiff competition from the private sector.

Government policy is to permit almost unrestricted access to the market. However, it has mandated the control of services to the Ghana Private Road Transport Union (GPRTU), which in effect has a monopoly with extremely wide powers. It vets new operators, allocates routes, and sets fares. It also collects income tax for the government and imposes fines for wrongdoings.

Generally, passengers are kept waiting until their bus is full before moving off: passengers en route have little chance of boarding. Many bus passengers face much walking and interchanging. Despite fares being very low, passengers are very dissatisfied with the services. An indication that many passengers would be prepared to pay more for a better service is the very large number using shared taxis at two to three times the bus fare. However, the 30 000 taxis add very substantially to serious congestion and pollution in Accra (Armstrong-Wright, 1993).

Currently the system seems to be designed for the benefit of the GPTRU and its members: government's liberal policy is not benefiting passengers. What seems to be needed is proper schedules, some larger buses, rational routing, and a number of premium services. The control of the system by a single union with vested interests should be replaced by some form of regulatory authority. While the operators should be free to set fares, they should be required to draw up and keep to schedules. In view of the economic and environmental benefits, operators should be given incentives to obtain better buses.

5. Santiago de Chile

In the course of the last 20 years bus services in Santiago have gone from being strictly regulated in the 1970s to being completely unregulated in the 1980s, and now in the 1990s routes are allocated by competitive tenders (Thomson, 1992).

Prior to deregulation, services were provided by a public sector company, ETC, private microbuses and taxibuses, and a few shared taxis. The state controlled fares, route allocation, frequencies, and bus imports. ETC, facing losses, a hard line on subsidies and increased competition, was unable to sustain operations and was closed down in 1981. By 1983 services were completely deregulated, which resulted in a massive increase in supply, improved frequencies, and reduced overcrowding. However, in the first 10 years, because of alleged price fixing, fares escalated in real terms up to 158% on microbuses and 87% on taxibuses. By far the most serious result of deregulation was very severe congestion and pollution.

To overcome these drawbacks, in 1992 the government introduced a system of licensing operators, setting fares and allocating routes by competitive tenders. To soften the bus operators' hostility to the arrangements, the government bought up older buses which became surplus under the new system. The effect of these arrangements has been a substantial reduction in buses operating in the city (down 35%), a lessening of congestion in the main streets, and a reduction in fares. Whether or not the initial improvements can be sustained will depend on compliance with the tender conditions, which calls for a strong administrative commitment by the authorities.

6. Kingston, Jamaica

At one time bus services in Kingston were successfully run for many years by a private company, Jamaica Omnibus Service (JOS). However, when fares set by the government became unprofitable, the services of JOS declined so badly that the government felt obliged to take it over. Productivity then dropped further, costs rose, and government lost over U.S. $1 million every month. So the government then leased the assets to the private sector.

The privatization process involved the bus routes in the metropolitan area being divided into ten packages with the franchisee of each package paying an annual fee. In turn the franchisees sublet the routes to individual minibus drivers, and exercised little control. This arrangement, with fares once again set too low, led to yet a further deterioration.

In an effort to rectify the situation the government commissioned a major transport study, which concluded the need for five exclusive franchises, more rigid control and fares set at commercial levels. At the time bids were invited for the new franchises, no decision had been made on fare levels. Also, very few bidders, if any, were able to meet all the required criteria. Nevertheless the franchises were let but without a new fare table. To overcome the obvious difficulties this caused, a system of temporary subsidies was introduced. Once again, the outcome was disappointing.

Clearly, the policy on commercial fares needs to be implemented; the private operators are not yet sufficiently developed and need both technical and financial support; and financial uncertainties need to be removed if overseas investors are to be attracted. Also the local authority will need to be strengthened if the proposed franchising system is to be pursued.

7. Hong Kong

The Hong Kong public transport system comprises a variety of bus and minibus services, rail mass transit, light rail transit, and tram and ferry services (Hong Kong Transport Department, 1999). None of these services are subsidized. Government is the major shareholder of the rail mass transit and light rail transit corporations and expects to receive a return on capital. All the bus, minibus, tram, and ferry services are privately owned and operated. Most are given operating rights by either franchises or licenses in some way.

Conventional bus services are provided by six franchised private companies, using some 6000 high-capacity buses, mainly double-deckers. They have operated generally on a district basis, though there is some overlapping of routes: in particular they share the high-demand cross-harbor tunnel routes. Three franchisees are exceptionally large for private concerns, with 4000 buses in one case and 700–8000 buses in each of the other two cases.

Fares, routes, and timetables for franchised services are determined by the Commissioner for Transport in consultation with operators and subject to government approval. The Transport Advisory Committee, made up of community leaders and transport officials, provides advice on these topics.

Hong Kong has recently introduced a computerized multitrip and multimode ticketing system. This makes use of smart cards ("Octopus cards") valid on competing bus, ferry, and metro services and some minibuses. The system accurately shares out the revenue and so overcomes the major barrier to use of multitrip tickets on competing services.

Franchises are let on a competitive basis and performance is closely monitored. When the franchise of a long-established but unsatisfactory operator expired in 1998, the network was put out to competitive tender. A franchise was granted to a new company. The criteria for selection of the franchisee included experience, resource capability, proposed service and fare schedules, age of vehicles, and availability of maintenance facilities. The result has been the introduction of quality services, including some air-conditioned buses, at competitive fares. The fare levels, including any changes during the franchise term, are subject to the approval of the Chief Executive in Council. Fare increases are based on well-established criteria such as the franchisee's financial situation and public acceptability.

A large number of minibus services (Green Public Light Buses: PLBs) have also been let by tender. Generally these are for routes unsuitable for high-capacity buses. There are about 2300 Green PLBs in operation.

In addition, there are 2000 minibuses which are only partly regulated. These are called Red Public Light Buses. They are authorized to carry passengers at fares, routes, and schedules determined by the operators. They are prohibited from operating in the new towns and in new housing estates. Within their operating areas, they are free to compete with franchised bus services. Standing is not permitted in Public Light Buses and the size of the vehicle is limited to 16 seats.

There are also a large number of non-franchised buses. These are mainly premium coach services designed to meet the needs of people living in new residential developments. These services are run by private operators under contract to the developers of the residential areas or to associations of residents. Although these services are regulated to avoid conflicts with the franchises of the main bus companies, the fares are not fixed by the government. Most of the residential coach services use medium-size buses, usually air-conditioned, each with a capacity to seat about 50 passengers. There are similar services for transporting factory workers and schoolchildren.

The franchising system allows sanctions to be imposed on operators who fail to perform satisfactorily. Apart from revocation and non-renewal of the franchise, financial penalties may also be imposed in case of non-compliance with the franchise terms by the operator. The mechanism is effective, as demonstrated by the recent case of a long-established but unsatisfactory operator being replaced by a new operator for services on Hong Kong Island. The Government also has powers to withdraw one or more routes (including profitable routes) from a franchise.

Much of the success of the Hong Kong public transport system is attributed to the inherent management and entrepreneurial skills of the local bus companies. However, there is little doubt that franchised services of this type require a high degree of planning, administration, monitoring, and control. In the case of Hong Kong, this has been possible because of the high standard of professional supervision displayed by staff of the Transport Department, working within a strong and sound administration.

8. Riga, Latvia

Public transport services in Riga are provided by three municipal transport enterprises owned by the Riga City Council (RCC), and a very limited private sector. None of the services is adequate and there is much public dissatisfaction. The low level of service arises because of the cost and difficulties in maintaining the existing, aging fleets. In the Soviet era there was no lack of spares at low, subsidized cost. But now suppliers require payment of the full cost in hard currency, which the operators cannot afford.

The revenue from the municipal public transport services is unable to cover more than about 67% of operating costs. This is mainly due to fares set by RCC still being too low despite recent increases, too many concessionary passengers, and widespread fare evasion. All the enterprises rely on high subsidies based on total vehicle/km, rather than passengers carried, providing little incentive to consider passengers or to attract more passengers.

There is a small fleet of 70 minibuses run by a state-owned company without subsidies and at fares roughly double RCC bus fares (Bushell, 1996–97). Clearly people are prepared to pay much more for a more convenient and comfortable service.

Private buses and minibuses have recently increased substantially, due mainly to fare increases in the competing public sector. However, interest is dampened by the lack of compensation for concessionary trips, and the high cost of credit for the purchase of vehicles. The lack of a suitable legal and regulatory framework discourages local and foreign participation and funding.

To overcome the shortcomings, RCC is pursuing a policy that will lead to the privatization of municipal transport enterprises. It has set up a Public Transport Unit to oversee the process. In the short term the municipal enterprises are being corporatized and supply services on the basis of contracts with RCC, which include operating subsidies and fares set by RCC.

The pace of privatization will depend on how soon the municipal enterprises become financially viable and the private sector becomes sufficiently developed and competitive. Major privatization problems may well arise because some of the land occupied by the municipal enterprises is owned by many different returning exiles ousted by the Soviet Union. They are now entitled to reclaim their land. Also, without a suitable legal framework, legitimate commercial interests are likely to be discouraged, with the risk that the vacuum will be filled by unscrupulous operators.

9. Nairobi, Kenya

Nairobi has a population of close to three million and an annual growth rate well in excess of 5%. The rapid expansion of the population and spread of the urban area of Nairobi in recent years has put a very considerable strain on the public transport system and the road network. As a result, bus services are overloaded and roads in the center and along commuter corridors are heavily congested during peak periods. By far the majority of commuters are low-income workers. At least 25% of all trips are made on foot – some of them being in excess of 10 km.

Public transport in Nairobi is provided by the private sector, comprising large buses operated by a franchised bus company, Kenya Bus Services (KBS), and a large paratransit fleet of small buses and converted pickups, called *matatus*. There is also a public sector bus system, Nyayo Bus Services, owned by the government.

While *matatus* have never been subject to much control, economic liberalization measures adopted by the government in 1994 in effect completely deregulated bus services. These measures also included the relaxation of restrictions on the importation of vehicles and gave rise to a massive increase in *matatus* and private cars.

Matatus operate informal paratransit services, with an element of route control exercised by operators associations. Originally *matatus* were mainly converted pickups, but in recent years these mostly have been replaced by 12 and 25 seat minibuses, usually with locally made bodies. Generally, *matatu* owners own only one or two vehicles, but a few own small fleets. They provide a very cheap and basic form of transport.

For many years KBS, using large buses, was owned and operated by United Transport International. In 1991 it sold out to Stagecoach Holdings, which in turn in 1998 sold its 95% shareholding to a consortium of local businessmen. Stagecoach Kenya Bus, in which the Nairobi City Council had a 5% stake, operated about 320 large buses, including some double-deck buses, in and around Nairobi. Under the terms of its franchise the company determined its own routes, fares and schedules without reference to any regulatory authority. This freedom, along with its management skills, allowed Stagecoach to provide services without any subsidies that initially were both popular and profitable. In particular, Stagecoach had been able to meet demands for more reliability, comfort, and convenience than that provided by *matatus* at roughly the same fares. As a result, Stagecoach was able to retain about 50% of the market despite cutthroat competition from some 2500 *matatus*. The fact that in areas of Nairobi where Stagecoach did not operate, *matatu* fare levels were higher is an indication of the effectiveness of its terms of franchise.

This type of franchise places very little administrative or supervisory burden on the authorities. In the case of Nairobi, far from being a financial burden, initially Stagecoach was in a position to make a small return for the City Council. It was thus in the city's interests to provide facilities to aid the smooth operation of the services. Ironically, one of the most serious impediments to cost-effective operation of Stagecoach's services was the poor state of the roads, and lack of road maintenance. In particular, services were affected by the massive road damage caused by the devastating rainstorms (El Niño) of 1997 and 1998. This led to Stagecoach making substantial losses and hence being sold to local interests. Also, services were increasingly hampered by severe congestion and a lack of effective traffic policing. Since the takeover, road rehabilitation programs (financed by the African Development Fund) have been initiated, bus operating costs have been trimmed, and the new company has substantially cut fares.

The government's Nyayo Bus Services, established in 1986, relies mainly on youth service volunteers as drivers. It has also lacked technical and managerial support. Despite heavy capital and operating subsidies, Nyayo has never provided effective services. Out of its fleet of some 250 new buses, it has rarely put into service more than 40–60. Thus it is not surprising to find that in the highly competitive and commercial environment that prevails in Nairobi, Nyayo has not survived: its buses are now rarely, if ever, seen operating in the city.

10. Lessons

The experience of economic deregulation and privatization of bus services has highlighted many problems and pointed to possible solutions in the cities concerned (Maunder and Mbara, 1996; Palmer et al., 1997). It may be possible to apply these solutions to other cities. Clearly in doing so some adaptation will be needed because of widely different local conditions that can be expected. For example: the economies; the institutional structures; the strength of the private sector; and the earnings, aspirations, and priorities of the inhabitants. However, in general terms, the following factors may be worthy of consideration.

10.1. Objectives

The objectives or expectations of deregulation and privatization of bus services generally have been to introduce competition and market forces so that the supply of bus services more efficiently meets demand. More specifically, the aims have been to

(1) reduce costs,
(2) improve revenue collection,
(3) produce services that are more responsive and innovative in meeting the needs of the public with regard to both quantity and quality,
(4) encourage investment in services,
(5) eliminate or at least reduce subsidies for the supply of services,
(6) encourage greater use of public transport (rather than cars).

10.2. The need for some regulation even in a "deregulated system"

While there should be fewer restrictions on operators wanting to take part in the provision of bus services, it is evident that a certain amount of control is necessary. This needs to be considered and in place, along with the setting up of an effective authority, before deregulation is launched. The authority should have sufficient powers and resources to

(1) oversee the deregulation process and monitor services,
(2) deal swiftly with malpractices,
(3) supervise any contracts or franchises that may be let,
(4) enforce vehicle and driver safety requirements,
(5) impose penalties for malpractices and award compensation to the victims of malpractices,
(6) register operators and their vehicles.

Franchising (on the lines of the London contract system), as an alternative to deregulation, is likely to require a much stronger authority. Most forms of franchising involve a very considerable administrative burden, and to be successful require experienced technical staff. The more relaxed type of franchise, similar to that in Nairobi, does not seem to have these drawbacks, but relies heavily on the commitment of the owners to provide a satisfactory service.

10.3. Encouraging public transport

It is evident that the chances of achieving favorable results with deregulation are improved where local authorities have adopted a policy of encouraging the use of public transport (e.g., in Oxford, U.K. – see Enoch, 1996). To this end they have provided bus priority measures, including special rights of way within city centers, effective park and ride schemes, and facilities for passengers. At the same time they have discouraged the use of private cars in city centers by exercising tough parking controls and imposing parking charges.

10.4. Decontrol of fares

All the evidence suggests that, where fares are set by operators in a competitive market, the objectives of deregulation have a better chance of being achieved. On the other hand, the deregulation process has been impaired in several countries because the authorities have found it difficult to allow operators to set their own fares. In some cities the public transport system has foundered because fares were too low to sustain viable services (Kingston provides a classic example). This has often given rise to heavy subsidies that could not be sustained. Fears that decontrol would lead to seriously escalating fares have generally been unfounded. In many places fares have had to rise above subsidized levels to achieve financial viability, but rises have not been excessive and are generally affordable. (Where fares have escalated, it is usually because anticompetitive practices have been allowed to go unchecked, as in the case of Santiago.)

10.5. Fare evasion

Revenue loss, in particular, fare evasion, seems to be far less of a problem for private operators than it is for public operators. Nevertheless, where this has become a problem, mainly on large buses, some operators have found it cost-effective to reintroduce conductors (e.g., in Riga and Nairobi).

10.6. Non-commercial routes

A frequent problem facing city authorities is that a number of routes that may be economically or socially desirable are not financially viable at affordable fares. As a result they may be ignored by operators in an open market. A solution adopted in a number of countries (U.K. in particular) is to put these routes out to contract on the basis of competitive tenders. Where there is a lack of competition, negotiated contracts may be necessary. In either case such routes may need and justify subsidization.

Contracts for non-commercial routes on the basis of gross cost contracts are likely to give rise to problems. This is because operators have little incentive to increase revenue or improve the quality of service. A more appropriate solution for developing countries would be some form of net cost contract with payments closely related to passengers carried. Payments could also be related to quality of services (such as London's quality incentive contracts), with increased payments for positive quality achievements and reduced payments when standards set out in the contract are not reached. Such contracts require a high standard of supervision, with regular detailed surveys and inspections, which may be difficult for most local authorities to achieve. The introduction of effective ticketing systems would greatly enhance the benefits and control of net cost contracts.

10.7. Commercial law

In a number of countries, for example those previously in the Soviet Union, such as Latvia, the existing law may not readily cope with the commercial environment needed to facilitate privatization. To encourage legitimate commercial interest in the provision of public transport, the legal framework needs to cover rights to property, company and foreign investment, contract law, liquidation and bankruptcy, antimonopoly provisions and enforcement, and resolution of disputes. This will be vital if the intention is also to seek the participation of foreign investors.

10.8. The capacity of the bus industry

For full deregulation to be successful the local bus industry needs to have sufficient capacity to meet demand and to provide effective competition. If the industry has limited capacity operators will concentrate only on the most lucrative routes. With a lack of competition, fares will be high and services less responsive to the needs of the public. As a result, there may be pressure to reintroduce controls, particularly on fare levels, which will probably make matters worse.

In many developing countries (e.g., Ghana, Jamaica, and Latvia) the private sector is held back by a lack of suitable vehicles and funding. Clearly they are likely to benefit from assistance with the purchase of buses and help with access to foreign exchange. Ongoing assistance could be in the form of fuel tax rebates. This would have the added benefit of giving bus travel a cost advantage over the use of private cars (which, in any case, should be restrained by higher taxation and parking charges, etc.)

In some developing countries, while the size of the private sector engaged in public transport may be very small, often there are many "own account" bus services operated by factories and collective farms, etc. Given the opportunity and encouraged by deregulation, these may well be interested in expanding into the provision of public transport bus services.

Where the public sector dominates and where private operators are unlikely to cope for some time, as an intermediate stage before deregulation, public operators should be placed on a more commercial and competitive basis. (e.g., the approach in Riga). At the same time, the private sector can be nurtured with assistance with funding as mentioned above, accompanied with some relaxation of regulations to stimulate interest.

10.9. Private monopolies

A problem that has occurred in a number of cities (in the U.K. in particular) is that deregulated services have been dominated by a few large operators, resulting in private monopolies and the risk of malpractices. An absence of competition is likely to lead to higher fares and a lack of response to the needs of the public. This problem may well be avoided by having in place procedures for dealing with anticompetitive practices before deregulation and privatization. One solution has been to require operators who blatantly undercut competitors to be required to continue with such low fares for a stipulated minimum period. Also, if the situation so demands, authorities should be in a position to see that the timetables of large competing operators are reasonably well spaced, and that they are required to keep to them as far as possible.

The process of dealing with anticompetitive practices has in some cases been very time-consuming. Thus, in a deregulated environment, the authorities need to be given adequate and swift powers for dealing with malpractices before they get out of hand. But at the same time they should not be allowed to interfere with healthy commercial practices, including stiff competition.

10.10. The reliability of services

It is evident that one of the factors that discourages passengers from using public transport is a lack of reliability (e.g., in Accra). This can be improved by requiring bus operators to publicize routes and timetables, and for lengthy notice to be given in respect of any changes. However, the introduction of measures to improve reliability is likely to put up administrative and operating costs. While those in low-income countries would benefit from more reliable services, this has to be balanced against their desire for the lowest possible cost of trips.

10.11. Multitrip ticketing

In some cities, deregulation has resulted in multitrip ticketing being discontinued because of the problems of allocating revenue between competing operators. This has been both inconvenient and costly for some passengers, and is one of the reasons why there has been a reduction in patronage in some cities in the U.K. Problems with revenue sharing can be overcome by regular and independent passenger surveys and inspections. Another, more effective solution is provided by electronic ticketing technology (e.g., the Hong Kong "Octopus" smart card system). This facilitates integration of competing operators, and provides accurate revenue sharing as well as other opportunities to improve revenue collection. Some developing countries are sufficiently advanced to introduce similar electronic ticketing systems, and it may not be too long before others are able to follow suit.

10.12. Concessionary travel

A similar revenue-sharing problem to that of multitrip ticketing has also arisen with concessionary travel passes. An additional problem is the natural resistance of private operators to concessions that undermine viability (e.g., in Riga). Where the state or city requires certain groups of passengers (e.g., students, the elderly, etc.) to be carried at concessionary rates, resistance has been overcome by compensating operators for any loss of revenue. Some operators (e.g., in Hong Kong and Nairobi) voluntarily provide concessionary travel based on their own commercial judgment (e.g., to make use of spare capacity in off-peak periods) and do not expect to be compensated.

10.13. Privatization process

Clearly, it is important to develop a detailed strategy in advance of privatization (as in the case of London) rather than dealing with problems as they arise (e.g., in Kingston). The following are examples of factors that may need to be taken into account:

(1) Private funds are often limited in developing countries, and there may be a need to phase the sales program to ensure sufficient competition in the bidding process. Most publicly owned bus enterprises are divided into fairly unique operating areas, each served by one or more depots. Where private funds are likely to be limited, it would be appropriate, as a first stage, only to privatize one of the most attractive operating areas. Subsequent stages would be undertaken in the light of experience with the first stage.

(2) There may be advantages if services up for sale are packaged to attract both large overseas and small local investors.

(3) Private purchasers of bus companies will need to be vetted for financial standing, administrative and operational experience, and ability. It will be

necessary to ensure that there is sufficient competition in the bidding process to attract satisfactory bids, and to avoid one private operator securing an extensive monopoly.

(4) To take advantage of their experience and skills, management and staff of the existing publicly owned bus companies should be encouraged to bid for the services up for sale. This, together with financial and business management assistance, may also help to reduce any redundancy problems likely to arise.

(5) The terms for the ownership of any land and buildings included in the sale will need to be carefully spelled out. To avoid asset stripping "property claw-back" needs to be a condition of sale. The valuation of assets needs to be undertaken before calling for bids in order to provide a benchmark for the evaluation of bids and for negotiations.

(6) To be successful the bidding process will need to be conducted in a fair, open and impartial manner. Bidders should be given full details of the operations, administration, finances, and assets of the companies up for sale, and the arrangements for dealing with existing staff.

11. Conclusion

While the results of deregulation and privatization of bus services have been very varied, there is little doubt that the process can be successful in providing affordable and efficient bus services at acceptable standards. The need for better bus services, already obvious in most Third World cities, is becoming increasingly important as city authorities grapple with the problems of rapidly growing populations and urban road congestion. All the indications are that for success, rather than quick fixes, there is a need to develop well-thought-out deregulation and privatization strategies having regard for local conditions, and the pitfalls and good results experienced elsewhere.

References

Armstrong-Wright, A. (1993) "Public transport in third world cities: State of the art review", No. 10, Transport Research Laboratory. London: HMSO.

Armstrong-Wright, A. and T. Thiriez (1997) "*Bus services: Reducing costs, raising standards*", World Bank, Washington, DC, Technical Paper No. 68.

Bayliss, D. (1997) "Bus privatisation in Great Britain", *Proceedings of the Institution of Civil Engineers, Transport*, 123(2):81–93.

Bushell, C. ed. (1996–97) *Jane's urban transport systems*. London: Jane's Information Group.

Comptroller and Auditor General (1995) *The sale of London Transport's bus operating companies: Report by the Comptroller and Auditor General*. London: HMSO.

Department of the Environment, Transport and the Regions (1999) *From workhorse to thoroughbred: A better role for bus travel*. London: HMSO.
Enoch, M. (1996) "Oxford and Darlington – The mess and success of bus deregulation", *Proceedings of the Chartered Institute of Transport*, 6(1):29–47.
Hong Kong Transport Department (1999) *Annual transport digest*. Hong Kong: Hong Kong Transport Department.
House of Commons Transport Committee (1995) *The consequences of bus deregulation*, Report of the House of Commons Transport Committee. London: HMSO.
London Buses (1993) *The road to privatisation*. London: London Buses Limited.
London Transport (1999) *The bus tendering process*. London: London Transport.
Maunder, D.A.C. and T.C. Mbara (1996) *Liberalisation of urban public transport services: What are the implications?* Crowthorne: Transport Research Laboratory.
Palmer, C.J., A.J. Astrop and D.A.C. Maunder (1997) "Constraints, attitudes and travel behaviour of low income households in two developing cities", Transport Research Laboratory, Crowthorne, Report 263.
Thomson, I. (1992) "Urban bus deregulation in Chile", *Journal of Transport Economics and Policy*, 26:319–326.

Chapter 18

TRANSPORTATION IN SPARSELY POPULATED REGIONS

ÅKE E. ANDERSSON
Royal Institute of Technology, Infrastructure and Planning, Stockholm

1. Introduction

Transportation planning and policymaking is highly dependent on the density of population and economic activity. While excessive congestion is a primary problem to be handled by policymaking in dense regions, inefficient use of capacity is the central problem of the sparsely populated regions. The mirror image of excess capacity is the problem of financing investments and reinvestments in the roads, railroads, and airports of these regions. Taxation of fuels and other charges related to the use of transport equipment as a means of covering infrastructural costs tend to reinforce the problem of excess capacity.

One of the central problems of the sparsely populated regions is the combination of economies of scale and low accessibility levels almost everywhere. The consequence of this combination is monopolistic pricing of transportation services as well as of other consumer and producer services. Improvement of the transport and communication systems in such regions can lead to changes in the market form, a consequence rarely taken into account in the cost–benefit studies of transportation investments.

Transportation investment as a means of furthering regional development has become a major issue of analysis. In econometric studies it has been demonstrated that investment in transportation infrastructure mostly has a significant development effect. A study of infrastructural investments in Sweden has, for example, shown that investments in roads and airports have a stronger impact on regional productivity than investment in railroads and port infrastructure.

2. Sparsely populated regions in developed countries

Among the economically developed countries, only a few could be reasonably classified as sparsely populated. To this group belong the countries of recent

Handbook of Transport Systems and Traffic Control, Edited by K.J. Button and D.A. Hensher

immigration – the U.S.A., Australia, Canada, and New Zealand. The other group is located in the northern periphery of Europe and consists of Iceland, Norway, Sweden, and Finland. Table 1 reveals some of the important statistical facts about these countries.

As can be seen from the table, the average density of population of all these countries falls well below 30 inhabitants per square kilometer, on average. However, the population distribution is very skewed, especially in the countries of late immigration. The population share of the three largest regions – Sydney, Melbourne, and Adelaide – is, for instance, approximately 50% of Australia's total population, while the corresponding share of the three largest metropolitan regions in Sweden is approximately 35% of the total population.

This implies that the sparsely populated countries are facing traffic system and congestion disequilibria as part of their transportation policy problems. At the same time they have to cope with equally complicated problems caused by transportation system indivisibilities, combined with a level of demand that is insufficient to cover long-run transportation costs.

3. The excess capacity problem of sparsely populated regions

Equilibrium modeling of transportation flows is quite a relevant approach for any given transportation network connecting the different parts of a sparsely populated region. The generalized prices of transportation can be assumed to be given and independent of transportation flows. Thus, the expected prices will correspond to the realized prices (disregarding the randomness caused by meteorological variations). Accordingly, each transportation plan will be optimized independently of other transport system users so as to maximize the profit or utility of the transport between the origin and destination. There are thus no congestion or other externality problems in analyzing short-term equilibria in the case of transportation and traffic within sparsely populated regions.

However, in the long run the problems of equilibrium solutions are notoriously complicated by the fact that most of the links will have excess capacity. This is of course due to the fact that there are unavoidable indivisibilities of investment in links, crossings, and terminals in any transportation system. These indivisibility problems tend to increase over time as a consequence of technological improvements of vehicles and increasing engineering standards. In earlier times, of stagecoaches and other simple means of transportation, indivisibility of infrastructure and excess capacity was a minor problem. Over time this has become the major problem in the provision of transportation infrastructure and services for sparsely populated regions.

In countries of low average density, investment in roads and railroads within and between sparsely populated regions with limited total demand and low

Table 1
Population, income, and transportation characteristics of sparsely populated countries

	Income (purchasing power parity) per capita	Population (million)	Population density per km^2	Vehicles per 1000 inhabitants	Air trips per inhabitant	Share of paved roads (%)
Australia	100	19.0	2.9	605	1.59	39
Finland	93	5.2	16.8	448	1.31	64
Norway	120	4.5	14.2	498	3.18	75
Canada	106	30.6	3.2	560	0.81	35
New Zealand	86	3.8	13.6	579	2.29	94
Sweden	95	8.9	21.5	468	1.34	78
U.S.A.	137	272.9	28.7	767	2.15	59
Netherlands	103	15.8	456.0	421	1.18	90
Japan	115	126.6	333.2	560	0.80	75
World		6 000	44.0	116	0.23	43
Year	1996	1999	1995	1996	1998	1996

Source: World Bank.

willingness to pay for transport infrastructure tends to come into budgetary and other political conflicts with the larger demand and willingness to pay for corresponding projects in densely populated regions. Traditional benefit–cost calculations can, in these cases, rarely support decisions to favor investment in the sparsely populated regions. Politicians therefore regularly resort to distribution rather than efficiency criteria to motivate their investment policies.

Roads, railroads, and even air transportation infrastructure are, as a consequence of a lack of demand and of willingness to pay, financed by general taxes as well as by car, truck, gasoline, and diesel fuel taxes. Car, truck, gasoline, and diesel fuel taxes are especially inefficient income sources when levied on households and firms in sparsely populated regions. These taxes are often a dominant relative part of the generalized price of travel and transport of goods and will therefore unreasonably constrain transport demand in regions with excess capacity. To the extent that these taxes are not motivated by purely fiscal or other political arguments, they ought to be differentiated between congested and uncongested regions in order to optimize relative use of transportation capacity.

The ideal way of financing transportation infrastructure is by taxation of land rents. Most of the rental value of land – in the absence of speculation, the price of land – is determined by variations in accessibility over the transportation networks. This close link between transport accessibility and land rents has been demonstrated theoretically by a large number of economists, e.g., von Thünen

(1826), Beckmann (1957), Mills (1972), Beckmann and Puu (1985), Andersson and Karlqvist (1974), Fujita (1989), and Sasaki (1987). Taxation of land rents in order to finance transportation infrastructure would also be consistent with a generalized version of Frank Ramsey's theory of optimal taxation (Sahlin, 1990), requiring taxes to be levied on price-inelastic factors and goods.

4. The problem of spatial monopolies

Perfect competition in goods and service markets can only be achieved in a world of costless transportation. In a world with considerable cost of transport, monopolistic competition is the generic market form. In the sparsely populated parts of the world, this market form easily degenerates into pure monopoly as a consequence of the combination of the high cost of transport and the sparse spatial distribution of demand. In sparsely populated and distantly located regions it is mostly not possible to have more than one train, airline, or even bus connection per day. The transportation companies controlling these means of transportation tend to charge a monopoly price or even full monopolistic discrimination prices. From the point of view of use of capacity, monopolistic price discrimination is more efficient than uniform monopolistic pricing. A price discrimination that ranged between a very high price charged to the least price-sensitive group down to a price corresponding to the marginal cost would lead to an optimal use of capacity. In contrast, a uniform monopolistic price would lead to a low use of transportation capacity. With perfect price discrimination the average price charged to the users of the transportation system is close to the uniform monopoly price, while the use of capacity corresponds to the point of equality between price and marginal cost. However, the reduced real income under monopoly pricing would then imply a crowding-out of other services by the transport producers.

The extent of monopolistic pricing by transportation firms can be illustrated by an example from 1999 in Sweden. The cost of traveling by SAS from Gothenburg on the Swedish west coast to Visby on the island of Gotland (a distance of approximately 500 km) was U.S. $850, while at the same time it was possible to buy a Stockholm–New York one-week return ticket at two-thirds of this price.

Very often cost–benefit analyses are carried out under an assumption of perfect competition or, alternatively, are a subject of some "second-best" calculation based on some other assumed market form. The latter procedure is the more reasonable one in most cases. However, as has been pointed out by Blum (1992), large transport network changes can lead to a change of market form. Improving the transportation possibilities can lead to a new and more competitive situation with reduced consumer prices as a consequence.

The opening up of e-commerce as a consequence of improved communication infrastructure is a case of special importance for the future. Many formerly protected suppliers of consumer and producer goods and services in sparsely populated regions are facing a new and much more competitive situation than before as a consequence of lowered information and other transaction costs, especially in the case of books, recorded music and other easily transported goods. For those consumers who are capable of handling the new technology, IT will undoubtedly mean an improvement of welfare both in terms of reduction of monopoly power and new possibilities of telecommuting jobs. For groups with insufficient education and other skills to cope with the information technology, life could become harder in the rural areas.

Benefit and cost calculations ought to be modified to reflect the endogeneity of the market form during changing transportation and communication conditions. Similar complications arise if substantial technological externalities are caused by improved efficiency of the networks.

5. The importance of accessibility

Accessibility to jobs and public and private services is central to transportation planning, especially in sparsely populated countries. An axiomatic foundation of accessibility measurement has been provided by Weibull (1976). An axiomatically consistent accessibility measure is, for example,

$$A_i = \sum_j f(d_{ij})C_j,$$

where A_i is the accessibility of location i, d_{ij} is the distance from location i to location j, $f(d_{ij})$ is the distance friction function, non-decreasing with increasing d_{ij}, and C_j is the capacity of, for example, jobs or public or private services in location j. In most sparsely populated regions, service accessibility is to be measured within daily commuting distances. This means that it is the density of population or rather the density of demand within the region that constitutes the accessibility of importance to the consumers of services.

The problem of such local service provision was first treated by Chamberlin (1946). He modeled a process in which a service firm initially enjoys a monopoly position because of protection of transport costs. If the service firm generates sufficiently large monopoly profits there is, however, an attraction to opening up other similar service firms, until a monopolistic competition equilibrium is reached, at which each firm will cover its average cost. The denser the region, the more firms will be established and the easier it will be for the consumers to find a competitive provider. The lowest monopolistic competition prices for any given service will be found in the most densely populated (metropolitan) regions at any given service quality. This is illustrated in Figure 1.

It is also possible that the density is so low that the monopolistic price is everywhere lower than the long-run average cost. At sufficiently low densities of demand there would not even be a possibility of covering the long-run average costs, even if perfect price discrimination according to the individual willingness to pay were possible.

The availability of different services varies accordingly with the population size and density of the region, as can be seen in Figure 2, estimated for Sweden.

Demand thresholds do of course differ between different types of services. There is a very strong relation between the population density of Swedish regions and the number of service branches represented in the regions, as can be seen from Figure 2. The consumers and producers in sparsely populated regions are thus highly dependent upon efficient transportation for their services.

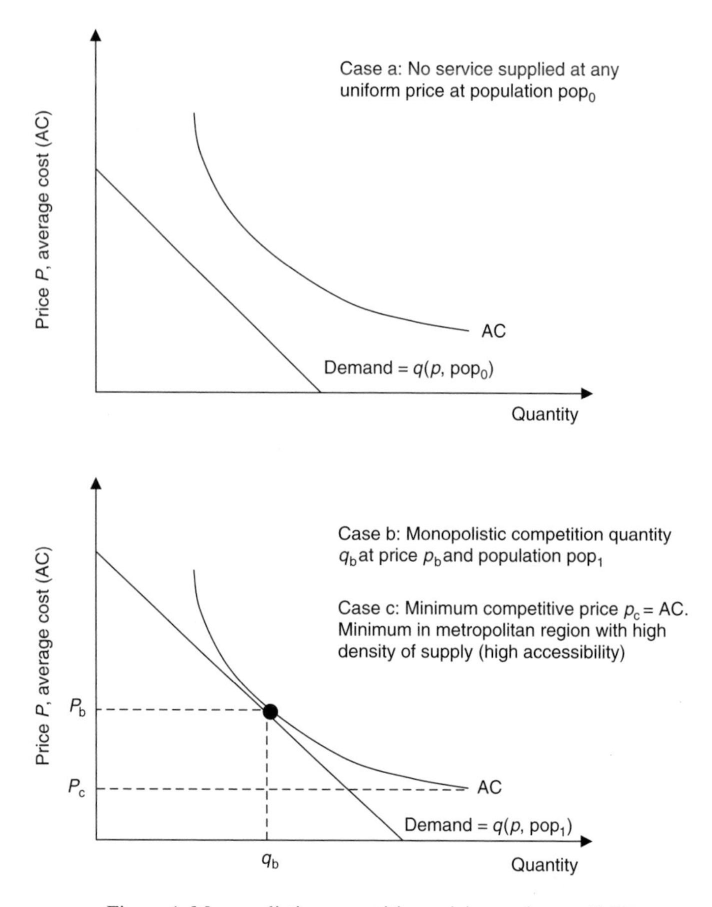

Figure 1. Monopolistic competition pricing and accessibility.

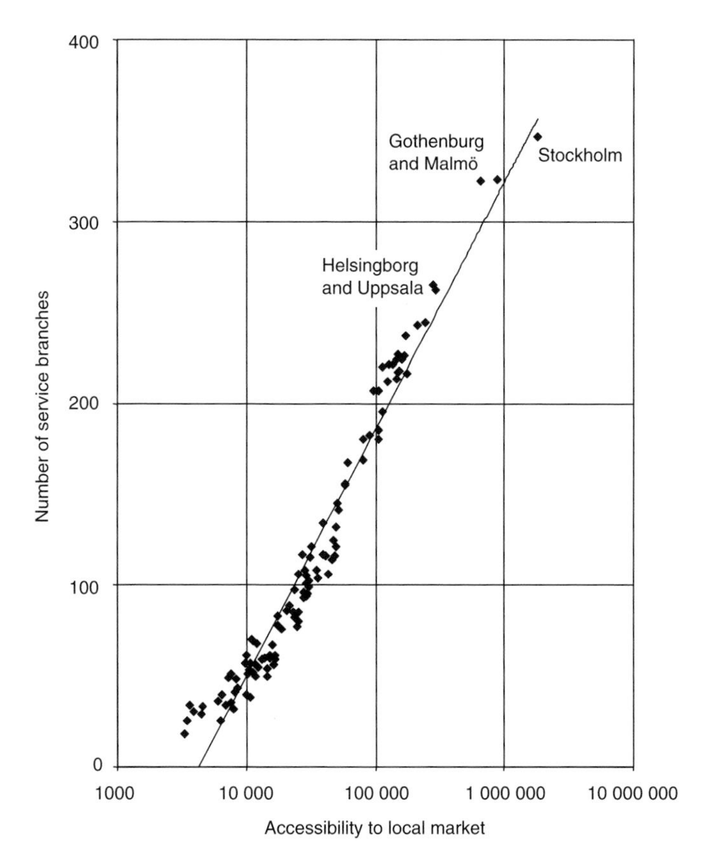

Figure 2. Service capacity and accessibility (*source*: Regioner, handel och tillväxt, 1998).

Many studies of transport demand have shown that rural mobility tends to be higher, when the differences in income, ethnicity, and sex have been accounted for. One example of such a study is included in the U.S. Department of Transportation *Annual Report* of 1997 (Department of Transportation, 1997), which gives the passenger miles per day in the U.S.A. for 1990 (Table 2).

People living in rural areas have to cope with a lower diversity of services than do similar residents of metropolitan regions.

6. Transport quality and optimal maintenance investment

Households and firms located in regions with sparse traffic flows regularly complain about the low quality of the transport infrastructure and the lack

Table 2
Miles of daily travel in the U.S.A., 1990 (age 16–64 years)

Population	Distance traveled (miles/day)
White men in rural areas	43
White men in urban areas	39
Black men in rural areas	30
Black men in urban areas	24

Source: Department of Transportation (1997).

of funds for proper maintenance of roads, railroads, and other transport infrastructure. Quality is a multidimensional concept, involving a number of different aspects such as snow clearance, water drainage, improvement of surface friction, curvatures, capacity to carry heavy loads, etc. The level and structure of maintenance (including reinvestment) will influence all of these different quality aspects to varying degrees.

This means that each infrastructure provider has to solve a complicated investment optimization problem.

The first part of this optimization problem is to find the technical relation between a certain quality and a given maintenance activity. Searching for these quality maintenance functions is at the core of much of infrastructure technology research (Robinson et al., 1998; Watanatada et al., 1987; Roney and McIlveen, 1991). But knowing the form of these technologically optimized (and thus efficient) maintenance functions is not enough for proper decision-making. It is equally important to know the utility to the users of the transportation system improvements in terms of the different quality dimensions. Using economic jargon, it is necessary to know the willingness to pay, or the "hedonic price," as a function of the different quality variables. It is of course also necessary to know the cost of maintenance as a function of the different maintenance variables and the traffic flow (possibly subdivided into different classes of traffic). Given that all this information is available, it is possible to formulate a (mathematical) economic optimization problem, within which the net discounted value is maximized with respect to the different types of maintenance activities.

The first rule of thumb says that each maintenance activity should be driven to the point at which marginal willingness to pay (marginal hedonic price), after summing over all quality dimensions affected, is equal to the marginal cost of the maintenance activity divided by the optimal flow of traffic. The optimal flow of traffic is, in turn, determined at the point at which the optimized willingness to pay corresponds to the marginal cost of the traffic flow. As can be seen from the first optimality condition, there is an advantage in a large flow. Optimal maintenance activities concentrate on those parts of the network that have a large flow of traffic.

By tradition, cost–benefit analyses of infrastructure investments have been confined to investment in new parts of the network. Recently there have been efforts at extending cost–benefit analyses in the directions indicated here. Some of these applications have been oriented toward optimal maintenance in sparsely populated parts of the transportation network.

7. Transportation investments and regional development

Over the last half century there have been a large number of theoretical and applied studies of the impact of transportation investment on regional growth, development, and productivity. It is almost trivially obvious that transportation accessibility or the expected transportation cost will influence the allocation of production capacity between regions, as well as realized regional production and household income. A higher transport cost per unit of output is equivalent to a lowered net price of the product. This means that the marginal revenue product as a function of the input of production capacity is lowered in proportion to the unit transport cost. At a given warranted rate of return, this implies that the share of the total production capacity allocated to different regions will decline with their expected cost of transportation, given a production technology that can be applied everywhere without losses of efficiency.

In 1995 the U.S. Department of Transportation summarized the international research on the impact of infrastructure and, especially, transport infrastructure on the level and growth of regional and national productivity (Department of Transportation, 1995).

According to that review, systematic studies of the economic impacts of infrastructure investment can be traced back to the 1940s and the contributions by Rosenstein-Rodan (1943), which analyzed the need for investment in infrastructure in backward eastern and southern European regions and the large productivity effects to be expected from it. However, as early as the beginning of the 20th century the economist Heckscher (1907) analyzed the impact of investment in railroad infrastructure on regional activity levels with a focus on the consequences for sparsely populated regions.

In the early 1970s, Koichi Mera (1973) found that investment in transport communication infrastructure in eight Japanese regions substantially increased production levels of the manufacturing and service sectors. The estimated output elasticities of infrastructure for the period 1954 to 1963 were 0.35 and 0.40, respectively.

During the 1980s and 1990s the number of studies of productivity of transport and communication infrastructure expanded at a rapid pace in the U.S.A. and Europe.

Some of the American studies were triggered by Aschauer's contribution (Aschauer, 1989) as an economist at the World Bank. Some of his claims were quite sweeping and thus triggered an intensive discussion about the interaction between economic growth and the supply of infrastructure (see, e.g., Aaron, 1991; Evans and Karras, 1994; Holtz-Eakin, 1992; Munnell, 1990).

In a recent paper, Aschauer (2000) has estimated a cross-section equation for state data relating the rate of growth of U.S. state products to the availability of network capital. He estimated the elasticity of network capital to be at a value of 0.25.

European studies have essentially been based on regional data. One influential example is Biehl (1986). His econometric analysis relating regional products per capita to the supply of transport and communication infrastructure for 139 European regions resulted in an infrastructural elasticity of 0.35. Similar results were reported for the German regions by Blum (1992).

Most of the studies referred to above are based on aggregated measures of the availability of infrastructure and of its impact on the level or growth of productivity. In a study by Andersson et al. (1990) and in Johansson (1991), infrastructure is subdivided into R&D capacity, airport capacity, intraregional road capital, railroad capital, interregional accessibility, accessibility to metropolitan regions, and the level of education. One important outcome of this study is an indication that railroad capital played a significant role in the determination of regional productivity for the cross-section data of 1970. By 1980 railroad and port capital had no significant productivity impact. Airport and road capacity had by then become the two types of network infrastructure having the largest impact on regional productivities. This development of infrastructure productivity is generally advantageous for the rural and other sparsely populated regions, because these regions can much more easily be supplied with adequate air and road transportation infrastructure than with railroads. However, in the long run the emerging postindustrial economy might favor densely populated areas, a factor that could counteract the advantages of investments in the road and air networks.

On the basis of a review by Rietveld (1995), Button (1998) concludes:

> The empirical studies implicitly looking at endogenous growth effects are certainly far from conclusive and can offer conflicting results. What does seem to emerge from the evidence offered ... is that previous surveys that have concentrated on the North American work may have been under-estimating output elasticities in many parts of the world. On the other hand, those who have advocated a greater public sector involvement through the upgrading of a region's infrastructure also face strong criticisms both from an empirical perspective and from a theoretical point of view.

References

Aaron, H.J. (1991) Discussion on D.A. Aschauer, "Why is infrastructure important?", in: A.H. Munnell, ed., *Is there a shortfall in public capital investment?* Boston: Federal Reserve Bank of Boston.

Andersson, Å.E. and A. Karlqvist (1974) "Population and capital in geographical space", in: J. Loz and M. Loz eds., *Computing equilibria. why and how.* Amsterdam: North Holland.

Andersson, Å.E., C. Anderstig and B. Hårsman (1990) "Knowledge and communications infrastructure and regional economic change", *Regional Science and Urban Economics*, 20:359–376

Aschauer, D.A. (1989) "Is public expenditure productive?", *Journal of Monetary Economics*, 23:177–200.

Aschauer, D.A. (2000) "Do states optimize? Public capital and economic growth", *Annals of Regional Science*, 34:343–363.

Beckmann, M.J. (1957) "On the distribution of rent and residential density in cities", in: Inter-Departmental Seminar on Mathematical Applications in the Social Sciences, Yale University.

Beckmann, M.J. and T. Puu (1985) *Spatial economics: Density, potential, and flow.* Amsterdam: North-Holland.

Biehl, D. (1986) *The contribution of infrastructure to regional development*. Brussels: Regional Policy Division, European Communities.

Blum, U. (1992) *The full costs and benefits of transportation*. Berlin: Springer-Verlag.

Button, K. (1998) "Infrastructure investment, endogenous growth and economic convergence", *Annals of Regional Science*, 32:145–162.

Chamberlin, E.H. (1946) *The theory of monopolistic competition*, 5th edn. Cambridge, MA: Harvard University Press.

Robinson, R., U. Danielsson and M. Snaith (1998) *Road maintenance management*. Houndsmill: Macmillan

Department of Transportation (1995) *Transportation statistics*. Washington, DC: U.S. Department of Transportation.

Department of Transportation (1997) *Annual report*. Washington, DC: U.S. Department of Transportation.

Evans, P. and G. Karras (1994) "Is government capital productive? Evidence from a panel of seven countries", *Journal of Macroeconomics*, 16:271–279.

Fujita, M. (1989) *Urban economic theory*. New York: Cambridge University Press.

Heckscher, E.F. (1907) "Till belysning av järnvägarnas betydelse för Sveriges ekonomiska utveckling" [The impact of railroads on the economic development of Sweden], Stockholm.

Holtz-Eakin, D. (1992) "Public sector capital and the productivity puzzle", National Bureau of Economic Research, Working Paper 4122.

Johansson, B. (1991) "Infrastruktur och produktivitet", Allmänna Förlaget, Stockholm, Expertrapport No. 9, till Produktivitetsdelegationen.

Mera, K. (1973) "Regional production functions and social overhead capital: An analysis of the Japaneses case", *Regional and Urban Economics*, 3:157–186.

Mills, E.S. (1972) *Studies in the structure of the urban economy*. Baltimore: Johns Hopkins University Press.

Munnell, A.H. (1990) "Is there a shortfall in public capital investment?", *New England Economic Review*, Sep./Oct.:11–32.

Regioner, handel och tillväxt (1998) "Regionplane- och trafikkontoret", Stockholm, Report 6.

Rietveld P. (1995) "Transport infrastructure and the economy", in: *Investment, productivity and employment*. Paris: OECD.

Roney, M.D. and E.R. McIlveen (1991) "Predicting the future needs for programmed track maintenance on CP Rail", in: Heavy Haul Railway, International Conference, 1991, Vancouver, Canada.

Rosenstein-Rodan, P. (1943) "Problems of industrialization in eastern and south-eastern Europe", *Economic Journal*, 53:202–211.

Sahlin, N.-E. (1990) *The philosophy of F.P. Ramsey*. Cambridge: Cambridge University Press.

Sasaki, K. (1987) "A comparative static analysis of urban structure in the setting of endogenous income", *Journal of Urban Economics*, 22:53–72.

von Thünen, J. (1826) *Der Isolierte Staat in Beziehung auf Landwirtschaft und Nationalekonomie*. Hamburg.

Watanatada, T. (1987) *The highway design and maintenance standards model*, vol. 1. *Description of the HDM-III model*. Baltimore: Johns Hopkins University Press.

Weibull, J. (1976) "An axiomatic approach to the measurement of accessibility", *Regional Science and Urban Economics*, 6:357–379.

299 - 319

Chapter 19

BIKING AND WALKING: THE POSITION OF NON-MOTORIZED TRANSPORT MODES IN TRANSPORT SYSTEMS

R41

PIET RIETVELD*
Vrije Universiteit, Amsterdam

1. Introduction

Biking and walking are rather neglected transport modes within transportation research. In terms of their contribution to the total number of kilometers traveled, their share is indeed small in most countries. However, their share in the total number of trips made is substantial almost everywhere. In less developed countries walking and biking are the natural transport modes for low-income households, but also in many countries with higher incomes they are considered important transport modes. Walking and biking remain attractive transport modes for a number of reasons:

(1) they provide door-to-door transport;
(2) biking and walking infrastructure usually has a very high spatial penetration;
(3) biking and walking do not lead to waiting, times compared with waiting at public transport stops;
(4) walking and biking have a favorable environmental performance;
(5) they are cheap transport modes;
(6) biking and walking are essential elements in multimodal transport chains;
(7) biking and walking are healthy activities (see for example Hendriksen, 1996)

Among the negative aspects of non-motorized (or slow†) transport modes are:

(1) their low speed (although in heavily congested areas their speed may be comparable to other transport modes);
(2) relatively high accident rates;

*I would like to thank Jasper Dekkers for his assistance in the literature search and Ken Button for constructive suggestions.
†We shall use both terms interchangeably.

Handbook of Transport Systems and Traffic Control, Edited by K.J. Button and D.A. Hensher

(3) low level of comfort (susceptibility to weather conditions);
(4) the physical effort that is required (depending, among other things, on wind, temperature, and gradients).

Walking has been the dominant transport mode for many centuries in all countries. Before industrialization took place there were some alternatives such as riding animals, horse-drawn carriages, and water transport, but walking must have been by far the most substantial transport mode. The Roman roads of 2000 years ago were predominantly used by pedestrians. Conflicts between pedestrians and wheeled traffic did occur, however, especially in urban areas. For example, Hass-Klau (1990) mentions that Julius Caesar banned chariots from the streets in Rome between sunrise and sunset to offer space to the pedestrian. Since the 19th century the development of railway and highway systems has led to dramatic changes in travel behavior towards motorized transport modes. In addition, the bicycle became available to the pedestrian as a possible substitute.

The bicycle started its development at the beginning of the 19th century, and it took about 80 years before it reached a level of quality and comfort that allowed massive adoption. Important steps towards the large-scale use of bicycles were the use of the inflatable rubber tyre, the introduction of the "safety" type, which was much easier to use, and the construction of extensive road networks of sufficient quality. It is interesting to note that the bicycle started its career as a means of sport and recreation for the upper class in the last part of the 19th century (Veraart, 1990; Dutch Ministry of Transport, 1999). Later on it developed into a transport means for a much larger public, for many transport motives. Filarski (2000) mentions the bicycle and the tram as the main transport modes in Dutch cities during the period 1910–1950. During the First and the Second World War the bicycle played a very important role in military operations. For example, the French and the Germans were reported to have a total of 150 000 bicycle troops. Similar numbers have been mentioned for the British and Turks. Bicycle troops played an important role in Japanese military operations in China and the Malay Peninsula in the 1930s and 1940s. In the Vietnam war the bicycle was an essential element in the logistical operations of the Vietcong.

The car has become the dominant transport mode since 1950 in most industrialized countries. Since then bicycle use has decreased substantially. In some countries it has returned to its marginal role, where it is mainly used for recreational purposes, but there are exceptions, especially in northern Europe. In the meantime, in several major developing countries, including China and India, the bicycle continues to play a very important role in transport.

There are a number of long-run trends that pose a threat to non-motorized transport modes. For example, with an increase of per capita income, leading to higher values of time and an increasing priority for comfort, a shift might be expected towards fast and comfortable transport modes. Also, the trends towards

Table 1
Annual sales and stocks of bicycles and cars (millions), world level

	Annual bicycle sales	Annual car sales	Bicycle stock (a)	Car stock
1950	11	8	65	53
1955	15	11	100	73
1960	20	13	150	98
1965	21	19	190	140
1970	36	22	230	194
1975	43	25	360	260
1980	62	29	500	320
1985	79	32	630	374
1990	94	36	840	445
1995	107	36	1000	477

Source: Worldwatch Institute, plus own estimates.
Note: (a)These figures are estimates based on the assumption that the average lifetime of a bicycle is 10 years (Bouwman, 2000).

urban sprawl and low-density construction provide a setback to non-motorized transport modes. Another trend is the increasing use of information and communications technology in motorized transport that may increase the quality gap between motorized and non-motorized modes. However, some studies indicate a rather stable position of non-motorized transport modes and even tendencies towards expansion. For example, the Organisation for Economic Co-operation and Development (2000) mentions a remarkable stability of the number of kilometers walked by Europeans of about 1 km per person per day during the last few decades. In addition, a renaissance in bicycle use in a number of European countries has been mentioned (Pucher et al., 1999; Dutch Ministry of Transport, 1999; Organisation for Economic Co-operation and Development, 2000), implying that the life cycle of the bicycle is taking an upward turn.

At the world level the bicycle appears to be an extremely vital transport mode. Its annual sales volume was of a comparable size to that of the car during the 1950s and 1960s but since then the total sales have increased at a faster rate than that of the car. This also holds true for the total stock. There are two main reasons why, nevertheless, during the past 50 years the car has become a much more dominant transport mode than the bike in Western countries. First, a major part of the growth in bicycle stocks took place in developing countries (see also Section 6). Second, the increase in ownership of bicycles does not necessarily mean that the bicycles are used more often. In the U.S.A. the share of the bicycle in transport is very low, but a large share of the households own at least one bicycle. This illustrates the position of the bicycle in several high-income countries, where it is only used occasionally as a medium for recreational purposes. Nevertheless, the figures presented in Table 1 demonstrate that the bicycle plays a much larger role in the world than is usually thought. Assuming that there are about 2 billion households in the world, Table 1 implies that there is on average one bicycle per two households in the world.

2. Demand for non-motorized transport modes

Most markets of transport services are studied in terms of demand and supply. In the case of walking and biking this distinction is not very helpful, however, since here the supplier and the consumer coincide. The consumer is usually producing the transport service him/herself. There are some exceptions in developing countries (palanquin, rickshaw), and these will be discussed in some detail in Section 6, but in industrialized countries non-motorized transport services are always self produced.

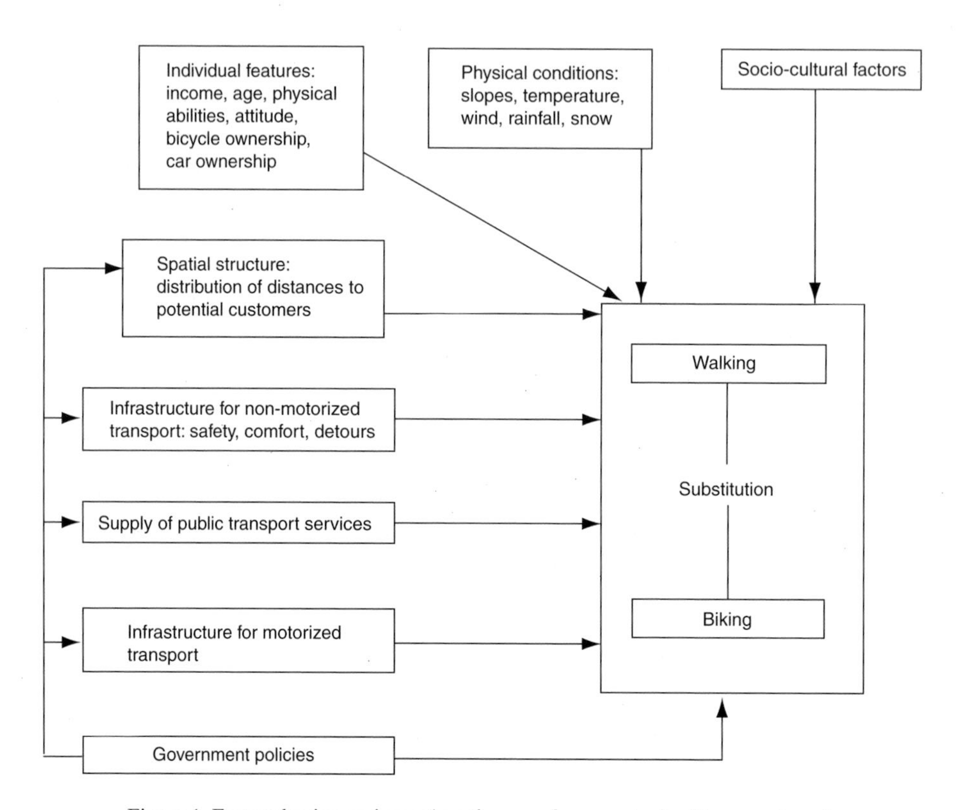

Figure 1. Factors having an impact on the use of non-motorized transport modes.

The fact that supplier and consumer coincide means that there are no monetary costs involved in the trip and that waiting for the transport service does not play a role. Thus, the cost structure differs strongly from that of most motorized transport modes. Factors which have an impact on the use of non-motorized transport modes include individual features, physical conditions, socio-cultural factors, and infrastructure. Also, the availability of other travel alternatives and government policies play a role. Figure 1 provides a schematic representation.

Relevant individual features are age, income, and physical abilities. For children and youngsters non-motorized transport tends to be relatively important. People having low incomes may not be able to afford a car, so that non-motorized transport modes are important alternatives. Students of high schools and academic institutions are known to be frequent bike users. In the U.S.A. university towns score relatively high in terms of the share of bicycle trips (see Gordon and Richardson, 1998; Pucher et al., 1999). In addition, physical conditions (gradients) play an important role. Dimitriou (1995) mentions that bicycles are not convenient for slopes higher than 4%. Of course, weather conditions (temperature, wind, rain, snow) are another group of determinants of modal choice. This holds true both at the strategic level and at the level of daily varying travel patterns (Golob et al., 1996; Khattak and de Palma, 1997). Infrastructure is also frequently mentioned as a factor: bicycle paths may be quite instrumental in improving the convenience and safety of bicycle trips (see also Pucher et al., 1999). In addition, detour factors in networks for motorized versus non-motorized modes play a role. The choice of non-motorized transport modes is, further, a matter of the composition of choice sets. In rural areas without public transport these modes will be chosen more often, since the car alternative is not available. Another factor is urban spatial structure: in areas with high densities where trip distances tend to be short, non-motorized alternatives often perform well (Tolley, 1997). Finally, social and cultural factors should be mentioned: non-motorized transport modes tend to be associated with low status (Dimitriou, 1995), but this is not always true. As we saw in the historical excursion in Section 1, the bicycle started as the transportation means of upper-class citizens, and present bicycle manufacturers do their best to develop fashionable models that are attractive to higher-income consumers. At a cross-country level, there appear to be large differences between the U.S.A. and Europe in the appreciation of bicycle use (Pucher et al., 1999). Within Europe, a clear distinction can be observed between northern and southern countries (high versus low status; Organisation for Economic Co-operation and Development, 2000)

Some data on the use of non-motorized transport modes in various countries can be found in Table 2. The table shows consistently high shares of these transport modes in terms of the number of trips (between 31 and 47%). Given the

Table 2
Shares of transport modes in various European countries (%)

Country	Year	Non-motorized share (trips)	Public transport share (trips)	Car share (trips)	Non-motorized share (km)	Public transport share (km)	Car share (km)
Austria	1983	40	19	42	8	34	58
Finland	1986	31	12	57	6	19	75
France	1984	41	8	51	8	17	75
Germany	1982	41	14	45	8	25	67
Israel	1984	37	31	32	–	–	–
Netherlands	1987	47	5	47	16	12	72
Norway	1985	35	11	54	6	31	63
Sweden	1983	38	12	50	5	20	75
Switzerland	1984	46	12	42	10	20	70
U.K.	1986	37	14	49	9	19	72

Source: Orfeuil and Salomon (1993), adjusted.

ways the questionnaires were designed, it is quite plausible that these shares have been underestimated. The design of many travel surveys lead to a neglect of short trips such as walking the dog, taking a letter to the mailbox, or calling at somebody nearby. In terms of total number of kilometers traveled this neglect will not have a large impact, but in terms of number of trips and travel time this may lead to a substantial underestimate.

Non-motorized transport modes tend to have high shares in countries such as The Netherlands and Switzerland. Low values are found in the Nordic countries (Norway, Sweden, and Finland).* Table 2 demonstrates that there seems to be a substantial substitution between non-motorized transport modes and public transport, leaving car use rather unaffected. This suggests that policies aimed at increasing the share of non-motorized transport modes do not necessarily lead to a decrease of car use. In this table non-motorized transport modes are treated as a combined mode. A more detailed treatment of the two modes reveals that the two are to some extent substitutes. For example, Gerondeau (1997) shows that countries with a small share of bicyclists tend to have a large share of pedestrians and vice versa. Thus, substitution between the two modes should be taken into account in the assessment of policies to stimulate one of the two modes. A more detailed picture of the mobility needs that are satisfied by the two non-motorized transport modes can be found in Table 3.

*The low scores for the Nordic countries are the result of very low shares for pedestrians and rather high shares for bikers, the first effect dominating the second.

Table 3
Average number of trips per person per day according to distance class and main travel mode, for The Netherlands, 1997

Distance class (km)	Car (driver)	Car (passenger)	Train	Bus/ tram/ metro	Moped	Bicycle	Walking	Other	Total
0–0.5	0.01	0.00	0.00	0.00	0.00	0.05	0.18	0.00	0.24
0.5–1	0.03	0.02	0.00	0.00	0.00	0.14	0.17	0.01	0.35
1–2.5	0.19	0.11	0.00	0.01	0.01	0.44	0.23	0.01	0.99
2.5–3.7	0.11	0.06	0.00	0.01	0.00	0.17	0.03	0.01	0.39
3.7–5	0.07	0.04	0.00	0.01	0.00	0.06	0.01	0.00	0.18
5–7.5	0.16	0.09	0.00	0.02	0.01	0.09	0.02	0.01	0.39
7.5–10	0.06	0.03	0.00	0.01	0.00	0.02	0.00	0.00	0.13
10–15	0.12	0.07	0.00	0.02	0.00	0.03	0.00	0.00	0.25
15–20	0.10	0.05	0.01	0.01	0.00	0.01	0.00	0.00	0.19
20–30	0.09	0.04	0.01	0.01	0.00	0.01	0.00	0.00	0.16
30–40	0.04	0.02	0.01	0.00	0.00	0.00	0.00	0.00	0.09
40–50	0.03	0.01	0.01	0.00	0.00	0.00	0.00	0.00	0.05
50	0.07	0.04	0.03	0.00	0.00	0.00	0.00	0.00	0.15
Total	1.07	0.59	0.07	0.10	0.03	1.01	0.63	0.06	3.57

Source: Central Bureau of Statistics (1998).

Table 3 presents a more detailed view of the share of walking and biking in trips as a function of distance traveled, in The Netherlands. The table shows that trips up to 1 km are dominated by pedestrians. Between 1 and 3.7 km, the bicycle is the most frequently used transport mode. Up to distances of 7.5 km, the non-motorized transport modes play a substantial role. In terms of the total number of trips, the share of the pedestrian is 18% and that of the bicyclist 28% implying a total modal share of slow transport modes of slightly less than 50%. It is also interesting to note that the average number of bicycle trips per person per day equals 1.01. Since a standard tour includes two trips, this means more or less that every day 50% of the average Dutch residents uses their bike. The table also demonstrates the potential for substitution between the two slow transport modes. If bicycling were not feasible for some reason, the share of pedestrian trips in the range up to 2.5 km might easily be much larger. For longer trips such a substitution would be less easy to imagine.

Shares of non-motorized transport modes vary according to the objective of a trip. In many countries non-motorized transport modes are over-represented in leisure trips and trips to educational institutions, whereas in shopping trips and commuting they are underrepresented (Gordon and Richardson, 1998;

Organisation for Economic Co-operation and Development, 2000). However, there are exceptions. For example, in The Netherlands an above average share of non-motorized transport modes is observed for shopping (the share of non-motorized transport in shopping trips is 51%).

Given the orientation of slow transport modes towards short distances, one may expect that high population densities are favorable. This is indeed confirmed by Dutch data (see Table 4), although the impact is smaller than one might expect. The share of walking indeed declines consistently with density. However, the link with bicycle use is much less clear. The reason is that in higher-density municipalities the quality of local public transport makes it a competitor of the bicycle. We conclude that our expectation of a clear link between urban density and non-motorized transport modes is confirmed for the pedestrian, but not for the bicyclist, a possible

Table 4
Average number of trips per person per day according to degree of urbanization and main transport mode, for The Netherlands, 1997

Degree of urbanization	Car	Public transport	Bicycle	Walking	Other	Total
Very high	1.26	0.36	0.95	0.82	0.09	3.48
High	1.63	0.16	1.04	0.67	0.09	3.60
Medium	1.75	0.13	1.04	0.61	0.09	3.62
Low	1.81	0.10	1.07	0.54	0.09	3.61
Very low	1.84	0.09	0.93	0.52	0.09	3.48
Total	1.66	0.17	1.01	0.63	0.09	3.57

Source: Central Bureau of Statistics (1998).

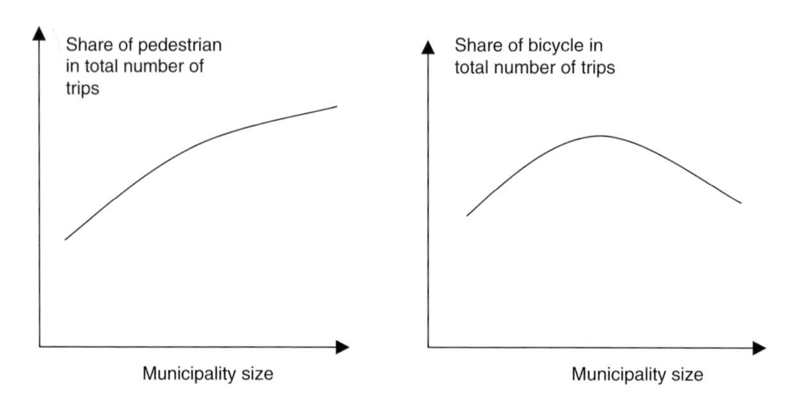

Figure 2. Share of non-motorized transport modes as a function of municipality size.

Table 5
Total number of trips per person per day made for various transport modes, taking into account multimodality, for The Netherlands, 1997

Mode of transport	Main trips	Transport on foot before/after	Transport by bicycle before/after	Total transport before/after inclusive
On foot	0.67			4.37
Bicycle	1.05			1.09
Moped	0.03			0.03
Car (driver)	1.13	2.26		1.13
Car (passenger)	0.60	1.22		0.60
Train	0.07	0.05	0.03	0.07
Bus/tram/metro	0.10	0.17	0.01	0.10
Other	0.06			0.06
Total	3.71	3.70	0.04	7.46

Source: Central Bureau of Statisics (1998), Voetgangersvereniging (1998), adjusted.

explanation for the latter being the competition from public transport (see also Organisation for Economic Co-operation and Development, 2000).

Given the rather direct link between degree of urbanization (measured via urban density) and city size, one may expect patterns as described in Figure 2. Information of this type is important for municipality governments that want to apply benchmarking approaches to non-motorized transport.

In the above presentation, non-motorized and motorized transport have been presented in terms of competing modes. However, in addition to substitution, complementarity of transport modes has to be considered. This aspect will be discussed in the next section.

3. The role of non-motorized transport modes in multimodal transport chains

Transport statistics are usually formulated in terms of "main" transport mode. This leads to a systematic underestimation of non-motorized transport modes. Even in the case of car trips, walking to and from the parking place is an inevitable element of the chain. The same holds true for walking and biking to the bus stop or the railway station. A consequence of this complementarity is that when the various trip elements are considered, the share of bike and walking is much higher. An example of a systematic inventory is given in Table 5.

The average number of trips made per person per day in The Netherlands is 3.7. Taking into account that for car trips and public transport trips walking or biking elements have to be included in order to give the trip a door-to-door character, the total number of trip elements rises to 7.5 per person per day. In terms of total distances traveled this has a rather negligible consequence. But in terms of total travel time, its consequence cannot be ignored. It leads to low travel speeds for short-distance motorized trips.

Although Table 5 indicates that pedestrians dominate the scene for multimodal chains, there is one subset where bicycles are also rather important: train-related public transport trips. As indicated by Rietveld (2000), the bicycle is the dominant access mode at the home end of trips by train (the modal share of bicycle is about 35–40%, of pedestrians it is about 25%). At the activity end these shares are 10 and 45%, respectively. These figures underline the complementarity of train and slow transport modes. They also point to the importance of co-ordination of physical planning and infrastructure planning. Concentration of new residential construction in zones up to some 3 km from railway stations offers favorable opportunities for train-based public transport chains. A similar point holds true at the "activity end" of train-based chains, but there is a difference. Since the bicycle is not available at the activity end, concentration of destinations within a 1 km radius is important to make train based chains attractive. The high share of slow transport modes in rail-based chains means that many travelers prefer the use of slow transport modes above bus, tram, and metro. The reason is that on relatively short-distance trips to railway stations, slow transport modes are often faster than bus or tram, especially when aspects like rescheduling costs and uncertainty costs are taken into account: because of their time-continuous character, the slow modes do not give rise to the risk of missing a connection in a chain. For railway companies, an important lesson to be drawn is that real estate developments in the immediate proximity of railway stations are of eminent importance for the patronage of railway services. In addition, investments in local infrastructure for slow modes, such as safe and convenient pedestrian routes to railway stations and parking facilities for bikes near railway stations, are necessary ingredients for a successful exploitation of a railway line.

The discussion thus far has focused on non-motorized transport modes for passenger transport. However, in freight transport they are also relevant. In many countries walking and biking are part of logistical chains to provide customers with the goods they need. Major examples are the postman and the paperboy, who provide an essential contribution to distribution in fine-meshed networks in urban areas: non-motorized transport modes often appear to be essential elements of chains dominated by motorized modes. Also, in the sector of express service deliveries, the importance of the bicycle has been recognized. Firms such as FedEx observed that the bike was the fastest transport mode in some congested urban centers and shifted their express delivery system to that mode. In Amsterdam barges are used for garbage collection to avoid blockages of the narrow streets. More on non-motorized transport can be found in Section 6, where the case of developing countries is discussed.

Table 6
Comparison of costs of various transport modes (1998; based on average trip length of each mode)

Transport mode	Space used for infrastructure (10^{-2} m^2/pkm)	Direct + indirect energy use (MJ/pkm)	Average costs paid by traveller (euros/pkm)	Travel time (min/pkm)
Petrol passenger car	0.55	1.79	0.170	1.34
Train	0.21	0.98	0.075	0.94
Bus, tram, metro	0.51	1.11	0.085	1.92
Bicyle	0.71	0.04	0.045	5.40
Walking	1.7	0.03	0.000	10.77

Source: Bouwman (2000).

4. Cost comparison of transport modes

Non-motorized transport modes are considered to have a favorable energy performance. This is based on their low direct energy input needed per passenger kilometer. For a more complete comparison between transport modes, indirect effects also have to be taken into account (the use of infrastructure and vehicles). The results for average costs per passenger kilometer (pkm) are provided in Table 6.

Table 6 shows zero monetary costs of walking, but the costs of biking are not as low as one might expect, because they include the costs of interest on the cost of the bike, plus maintenance. Note that Table 6 is in terms of average costs; the marginal costs of modes where the user owns the vehicle (car and bike) are much smaller than the average costs. The table further underlines the slow nature of the non-motorized transport modes and their favorable energy performance. Somewhat surprising may be the high levels of space consumption for the non-motorized transport modes, because these are often reported to be highly space-efficient given their low demands for parking space. The reason for this outcome is that sidewalks and bicycle paths occupy a substantial surface area but are not used in a very intensive way. Given the low speed of pedestrians, the space needed per kilometer traveled annually is relatively high. The same holds true to some extent for bikers. One must be aware, however, that sidewalks also fulfill other functions in residential areas: they function as buffers between motorized traffic and residences, they provide playing space for children, and they provide space so that light can enter the houses. In shopping areas the sidewalks provide space for pavements cafés and traders; in addition, they allow consumers to walk past the shop windows to inspect the goods on offer. Sidewalks can be interpreted therefore as inputs in a multiple-output production function, so that the space requirements in the table are overestimated.

Table 7
Comparison of traffic casualties across transport modes, for the European Union, 1995

Transport mode	Number of casualties per 100 million person-kilometers	Number of casualties per 100 million person-hours
Car	0.80	30
Train	0.04	2
Bicycle	6.3	90
Walking	7.5	30

Source: European Transport Safety Council (1999).

The figures in Table 6 give a comparison of transport modes per passenger kilometer given the average trip length per transport mode. If one wants to compare the suitability of the modes for a given trip length, these figures will change. For example, for very short-distance trips the effective speed of car, train, and bus will be rather low. This explains, of course, why slow transport modes are most frequently adopted for short trips.

Another point of importance when one compares costs of transport modes is that costs per kilometer are not necessarily the most appropriate way of standardization. For example, in the context of residential choice, the appropriate comparison may be between a residence at 15 km distance (implying a 20 minute trip by car) and a residence at 5 km distance (a 20 minute trip by bicycle). In examples like this the appropriate standard of cost comparison is the cost of the total trip, instead of the per kilometer cost. This tends to make slow transport modes more attractive, because their length is lower, so that multiplication of cost per kilometer by length in kilometers yields relatively low costs. Gerondeau (1997) emphasizes that such a comparison on the basis of trips of equal duration instead of equal length makes sense given the observed (quasi-) constancy of travel budgets.

Traffic accidents deserve special attention when discussing non-motorized transport modes. As indicated in Table 7, the risks of a fatal accident per kilometer differ strongly among transport modes. The non-motorized transport modes in Europe have casualty rates that are much higher than that of the car, the difference being almost a factor of 10. Also, for the U.S.A., bicycling dangers are reported to be a major obstacle to bicycling. These figures vary strongly between continents. In Asia and Africa accident risks are much higher than the figures reported in this table.

There are two perspectives that lead to a different view of accident rates of slow transport modes. First, when the risks are computed per travel hour, the differences are much smaller. As Table 7 demonstrates, risks per hour are quite similar for the pedestrian and the car driver. Thus when exposure is measured in terms of minutes instead of in kilometers, slow transport modes perform much better. Second, the above figures relate to the traveler's own risk. When risks for other road users are taken into account, non-motorized transport modes have a more favorable performance than the motorized ones because of the low degree of externalities involved (Persson and Odegaard, 1995; Dutch Ministry of Transport, 1999).

5. Government policies

Governments have various means to stimulate non-motorized transport modes. We shall use the following classification:

(1) physical planning,
(2) infrastructure planning,
(3) regulation of transport,
(4) technological development,
(5) financial instruments,
(6) organizational measures (stimulation of other actors), and
(7) policies with respect to other transport modes.

As indicated in Section 3, physical planning is potentially important because spatial structure has an impact on the use of non-motorized transport modes, especially on walking. It is no surprise, therefore, that compact solutions are often proposed to achieve green urban transport. Examples of such compact solutions can be found in several European countries, where governments provide guidelines on minimum densities of residential construction and on the choice of locations where residential construction is allowed or not allowed.

Infrastructure planning is another field where governments can stimulate non-motorized transport modes by providing space for bicycle paths and sidewalks. This separation of motorized and non-motorized transport also strongly improves traffic safety. A historical account of transport infrastructure planning and traffic calming can be found in Hass-Klau (1990). When separation of fast and slow transport modes is included in the original design of transport networks, it is often rather inexpensive to create an adequate quality level. Once infrastructure networks have been made without proper attention to the slow transport modes, however, the attainment of adequate quality levels tends to be more expensive.

An interesting development is the conversion of old rail tracks to cycleways in countries such as the U.K. and the U.S.A. and the construction of special walking routes for pedestrians. These initiatives usually address leisure activities taking place in rural areas.

One particular aspect that deserves attention in transport networks for slow transport modes is that railways, canals, and highways often have substantial barrier effects on other transport modes, and in particular on non-motorized transport. Tunnels and bridges may help to overcome this problem. For a review of options the reader is referred to Dutch Ministry of Transport (1999).

Another special point of attention is the layout of transport networks so that non-motorized transport gets a strong advantage above motorized modes by creating large detour factors for the latter. An example where the latter approach is followed is in the city of Houten (30 000 inhabitants, The Netherlands). From residential areas to the city center, slow transport modes typically bridge distances that are only one third of the distances of car users. This leads to very high shares of these transport modes. The other side of the coin is that the detours lead to extra car kilometers. Thus, although the performance in terms of modal share looks quite good, the performance in terms of car kilometers generated is less favorable. The opportunities to create transport networks that are favorable for slow transport modes are usually favorable in new towns. This demonstrates the importance of an integrated approach to physical planning and to the planning of transport networks.

At a micro level, the design of transport networks is important as well. Safety concerns call for a design of roads and crossings so that conflicts between fast and slow traffic is minimized. One of the means is the introduction of "*woonerven*" and 30 km zones in residential areas. In addition to such safety-oriented measures, convenience-oriented measures are important for non-motorized transport modes. Examples are the quality of the road surface of sidewalks and bicycle paths, protection against wind or sun, removal of snow and ice, preferential treatment at crossings for slow transport modes, etc. A nice example is given by Garder and Leden (2000), who mention that residents of the city of Helsinki (Finland) make extensive use of the bicycle, but only during the summer period, an important reason being that winter maintenance procedures favor automobile traffic (see also Lahrmann and Lohmann-Hansen, 1998).

An interesting development is the regulation of transport via the introduction of pedestrian zones in urban centers. Well-known European examples are the cities of Freiburg, Nuremberg (both in Germany) and Groningen (The Netherlands). In these historic cities relatively large areas have been made relatively car free, leading to networks of up to 10 km of pedestrian streets. A useful review is given by Monheim (1997). The concept has spread to many other cities. An important question is how the retail sector in these areas responds. This appears to depend on a number of factors, such as government policies to allow development of large-scale retailing at urban fringes, the existence of edge cities with their own retailing activities, location and pricing of parking places, public transport services for visitors from the wider region, etc. In general terms, however, it appears that in many cities these pedestrian zones did not hurt the position of the shopping area or even improved it. However, not all sectors will respond in the same manner; the more specialized a shop, the more it may be expected to benefit (see Monheim, 1997). An important factor appears to be the overall quality of the city center for the visitor. It is especially the cities with a historical character that are considered as attractive destinations for tourists and leisure shoppers. Making these city centers relatively car free will reinforce the perceived quality.

Safety is an important issue in policies to stimulate the use of non-motorized transport modes. This holds true in particular for children. Many parents function as private taxi drivers for their children to bring them to school or other destinations. Safer transport networks would yield opportunities for substitution towards non-motorized modes. As has been mentioned above, there are a number of ways to improve safety of transport networks by providing adequate infrastructure. Legal aspects also play a role. In some countries a tendency can be discerned where liability for traffic accidents is shifted away from non-motorized travelers to motorized ones. This will reinforce the legal position of non-motorized travelers and may induce a more careful driving behavior from their side, a possible drawback being a more risky travel behavior of the non-motorized traveler, providing a clear example of risk compensating behavior.

One of the means that is sometimes promoted to improve safety of bikers is the use of helmets. However, there are some concerns here because Australian research has demonstrated that requiring the use of bicycle helmets may lead to a reduced bicycle use (Finch et al., 1993). Wardlow (2000) claims that the health benefits of biking are so large that they exceed by far the accident risks. Therefore, the requirement for the use of bicycle helmets, if it leads to a decrease in bicycle use, would have adverse health effects. Another potential problem is that bikers may be induced to take more risks given the protection of the helmet. Garder and Leden (2000) report figures about different rates of adoption of bicycle helmets in various parts of Scandinavia and conclude that much depends on promotion activities to support people who want to use a helmet.

Another means to promote safety of bicyclists is the requirement to equip the bike with lights and reflectors. This is important because a relatively large share of bicycle accidents takes place at twilight and nighttime, one of the possible risks being collisions between bicycles.

In addition to traffic safety there is also social safety. In many cities the perceived lack of social safety discourages the use of non-motorized transport and of public transport. It is here that local governments can do much to improve the situation. Policies may range from increased police surveillance in general, with special attention paid to black spots, to improved lightning of bicycle paths and footpaths. Also, the introduction of social safety as a design criterion in the planning of infrastructure for non-motorized transport modes will improve social safety problems.

Bicycle theft and vandalism is a serious problem in some countries. For example, the probability that a bicycle is stolen is estimated to be 5% or more per year in The Netherlands (Dutch Ministry of Transport, 1999). Among the solutions are insurance (but premiums are high), guided parking, police surveillance, and ICT applications as indicated below.

A point of special importance is the application of advanced technologies such as ICT in transport. ICT is gradually increasing the quality of motorized transport modes, making them more comfortable, and providing the traveler with information. An aspect that is easily overlooked concerns the potential contribution of ICT to improving the situation of non-motorized travelers. Governments have a responsibility in this respect because, in the case of non-motorized transport, there are no strong parties that have an interest in promoting ICT applications (in motorized transport, car manufacturers and public transport operators obviously do have such an interest). Several opportunities exist to avoid a pro-motorized transport bias in ICT applications. For example, ICT could be used to impose strict adherence by car drivers to speed limits in residential areas. Similarly, ICT could also be applied on a larger scale to provide pedestrians and bicyclists with fair treatment at traffic lights. Also, ICT may be a promising solution to the problem of bicycle theft since it makes the tracing of stolen bicycles much easier.

One of the policy opportunities for governments is to put non-motorized transport modes on the agenda of employers. Bicycle use tends to be an exception for commuters in the U.S.A. (the modal share is far below 1%). Yet even in a low-density country such as the U.S.A., the share of commuters living within a distance of 5 km from their place of work is no less than 30%, implying that for a considerable share of commuters the non-motorized transport modes would be feasible alternatives. Possible ways to stimulate bicycle use by commuters are the provision of adequate bicycle parking facilities at the workplace and a subsidy for workers that use the bike (it saves employers expenditure on expensive car parking space). Note that the provision of free parking space also implies a subsidy by the employer. Obviously such subsidies may have fiscal implications. Touwen (1997) surveys several case studies of policies by employers. In one case the share of bicycle use by commuters at a certain firm increased from 14 to 21% after the introduction of a number of stimulating measures, despite the fact that in this case the number of employees having a company car was no less than 33%.

Financial measures play a small role in policies to promote non-motorized transport modes. A relationship exists with the above issue of the fiscal treatment of employers who give a bicycle to their personnel. Of marginal importance is a measure in The Netherlands where business trips that are made by bike yield an opportunity for a tax deduction of 6 Eurocents per kilometer.

Of much more importance for the non-motorized transport modes are policies with respect to motorized transport modes. Some of them have already been mentioned above, such as traffic calming of car traffic, which affects their speeds or forces cars to make large detours. Other measures to promote non-motorized transport concern parking policies. As the charges for parking increase (especially in high-density areas), non-motorized transport modes become interesting alternatives for short-distance trips.

A final point of importance concerns the stimulation of non-motorized transport modes as access modes to public transport. For pedestrians, the construction of safe and convenient footpaths to railway stations and bus stops is important. For bikers, similar measures are needed, and also the provision of adequate bicycle parking facilities near railway stations and bus stops.

6. Non-motorized transport modes in developing countries

In developing countries, non-motorized transport modes often strongly dominate travel patterns. Also, for freight transport they may play an important role. This holds true for both rural and urban areas. In rural areas in Indonesia, infrastructure may be so bad and vehicle ownership so low that almost 100% of the trips within a village take place by slow modes. For trips leaving these villages, between 80 and 90% of the trips are made by non-motorized modes (Rietveld et al., 1988). Similar figures are observed in African villages. Also, within urban areas the share of slow transport modes is important. Although urban residents tend to have higher incomes than their rural compatriots, ownership of motorized vehicles is still low. In addition, the layout of road networks in residential areas often does not allow minibuses and cars to enter (Dimitriou, 1995).

China and India are examples of countries with a very strong presence of non-motorized transport in urban areas. Yang (1985) reports that in Chinese cities the share of bicycle trips generally varies between 30% and 60%. Average trip lengths are considerable: for commuting the average bike trip length is about 9 km for males and 5 km for females in large Chinese cities. The popularity of the bike in the large cities leads to extremely large flows. Some intersections are reported to have flows of some 20 000 bicycles per hour. Speeds of busses and cars are not far above the speed of the bikes in the most busy parts of the cities. In Beijing policies of staggered working hours have been introduced to spread bicycle flows over time. For India, Pendakur (1988) reports bicycle shares of 10–20% for trips in large urban areas. The share of pedestrians varies between 15 and 45%.

An important difference between developing and industrialized countries is that slow transport modes play a substantial role in freight transport. Pushcarts, horse-drawn carriages, and rickshaws are frequently used for this purpose. These transport modes appear to be suitable for the very small-scale trading and manufacturing enterprises. For example, street vendors and small-scale food manufacturing have a large market share in the urban economy in low-income countries. The narrowness of roads in residential areas gives non-motorized transport modes a natural advantage. Thus, in the marketing channels of inputs and outputs of small- and medium-scale enterprise within these countries, non-motorized transport modes cannot be missed. Also, bikes play an important role in freight transport in many developing countries, for example in transporting agricultural products from the villages to the nearby marketplaces.

Another important difference between developing countries and industrialized countries concerns the degree of self-production of non-motorized transport services. Whereas in industrialized countries the producer and consumer of non-motorized transport services usually coincide, they are often different persons in developing countries. In some cities a substantial part of the working population earns an income as rickshaw drivers, or as workers with pushcarts in market areas. The quality of these services is rather high: they provide flexible, personalized door-to-door service in very fine-meshed transport networks, and, given the large supply of these services the waiting times are usually low. The charge per kilometer of these services is really high, however. A one-way trip may be as high as the daily wage of a laborer (Kartodirdjo, 1981). In terms of charge per kilometer, minibuses are much cheaper. Dimitriou (1995) reports that, per passenger kilometer rickshaws are 5–10 times more expensive than minibuses in Indonesian cities. This means that rickshaws and other non-motorized services are primarily consumed by higher-income residents in these cities.

In rural parts of developing countries, animals are often used extensively in freight transport. The tractor has not yet captured agricultural activity in all places, so that in the rural parts of many countries one may still observe a large variety of animal traction. Examples are buffaloes, horses, oxen, mules, donkeys, and camels, all involved in various agricultural and transport activities. Also, elephants are sometimes used in freight transport. It is good to remember that it was not too long ago that horses played a vital role in rural freight transport in Europe and North America (Ausubel et al., 2000).

Infrastructure problems have a strong impact on non-motorized transport in developing countries. The shortage of road infrastructure means that all transport modes make use of the same road. This leads to two problems. First, congestion in urban areas is very high. Separation of slow and fast traffic will be one of the means to make this problem manageable. Pendakur indicates that planning practices in Indian cities are heavily biased in favor of motorized transport. One of the issues is the provision of "proximity," allowing poor urban residents to find appropriate destinations of their trips within walking distance. The other problem relates to safety. The number of casualties per kilometer driven in developing countries is extremely high. The fatality rates in transport per motorized kilometer may be a factor of 100 higher in developing countries compared with industrialized countries. Separation of fast and slow transport modes will again be one of the tools to ameliorate this situation.

Given the strong presence of the bike in many developing countries, it is no surprise that bicycle production is an important economic sector in several of them. The major world bicycle-producing countries are located in Asia. Annual production figures in 1995 are 45 million in China, 10 million in India, 8 million in Taiwan, and 6 million in Japan. The U.S.A. is the major producer outside Asia (8 million), but a quite substantial share of the components (for example, drive trains, hubs, and brakes) of U.S. and other non-Asian producers are imported from Asia. In addition to bicycle manufacturing, bicycle repair is an important economic activity in some of these Asian countries. In Africa, biking plays a much less important role than in Asia, but in certain countries, such as Zimbabwe and Tanzania, there are pockets of high bicycle use (Muller, 1988). Bicycle production is negligible in this continent. A major bottleneck for bicycle use in many African countries is the availability of spare parts.

7. Conclusions

With the increase of incomes in most countries, the share of motorized transport modes has increased, allowing people to travel longer distances. Thus, in terms of the contribution of non-motorized transport modes to the total number of passenger kilometers, one can observe a continuous decline. In terms of the total number of trips, however, non-motorized transport modes have retained high shares (Table 2 mentions shares of 31–47% in a sample of European countries).

For the quality of life in cities, a substantial share of the non-motorized transport modes appears to be an unavoidable and essential element. Every trip contains non-motorized elements (for example, walking from the parking place to the final destination, biking to the railway station, etc.). The contribution of non-motorized transport modes to the urban quality of life is gaining increasing attention at the level of national and local governments in many countries (Tolley, 1997; Dutch Ministry of Transport, 1999; Pucher et al., 1999; Organisation for Economic Co-operation and Development, 2000).

Among the factors that may stimulate non-motorized transport modes in the future are:

(1) Improved image: health considerations are becoming more and more important in consumer behavior and even limited use of non-motorized transport modes appears to reduce the risk of heart problems (Hendriksen, 1996). Health considerations were mentioned as the major reason for participation in a bicycle use program in Denmark (Lahrmann and Lohmann-Hansen, 1998).

(2) Introduction of new types of bicycles (for example, the mountain bike led to an improved popularity of biking among youngsters).

(3) Consistent government policies to remove barriers against non-motorized transport modes, for example traffic calming.

Thus, there seems to be a potential for substitution of trips from motorized ones towards non-motorized ones. It appears indeed that in the range of trips of distances up to about 7 km where non-motorized modes are potentially attractive, there is substantial room for substitution. However, a closer examination of the data yields several indications that substitution between non-motorized transport modes and public transport is stronger than between non-motorized transport modes and the car. Thus, stimulation of non-motorized transport does not necessarily lead to less car use in cities.

An important implication of the above is that policies aiming at lower levels of car use in cities in order to improve the environmental quality by offering cheap or improved public transport may appear rather ineffective. The most notable effects that can be expected are "new demand" (new public transport trips that do not replace other trips, and long public transport trips that substitute short public transport trips) and substitution of non-motorized trips by public transport trips. These two effects do not yield the anticipated target of green transport in cities.

The above discussion on substitution between transport modes should not lead one to forget that complementarity is also important. This chapter has demonstrated the importance of non-motorized transport modes for the proper functioning of multimodal chains, both car-oriented and public-transport-oriented ones.

References

Ausubel, J.H., C. Marchetti and P.S. Meyer (2000) "Toward green mobility", *European Review*, 6:137–156.

Bouwman, M. (2000) *Tracking transport systems, an environmental perspective on passenger transport modes*. Groningen: Geo Press.

Central Bureau of Statistics (1998) *De mobiliteit van de Nederlandse bevolking in 1997*. The Hague: Central Bureau of Statistics.

Dimitriou, H.T. (1995) *A developmental approach to urban transport planning*. Aldershot: Avebury.

Dutch Ministry of Transport (1999) *The Dutch bicycle master plan, description and evaluation in an historical context*. The Hague: Dutch Ministry of Transport.

European Transport Safety Council (1999) *Exposure data for the assessment of risks: Use and needs within and across the transport modes in the EU*. Brussels: ETSC.

Filarski, R. (2000) *Opkomst en verval van vervoersystemen*. Delft: PAO-VV (Post-academic Education in Transport).

Finch, C.F., L. Heiman and D. Neiger (1993) "Bicycle use and helmet wearing rates in Melbourne: the influence of the helmet wearing law", Monash University, Report 1993/02.

Garder, P. and L. Leden (2000) "Promoting safe walking and cycling", *Nordic Road and Transport Research*, 12(3):12–15.

Gerondeau, C.(1997) *Transport in Europe*. London: Artech House.

Golob, T.F., K. Seyoung and R. Weiping (1996) "How households use different types of vehicles: A structural driver allocation and usage model", *Transportation Research*, 30A:103–118.

Gordon, P. and H.W. Richardson (1998) "Bicycling in the United States: A fringe mode?", *Transportation Quarterly*, 52:9 -11.
Hass-Klau, C. (1990) *The pedestrian and city traffic*. London: Belhaven.
Hendriksen, I. (1996) "The effect of commuter cycling on physical performance and coronary heart disease risk factors", Ph.D. thesis, University of Yogyakarta: Amsterdam.
Kartodirdjo, S. (1981) *The pedicab in Yogya karta*. Yogyakarta: Gadjah Mada University Press.
Khattak, A.J. and A. de Palma (1997) "The impact of adverse weather conditions on the propensity to change travel decisions", *Transportation Research*, 31A:181–203.
Lahrmann, H. and A. Lohmann-Hansen (1998) *A sustainable transport system: From cars to bicycles via incentive motivation*. Aalborg: Aalborg University.
Monheim, R. (1997) "The evolution from pedestrian areas to 'car free' city centres in Germany", in: R. Tolley, ed., *The greening of urban transport*. Chichester: Wiley.
Muller, A.K. (1988) "Bicycle transport in rural east Africa (1988)", in: Transportation Research Board, 67th Annual Meeting, Washington, DC.
Orfeuil, J.P. and I. Salomon (1993) "Travel patterns of the Europeans in everyday life", in: I. Salomon, P. Bovy and J.P. Orfeuil, eds., *A billion trips a day*. Dordrecht: Kluwer.
Organisation for Economic Development (2000) *Transport benchmarking, methodologies, applications and data needs*. Paris: OECD.
Pendakur, V.S. (1988) "Non-motorised urban transport in India", in: Transportation Research Board, 67th Annual Meeting, Washington, DC.
Persson, U. and K. Odegaard (1995) "External cost estimates of road traffic accidents, an international comparison", *Journal of Transport Economics and Policy*, 29:291–304.
Pucher, J., C. Komanoff and P. Schimek (1999) "Bicycling renaissance in North America? Recent trends and alternative policies to promote bicycling", *Transportation Research A*, 33:625–654.
Rietveld, P (2000) "The accessibility of railway stations: The role of the bicycle in The Netherlands", *Transportation Research D*, 5:71–75.
Rietveld, P., Sadyadharma and Sudarno (1988) "Rural mobility in Java: The village economy and the rest of the world", *Singapore Journal of Tropical Geography*, 9(2):112–124.
Tolley, R., ed. (1997) *The greening of urban transport*. Chichester: Wiley.
Touwen, M. (1997) "Stimulating bicycle use by companies in The Netherlands", in: R. Tolley, ed., *The greening of urban transport*. Chichester: Wiley.
Veraart, C.F.A. (1990) *Geschiedenis van de fiets in Nederland, 1870–1940*. Eindhoven: Technische Universiteit.
Voetgangersvereniging (1998) *Het voetgangerscijferboek*. The Hague: Voetgangersvereniging.
Wardlow, M.J. (2000) "Three lessons for a better cycling future", *British Medical Journal*, 321:1582–1585.
Yang, J.-M. (1985) "Bicycle traffic in China", *Transportation Quarterly*, 39:93–107.

Chapter 20

TRAFFIC CALMING

RAY BRINDLE
Eldamar Research Associates, Kyneton

1. Scope of this chapter

In the broadest sense, *traffic calming* is the process of reducing the physical and social impacts of traffic on urban life, principally through the reduction of traffic speeds and volumes. Its main objectives are to reduce accidents and help to improve urban amenity.

In principle, traffic calming may be directed at neighborhood streets and areas, urban or rural corridors, or whole cities. Programs may involve direct intervention using physical devices or other management techniques (including emerging technologies), or more fundamental social changes that result in different travel choices and driver behavior. This can mean, ultimately, reduction in the total levels of traffic in cities.

Nevertheless (as the *Shorter Oxford English Dictionary* indicates), most current examples of traffic calming involve "the deliberate slowing of road traffic, especially through residential areas, by narrowing or obstructing roads." In this chapter, we focus on this conventional use of traffic management techniques in neighborhoods and along corridors. Measures to reduce the total level of traffic in cities are discussed in Chapter 12 (on travel demand management). Reduced traffic levels alone may not meet all the objectives of traffic calming and may still require supplementary traffic management measures, however.

Further material and references to a wide range of sources can be found in O'Brien and Brindle (1999).

2. Traffic calming in context

Concern about the impacts of traffic on urban life is not new. Even before the widespread use of motor vehicles, and in fact dating back into ancient times, there were examples of legal and physical attempts to restrict vehicle speeds and numbers (Lay, 1992).

Handbook of Transport Systems and Traffic Control, Edited by K.J. Button and D.A. Hensher

The issues creating the demand for traffic calming were probably first formally addressed in the U.K. report *Traffic in Towns* (Department of Transport, 1963), which pointed out the relationship between the quantity of traffic, urban amenity, and the level of investment in the land use–transportation system.

Subsequently, the first tentative programs of local traffic restraint were established in the U.K. and elsewhere in Europe in the late 1960s and early 1970s. These programs were based on the assumption (from *Traffic in Towns)* that the "problem" was caused by intruding non-local traffic exploiting excessive connectivity in local street networks. At that time, the common English term for these actions was "environmental traffic management." The principal aim was to alter "grid" street networks (using street closures, one-way links, and so on) to make the streets less connective for through traffic, and to create (or reinforce) a road hierarchy.

There was mixed success with these techniques of network modification (e.g., in the U.K., U.S.A., and Australia). These changes had the inevitable effect of changing local access patterns, leading to opposition by some residents. In addition, it became clear that in many neighborhoods the removal of non-local traffic did not remove the core problems. This caused some reconsideration in the 1970s. About the same time, concern was growing about the large number of casualty crashes that were occurring in local streets (typically between a quarter and a third of all reported casualties in urban areas), which had up to then not previously received much road safety attention. The emphasis shifted from changes in the nature of the local street network to the modification of the behavior of all vehicles that used the street. 30 km/h residential zones were beginning to be introduced in Germany. A radically new model had been offered by the emergence of the *woonerf*, or low-speed shared zone, in Delft (The Netherlands), which required a different understanding of the mutual relationship between vehicles and other road users. Derivatives of this model evolved in Germany and Denmark. In several countries, tools were sought that influenced a reduction of vehicle speeds, and the creation of opportunities for streetscaping to change the character of the street.

Thus, by the end of the 1970s, various techniques for both network modification and speed management had gained widespread use in Europe and Australia, and were being promoted in the U.S.A. In Australia, small roundabouts at local street intersections were already numerous and set an example that Europe and the U.S.A. were later to follow. 100 mm high, 3.6 m long round-profile speed humps became the subject of careful research in the U.K. during the 1970s and subsequently in Australia into the 1980s. This research encouraged rapid expansion of humps in local streets in Australia, while their use became less common in the U.K. as a result of perceived legal and administrative constraints.

The term "traffic calming" is a translation of *Verkehrsberuhigung*, describing the physical measures that had been introduced into German streets during the 1970s as part of the evolving "shared zone" and "30 km/h" philosophy. Its earlier English equivalents (such as "traffic pacification") were used in the 1980s to describe the local traffic management techniques that had already become widespread elsewhere. But the term "traffic calming" grabbed both the public and the professional imagination as if it were a new concept. During the early 1990s the pursuit of traffic calming spread to the U.S.A. and Canada, although several cities had initiated schemes in the mid 1980s, following on from the Berkeley, CA, trials of the 1960s and 1970s (Ewing and Kooshian, 1999).

Although the most familiar forms of traffic calming action worldwide thus involve the use of physical treatments at the local street level, international traffic calming practice is not limited to low-volume neighborhood streets; it may also describe traffic management in busier streets and corridors. Sometimes, it has been appropriate to impose forms of traffic restraint on higher-order roads (carrying perhaps up to 20 000 vehicles per day) in response to the competing functions of the street space. Sometimes the partial or full exclusion of traffic from town centers (which has been a long-established practice for city-center rejuvenation) is promoted as a form of traffic calming. More recently, concepts of city-wide traffic reduction, changes in travel choice, promotion of other modes of transport, and even manipulation of the land use–transport system have been interpreted by some as the ultimate form of traffic calming. This would shift the focus from changing driver behavior to inducing more fundamental social and attitudinal changes that would be reflected in travel behavior.

We could summarize this by concluding that "traffic calming" describes a *process* and a desired outcome; the installation of physical speed control devices is still by far the most common means towards that end, but it is only one of the possible tools.

3. Planning for traffic calming

In both existing and proposed local networks, traffic calming can focus particularly on three aspects of planning (as distinct from specific engineering treatments or details):

(1) local traffic as a planning rather than inherently an engineering issue;
(2) the need to see neighborhoods as systems; and
(3) following a systematic planning process when designing or, especially, redesigning a locality.

3.1. Local traffic as a planning issue

Traffic problems in neighborhoods may arise from the inherent characteristics of the local land use/network pattern, from changes in the nature of intruding traffic, from sudden changes in the nature of traffic demand affecting the area (such as traffic generated by a new commercial center nearby), or combinations of these. Yet planning decisions are often made without regard for the local traffic consequences, on the implicit assumption that "traffic calming" will fix any problems that may arise. The very success of many physical traffic control measures in neighborhoods thus helps to divert attention away from the land use/traffic system as the underlying cause. Many of the situations that traffic calming tries to resolve could be avoided by proper land development and planning decisions in the first place. Thus, local area traffic management, even traffic calming as a whole, is not, at its root, essentially an "engineering" problem. Rather, local traffic calming could be seen as an engineering solution to a planning problem.

It is not coincidental that most local speed management and safety measures exist in the more connective local street systems, since it is there that we tend to find higher levels of intruding traffic, higher speeds, more intersections, and more accidents. Experience has borne out the research finding that neighborhoods based on culs-de-sac and other forms of low-connectivity street have greater overall levels of safety. Possible conflict with current urban design preference for connective local street systems with no culs-de-sac will be noted. The street forms themselves, fixed at the time of planning approval and construction, can leave an imprint on subsequent traffic and accident experience that can be hard to change.

3.2. Neighborhoods as systems

A neighborhood traffic calming plan should be more than a catalogue of works; the effective area-wide plan is truly greater than the sum of its component treatments. There are two reasons for this:

(1) streets are part of networks, and
(2) movement networks are only one part of the urban system.

Streets within networks

The adaptability of networks is well known to traffic engineers, and soundly based local traffic calming schemes will have regard for the effects of the proposals on travel decisions and driver route choice, and hence on traffic displacement and reduction. There is sometimes an unduly optimistic expectation of the extent to which local traffic calming will reduce total travel, but the effects of street changes on travel and route choice are well established. If the diversion of traffic to other routes is not anticipated and carefully analyzed, there may be adverse community response. In Australia, the term for local street traffic calming – Local Area Traffic Management – was coined in the late 1970s specifically to emphasize the need for such an area-wide approach. Analytical techniques are available to assist practitioners estimate the trip change effects of treatments. In the absence of these techniques, experienced practitioners can make reasonable estimates of the range of changes that could occur. Network effects, including diversion of traffic to nearby local streets and effects on arterials, should always be considered by one means or the other.

A more familiar aspect of streets as part of networks is the matter of "road function" and "hierarchy." Getting community consensus on the application of a conventional road hierarchy to the real network can be controversial. In most cases, this difficulty is unnecessary. The application of a local traffic calming scheme presupposes that there is a community agreement on at least one fundamental point: that the streets in which these actions are to be taken are different in nature and purpose from other roads where traffic is expected to pass without such constraints. Thus, local traffic calming programs imply the existence of a road hierarchy comprising at least two categories:

(1) those elements which exist to carry traffic reasonably efficiently, on which severe traffic restraint is inappropriate (i.e., "traffic roads"); and
(2) those elements on which living and environmental conditions predominate, and on which physical speed management may be considered (i.e., "local streets").

Reaching consensus on this point should present fewer problems.

There may be good reason to be concerned about the traffic and safety conditions on some "traffic roads." Rather than let the road classification drive traffic management actions in these cases, and to overcome the problem artificially created by slavish adherence to hierarchical definitions, traffic planners have explored ways to reconcile traffic importance with local sensitivities. Traffic calming action may not only be taken on local streets; it may also be directed towards creating moderated speed conditions along traffic routes passing through various types and intensities of community activities (e.g., strip retail centers, and roads through small country towns and villages). This requires different sorts of measures, based on the broad principle that, even on some traffic routes, the desirable speed environment may be moderately low. Clearly, a conventional approach to road classification would inhibit such a proposal. Traffic calming on traffic roads thus is being introduced via two generalized strategies:

(1) the adoption of a road type definition that recognizes a form of subarterial (or "distributor road," in the U.K. terminology) on which the traffic function is restrained; and/or
(2) varying the physical form of traffic roads along their length to reflect the adjacent land use and level of conflict. (For example, a road may be managed to provide a good level of service along most of its length, but through a retail precinct it may have its traffic function lowered to allow some priority to parking and pedestrian movements.)

Networks in the urban system

The place of traffic calming within the urban system is more elusive. One way to approach this is to consider what is the root cause of the problem and if in fact physical traffic management treatments are the only way to resolve it. Without a clear definition of problems, appropriate solutions are difficult to select and there are inadequate criteria by which to measure their performance. The devices become the focus of attention, from concept to implementation and public debate, and often become ends in themselves.

At the very least, an attempt should be made to see problems and solutions in the context of the locality (neighborhood or "main street," for example) as a functioning unit, not just as a site-specific traffic matter. The solution to traffic problems in a residential precinct, for example, could lie in finding ways to modify the form or operation of a nearby employment node.

3.3. A systematic planning process

Typically, the selection, placement, and design of traffic calming devices are arbitrary and respond more to local politics and practical constraints than to logical traffic planning. In order to maximize the efficiency of a given program, a suitable process or framework for making planning decisions about traffic calming first needs to be established.

The key to consistent, logical, and effective planning for traffic calming at any level is to adopt a strategic decision-making approach.

In essence, the strategic decision-making approach forces attention to be focused on the desired outcomes to be achieved, and the effectiveness of the adopted actions towards that end. This is especially important in the selection and placement of devices for neighborhood and road corridor traffic calming.

The strategic decision-making process can be understood in terms of a series of questions, such as:

(1) What are we hoping to achieve? *Outcomes*: Description of future states or changes.
(2) What are going to be our measures of success towards that end? *Objectives*: Description of measurable targets to achieve the desired outcomes.
(3) What are the most effective ways by which we can move towards those targets? *Strategies*: Selection of the general methods of approach.
(4) How best can we implement the chosen approaches? *Actions/schemes*: Location and design of treatments.

The implicit question linking each of these is the word "why?":

Why is a road hump proposed to be placed here?
Because it is an essential part of a road hump implementation program for this area.

Why are road humps proposed for this area?
Because that is considered to be the most effective approach (strategy) to speed reduction in the area.

Why are we trying to reduce speeds?
Because that is an effective means by which accident potential will be reduced and living conditions improved in the area.

The validity of each of these "why–because" links, and thus the whole process, depends on each step in fact satisfying its implied rationale. For example, if there were no established connection between speed reduction and accident reduction, then the adoption of speed reduction as an objective towards accident reduction would be questionable. So performance measurement, or anticipation of performance from practice and experience elsewhere in the case of project planning, is a vital part of planning for traffic calming schemes. This continuous background checking of the links between each stage in the process of project development can be called validation. Familiarity with data from empirical research and practice is clearly important.

Validation suggests a means by which guidance can be given to the selection of treatments, by effectively reversing the "why?" question to become an "if ... then" statement. In other words, experience and research can be accumulated to allow us to say: "*If* you want to achieve x, then consider doing y (and/or z)." Validation seems to be the most important missing component in the local traffic calming planning process. If it has authoritative guidelines that validate the links between various strategies, objectives, and desired outcomes, local government can proceed more confidently. This simple concept forms the basis of a consistent framework for selection of strategies and installation design, and allows the practitioner and decision-maker to make informed judgments about the many traffic calming options available to them.

4. Designing for traffic calming

4.1. Creating the preferred traffic character of the street

Some rules of thumb can be suggested for the design of lower-speed street systems, and, by extension, for the planning of local traffic calming measures, based on research and observation:

(1) Get the network right in the first place. Direct and connective pedestrian networks do not require traffic networks also to be connective, and "spreading the traffic" may mean spreading the problem.
(2) Keep the total length of lower-speed travel down to an acceptable level. One minute (e.g., 500 m at an average of 30 km/h) is ample to encompass most neighborhoods, even at lower densities. If motorists are expected to travel too far at lower speeds, they will increasingly try to travel at the upper limit allowed by the road geometry.
(3) Conversely, minimize local trip lengths on streets on which you want to restrain speeds. Network design should be such that traffic on any low-speed access street is not more than, say, 500 m from its origin or destination.
(4) Do not expect motorists to readily accept restrained-speed conditions on a "connector" (a local street that provides a direct connective link in the traffic road network). Conversely, if you want a neighborhood street to operate at a lower speed, do not make it a connector unless it is very short.
(5) Be cautious about the number of intersections, the lengths of street segments, sight line impedances, and the mixture of extraneous traffic with living activities.
(6) Most importantly, limit the length of unrestrained street sections. It is generally agreed that 200 m is an upper limit for the length of access street sections where speeds are hoped to be moderated, and well below 100 m where speeds 30 km/h or less are targeted. Street sections can be defined as the length between terminations (at T junctions), or between other points where the speed is very low (say, 20 km/h or less). This is supported by parallel evidence on the spacing of traffic calming devices. Note that many devices and curves (or bends) have exit speeds of 30 km/h or more, and thus cannot be expected to contribute to speed environments below 40 km/h if spaced such that vehicles can accelerate to above that level between constraints.
(7) Do not rely on narrower pavements alone to reduce speeds; streets that are long and narrow are likely to combine moderately high speeds with lower sight distances, which is just what is not needed. If the street lengths are limited, width as such is less critical. Speeds on a short, relatively wide street are likely to be restrained, but those on a long, narrow street may not be. Thus, concern for neighborhood safety does not necessarily mean that spacious streets are prohibited. What must be avoided are wide streets that are also long, if speeds are to be restrained.
(8) Nevertheless, constrictions or impedances are important traffic calming tools. A reduction of hindrance between opposing vehicles is not a logical design criterion if you want to restrain speed. This means not providing more width than you really need for parking and reasonable passage, even if it means building in artificial width constraints such as parking protectors.

A 6 m street on which there is never any parking is functionally as "wide" as an 8.5 m street with parked cars on one side, or a 10–11 m street with parking on both sides.

(9) There is no justification for providing carriageway widths which permit overtaking in residential areas. Slower and turning vehicles become "speed control devices" if they cannot be overtaken.

(10) Avoid long and wide sight lines, while being careful that stopping sight distances are satisfied. The "design speed" for vertical sight distance should not be less than that for horizontal sight distance, and should certainly be greater if horizontal sight distance is towards the minimum.

(11) Avoid free-flowing alignments which introduce indirectness into the pavement but which allow the driver to see what is beyond the curve.

(12) Conventional "design speeds" for the alignment of curves tend to understate the actual speeds that drivers will adopt, so bends and curves that are introduced for speed management need to be conservatively designed (i.e., to design speeds below the target speed). In particular, do not rely on 90 degree bends to yield spot speeds below 25 km/h or street speeds below 40 km/h.

(13) Use occasional interruptions to the parking lanes (such as planting areas) to constrain the "optical width" of the street, again being careful to protect stopping sight distances. This would also remove parking from the traveled way, thus reducing the collision risk presented by parked vehicles.

(14) At the small scale, encourage innovative total design to produce speeds well below those in conventional estates. Total design of the street, combining considerations of length, visibility, texture and materials, cross section, edge treatments, activity, roadside development, and planting, is required to make sure that all the variables work in concert to produce a safe, low-speed environment. Perhaps a useful image is that of a landscaped parking access way rather than a traffic route.

4.2. Selection of street modification treatments

The first step of traffic calming design is the selection of the types of measures that are appropriate to the objectives being sought, using a strategic decision process as described previously (Hawley et al., 1993; Ewing and Kooshian, 1999; AustRoads, 1988; Transportation Association of Canada and Canadian Institute of Transportation Engineers, 1998; Road Data Laboratory, 1993; CROW, 1988).

The next step is to select the candidate treatments for the site in question. No foolproof, automatic treatment selection process exists, because at this stage all the site and community factors that may affect the choices require careful consideration. Reference materials are available to suggest the performance of

Table 1
Actions that support traffic calming strategies

Strategies	Types of actions ("devices," treatments, and other measures)
Speed control	
Vertical speed control devices	Round-profile hump Flat-topped hump/plateau hump Raised intersection Raised pavement (non-intersection) Raised pedestrian crossing
Horizontal speed control devices	Single-lane slow point Single-lane angled slow point Two-lane slow point Two-lane angled slow point Diamond slow point
Non-intersection path deflections	Midblock island Alternated parking arrangements ("axial shift")
Intersection treatments	Roundabout Central deflector ("impeller") Other intersection deviations Modified intersections (including priority changes)
Changes to the form and use of the street	
Changed street form/ changed visual signals to road users	Total street reconstruction/changed street form Reduced carriageway width/narrowed pavement Curvilinear carriageway construction Reduced visual length of street Kerb extensions
Streetscape changes	Varied paving materials Review signs and other street furniture Modify planting and materials
Reallocate street use	Review parking provision and policies Legislated "rights" and obligations of street use
Zone-based speed conditions	Shared zone (e.g., *woonerf*) Lower area speed limit Perimeter (threshold) treatment
Non-physical speed management	Police presence Electronic speed management and/or enforcement Community speed watch (etc.)

Table 1
Continued

Strategies	Types of actions ("devices," treatments, and other measures)
Traffic network connectivity change	Street closure Half-closure Diagonal closure Bus- and service-vehicle-only link One-way street Driveway link/low-speed connection Improved connectivity of arterial alternatives
Improve pedestrian/cycle access and connectivity	Median islands Bike-modified treatments Carriageway narrowing Raised crossings Bicycle lanes Controlled road crossings Separate footways Segregated ped/bike facilities Ped/bike links Bicycle bypasses of treatments
Improve public transport accessibility and connectivity	Bus-only links Bus-modified traffic control devices Bus bypasses of treatments

the alternatives. Table 1 outlines the common strategies and corresponding types of measure that might be taken. Note that the possibilities are increasingly including technological rather than physical measures.

4.3. Changing the speed profile of the street

The purpose of physical speed control devices is to lower the profile of vehicle speeds along the streets, that is, to reduce the difference between the actual speeds and the desired (or "target") speed. The traffic planner can estimate the typical speeds of vehicles along the street, using known acceleration and deceleration rates and information about the effectiveness of various physical devices in reducing vehicle speeds (Figure 1). Arbitrary location of speed control devices that does not take account of their effects on the speed profile may lead to disappointing outcomes. There may be a temptation (either for cost reasons or to placate residents) to opt for treatments that are too far apart to be fully effective. A better approach is to treat the street section as a whole rather than as a series of isolated devices.

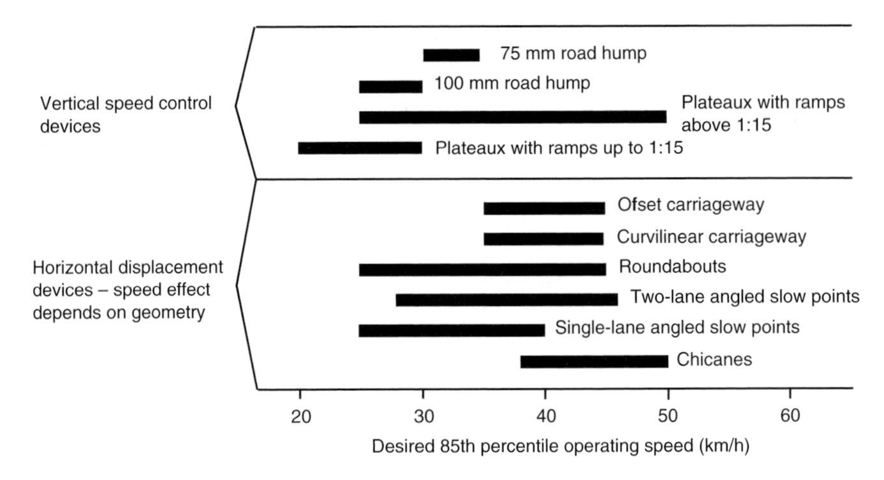

Figure 1. Reported operating speed ranges for selected device types.

4.4. The specific treatments

The design of the speed control devices (such as those listed in Table 1) is a detailed engineering and urban design task. There are now many design aids for the practitioner, and local codes and requirements may also prescribe the form and detail of devices (including their signage). The form of a device affects the speed at which vehicles will traverse it. This can vary widely for a given type of treatment, as Figure 1 shows.

4.5. Designing for special users

The detailed design of treatments should meet the general requirements of function, appearance, and safety. In addition, the selection, placement, and design of treatments should have regard to the needs of special road users such as buses, emergency vehicles, and bicycles.

An underlying principle of traffic calming is that conditions should be made better for pedestrians and cyclists (Cleary, 1991). A well-designed traffic calming scheme should create an improved cycling environment, but specific devices are often criticized for increasing rather than decreasing risks to cyclists. Thus, it appears to be a matter of how things are done rather than simply what is done. There are several key sources that can be referred to for guidance on design of traffic calming schemes in ways that acknowledge cyclists as an integral part of the design process rather than being treated as a supplementary or postdesign check (summarized by O'Brien and Brindle, 1999). These cover such matters as

(1) bicycle/vehicle conflict, bicycle/pedestrian conflict, and bicycle service and comfort;
(2) the quality of the riding surface (especially at vertical speed control devices);
(3) Avoidance of squeeze points where vehicles conflict with bikes, unless these treatments can be integrated into a total scheme that creates a slower-speed shared environment, or unless cycle bypasses of the squeeze points can be provided;
(4) quality of delineation, signage, visibility, and lighting;
(5) provision of cycle paths through street closures to maintain route connectivity and continuity;
(6) adequate design of islands and refuges so they can shelter bicycles;
(7) conditions in local streets in general should permit safe sharing of the street space, without the need to go overboard in providing separately for cyclists.

5. Final comment

Planners and traffic engineers should note that "traffic calming" based on traffic restraint devices is essentially negative in purpose. Quality streets, on the other hand, are about creating public, private, and interface spaces which are attractive to look at, good to use, and amenable to be in. Local traffic calming techniques, as contributors to the streetscape, are likely to focus increasingly on integrated design of the street, and emphasize materials, planting, and form. This does not mean, as is often implied, that environmental design of streets will in future exclude traffic specialists. Safe and effective reallocation of the street space for other uses will require traffic management (i.e., functional) skills in the planning and design process. Human-scale streets will require more, not less, traffic management and design skill, even if the use of electronic speed control increases.

Finally, some cautionary notes about the limitations of conventional traffic calming. Experience over the past twenty years or so has emphasized the need to be clear and open about the potential negative consequences of physical speed control as well as their advantages, and to educate the community about the need to make intelligent trade-offs. Safety and amenity gains may come at the price of driver resistance and localized noise and inconvenience. There are established programs and techniques for community involvement in traffic calming. These should be a normal part of any local traffic management scheme. However, a community may have to accept that sometimes the conflicts are not fully resolvable.

Regardless of the many benefits that engineering-based traffic calming techniques can bring, "sustainable cities" will not be created through engineering treatments alone. Already, there are developments in electronic speed control (as noted in Table 1) that may supplement or even replace some physical treatments.

However, the achievement of "traffic calming" on a large scale, with or without physical treatments or electronic speed management, requires widespread and fundamental changes in the community's attitudes to urban development, travel mode, and how they behave as drivers. We have enough experience to verify that traffic management at a significant level cannot get too far ahead of social attitudes, and that we cannot expect to bring about cultural change through traffic engineering alone.

References

AustRoads (1988) *Guide to traffic engineering practice*, Part 10: *Local area traffic management*. Sydney: AustRoads.

Department of Transport (1963) "Traffic in Towns", HMSO, London, Report of the Working Group appointed by the Minister of Transport.

Cleary, J. (1991) "Cyclists and traffic calming", Cyclists Touring Club, Godalming, CTC Technical Note.

CROW (1988) *Recommendations for urban traffic facilities*, 3rd edn. Ede, The Netherlands: CROW [Center for Legislation and Research in Geotechnical, Civil and Traffic Engineering].

Ewing, R. and C. Kooshian (1999) *National traffic calming report*. Washington DC.: Institution of Transportation Engineers.

Hawley, L., C. Henson, A. Hulse and R. Brindle (1993) *Towards traffic calming*: *A practitioner's manual of implemented local area traffic management and blackspot devices*. Canberra: Federal Office of Road Safety.

Lay, M.G. (1992) *Ways of the world*: *A history of the world's roads and of the vehicles that used them*. New Brunswick, NJ: Rutgers University Press.

O'Brien, A.P. and R.E. Brindle (1999) "Traffic calming", in: J.L. Pline, ed., *Traffic engineering handbook*, chap. 9. Washington, DC: Institute of Transportation Engineers.

Road Data Laboratory (1993) "*An improved traffic environment. A catalogue of ideas*", Danish Road Directorate, Road Data Laboratory, Road Standards Division, Herlev, Denmark, Report 106.

Transportation Association of Canada and Canadian Institute of Transportation Engineers (1998) *Canadian guide to neighbourhood traffic management*. Ottawa: Transportation Association of Canada/Canadian Institute of Transportation Engineers.

Chapter 21

SPECIALIZED TRANSPORT

DAVID GILLINGWATER
Loughborough University

NICK TYLER
University College London

1. What is "specialized transport"?

"Specialized transport" can be defined as any form of public passenger transport where eligibility rules apply: people who wish to use such a service have to meet certain criteria for ridership which apply to them and for which that transport is then made available. Such provision embraces a wide range of passenger transport service use (Nutley, 1990), including "school transport" – services to transport children to and from school; "patient transport" – services to transport non-emergency patients to and from hospital or medical-center appointments; "statutory sector transport" – services to transport individuals (typically those with physical disabilities or learning impairments) to and from facilities such as day centers; "socially necessary transport" – where services used by social groups (typically elderly and retired or unemployed people) are subsidized; and "community buses" – services to transport particular individuals and/or social groups (typically those living in communities experiencing social exclusion) to enable them to meet their social and welfare needs. As governments have been trying to tackle problems of social exclusion, the needs of people living in isolated locations or of those who find it difficult for some reason to use conventional public transport have sometimes yielded opportunities for services which are specifically designed to be easy to use. Such services are designed to include people in a way that removes the need for eligibility criteria. Thus conventional public transport is becoming more "specialized" by including in its basic design facilities for people who experience physical, sensory, or cognitive disabilities.

Each example serves to demonstrate the basic rationale which underpins the provision of "specialized transport": that is, some organization somewhere – usually identified with a state agency, government service, or local community – deems it to be important that eligible individuals and social groups merit or need

Handbook of Transport Systems and Traffic Control, Edited by K.J. Button and D.A. Hensher

access to public passenger transport. The key word here is "access" – indeed, in many quarters such transport provision is often referred to as "accessible transport" (Sutton and Gillingwater, 1995). The argument can be simply put: as transport is essentially a derived demand (in the sense that very little transport takes place for its own sake – rather it is undertaken in order to meet other objectives), then any barriers which prevent individuals and/or social groups from being able to participate in activities which most people take for granted (such as getting to and from work, being able to shop, getting to and from the post office or bank for personal business, etc.) need to be identified, addressed, and ultimately removed. Thus the provision of a "specialized transport" service is often identified with meeting the objectives which underpin a particular social policy like "social exclusion" (Department of the Environment, Transport and the Regions, 1998).

This emphasis on accessibility – and its corollary, barriers to access – is becoming an increasingly important public interest issue for many countries around the world. There appears to be a strong correlation between increasing economic prosperity and the spatial concentration of activities, resulting in complex divisions in land use reflecting the impact of changing modes of personal transport – especially the increasing significance of car ownership and use (Vuchic, 1999). The "taken for grantedness" of, for example, accessing an edge-of-town shopping mall is for most people based on taking the car. The perception which underpins such a journey – the chain of events – consists of leaving home, jumping into the car, driving to the mall, finding a parking space, visiting outlets, returning to the car, and driving home. The journey chain is therefore relatively seamless. Contrast this with accessing the same mall by someone without access to a car. Can they make the same journey – to achieve the same desired objective – using a non-car means of transport? The answer is, more than probably, not or with much greater difficulty. Add to this the problem of getting to activities such as health, education, work, or voting, and the need for transport to be available to everyone becomes an issue of fundamental human rights and not just a matter of convenience. Those people for whom the "car-norm" does not apply (whether through health, age, or economic reasons) are effectively excluded from taking a full part in society unless transport is truly available for them. "Specialized transport" provides this for that part of the population that is excluded from conventional public transport. Consider how the same journey might be made by public transport.

The public transport journey chain is far from comparable to its car-based counterpart or as seamless (Frye, 1996): there is the walk to the bus stop, the wait for the bus, the time taken to transfer from the bus to another which serves the mall and then the return process. Although there may be barriers to our car driver accessing the mall (e.g., road works, traffic congestion, lack of driver information, etc.), these are qualitatively different from those barriers facing our non-car traveler. There may well be physical barriers – the pavement to and from the bus stop may be uneven; the bus stop may be on the other side of a busy, dual-carriageway road, necessitating finding a suitable crossing point. Once at the bus stop, there may be no suitable seating or waiting area or accessible timetable display. When the bus arrives, there may be a gap between the kerb edge and the vehicle entrance and the design of the vehicle may require a steep step up to get on and then down to get off the bus.

All of this assumes that it has first been possible to find out, understand, and act on information about the service in order to know that it exists and how to use it. It also assumes our traveler is fit and well and more than able to cope with such arduous tasks. Evidence suggests otherwise. For example, how feasible would such a trip be for a person in a wheelchair? This raises another type of accessibility barrier – not so much physical, sensory, or cognitive but rather more social – that can affect in particular people who are in some sense vulnerable and dependent on others to help them meet their personal needs. Frail elderly people are an obvious case of what is called "mobility impairment," together with those suffering severe health problems who may use a wheelchair to move around. But there are many other groups, including, for example, young children, people on income support or very low incomes, single parents, people from ethnic minorities, etc. Each of these people needs to travel in order to engage in those activities which most people take for granted, and yet these barriers serve to exclude them from such participation. In terms of public passenger transport provision, the examples of "specialized transport" cited earlier have each been developed in response to a lack of provision to tackle one or more of these physical and/or social barriers to accessibility. More often than not this is as part of a wider aspect of the kind of social policy associated with vulnerability, dependence, and exclusion (Banister, 1995).

2. The structure and organization of "specialized transport"

Figure 1 is an attempt to devise a structure for the organization of "specialized transport" (Sutton, 1988; Sutton and Gillingwater, 1995). It shows that "specialized transport" functions in a three-level hierarchy of service provision – primary, secondary and tertiary – mapped against four types of service – fixed-route, variable-route, contract hire, and demand responsive characteristics.

The primary level consists of conventional public passenger transport (commercial, for-profit operations), services that are universally available to anybody who wishes to use them – subject to their ability to pay. These dominate service provision and account for the vast majority of services – whether by number of passenger journeys, passenger-miles, or vehicle-miles (in the U.K., well in excess of 95% of all public passenger transport journeys). These include conventional, fixed-route, stage carriage bus operations undertaken by public

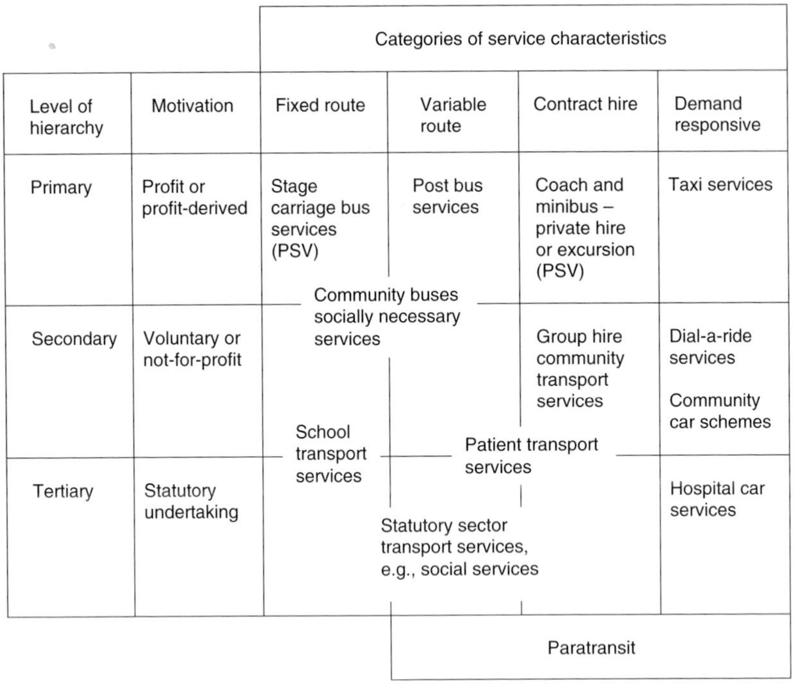

		Categories of service characteristics			
Level of hierarchy	Motivation	Fixed route	Variable route	Contract hire	Demand responsive
Primary	Profit or profit-derived	Stage carriage bus services (PSV)	Post bus services	Coach and minibus – private hire or excursion (PSV)	Taxi services
		Community buses socially necessary services			
Secondary	Voluntary or not-for-profit			Group hire community transport services	Dial-a-ride services Community car schemes
		School transport services	Patient transport services		
Tertiary	Statutory undertaking		Statutory sector transport services, e.g., social services		Hospital car services
			Paratransit		

Figure 1. The structure and organization of "specialized transport."

service vehicles (PSVs) as well as variable-route operations (e.g., rural post bus services), private contract hire and excursion operations with PSVs, and conventional taxis operating demand-responsive services. These primary-level services, by definition, do not incorporate "specialized transport," although recent legislation is enabling them to carry people who would previously have had to use the specialized transport to be found primarily in the secondary and tertiary levels of the hierarchy. That said, there is clearly an element of "socially necessary transport" which comes under the umbrella of "specialized transport." This consists of the payment of subsidies to commercial, fixed-route operators for services which an operator regards as not viable (e.g., late night or weekend off-peak services) but which a local government might deem desirable on social-policy or public-interest grounds to subsidize. In the U.K., such services are usually open to competition by tender.

The tertiary level consists of transport services operated predominantly by or on behalf of statutory undertakings such as local government or health or other public agencies responsible for implementing and delivering social policies formulated nationally. These services more often than not have to be provided by law, and in the U.K. reflect the "free at the point of delivery" philosophy of intervention for, for example, health and education. Although there may be some notional cost which users pay, the full cost is met either through general taxation or subsidy or through virement of local resources. Such services typically include fixed-route school transport, variable-route patient transport, contract hire statutory sector transport purchased through an "internal market," and demand-responsive hospital car services. The whole thrust of transport provision at this tertiary level is to ensure that people – "clients" – get to destinations at times which suit the agency so that their statutory obligations may be discharged. In this sense, client mobility (rather than profitability at the primary level) is the driving force behind the deployment of transport resources, as one part of the delivery of public services through a process of administrative rationing.

Until recently – in the past 20 years in the U.K. and North America – public passenger transport consisted of just two tiers, primary and tertiary, with little or no interaction between them. This has now changed with the advent of a secondary level which has come to be directly associated with "specialized transport." This secondary level consists of transport services operated predominantly by voluntary and/or "not-for-profit" organizations with strong roots in local communities. The modus operandi of these services is to provide accessibility for people with a mobility impairment. This is qualitatively different from the goals of profitability at the primary level or delivery to statutory service provision at the tertiary level. These services include "socially necessary transport" and "community bus" operations: both fixed- and variable-route (urban and rural), as well as contract hire (usually referred to as "group hire" operations) and demand-responsive, "dial-a-ride" operations (both PSV and car-based "community car schemes").

In a recent study of this secondary level of public passenger transport in the U.K. (Department of the Environment, Transport and the Regions, 1999), it has been estimated that there are some 5300 voluntary and "not-for-profit" organizations which provide services for about 21 million trips per year. Within this sector it is possible to identify three types of service provision. The first group consists of about 40 very large and well-established transport operators – most of whom are either based in very large cities like Birmingham or Manchester or operate within certain parts of London. These 40 account for almost 25% of the 21 million trips per year. In a few cases, these organizations operate vehicle fleets which are roughly the same size as their primary-level commercial counterparts. The second type of service provision consists of about 5000 usually very small, well-established and self-contained transport operators – most of whom are based either in rural settlements (where they rely on volunteer drivers using their own vehicles) or in small towns (where they operate typically fewer than five vehicles). However, these small operators account for over 40% of the 21 million trips per year. Finally, there are an estimated 300 small and comparatively new transport operators, both urban- and rural-based, providing the remaining 33% of annual trips and who share characteristics similar to the previous group.

Although these 21 million trips account for less than one per cent of the trips provided by the primary and tertiary sectors combined, what makes this sector of interest is its rapid growth and the way in which it develops innovative services which challenge both primary and especially tertiary service providers. These are illustrated in Figure 1 by examples of services provided at the secondary–primary and secondary–tertiary interfaces. This shows how the secondary level has been nibbling away at its boundaries with more conventional operations by both the primary and tertiary providers.

Under U.K. legislation, it is now feasible under the Transport Act of 1985 for a voluntary or "not-for-profit" operator to provide local bus services, which until recently have been the prerogative of commercial operators. In "specialized transport" terminology, these constitute "socially necessary transport" services (typically those regarded by a commercial operator as unprofitable and in need of subsidy or unviable for subsidy by the local authority) and "community bus" services (typically services where a commercial operator has withdrawn or where no commercial service has ever operated). However, it is at the secondary–tertiary interface where such operators have really had a major impact on service provision. Where transport services have typically been provided "in house" by a public agency (e.g., a local authority social services department), recent legislation has required such services to be market-tested and in many cases put out to competitive tender. As a result of this what has been called "contract culture," many secondary-level transport providers have become the preferred and sometimes even the only supplier of tertiary-level services.

One of the key reasons for this impact has to do with the way in which these providers have developed their services around the concept of accessible transport. With their strong roots in local communities and because providing accessibility for people with a mobility impairment is an important driver, these operators also find themselves in tune with legislative changes aimed primarily at "equality of opportunity" antidiscrimination measures for people with disabilities. In the U.S.A. the Disabilities Act (ADA) of 1990 has been instrumental in this "human rights"-based approach to transport provision, and in many cases has prefigured similar legislation elsewhere, including the U.K.'s Disability Discrimination Act (DDA) of 1995 (Brown, 1998). In many cases, these secondary-level transport providers have been the only operators to have vehicles and facilities which have met the basic requirements of this type of legislation (e.g., wheelchair access and restraints, driver assistance, etc.) prior to its implementation.

The following three case studies illustrate the variety of innovative services which these secondary-level operators provide. Although they are U.K. based, they do embody principles and approaches which have been developed in other countries and which are applicable to other countries' experiences too. In addition they are examples where the authors have themselves been directly involved in service development. The first is a case study of Barnsley Dial-a-Ride and Community Transport – an essentially urban-based organization providing a full range of "specialized transport" services. The second is a case study of a highly innovative approach to transport provision in a sparsely populated rural area – the Cumbria Plusbus. The final case study is an example of a highly innovative approach to a particularly complex problem facing many providers – the co-ordination of "specialized transport" services in one area of London.

3. Case studies of "specialized transport" services

3.1. Barnsley Dial-a-Ride and Community Transport

Barnsley Dial-a-Ride and Community Transport is a long-established "not-for-profit" organization which provides "specialized transport" services in a largely urban area to the north-east of the South Yorkshire conurbation around the city of Sheffield. Although it employs 35 paid staff, the organization relies heavily on numerous volunteers to operate its 17 vehicles, ranging from wheelchair-accessible small vehicles used for its community car service (essentially a "not-for-profit" taxi service) to wheelchair-accessible people carriers for its dial-a-ride service and up to 16-seater wheelchair-accessible PSV minibuses for its group hire and "community bus" services. Nominal rather than full-cost user charges apply for most of these services.

These services provide around 100 000 passenger trips per year, paid for by the South Yorkshire Passenger Transport Authority under a service level agreement, which has a 30% performance payment related to the number of passengers actually carried. A further 20 000 passenger trips are provided annually by two subsidiary operations – 15 000 by Coalfields Community Transport, and 5000 by Penistone Rural Ride, a service which operates near to the edge of the Peak District National Park. In addition, 28 000 trips a year are provided under service level agreements with Barnsley Metropolitan Borough Council and a number of support services, which complement services for the education and social services departments. A further 7000 trips per year relate to "Shopmobility," a loan service of 17 battery-powered scooters and 20 wheelchairs for use in the main shopping center – a service which is also seen as contributing to the regeneration of the economic and social wealth of the locality. These 155 000 trips contrast with a total of just 1000 trips when the organization was first operational in 1983.

The dial-a-ride operation provides half of all such services in the whole of the South Yorkshire area, mainly because of the very high incidence of people with disability and mobility problems in the area around Barnsley: almost 18% of the population – that is, 44 000 out of 250 000 – have a definable mobility impairment according to the local authority's community care plan. Of these, over 7000 are registered to use the services within the parameters of the various eligibility schemes, delivering over 350 000 miles of passenger trips a year at a unit cost in the region of £4.25 (U.S. $6.50) per trip – £3.35 (U.S. $5.00) net if user charges are included.

3.2. *Cumbria Plusbus*

The Cumbria Plusbus is a public local bus service operating in the county of Cumbria in north-west England – an area associated with the Lake District National Park. The service is unusual in many respects, but an important feature is that it forms part of an action research project covering not only transport issues but also the way in which the community interacts with such a service and vice versa.*

The service joins 12 villages and the small market town of Kirkby Stephen. The study area was chosen carefully to allow observation of the connectivities opened up by the service. The villages are fairly isolated, yet close enough together to permit a bus service to operate between them at a reasonable frequency. Before the Plusbus entered service, most of the villages had no bus service at all (for example, people were walking 5 km to go shopping). About 4500 people live in the whole area (the average population density is about 0.6 persons/hectare), of which Kirkby Stephen accounts for a little less than half. Although the bus operates on a "hail and ride" basis, there is at least one formal bus stop in each village. This provides an information point and an indication of where the bus will be found. The service headway is three hours, providing four journeys in each direction per day on six days per week.

The vehicle (part funded by the Countryside Agency) was specified by University College London (UCL) on the basis of previous experience with other small, accessible buses and consultation with people in the project area. The bus has a low floor with no steps inside the vehicle and an aisle wide enough to allow

*The Plusbus is a local service which operates as a public bus service but at a much higher frequency than would normally be the case in such a rural operating environment. The vehicle is small and designed to be fully accessible. In the cases cited in this chapter, the service has a fixed route and schedule. The service forms part of an action research project funded largely by the U.K. Engineering and Physical Sciences Research Council to investigate the outcomes of providing a fully accessible, frequent bus service in a remote rural area. The project – called "Accessible Public Transport in Rural Areas" (APTRA) – is being carried out by researchers at the Centre for Transport Studies at University College London and covers not only the transport issues of the specification and design of the vehicle, route, schedule, and information systems, but also the way in which the community interacts with such a service and vice versa.

wheelchairs to pass along the bus. In the bus there are three possible places for a wheelchair user to choose from. The bus seats up to 13 passengers, including up to two people in wheelchairs. The bus is fitted with high-visibility non-slip stanchions, handrails and grab handles, and an induction loop for people with hearing aids. The vehicle has a wide side door at the front and at the rear of the vehicle (to cater for the event that the road is too narrow to permit boarding or alighting via the front door).

Information about the schedule is provided at each bus stop in large print with a schematized route map. In addition, leaflets are available with timetable and fares information and a route map, including a detailed map of each village showing the location of the bus stop and other features. In all cases attention has been paid to make sure that the information is legible (through a careful choice of typeface and font size). A document is also available containing a tactile map with the timetable and information in Braille, and for those who do not read Braille, information and the timetable are available in audible format on compact disk.

After one year, the service carries between 280 and 300 passengers per week and distributes bulk newspapers to the villages. A recent survey of users showed that about 40% are under 16 years old and 21% are over 59 years of age. 62% of the users say they have increased the amount of travel because of the bus, often making three to four shopping trips per week by bus instead of one by car – 41% of users are ex-car passengers and 29% are now using the bus where they used to drive a car. The operation is carried out under contract to UCL by a local bus company. An interesting element of the contract is the insistence on enforceable quality standards in terms of performance and driver training. The project has helped the local community to start a management company to take over the management of the bus service and this company will supervise the contract with the operator (and thus enforce the quality standards) in the future.

3.3. Transport co-ordination center in a London borough

It is important to realize that not only does the design of "specialized transport" services have to be addressed, but also the institutional arrangements by which they are organized. In one London borough, a review of the "in-house" transport service showed that there was a remarkable disparity between the "style" of the transport operation and the needs of the providers of social services and special education. The gap had become worse as a result of the borough's attempts to implement the National Health Service and Community Care Act 1990 by providing service users with a more individual care package. This meant that, for example, the service providers wanted to enable people to travel for individual activities (e.g., a visit to a swimming therapy session) as part of their care plan. Also, a user could be in the position of having services provided by more than one agency (e.g., health and social services). This meant that, because of the different budgets involved, transport was organized by each agency to the exclusion of any possibility of linking trips. Thus a user could not travel from a health clinic to a social services appointment without first being returned home. The transport service was organized to collect people and deliver them en masse to and from their activities since individual trips were prohibitively expensive to provide. But different service providers had different eligibility criteria so that often the only way to transport someone was as the only passenger in a vehicle. Funds for transporting a user went with the vehicle, tending to make the transport service suit the vehicle rather than the user. The quality of care was suffering due to the inflexible nature of the transport service.

The approach adopted to resolve this problem was to set up an independent transport co-ordination center (TCC) so that it would be independent of the borough yet could act as an agent of the borough in contracting transport services from contractors who could satisfy a strict set of performance and quality criteria. These criteria were designed on the basis of consultation with service users and the social services and education departments. The TCC is required to obtain the most appropriate, least-cost transport that meets the user's and service provider's needs at the time of booking. The contract between the TCC and the borough sets out a number of strict service standards and criteria on which the TCC will be judged. These standards are passed to the contractors by the TCC as part of the conditions of contract. In this way, quality is made a contractual issue. Importantly, the TCC works on the basis that users have allocated to them a "transport budget" as part of the total cost of their care package. The transport money therefore follows the user rather than the vehicle, thus helping to ensure that the most appropriate transport is obtained.

Transport can be booked at any time (both the TCC and transport contractors are required to provide transport within set time limits). Cost savings can be expected as a result of being able to mix use of vehicles, to make better use of down time, and to co-ordinate the transport of different people. This includes education, social service, and leisure trips being carried out at the same time on the same vehicle. Eventually some users will be allocated their own transport budget so that they can order their own transport from the TCC independently of the provider or their other services.

The TCC is a registered company (but with no share capital and for the benefit of the community) as a first step towards being registered as a charity. The reason for this is to ensure that the TCC is responsive to the end-users and is transparent in its relations with contractors and the borough. Also, as an independent body, the TCC can examine and test new and progressive ideas about meeting transport needs (e.g., the possible introduction of transport mentors and travel training): this would be much more difficult to do as a department of the borough council.

The structure of the TCC is designed to work towards enabling people to travel as independently as possible. Training people to travel on public transport (independence training) will free up space and resources used on specialized transport as well as providing a more sustainable life change for the user. Additionally – and perhaps more importantly – the quality of transport provision can be expected to improve as a result of the contractual requirements on vehicle standards and equipment, crew quality, and training and reliability.

The TCC was put into action for a trial period to procure approximately 10% of the borough's transport provision as well as all taxi work, through a tendering process. This amounts to approximately £1 million (U.S. $1.5 million) per annum out of a total spend of £3.6 million (U.S. $5.4 million). The tendering process has brought about a reduction in costs to the borough as a result of the competition between potential operators. Although some further reduction in costs is expected from further tendering, the greater savings – and improvements in quality – will come from the co-ordination of transport and people.

4. Key issues in "specialized transport" service provision

Each of the three case studies outlined above illustrates in one or more ways the core principles which underpin the design of "specialized transport" services operating within or at the boundaries of the secondary level of the hierarchy of transport provision. The emphasis on accessibility – and the removal of barriers to access – lies at the very center of these core principles, as does the assumption or belief that it is in the best interests of all people (with or without mobility difficulties) to be able to make their own informed travel decisions (Tyler and Caiaffa, 2000). The concept of independence – the ability to ascribe travel need to oneself rather than have some administrator or bureaucrat ascribing travel need on that individual's behalf – demonstrates one of the reasons why community-based, voluntary, or "not-for-profit" transport provision has grown so sharply in recent years in countries like the U.K. and U.S.A. (Gillingwater, 1995). These providers have struck a chord in showing that "specialized transport" services can be delivered in ways which challenge the traditional suppliers of tertiary-level services as well as delivering highly innovative services which challenge the mind-set of those providing conventional public transport.

The relationship between the ascription of travel need and the provision of accessible transport services has, until comparatively recently, been the source of much tension between the providers in the three tiers. Each has tended to argue for the defence of their own services as representing the "best" way to provide "specialized transport," primarily on the basis that "we know our own markets best." This segregationist approach has been the subject of increasing criticism in recent years, on the basis that individuals should be able to make up their own

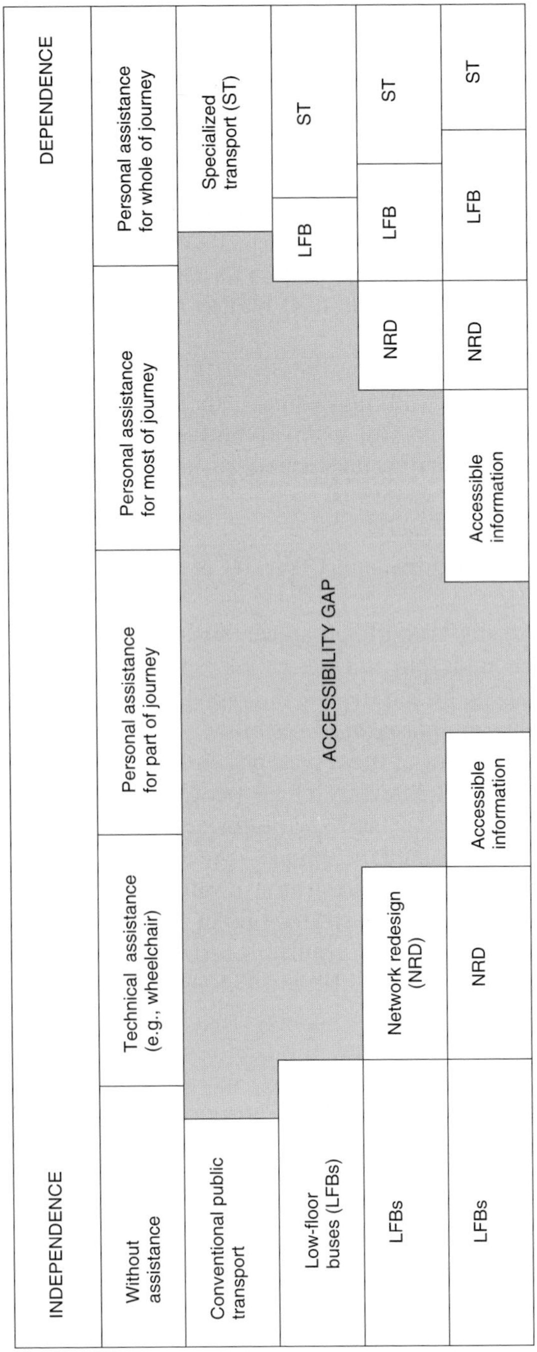

INDEPENDENCE						DEPENDENCE
Without assistance	Technical assistance (e.g., wheelchair)	Personal assistance for part of journey		Personal assistance for most of journey		Personal assistance for whole of journey
Conventional public transport	ACCESSIBILITY GAP					Specialized transport (ST)
Low-floor buses (LFBs)					LFB	ST
LFBs	Network redesign (NRD)			NRD	LFB	ST
LFBs	NRD	Accessible information		Accessible information	NRD	LFB ST

Figure 2. Closing the accessibility gap.

minds as to which service best suits their needs. As a result, there is a belated recognition on the part of service providers that an integrationist or inclusive approach to a public passenger transport network which is accessible to all offers a better way forward. The argument runs as follows: conventional public passenger transport is becoming more accessible (for example, through the adoption of low-floor buses together with better schedule and network design) and thus can include some people currently restricted to "specialized transport." This does not imply the demise of "specialized transport." It means that "specialized transport" services operating at the secondary level can become more focused on those who – even with the advent of more accessible primary-level public transport – cannot use primary-level services.

The key issues facing "specialized transport" providers – whether at primary, secondary, or tertiary level – are therefore how best to adapt their operations such that each can optimize the "fit" between the users' needs and service provision. The principles that underpin this adaptation process, which can be referred to as closing the accessibility gap, are illustrated in Figure 2.

Figure 2 suggests that it should be possible to assess the public transport journey chain in terms of the number of people for whom each element is a barrier and the degree of independence it permits amongst potential users. Different people encounter barriers at different points in the journey chain (Tyler, 1999) and this constitutes the main way in which independence or dependence – "self-ascription" or "other-ascription" – can be assessed (as far as public passenger transport journeys are concerned). The degree of independence is simplified to five levels, based on the possibility of an individual completing the whole journey chain with no assistance (thus implying maximum independence) through to the situation where a journey cannot be completed (thus implying maximum dependence). The five levels can therefore be conceived as defining step changes in dependence from zero dependence, to dependence on equipment (e.g., a wheelchair or hearing aid), to occasional dependence on others (e.g., help in boarding a bus), to dependence on being accompanied throughout the journey (e.g., a fully escorted service), to complete dependence for all needs. Each step change in dependence results in a threshold that determines the accessibility of the system.

Figure 2 shows the progressive reductions in the accessibility gap that might be anticipated due to changes made to a primary-level conventional bus operation. These reductions follow, first, the replacement of the bus fleet with new "low-floor" buses, second, the introduction of fully accessible information systems, and, finally, improvements to the design of the network. The core market for conventional public transport is represented by those requiring no assistance with their journeys, whereas those requiring assistance all through the journey or who could not complete a journey unaided have to rely on some form of "specialized transport" – with the consequent accessibility gap. People who are unable to use conventional public transport but who are not eligible for "specialized transport" come into the accessibility gap. Overall, this represents the situation before the introduction of service improvements. The next row shows the impact that could be expected with the introduction of "low-floor" buses. These vehicles offer travel opportunities to those who previously could not access conventional buses because they required some basic assistance. This expands the market for the bus operator. At the same time, some people previously restricted to using "specialized transport" might be able to use the new buses. This would free time and space on "specialized transport" services for others requiring assistance to travel so they might travel more frequently or at times which better suit their needs, thus improving the level of service to those users. The impact overall is a reduction in the accessibility gap.

The final two rows follow the same logic, showing the possible impact on the accessibility gap of introducing fully accessible information systems and improving network design. In both cases, it could be anticipated that ridership on primary-level public transport services might increase because of the transfer of a proportion of those previously using "specialized transport." As a result, the pressures on those "specialized transport" services might reduce. This would allow improvements to be made to the overall levels of service offered by those secondary-level service providers – in the sense of allowing eligible people to travel more frequently, at times more suited to their needs and to destinations of their choosing. As with the introduction of the new "low-floor" vehicles, the impact overall of these two additional improvements should lead to further reductions in the accessibility gap – reductions which, at the end of the day, constitute improvements for all those traveling by public passenger transport. In this context, "specialized transport" should not be seen as a marginalized or "special" form of public passenger transport provision but rather as part of a bigger and wider "patchwork quilt" of services – contributing to the development of a fully accessible network of public passenger transport service provision for all. That is its ultimate goal.

References

Banister, D. (1995) "Equity and efficiency in the evaluation of transport services for disadvantaged people", in: D. Gillingwater and J. Sutton, eds., *Community transport: Policy, planning, practice.* Luxembourg: Gordon & Breach.

Brown, N.L.C.K. (1998) "The effects of the ADA and DDA on urban bus services and social inclusion", University of London Centre for Transport Studies, University College London, Working Paper.

Department of Transport, Environment and the Regions (1998) *A new deal for transport: Better for everyone*, Cm 3950. London: The Stationery Office.

Department of the Environment, Transport and the Regions (1999) *Review of voluntary transport.* London: Mobility Unit, DETR.

Frye, A. (1996) "Bus travel – a vital link in the chain of accessible transport", in: *Proceedings of Bus and Coach '96*. London: Institution of Mechanical Engineers.

Gillingwater, D. (1995) "Resourcing accessible and non-conventional public transport", in: D. Gillingwater and J. Sutton, eds., *Community transport: Policy, planning, practice*. Luxembourg: Gordon and Breach.

Nutley, S. D. (1990) *Unconventional and community transport in the United Kingdom*. Luxembourg: Gordon and Breach.

Sutton, J. (1988) *Transport co-ordination and social policy*. Aldershot: Avebury Press.

Sutton, J. and D. Gillingwater (1995) "The history and evolution of community transport" in: D. Gillingwater and J. Sutton, eds., *Community transport: Policy, planning, practice* Luxembourg: Gordon and Breach.

Tyler, N. (1999) "Measuring accessibility to public transport: Concepts", University of London Centre for Transport Studies, University College London, Working Paper.

Tyler, N. and M. Caiaffa (2000) "Accessibility and independence", University of London Centre for Transport Studies, University College London, Working Paper.

Vuchic, V. (1999) *Transportation for livable cities*. New Brunswick, NJ: Rutgers University Press.

Chapter 22

WOMEN AND TRAVEL

LAURIE A. SCHINTLER
George Mason University, Fairfax, VA

1. Introduction

Transportation planning and engineering have not historically been gender neutral. It has traditionally been assumed that travelers are relatively homogeneous in their travel needs, values, preferences, and behavior. The typical traveler has been characterized as a commuter, who seeks to minimize travel time to work and who values time as forgone leisure time. Highway and transit facilities have often been designed to achieve some desirable level of mobility for this "representative" traveler. These assumptions are becoming increasingly unrealistic and outdated, however, as women become increasingly motorized.

At an aggregate level, women's travel patterns appear to be converging to those of men, at least in terms of per capita distance traveled. An increase in female labor force participation is largely responsible for this trend. In the U.S.A., for example, the number of women in the workforce has increased by 122% since 1969, which is in sharp contrast to men, whose numbers in the workforce increased by only 47% during the same time (Hu and Young, 1999). Women's role as primary caretakers of household and family obligations, their concerns for personal safety, and, more generally, the geographical dispersion of population are also contributing to a growing motorization of women. At micro level, the travel patterns and needs of women tend to differ from those of men. Gender differences exist in the reason for travel, trip lengths and frequencies, mode choice, and the complexity of trip making.

Effective transportation planning and engineering require an understanding of women's travel behavior and patterns. They are also important for helping to ensure that the needs and desires of women are reflected in transportation investment. This chapter explores some of the differences between men and women in terms of travel behavior, and their implications for transportation planning and engineering. The focus is primarily on urban travel patterns, and not those relating to rural, intercity, or international travel. Further, the conclusions drawn in this paper pertain mainly to developed countries in the western hemisphere, where data on women's travel patterns are readily available.

Handbook of Transport Systems and Traffic Control, Edited by K.J. Button and D.A. Hensher

2. The responsibilities and needs of women

Transportation is a derived demand, stimulated by the need to perform some activity at a particular destination. Needs range from the most basic, such as those to maintain human existence and safety, to those that are more complex, such as promoting a sense of belongingness, self-esteem, and self-actualization. Individual needs and responsibilities play an important role in shaping the determinants of travel behavior. Women's needs are different from those of men, and include factors such safety, comfort, and accessibility to opportunities. Travel time may also be important to women; however, the value placed on time is likely to differ from that placed by men due to gender-specific responsibilities and needs.

Traditionally, women have been the primary caretakers of household and domestic responsibilities. The responsibilities of women now extend beyond the domestic sphere, due largely to the increase in female labor force participation. Labor force participation in the U.S.A. has increased regardless of marital status. In fact, the increasing number of married women in the workforce largely explains the growth in female labor force participation.

Because women have responsibilities different from men, their valuation of time is likely to be different. This distinction, however, is rarely made in transportation analyses. The value of travel time (the shadow price of travel) is often based on some "representative" traveler who trades work for leisure time. In theory, this parameter is derived from the consumer welfare maximization problem and the Lagrange multipliers resulting from this optimization. Individuals or households seek to maximize welfare, or the consumption of market goods and services, leisure time, and time for other non-work-related activities (e.g., travel), subject to some set of constraints. These constraints typically relate to the household or individual's budget and limits on the time available for various activities. In the simplest formulation, the disutility of time is equivalent to the value of forgone leisure time (Becker, 1965). In this example, consumers seek to maximize their consumption of market goods and services and leisure time subject to budget and time constraints. Commuting reduces the amount of time available for leisure.

The value women place on time is likely to be more complex than that based solely on forgone leisure time. In theory, the value of time for an individual depends on his/her wage or income level, the number and variety of activities in the consumption set (e.g., shopping, recreation, childcare), and any time constraints relating to these activities (Small, 1992). Since men and women differ in terms of all of these factors, the value they place on time is likely to vary as well. For instance, their value of time might be affected by the need to attend to household and domestic responsibilities or the need to care for children. Since many of women's journeys are likely to combine "work" and "non-work" functions (e.g., shopping on the way to work), then time valuations need adjustment to reflect this.

In addition, the sole concern of women may not be to minimize travel time as assumed in transportation planning models. Factors like security may be an important concern for many women. Women are generally physically smaller than men and more often travel with vulnerable children. This poses real fears of attack and affects perceptions of the safety of travel. Many women may feel unsafe when traveling in the evening, particularly when waiting for underground trains, walking to a car in open or multistoried car parks, etc. This is an issue as more women are now traveling at hours later in the evening. Currently, about half of all "moonlighters" in the U.S.A., for example, are women. Furthermore, for many women who work during the day, trips for household and domestic duties must be completed in the evening hours.

3. Travel activity patterns

The needs and responsibilities of women play an important role in shaping their travel activity patterns, specifically, in their impact on trip purpose, frequency and distance of travel, mode of transportation used, and complexity of trip making. Transportation planning models are not designed to capture these differences.

3.1. Trip purpose

Women's entry into the workforce, along with their continued role as primary caretakers of domestic responsibilities, has also led to the emergence of "knock-on" trips, or trips generated by the substitution of home production for market production. Statistics show that women are more likely than men to make these types of trips. Approximately 50% of all person trips made by women in the U.S.A. are for family and personal business and two thirds of the trips women make are to take someone else someplace (U.K. Department of the Environment, Transport and the Regions, 1997). In 1994/96 women in the U.K. made 28% fewer commuter journeys and 68% fewer trips during their work than men. Women are making 65% more "escort education" journeys (taking children to school) and approximately 30% more shopping trips than men (U.K. Department of the Environment, Transport and the Regions, 1997).

3.2. Trip length and frequency

At an aggregate level, women are becoming increasingly motorized. In the U.K. for example, women in all age groups are increasing the distances that they travel by mechanized transport and, especially, as car/van drivers (U.K. Department of the Environment, Transport and the regions, 1997). Over the last several decades, the mobility of women in the U.S.A. from all age groups has also improved. Between 1969 and 1995, the average annual person trips taken by women increased by 11% (Hu and Young, 1999). This rate of increase was less for men, despite the fact that the population growth rates of men and women were both about 40% over the same time period.

Despite their increase in automobile use and enhanced mobility, women still travel shorter distances than men. In the U.S.A. women drive only 60% to 70% as many miles as men (Hu and Young, 1999) and on average they travel 27.8 person miles a day, which is slightly less than the 35.2 person miles a day that men travel. Women also tend to have shorter commutes (Hanson and Hanson, 1980).

These patterns arise from the time constraints imposed by women's responsibilities, such as the need to conduct several activities (e.g., picking up children at daycare and dropping of clothes at the dry-cleaner) within a fixed amount of time. In the U.K., there appears to be a correlation between the distance traveled by women and the need to attend to childcare, caring for the elderly, and domestic responsibilities. In this sense the pattern has changed little in recent years. These patterns are not unique to the U.K. or U.S.A.: Hanson and Hanson's (1980, 1981) work, for example, offers insights into the importance of married women's employment for travel patterns in Sweden, while Dyck (1989, 1990) covers similar issues in Canada.

3.3. Complex trip-making

Complex travel behavior such as trip chaining is also more common for women than men even when both males and females are in employment (Root et al., 2000). Women stop more for running household errands than do men, on both inward and outward commutes and irrespective of the number of persons in a household or its structure. On average two in three American women make stops on their way home and 25% make more than one stop. The places visited differ, with women tending to visit schools, day-care centers and shops more than men; the latter are twice as likely to go to a restaurant or bar (U.S. Department of Transportation, 2000). The trend for more complex commuting patterns is upwards. In the U.S.A. the number of intermediate stops on the way to work has grown by about 50% since 1980 and the number on the way home by about 20% (Hu and Young, 1999).

3.4. Mode choice

In comparison with men, women have tended to travel on what are generally considered less prestigious modes of transport. In the period 1994–96, women traveled less far than men in the U.K. by all modes apart from buses, walking, and as passengers in private cars. Statistics on aircraft, ship, and Channel Tunnel journeys according to gender are not available for the U.K., but they would probably tell the same story. In addition, women are passengers for about half of the travel they do, while men are for only one fourth of the time (Hu and Young, 1999). In 1975/76 women drove about a fifth of the miles driven by men but by 1994/96 the gap had closed somewhat, and women drove about two fifths of the miles that men drove (U.K. Department of the Environment, Transport and the Regions, 1997).

These patterns are changing, though. In the U.S.A. and U.K., women's use of local bus services in terms of distance traveled per capita has declined over the last few decades while for automobile it has increased (U.K. Department of the Environment, Transport and the Regions, 1997). One explanation for this is that women face certain constraints that limit the feasibility or desirability of using modes of transportation other than the automobile. "Coupling" constraints, or the need to plan travel around someone else's schedule, are common for women, as indicated by the high degree of "escort" trips made by women. These constraints, along with the need to trip chain, eliminate bus or rail as a feasible mode of transportation. Security concerns may also limit the feasibility of some modes at certain times of the day. These trends raise security issues for women, particularly for those who choose to travel by public transit, and will increasingly be concerns as more women demand transportation. Given security concerns, transit may not be a viable option for women "moonlighting," evening employment may not be possible without access to a dependable vehicle, or a certain route may not be feasible due to its perceived danger.

Demographic trends indicate further improvements in mobility for women and continued growth in the number of women drivers on the road. Many of these drivers are likely to fall in the older age cohorts (e.g., 65 years of age). In the U.S.A., between 1969 and 1995, the highest rates of growth in population were in these cohorts. In 1995, almost 65% of the population 75 years old and over were female (Hu and Young, 1999). Currently, elderly women are predominantly not driving license holders (Zauke and Spitzner, 1997) and consequently have only limited use of cars as passengers. This picture will change dramatically when the current cohorts of younger and middle-aged women begin to enter the older age categories.

As these trends continue, the share of elderly women on the road is likely to increase dramatically. Many of these women will be highly educated, wealthy, licensed to drive, and as a result extremely mobile. Furthermore, life expectancy figures suggest that because women are outliving men, they will wish to be mobile for longer. Women's life expectancy is greater than that of men in both North America and the E.U. Additionally, if, as demographers suggest, the whole population ages, the proportion of very elderly women who still have mobility requirements will increase further.

4. Decision-making and cognitive processes

The unique circumstances and psychology of women may lead them to very different rules of decision-making than for men, and in fact this behavior cannot be accurately reflected in travel demand models based on rational behavior and utility maximization. Two psychological attributes that differ by gender are risk propensity and the tendency for ambivalence (Root et al., 2000). Although not studied conclusively, there is some evidence to suggest that women are more risk averse than men when it comes to making travel-related decisions. In studies of traveler information services, women are often less prone to switch routes after receiving travel information on alternative routes (Abdel-Aty et al., 1996). In a stated-preference survey of Los Angeles travelers, women tended to be more conservative in their selection of travel alternatives. When given the hypothetical choice between two routes, for example, one that is relatively long but predictable in terms of travel time and another that is less predictable but potentially shorter in terms of travel time, women tended to select the former and men the latter. In addition, risk aversion may affect women's travel decisions, when security and safety are a concern.

The propensity for risk may lead to cognitive anomalies, or departures from rationality on the part of women. For example, in selecting a route from work to day care on a given day, a woman may show risk aversion for alternatives with lower travel times than her normal route (i.e., the reference point) and risk preference for losses. These points imply that the shapes of women's utility functions are likely to be different from those of men due to differences in risk propensity and that multiple functions may be necessary to capture risk propensity under different circumstances.

In addition to attitudes about risk, women also tend to be more ambivalent than men. Feelings are never simple, nor unequivocal in their linkage to behavior, but it is appropriate to discuss ambivalence in the context of transport choice. Women are, in general, considered to be more exposed to ambivalent feelings than men in their roles a parent. It would appear that many social groups, but particularly some groups of women, are especially prone to ambivalent feelings, but that their analysis of these can prompt leaps in thought and creative solutions to problems. This suggests that gaining a better understanding of the ways in which women use ambivalence might contribute to further insights into why women are, currently, choosing to travel more ecologically than men.

5. Conclusions

Existing transportation planning models fail to capture such variations in trip purpose. In aggregate models, trips are categorized simplistically as either work or non-work. Furthermore, these models predict travel demand on the basis of the characteristics of zones, rather than those of the individual. It is important to understanding the travel characteristics and trends of women specifically, for the development and implementation of more efficient and equitable transportation programs. Traditional transportation planning models do not characterize gender differences in transportation. These models assume that each traveler's primary concern is to minimize travel time or cost. One possibility highlighted in this chapter is that other factors such as safety, comfort, and accessibility to opportunities may be more important than travel time to many women. In addition women tend to have different cognitive processes or psychology from men.

Understanding how risk influences women's travel decisions is also important. The effectiveness of traveler information services in reducing congestion, improving air quality, and promoting sustainable transportation, for example, depends largely on the reactions of travelers to information provided through these services. Traveler information services are increasingly being seen as a tool for travel demand management. Current modeling efforts are focusing on developing methods for simulating the effects of traffic information on individuals' choices of departure time, route, mode of transport, and destination.

Gender differences in decision-making processes, such as those relating to the way that sets are formulated and how alternatives are selected, should be recognized. This information could be incorporated into newly emerging modeling frameworks that assume travelers to be boundedly rational and satisficers. Nonlinear utility functions that reflect women's attitudes toward risk should be explored.

References

Abdel-Aty, M., R. Kitamura and P. Jovanis (1996) "Investigating effect of travel time variability on route choice using repeated-measurement stated preference data", *Transportation Research Record*, 1493:39–45.

Becker, G.S. (1965) "A theory of the allocation of time, *Economic Journal*, 75:493–517.

Dyck, I. (1989) "Integrating home and wage workplace: Women's family lives in a Canadian suburb", *Canadian Geographer*, 4:329–341.

Dyck, I. (1990) "Space, time, and renegotiating motherhood: An exploration of the domestic workplace", *Environment and Planning A*, 22:459–483.

Hanson, S. and P. Hanson (1980) "Gender and urban activity patterns in Uppsala, Sweden", *Geographical Review*, 70:291–299.

Hanson. S. and P. Hanson ((1981) "The impact of married women's employment on household travel patterns: A Swedish example", *Transportation*, 10:165–185.

Hu, P.S. and J. Young (1999) *Summary of travel trends: 1995 nationwide personal transportation survey*. Washington, DC: Federal Highway Administration, U.S. Department of Transportation.

Root, A., L. Schintler and K.J. Button (2000) "Women, travel, and the idea of 'sustainable' transport", *Transport Reviews*, 20:369–383.

Small, K. (1992) *Urban transportation economics*. Chur: Harwood.

U.K. Department of the Environment, Transport and the Regions (1997) *Transport statistics Great Britain 1997 edition*. London: HMSO.

U.S. Department of Transportation (2000) *United States Nationwide Personal Transportation Survey*, 1995. Washington, DC: DOT.

Zauke, G. and M. Spitzner (1997) "Freedom of movement for women: Feminist approaches to traffic reduction and a more ecological transport science", *World Transport Policy and Practice*, 3:17–23.

Chapter 23

CROSS-BORDER TRAFFIC

KENNETH J. BUTTON
George Mason University, Fairfax, VA

1. Introduction

The globalization and internationalization of production are inevitably leading to a growth in the demand for international transportation. This is compounded by a rapid increase in tourism as in the industrialized world people enjoy rising personal incomes, better health, and longer retirements and are more adventurous in their leisure pursuits. The result is increased amounts of international traffic and more international personal mobility. The forecasts are that the trend is far from being transient but that longer-term growth in international trade and in international travel will continue into the foreseeable future.

Transportation represents an impediment to international trade. Part of this is a natural function of the costs of distance but there are also institutional issues. In some cases national (or state) governments manipulate transportation rates or infrastructure provision to favor their own exporters. Indeed, one of the major initiatives by the European Coal and Steel Community (ECSC) after its foundation in 1951 was to act to try and remove such distortions, and it has also been a priority of the Common Transport Policy of the European Union. This chapter is less concerned with this type of problem than with the more direct one of ensuring that transportation is provided across actual borders in as frictionless a way as possible.

Borders have traditionally proved to be bottlenecks in the international transportation system. In general there are few border crossing points and traffic must funnel through them. At the very least they provide relatively easy locations to control trade and collect revenues. They can also serve as a means of meeting other non-economic objectives such as the detection of illegal drugs or the prevention of disease or harmful insects entering a country. In many cases the bottlenecks are, therefore, deliberate and deemed to be an effective way of meeting explicit non-transportation objectives. This does still, however, raise questions about the efficiency with which these border activities are conducted. In addition, border constraints can serve as non-tariff barriers to trade of a less explicit type.

Handbook of Transport Systems and Traffic Control, Edited by K.J. Button and D.A. Hensher

This contribution covers three main topics: the rationale and nature of cross-border impediments, the benefits of freer cross-border traffic movements, and some examples of how cross-border traffic flows are being freed up. It focuses mainly on what may be called macro-borders – those between countries – but these are not always the important borders. In some cases there are meso-level borders between, for example, states within a country (differing weight limits between U.S. states and regulatory structures between Canadian provinces have created border effects). There are also micro-level borders involving, for example, cities limiting entry of vehicles of certain types (e.g., carrying hazardous material), or to certain links on the network (e.g., lorry routes) or at specified times of the day (e.g., bans on nighttime deliveries). This chapter also focuses on the purely transportation aspects of border traffic and excludes land use and local production implications of traffic congregating at border areas – see, Corvers and Giaoutzi (1998) and Ratti (1995).

The chapter also covers both person and freight transportation. In many cases the issues surrounding both types of movement across borders are very similar, for example in terms of the need for adequate and co-ordinated infrastructure on both sides of a border. In other cases there are differences which pose slightly different challenges. Whilst most freight is moved in an unidirectional manner (raw materials go across borders to be refined and consumed there), the majority of person movements involve two-way trips (tourists return home after vacationing). This has institutional implications in terms of documentation and the institutions involved. Also, people are not normally happy to remain in limbo at one side of a border for long periods as formalities are dealt with, while warehousing and holding of freight at border crossings is relatively common.

2. Reasons for limiting free cross-border traffic and methods used

Some borders present natural problems for transportation. For a variety of reasons many borders reflect natural barriers – e.g., seas or mountain ranges. In this case the cost of cross-border movements is inevitably higher than for comparable movements within a country. They may involve additional trans-shipments and often there are limitations to the technology available (e.g., the number of deep-water ports). These are inevitable problems and just reflect the naturally higher costs of some transportation activities that here, by chance, coincide with borders.

There are also situations where there are legacy effects that mean the technology on either side of the border is not compatible. In the case of railways this can be in gauge width or power supply, and with canals may involve depth and lock standards. Whilst these are in a sense artificial barriers to cross-border traffic, they are a function of past decisions and not necessarily a reflection of what current policy-makers support. These types of barrier are again not dealt with here in any detail.

There may also be a number of legitimate reasons for having artificial border crossing controls. Some of these have little to do with economics but are based in concerns about national security or to meet social concerns of a local population (e.g., to do with illegal immigration, drug smuggling, or the spread of pests or disease). There may also be environmental and safety concerns about the vehicles used in the transportation process. Economic concerns are more to do with smuggling and avoidance of taxation or other duties. What the border crossing impediment amounts to in this context is an additional transaction cost in business that must be set against the benefits that are gained by having a "secure" border.

Other motivations for impeding the free movement of traffic across borders are less strongly founded in conventional trade theory. In some cases the aim is to protect domestic production or jobs in particular industries. From a national perspective this may generate short-term gains but trade theory indicates that restricting the movement of commodities from more efficient producers into a country ultimately is wasteful of its own resources that could be more fruitfully employed than in import substitution. As with direct tariffs they inhibit a country from fully exploiting its comparative advantages.

There are many forms of cross-border traffic restrictions, some explicitly imposed at the border but others that go further back in the transportation chain. There are controls that are outside of the immediate transportation field, such as product regulations that prevent goods produced in one country from being sold in another country. This can extend more directly to transportation when there are inconsistencies in technical and safety-related transportation regulations, such as vehicle size and weight restrictions, that determine whether a transportation supplier on one side of a border can move across a border. In many cases, even if there is technical consistency, there are simply regulatory prohibitions. This leads to high trans-shipment costs at borders that can be compounded if a double trans-shipment is required to transport goods across a "neutral" trade zone.

Differing regulations regarding such things as fares or carriage rates on either side of a border can pose problems. This may involve explicit discrimination against foreign goods and carriers but it may also be seen in rate structures (e.g., in terms of differential point-to-point rates or differing tapers) on one side designed to help that country's exporters. There may be specific border-crossing fees, in addition to any forwarding fees the private sector may introduce for its services as a facilitator.

Time is important in many transportation activities (see Gunn, 2000) and borders almost inevitably impose time costs on travelers and transporters. These involve document, vehicle, and consignment inspections. The costs are not simply those incurred during the inspection but are usually seen in much greater magnitude further back in the chain in terms of acquiring passports, visas, and trade documents. They are, nevertheless, an additional cost of a cross-border movement. The volume and nature of documents and the number of checks at a border impose restrictions on movement but some of these can be reduced with the introduction of new information-handling and inspection technologies.

It has been assumed that all cross-border transportation restrictions stem from public policy actions. This is not always the case. In some instances there may be market manipulation by suppliers to limit supply of cross-border services. In shipping, for example, the formation of conferences to provide regular sailings, and more recently consortia and alliances, could be seen as means for preventing other shipping companies from providing international services. Similar structures exist in the railroad and air transportation sectors.

3. Benefits of freer cross-border traffic movements

In economic terms, the existence of borders can act as a non-tariff trade restriction. Border controls of any kind push up the cost of transportation above the cost of comparable movements within a country.

The implications of this can be seen in Figure 1. This looks at the simple case of two countries that produce a common product. It shows a back-to-back diagram, where S_i and D_i are the supply and demand schedules for the commodity in country i, and S_j and D_j the supply and demand in country j. The demand for imports and the supply of exports are obtained by subtracting horizontally the domestic supply from demand. The demand for imports (exports) at each price is the difference between the quantities supplied and demanded assuming domestic and import commodities

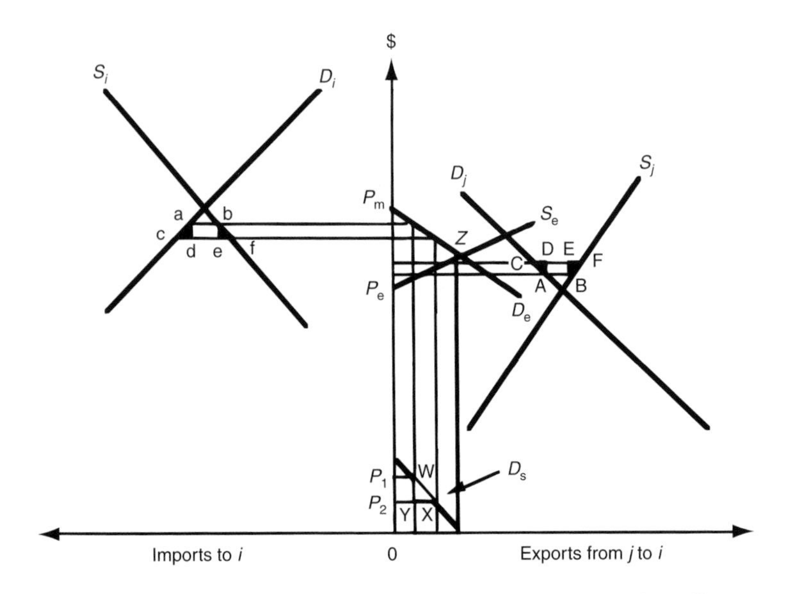

Figure 1. Welfare gains from reducing transborder transportation impediments.

are perfect substitutes. D_e and S_e in the figure are derived in this fashion – the vertical difference between these curves then represents the demand for international trade, shown as D_s. (If transportation charges were zero, for example, then the free-trade equilibrium would be *Z*.) Suppose actual shipping rates when there are border crossing impediments are P_1^1 then at that rate the price of imports from country *i* confronting country *j* is seen to be P_m (the CIF price), while the cost of exports to country *j* would be seen in country *i* to be P_e (the FOB price). Country *i* would then import an amount ab equal to country *j*'s exports of AB.

There is now a reduction in border impediments (e.g., less documentation, longer opening hours for customs clearance, or reduced needs for trans-shipment). The effect is that shipping costs fall to P_2, resulting in exports from country *j* rising to CF to match the higher imports of cf into country *i*. The benefit of this trade to the two countries is represented by areas in the figure. Area adc is the extra consumption enjoyed by the importing country as a result of the fall in the CIF price, while bef is a positive production effect resulting from a contraction of country *i*'s relatively high-cost industry. The areas ADC and BEF are the symmetrical benefits to the exporting country. (Interestingly, the sum of these benefits can be measured directly as the area WXY under the demand curve for cross-border transportation services.)

4. Basic requirements for freer cross-border transportation

Although this chapter is written in English, it is perhaps helpful to resort to the bastardized jargon of the European Union (E.U.) to consider the requirements that reduce cross-border impediments to transportation. The effort to remove cross-border problems has led to the isolation of three key ingredients. There should be adequate "interoperability," "interconnectivity" and "intermodality." These are interrelated, multidimensional concepts that embody institutional as well as technical considerations.

Interoperability means that the operating equipment (trucks, trains, ships, etc.) can operate on either side of the border equally efficiently. This means common technical specifications, or at least sufficient flexibility in specifications, but it also means common institutions such as licenses, insurance, waybills, computer and information systems, safety standards, and labor laws and practices. Without these features, there is the need for consignments or passengers to change carrier at the border even if the same mode is used on either side.

Interconnectivity is largely, but not exclusively, an infrastructure matter. Railways require the same gauge on either side of a border to be efficient and, with electric locomotion, the same power system. Roads must be of comparable quality to carry heavy trucks. The quality of a cross-border air service is only as good as the worst air traffic control system on either side of the boundary. But there are also operational considerations. Timetables for public modes of transportation, for example, need to be co-ordinated across boundaries for full efficiency.

The idea of intermodality is not strictly only a transborder concern but involves the more generic issue of being able to switch between transportation modes at minimal generalized cost. It concerns efficient interchange between modes. In some cases this has little to do with cross-border traffic but does become particularly relevant when seaports and airports are important elements in cross-border traffic. If these are the main gateways into a country, irrespective of how far they may be from the legal border, then these are the de facto places where goods and people encounter a cross-border situation. To reduce friction at these points where a modal change is frequently required there is a need for efficient consolidation and trans-shipment facilities and procedures.

5. Removing border constraints

There have been a number of efforts to reduce the impediments associated with border crossings. While formalization of the above concepts is relatively new, there has been a steady movement to improve interoperability, interconnectivity and intermodality. A major innovation that has reduced some of the technical problems of interoperability (and also interconnectivity and intermodality) has been containerization. The introduction of passports is perhaps the most important of the innovations concerned with individual travel. Further, transborder institutions such as the International Civil Aviation Organization have been created to harmonize technical standards, in this case involving aircraft, as a move towards greater interoperability.

More recently the emphasis has been on removing border crossing restrictions within blocks of countries. The most notable of these have been the initiatives of the European Union and that of the North American Free Trade Area (NAFTA). These are discussed in some detail.

5.1. The European Union[*]

The upsurge of interest in supply chain management, just-in-time production, and the like has led to a wider appreciation of the general need to enhance the efficiency of European transport if the region as a whole is going to compete successfully in the global economy. The concern is that the effectiveness of

[*]The term "European Union" is a relatively new one. In the past there have been terms such as "European Economic Community" and later "European Community." For ease of exposition, "E.U." is used throughout here.

transport logistics in the E.U. area should be at least comparable with that elsewhere to ensure that the labor, capital, and natural resources of member states can be exploited in a fully efficient, economic manner.

It was against this broad background that the E.U. initially sought to develop a transport policy, of which the Common Transport Policy (CTP) has been but one element, designed to reduce artificial friction. It has taken time for the CTP and other elements of transport policy to come together to represent anything like a coherent strategy. The process has not been smooth and has involved a number of almost completely discrete phases (Button, 2000). The focus here is on the most recent initiatives from about 1990.

A simple examination of a map of the E.U. provides guidance on some of the problems of devising a common transport policy (Button, 1984). At the macro level the E.U. does not geographically conform to an efficient transport market. Ideally, transport functions most effectively on a hub-and-spoke basis, with large concentrations of population and economic activity located at corners and in the center and with the various transport networks linking them (see Chapter 6). The overall distribution of economic activities, the geographical separation of some states, and the logical routing of traffic through non-member countries do not conform to this. Further, while some countries sought to ensure their transportation was provided efficiently (the Anglo-Saxon philosophy), others treated transportation as a way of meeting wider political and social goals (the Continental philosophy). This led to divergent policies on different sides of many borders. Remarkably, given this background, many border crossing barriers have now been removed.

Road transportation is the dominant mode for freight and passengers in the E.U. Initial efforts to develop a common policy regarding road transport, however, proved problematic. Technical matters were more easily solved than those of creating a common economic framework of supply, although even here issues concerning such things as maximum weight limits for trucks tended to be fudged over. Economic controls lingered on as countries with less efficient road haulage industries sought to shelter them from the more competitive fleets of countries such as the U.K. and The Netherlands. There were also more legitimate efficiency concerns over the social and environmental costs of road transportation as well as narrower infrastructure utilization questions.

The Single Market initiative from 1987, also later influenced by the potential of new trade with the post-Communist states of eastern and central Europe (Button, 1993), has resulted in significant reforms to economic regulation. Earlier measures had helped expand the supply of international permits in Europe (the E.U. quota complementing bilateral licensing) and reference tariffs had introduced a basis for more efficient common rate determination. The 1990s were concerned with building on this foundation and, as part of the 1992 Single Market initiative, a phased liberalization was initiated that gradually removed restrictions on trucking movements across national boundaries and phased in cabotage.

Passenger road transportation policy has largely been left to individual member states, although in the late 1990s the European Commission began to advocate the development of a "citizens' network" and more rational road charging policies. Perhaps the greatest progress has been made regarding social regulations on such things as the adoption of catalytic converters in efforts to limit the environmental intrusion of motor vehicles. It has taken time to develop a common policy regarding public transportation, despite efforts in the 1970s to facilitate easier cross-border coach and bus operations.

Rail transportation is an important freight mode in much of continental Europe and provides important passenger services along several major corridors. Much of the important economic reform of European railways was undertaken in the early phase of integration by the ECSC with actions on such things as the removal of discriminatory freight rates. Recent initiatives have been concerned less with issues of economic regulation and with operations and more with widening access to international networks and with technological developments, especially regarding the development of a high-speed rail network as part of the Trans-European Networks (TENs) initiative.

The earlier phase had initially sought to remove deliberate distortions to the market that favored national carriers, but from the late 1960s and 1970s had shifted to the rationalization of the subsidized networks through more effective and transparent cost accountancy. The Union has also instigated measures aimed at allowing the trains of one member to use the track of another with charges based upon economic costs. The implementation of the open-access strategy has, however, been slow, with limited impact.

The E.U. has traditionally found it difficult to devise practical and economically sound common pricing principles to apply to transport infrastructure. With regard to railways, the gist of the overall proposals is that short-run marginal costs (which are to include environmental and congestion costs as well as wear on the infrastructure) are to be recovered. Long-run elements of cost are only to be recovered in narrowly defined circumstances and in relation only to passenger services. This clearly has implications, especially on the freight side, if genuine full-cost-base competition is to be permitted with other modes over a complete E.U. system.

Rail transport has also received considerable support from the Commission as an integral part of making greater use of integrated, multimodal transport systems. Such systems would largely rely upon rail (including piggy-back systems and kangaroo trains) or waterborne modes for trunk haulage, with road transport used as the feeder mode. This is seen as environmentally desirable and as contributing to containing rising levels of road traffic congestion in Europe.

The difficulties that still remain with cross-border rail transport reflect technical variations in the infrastructure and working practices of individual states that are only slowly being co-ordinated. Some countries, such as The Netherlands,

Sweden, and the U.K., have pursued the broad liberalization philosophy of the E.U. and gone beyond the minimal requirements of the CTP, but in others rigidities remain and the rail network still largely lacks the integration required for full economies of scope, density, and market presence to be reaped (see Chapter 5).

Inland waterways had already been an issue in the early days of the E.U. This is mainly because they are primarily a concern of two founder member, The Netherlands and Germany. Progress has tended to be slow in formulating a policy, in part because of historical agreements covering the Rhine navigation (e.g., the Mannheim Convention) but mainly because the economic concern has been one of overcapacity. Contraction of supply is almost always inevitably difficult to manage, both because few countries are willing to pursue a contraction policy in isolation and because of the resistance of barge owners and labor.

As in other areas of transport, the E.U. had begun by seeking technical standardization, and principles for social harmonization were set out by the Commission in the 1970s, but economic concerns have taken over in the 1990s. In 1990 the E.U. initiated the adoption of a system of subsidies designed to stimulate scrappage of vessels. Subsequent measures only permitted new vessels into the inland fleet on a replacement basis. Labor subsidies operated in The Netherlands, Belgium, and France (the rota system that provides minimum wages for bargemen) have also been cut back in stages, to be removed entirely in 2000. This was coupled with an initiative in 1995 to co-ordinate investment in inland waterway infrastructure (the Trans-European Waterway Network).

Efforts to develop a common policy on maritime transport represent one of the broadenings-out of E.U. transport policy from the mid 1980s, because initially maritime transportation was excluded from the CTP. Much of the emphasis of the E.U. maritime policy in the late 1990s has been on the shipping market rather than on protecting the Union's fleet. In other words, user rather than supplier driven. In the 1990s the sector became increasingly concentrated as consortia grew in importance, mergers took place, and then the resultant large companies formed strategic alliances. An extension of the 1985 rules to cover consortia and other forms of market sharing was initiated in 1992.

In 1994 the E.U. took action to ban the Transatlantic Agreement that had been reached the preceding year by major shipping companies to gain tighter control over the loss-making North Atlantic routes. It did so on grounds of capacity manipulation and rate manipulation and because it contained agreements over pre- and on-carriage over land. In the same year it also fined 14 shipping companies that were members of the Far East Freight Conference for price fixing. The main issue was that these prices embodied multimodal carriage and, while shipping per se enjoyed a block exemption on price agreements, multimodal services did not.

Ports have also attracted the attention of the E.U. Technology advances have led to important changes in the ways in which ports operate and there has been a significant concentration in activities as shipping companies have moved towards hub-and-spoke operations. The main E.U. ports have capacity utilization levels of well over 80% and some are at or near their design capacity. This may be a function of a genuine capacity deficiency or it may reflect inappropriate port pricing charges. The Commission has produced new proposals for co-ordinating investment in port facilities.

Liberalization of E.U. air transportation may be considered one of the successes of the CTP in the late 1980s. The final reform – "Third Package" – came in 1992 and was phased in from the following year. The measures removed significant barriers to entry by setting common rules governing safety and financial requirements for new airlines. Since January 1993, E.U. airlines have been able to fly between member states without restriction and within member states (other than their own), subject to some controls on fares and capacity. National restrictions on ticket prices were removed, with only safeguards if fares fall too low or rise too high.

Consecutive cabotage was introduced, allowing a carrier to add a "domestic leg" on a flight starting out of its home base to a destination in another member state if the number of passengers on the second leg did not exceed 50% of the total in the main flight. Starting in 1997, full cabotage has been permitted, and fares are generally unregulated. Additionally, foreign ownership among Union carriers is permitted, and these carriers have, for E.U. internal purposes, become European airlines. This change does not apply to extra-Union agreements, where national bilateral arrangements still dominate the market. One result has been a considerable increase in cross-share holdings and a rapidly expanding number of alliances among airlines within the Union.

5.2. The North American Free Trade Area

The NAFTA went into effect on January 1, 1994 with the aim of opening the borders separating Canada, Mexico, and the U.S.A. to the free exchange of goods and services. It is a very comprehensive agreement, covering not only tariff elimination, but a number of highly contentious issues including non-tariff barriers, direct foreign investment, trade and services, government procurement, and intellectual property rights. In transportation, NAFTA sought to equalize the U.S.–Mexico transborder operations to those practiced between Canada and the U.S. Reciprocal entry in the trucking industry was to be permitted initially to zones in border states, later to border states, and in seven years to all states and all over Mexico. Yet half a decade into NAFTA, there remain many subtle and not so subtle barriers to cross-border movements.

Table 1
Maximum gross vehicle weights (kilograms) in the NAFTA countries

Truck type	U.S. federal	U.S. state maximum	Canada provincial minimum	Canada provincial maximum	Mexico
Tractor semitrailer (5 axles)	36 288	39 917	39 500	41 500	44 000
Tractor semitrailer (6 axles)	36 288	45 360	46 500	53 000	48 500
A train double (5 axles)	36 288	43 092	38 000	43 500	47 500
A train double (6 axles)	36 288	48 082	47 600	48 000	56 000

Source: North American Free Trade Agreement Land Transportation Standards Subcommittee (1997).

One of the most important problems is inconsistency in truck size and weight regulation across national borders (Table 1). Size regulation refers to limits on the width of the truck and on its overall length and the lengths of its component parts (tractors, semitrailers, and trailers). Weight regulation refers both to the gross vehicle weight (GVW) and to the distribution of weight across axles. Truck size and weight regulations are imposed to avoid excessive wear and damage to road and bridge infrastructure; to ensure consistency with the geometric design standards of roads; and to promote safety, especially in relationship to the interaction of trucks and automobiles in the traffic stream.

Inconsistencies in these regulations can add significantly to the cost of cross-border transportation. For example, suppose that the truck configuration typically used to ship lumber within Canada is not legal on U.S. roads. Shipments going from Canada to the U.S.A. must then either be transferred from one truck to another at the border, or be shipped via some lowest-common-denominator truck configuration that is legal in both countries. In the first case, considerable extra costs in terms of labor and delay are incurred. In the second, it may be necessary to use a truck configuration that is less efficient than the best option for shipping lumber in either country. Either way the outcome is the same – transport costs are higher than the costs of shipping the same load a similar distance within a single country.

Harmonization of truck size and weight regulation is necessary in order to achieve the full trade creation potential of the elimination of tariffs under NAFTA. This effort is retarded by two factors. There is the complexity of truck size and weight regulation, requiring agreement on a wide range of engineering and safety issues. The second is the problem of jurisdictional fragmentation. In each of the three NAFTA partners, state or provincial governments have some latitude in setting their own regulations. This means that, in principle, a total of 64 jurisdictions are involved in the harmonization process. Given these problems, a complete consensus on regulations is not seen as a realistic goal. Instead, a set of agreements and procedures that will minimize the impact of regulatory inconsistencies on cross-border traffic is sought.

There have, however, been fears that entry by large, well-financed U.S.-based truckers into Canada and Mexico poses serious threat to domestic trucking industries. Chow and McRae (1989) found that government policy in the form of nine non-tariff barriers minimally affected the competitive advantage in the U.S.–Canadian case. Only the long-haul truckload Canadian carriers had disadvantages, because of driver and equipment restrictions on domestic movements and an unfavorable spatial distribution of industry in the two countries. Chow and McRae also noted that "the level playing field" problem was exaggerated by the Canadian trucking industry. The disadvantages of operating as a Canadian-domiciled carrier were more than compensated by the lower input costs than U.S.-domiciled carriers, and even after accounting for marginal tax differences there was a cost advantage for Canadian operators. Since Mexican industry also fears unequal competition from U.S. carriers, there have been delays to NAFTA-agreed changes in Mexican regulations to improve cross-border flows.

Another transportation barrier is the existing U.S. restrictions on trade in domestic water transportation. In the large, multicoastal U.S. economy, foreign participation in trade is restricted by the 1920 Jones Act, which reserves cabotage traffics to U.S.-built and U.S.-registered ships that are predominantly owned and crewed by U.S. nationals. This remains outside of NAFTA.

Border crossings can be subject to long delays. This is partly because the frontiers are crossed by a relatively small number of road and rail links, resulting in bottlenecks. Furthermore, inspection and documentation activities that occur as vehicles cross the border are time-consuming. If delays at borders are long enough they can add significantly to transport costs. Immigration rules on the nationality of labor used in repositioning trucks in the other country after unloading an international shipment are essentially similar, except that Canada permits more flexibility. The Customs rules on the use of equipment to move goods in the other country are being harmonized and there are efforts at similar enforcement of rules across borders.

There have been also a number of initiatives by the public sector in Canada and Mexico and by the transportation companies in all three countries to expand the north–south NAFTA networks. In 1995 Canada privatized Canadian National Railways (CN) and removed subsidies, and the Ferrocarriles Nacionales de Mexico (FNM) was broken up after seven decades of government ownership and privatized into several concessions. In 1997, TMM (the largest marine transportation company in Mexico) and Kansas City Southern Railroad (KCSR) purchased the Laredo–Mexico City line. Meanwhile, CN merged with Illinois Central (IC). Jointly with the Canadian Pacific–Soo line and the KCSR–TMM line, CN-IC can provide single-line service. avoiding switching costs and delays, from Canada to Mexico.

Canada and the U.S.A. have long traded large volumes of goods, and in the process both governments have worked to develop relatively efficient border crossing routines. The situation along the U.S.–Mexican frontier is quite different. These border crossings generate long delays and many Mexican trucks are sent back due to violations of various U.S. regulations. The problem is not just the newness of the U.S.–Mexico border situation; there are also issues of illegal immigration and the transportation of drugs. The Mexican truck fleet is in a relatively poor state and Mexican carriers and drivers are often badly informed about U.S. regulations.

The situation along the Mexican border still presents an impediment to full implementation of NAFTA provisions. NAFTA specifies a timetable for providing full freedom of truck movement across the U.S.–Mexico border. Initially, Mexican trucks were only allowed to operate in a relatively small commercial zone extending just a few miles into the territory of the four states that border Mexico (Mexican goods bound for destinations outside this zone being transferred to U.S. trucks). By December 1995, Mexican trucks were to be allowed to make deliveries throughout the territories of the border states and U.S. trucks were to have similar access to Mexican border states. By 2000, Mexican trucks should have been able to travel throughout the U.S. and American trucks should have been able to travel throughout Mexico. Cabotage restrictions notwithstanding, Mexico and the U.S.A. were to have a similar arrangement to the one that now exists between Canada and the U.S.A. By 2000, however, not even the access for Mexican trucks that was planned for 1995 had been realized.

The U.S. government and the governments of the bordering states still fear that Mexican trucks will not meet U.S. regulations. This would not be a problem if effective surveillance could be applied. The inspection process, however, is highly complex because various federal agencies all have concerns about what may cross the border. Inspection of the trucks themselves (as opposed to their contents or personnel) comes under the jurisdiction of state Departments of Transportation, which receive some limited assistance from the U.S. Department of Transportation. State officials check trucks for size, weight, and safety violations. Since a relatively small number of inspectors are assigned (see Table 2) and because facilities are limited, it is only possible to conduct spot inspections. In these spot checks, roughly 50% of the trucks inspected have been put out of service due to a violation. State officials are therefore reluctant to allow Mexican trucks to travel further into their territories until a more stringent inspection process can be put in place or there is a much lower rate of violation.

There is potential for new information and communication technologies that come under the general heading of intelligent transportation systems (ITS) to speed border crossings. These can eliminate much of the need for paper handling by remotely reading truck identification and cargo information, and conducting certain basic checks on weight, length, height, and width while the truck is in

Table 2
Trucks from Mexico to the U.S.A. and inspection personnel at the seven busiest crossings (1996)

Crossing	Trucks 1996	Percentage of total	State inspectors	Federal inspectors	Total inspectors	Trucks/s
Ota y Mesa, CA	520 908	17	28	1	29	17 962
Calexico, CA	169 403	5	19	1	20	8 470
Nogales, AZ	225 274	7	7	2	9	25 030
El Paso, TX	577 152	19	9	2	11	52 468
Laredo, TX	899 754	29	8	2	10	89 975
McAllen, TX	198 260	6	5	0	5	39 652
Brownsville, TX	224 537	7	7	2	9	24 948
Subtotal	2 815 288	90	83	10	93	30 272
Others	297 803	10	10	1	11	27 073
Total	3 113 091	100	93	11	104	29 934

Source: U.S. General Accounting Office (1997).

motion. Also, electronic databases can be used to identify trucks and drivers with previous violation histories so that inspection efforts can be concentrated on them. The introduction of ITS is only taking place gradually.

6. Conclusions

There are many different types of border that can impact on the efficiency of transportation systems; some are natural but many are artificial. Borders can fragment transportation networks and reduce the cost gains enjoyed through economies of density and scope and the additional revenues generated from economies of widespread market presence. Impediments to trade restrict the economies that can come from countries or regions being able to fully exploit their natural comparative advantages. There are, however, increasingly moves to reduce the problems that can result from such inefficiencies, be they technical or institutional, at border points. In some cases these are bilateral actions involving just a pair of countries, but with the emergence of the E.U. and NAFTA, there are also multilateral changes taking place. The gradual increase in powers being given to global bodies such as the World Trade Organization may represent the next movement forward.

References

Button, K.J. (1984) *Road haulage licensing and EC transport policy*. Aldershot: Gower.
Button, K.J. (1993) "Freight in a Pan-European context", in: D. Banister and J. Berechman, eds., *Transportation in unified Europe: Policies and challenges*. Amsterdam: Elsevier.

Button, K.J. (2000) "Transport in the European Union", in: J.B. Polak and A. Heertje, eds., *Analytical transport economics*. Cheltenham: Edward Elgar.

Chow, G. and J.J. McRae (1989) "Non tariff barriers and the structure of U.S.–Canadian (transborder) trucking industry", *Transportation Journal*, 30:4–21.

Corvers, F. and M. Giaoutzi (1998) "Borders and barriers and changing opportunities for border regional development, in: K.J. Button, P. Nijkamp and H. Primus, H, eds., *Transport networks in Europe: Concept, analysis and policies*, Cheltenham: Edward Elgar.

Gunn, H.F. (2000a) "An introduction to the valuation of travel-time savings and losses", in: D.A. Hensher and K.J. Button, eds., *Handbooks in Transport 1. Transport modelling*. Oxford: Pergamon.

North American Free Trade Agreement Land Transportation Standards Subcommittee (1997) "Working group 2 – vehicle weights and dimensions, harmonization of vehicle weight and dimensions regulations within the NAFTA partnership".

Ratti, R. (1995) "Dissolution of borders and European logistic networks: Spatial implications and new trajectories for service performers, border regions and logistic networks", in: D, Banister, R. Capello and P. Nijkamp, eds., *European transport and communications networks: Policy evaluation and change*. New York: Wiley.

U.S. General Accounting Office (1997) *Commercial trucking: Safety concerns about mexican trucks remain even as inspection activity increases*. Washington, DC: GAO.

Chapter 24

PARKING AND TRAFFIC CONTROL

WILLIAM YOUNG
Monash University, Melbourne

1. Introduction

Parking has an important role to play in controlling transport usage. Parking policy related to the supply and pricing of parking will influence drivers' ability to store their vehicles and move on to destinations. It can influence the usage of transport modes, congestion, the quality of the urban environment, and the location decisions of households and firms. Parking guidance systems can influence parking usage, route choice, and the efficiency of the traffic network. On-street parking provision will directly influence the capacity of roads and therefore the movement of traffic in urban and rural areas.

Parking is fundamental to the performance of urban and transport systems. It can influence and is influenced by the environmental, social, and economic dimensions of cities and rural areas. To understand the interrelationship between parking and traffic control it is necessary to understand these influences.

McShane and Meyer (1982) highlight the relationship between urban areas and parking strategy through a discussion of urban goals. They point to the role of parking in achieving the goals of a healthy economic climate; efficient use of the existing transportation, land use, and public resources; ease of mobility and accessibility of resources; equity of resource distribution and preferential allocation of some resources; environmental goals; and enhancing amenity. This chapter will investigate traffic control within the context of the broad influence of parking on these urban goals. More specifically, it will introduce parking management, parking information, on-street parking, and rural parking. A case study of Melbourne, Australia is used to illustrate parking policy's impact on urban goals.

2. Parking management

The determination of the demand for parking is the first step in the design of parking facilities. The demand is a function of the level of activity in an area. This

Handbook of Transport Systems and Traffic Control, Edited by K.J. Button and D.A. Hensher

Table 1
Methods of parking restraint

Control	On-street	Off-street
Price control	Install meters Increase meter rates Street parking permit with fee Automatic payment systems	Parking tax Rate structure to discourage long-term use
Quantity control	Ban parking, totally or at specific times Ban parking except to specific groups (e.g., residential) Adjust meter times Relocate parking	'Freeze' new parking Reduce existing parking Control future parking Vary time of opening Relocate parking

is generally related to the area's land-use zoning. Questions like "Where should parking be located?," "Should the full parking demand be catered for?," or "Can the parking demand be changed?" are part of parking management. They are related to transport and land use policy decisions.

Parking management has been seen as a powerful tool in controlling transport demand. In general, parking can been seen in terms of four objectives:

(1) *Moving vehicles*. The use of roads for moving vehicles should take precedence. Parking management strategies can assist the movement of vehicles. Hence parking design can have a direct influence on traffic movement.
(2) *Public transport*. Parking management can be used to support public transport (Morall and Bolger, 1996; Finstad, 1996). Good public transport and its utilization reduce the overall need for parking. Similarly, parking spaces at local train stations are needed to support public transport policy. Control in the location, supply, and price of parking can directly influence the decision to travel and choice of mode. For instance, the parking supply at remote lots may have a small impact on shifting people out of cars but may have a sizable impact on congestion by distributing vehicles spatially.
(3) *Local environment*. Parking management can aid in improving local environmental conditions. Limitation of parking can improve the esthetics of an area (McCluskey, 1987), reduce the amount of traffic attracted to an area, and ensure parking for local residents. This can have an indirect influence on traffic movement through its impact on the desirability of an area and the desire to travel to these destinations.
(4) *Urban development*. This can be affected by making some developments more or less costly. The supply of parking can make some areas more attractive. Changes in the type of development in particular areas will influence the generation of trips and hence control traffic movement.

Parking control relates to the restraint of parking. Parking restraint can be introduced through the regulation of price and provision of parking. Traffic engineers more often than not have control over the quantity and location of parking rather than the price mechanism. Change in the provision of parking can be applied to on-street parking, and off-street parking as shown in Table 1. Developments in technology have now allowed payment of parking to be streamlined, and in some areas moves to integrated parking payment systems and road pricing systems are being considered.

2.1. Parking demand

The demand for parking is related to the type and level of activity in an area. This is clearly related to the level of economic activity, the mobility of the community, and the cultural attractiveness of the region. Complex modeling approaches (Young, 2000) can be used to determine the demand for parking. For practical purposes, the demand for parking is generally related to land-use zoning and the specific land use served. In many countries land-use zoning ordinances (Weant and Levinson, 1990) specify the number of parking spaces required for new construction or major building modifications. Zoning requirements can be used as a benchmark for planning and to allow planners to assess the adequacy of parking in new developments. Examples of rates (Weant and Levinson, 1990) are two spaces per single-family dwelling, 2.2 spaces per 1000 ft^2 gross floor area of general retail space, or 1.5 spaces per school classroom. Refinements on this concept could reflect the level of public transport provision. The Ministry of Planning and Environment (1987) in Victoria, Australia provides a range of parking rates. Low public transport provision would require 1 parking space per bedroom in a single-family residential dwelling and 0.25 spaces per bedroom where there is a high level of public transport provided. The zoning concept can also be used to define safe traffic operations by dealing with safe right-of-way needs, property setbacks, development densities, and access control.

Increasingly, urban areas are becoming more complex, and parking may provide access to a variety of land uses. These areas have been termed mixed-use or multiuse areas and tend to exhibit different characteristics from single-use developments. Where complex land uses exist, the parking needs are not calculated by adding the individual land use needs. Procedures for calculating the need are outlined in Urban Land Institute (1983) and Barton-Aschman Associates Inc. (1983). Calculation of the needs for mixed-use parking may be difficult because of the lack of data. The peak demand for parking in shared parking facilities must take into account the parking demand for each generator, displacing the generators temporally to determine the maximum demand.

If the provision of parking equals the demand then there is readily free access to an area and parking does not control traffic flows. Traffic flow is, rather, controlled by the traffic system. Appropriate provision of parking may make some areas more attractive and hence increase traffic movement to those areas.

2.2. Supply of parking

Parking that meets the demand may not be provided because of the availability of suitable land, because of its cost, or in order to control the movement of vehicles into particular parts of urban areas. Further, the supply of parking may be focused on particular groups who wish to partake in the activities in designated areas. Some of these activities may be short term in nature, while others may require a longer stay. Parking restraint, parking location, and parking duration decisions will all influence the movement of traffic into an area.

Parking may not be provided in close proximity to the final destination. Some parkers may have particular difficulties that make it difficult for them to walk a considerable distance. The definition of "convenient walking distance" varies with a number of factors, including trip purpose, or activity being undertaken, the duration of stay, and the size of the urban center. Burrage and Mogren (1957) quote data from a number of American cities of population less than 500 000 which indicate an average walking distance of 213 m for parkers whose trip purpose is work, 200 m for those who are shopping, and 128 m for sales or service trips. If parking is not placed at appropriate locations the area will be less desirable and people will use other modes or choose other destinations.

The parking duration can be limited to facilitate short-duration shopping trips or longer-duration business trips. It may be desirable to consider imposing time limits on the duration of parking. The successful application of time limits relies on compliance by all parkers, otherwise such limits lose their value. Therefore, any time limit should be supported by reasonable, regular, and effective enforcement.

Parking may not be able to be provided close to destinations, and a combination of parking and use of high-occupancy vehicles may be adopted. Park and ride provides an opportunity to control the flow of cars while maximizing the use of efficient modes of travel.

2.3. Pricing

The control of parking through price is common in many cities. The price drivers pay for parking influences their choice of destination and traffic movement. Parking price gradients and the level of pricing can encourage or discourage the use of particular areas and directly influence traffic distribution and flow. Further, the amount of parking provided will influence the level of competition for parking and hence influence the price levels as well as the rate structure.

The price of parking influences its demand. Shoup (1999) points out that the minimum parking requirements discussed in Section 2.1 "are based on two highly unreasonable assumptions: (1) the demand for parking does not depend on price, and (2) the supply of parking should not depend on its cost." The need to recognize the cost of parking and the impact of parking price on demand are crucial in the control of traffic demand.

In some situations, the true charge for parking may not be passed on to the driver and hence, there, travel decisions may not be as expected. For instance, the level of parking subsidies to employees, either directly by the provision of parking, or through the use of valet parking can influence the impact of parking pricing strategies since the employee is not directly impacted by the parking charge and will not consider alternative travel options. In the U.S.A., nine out of every ten commuters who drive to work do not pay for parking (Shoup and Wilson, 1992). Shoup (1997) indicate the possibility of paying out these subsidies in order to get more realistic market responses to parking policy initiatives. Further, some entities (e.g., shopping centers and educational establishments) may not charge the true cost of parking to encourage the use of their facilities. These market distortions can influence the outcome of parking pricing policies.

2.4. *The impacts of parking control*

Some methods of parking restraint are outlined in Table 1. They may have a direct or indirect influence on the movement of traffic since any management decision to locate parking supply away from a driver's destination or to increase the price of parking will impact users' perception of the desirability of the final destination, and/or the attraction of using the car to get to it. If the change is large enough then the driver may change the mode of travel or the destination, or may forgo participation in the activity.

The impacts of parking management decisions can be both direct and indirect. The parkers feel the direct impacts, while their reaction is related to its fairness and enforcement. Rules and incentives governing deliveries during evening or night hours would influence traffic flow and deter double parking and the reducing of lane capacity. Parking restraint can have indirect impacts on many groups. The most obvious and often the most vocal are the local traders in a commercial area where parking restraint is to be implemented. They fear loss of trade. This concern should be taken into account and alternative parking provided where possible. A balance between parking and road capacity should be obtained.

Parking restraints may also have spillover effects. For instance, time constraints in a parking area may increase parking in adjacent areas. Similarly, time limits in local areas may act as deterrents to residents. Thus, areas with parking restraint should be surrounded by a "buffer zone" to act as a transition between restraint and non-restraint areas.

Parking restraint may also be partially responsible for encouraging commercial, industrial, and retail activities to leave or not locate in particular areas. Many developments do not provide parking for all users. Further, certain types of parking may be discouraged. In some district centers, shoppers may be given priority while people working in the area are asked to park in less convenient parking places or encouraged to use public transport. In the short term, the people must adjust to the policy, in the long term, they may relocate their workplace (or other activity). Changes in the location of homes and workplaces will influence the long-term travel patterns in cities.

It is also important to consider the total impact of parking management before its inception. The indirect impacts may result in the parking management plan being an overall failure. For example, the reduction in the provision of parking in a central activity district (CAD) to below the demand may be directed at increasing public transport usage. In order for such a scheme to work, appropriate parking and access services must be supplied at public transport stations and the public transport system must be able to provide the appropriate service. If this overall view is not taken, then parking restraint alone is unlikely to produce the required result. When using parking to influence public transport usage, it should be noted that although there is a relationship between parking restraint and public transport usage it is not a simple relationship and other urban policies and decisions may influence the magnitude of the effect.

Another area where parking has a considerable impact on the efficient running of the transport system is when special events take place. On most weekends in winter hundreds and thousands of people converge on football grounds. Every four years we have the Olympic Games. Special events usually require people to enter and exit specific parts of urban regions in very short periods of time. This requires astute design of the parking system and control of traffic movement. Park and ride systems have been developed to support this concentration of activity (Trout and Ullman, 1997).

2.5. Administration and planning of parking

Parking management in many cities comes under the control of local government. State authorities may provide general guidance on parking rates and land use activities through their zoning ordinances or building requirements.

Within local government areas, private companies as well as government organizations may operate parking lots. Co-ordination of parking management plans therefore entails co-operation with the parking operators. This co-operation relates to such things as co-ordination of information on the availability, type, and price of parking and the provision of parking, and may include negotiation over prices.

Parking supply can change significantly in some areas depending on the rapidity of development. In some cases private parking is supplied because land is held open waiting for development. With changes in the market this land may be developed sooner.

Many private developers supply parking as a service required in commercial buildings or retail areas. Government can encourage the development of parking lots for general use. One method of doing this is through "cash in lieu" policies, where developers can pay a sum of money related to the cost of providing parking to the local authority. The authority can then use this to provide parking at another location or improve other aspects of the transport system (e.g., public transport provision).

3. Parking information systems

Parking systems cannot work efficiently without appropriate information on the location, type, and availability of parking. This information can be used to control the movement of traffic and reduce search times, hence improving the transport system's performance.

The management of parking is primarily achieved by signing. Such signing falls into three categories: the regulation of curbside and center-of-the-road parking by linear controls specific to each length of kerb, the regulation of parking on an area-wide basis using the signs to indicate the boundaries of the controlled area, and advisory signing to direct drivers to locations where parking is available, either on or off street.

By far the most common method of regulating or managing on-street parking is by linear control, i.e., parking signs which indicate parking regulations along a length of road (Standards Australia, 1988a,b). Pavement markings are commonly used to supplement these controls, by the marking of parking spaces on the pavement in accordance with the design guidelines (Standards Australia, 1988a). In most controlled areas it will be found that unless bays are marked, it will be difficult to enforce parking layouts, and inefficient usage of spaces will result.

3.1. Area control of parking

For suitable areas, a more economical method of parking management is area parking control (Standards Australia, 1988a). Area parking control can be implemented as follows:

(1) an area is identified in which a particular level of permissive parking (e.g., 1 hour) can apply to most of the available curbside space;
(2) the area is delineated by means of the signs at each entry and exit point in the area, together with such other reminder signs as may be required within the area; and
(3) additional linear control signs, overriding the general area control, are provided as necessary in those locations where more restrictive controls are needed, either full time or part time.

3.2. Advisory signs

Advice on the location and availability of parking can be presented through the use of information boards or radio broadcasts, but the most common method is the use of signs. These signs range in sophistication from static information systems to dynamic electronically controlled systems (Young, 1987).

Advanced guidance signs in rural areas are required for such places as lookouts, picnic areas, information bays, areas for heavy vehicles, etc. These signs should also be placed at the parking facility, directing the driver to the entrance and exit. In order to provide the driver with enough information to make a rational decision the signs should contain information on the types of facilities available (toilets, barbecues, shops, etc.). Secondary information on the location of the next parking place can also aid the motorist. As with other service information, these signs should have a white legend on a blue background.

Direction signs to off-street parking facilities are also useful in business and shopping centers. These enable parkers to locate parking places quickly and reduce circulation, traffic congestion, and pollution (Young, 1987). The information that may be presented on these signs relates to the location of, direction to, and availability of parking.

The location of parking can be presented on a sign by using a place name, number, or map of the region. This information can be self explanatory (as in the case of the place name or the map) or can be associated with a hand-held map that locates particular parking places using corresponding names or numbers. Many off-street parking facilities provide information on current parking availability at their entrance.

In some cities, information is provided to parkers via dynamically updated signs located throughout the region. These dynamic information systems require information on the availability of parking in an area. This can be collected through the use of detectors that monitor entry and exit flows at each car park. The variable-message signs present information on whether a parking facility is full, almost full, or less than a certain percentage full (say 90%). These approaches allow the control of traffic movement and parking utilization dynamically.

4. Design of parking systems

The control of traffic movement is also influenced by the design of parking at the road network level.

A traffic-related aspect of parking location is the accessibility of parking sites to the road system. Parking should be close to arterial roads to reduce traffic intrusion into local streets. Sites will also differ with respect to entrance and exit conditions. Interference with traffic flows and delays resulting from difficulties in finding gaps should be taken into account in the design process. In general, it is advantageous to have parking facilities surrounding developments, with good access to the road system. Locating the pedestrian walkways inside the ring of parking facilities reduces pedestrian–vehicle conflict.

Road networks can be designed to facilitate movement into a destination through the astute use of transport and parking systems. Since people must park their car before entering a facility the location of parking will influence the movement of vehicles. The location of parking on major arterial roads between the final destination and the road system can ensure easy movement and reductions in unnecessary traffic movement.

On-street parking can directly influence the capacity of roads (Taylor et al., 2000) and hence traffic flows. Clearways remove vehicles from roads at peak times and control traffic movement. Yousif and Purnawan (1999) highlight the effects of door opening and vehicle parking maneuvers on delay and congestion. A second vehicle-to-vehicle conflict is associated with parking and unparking. The parking space angle (30–60°) could be used to reduce parking and unparking times and hence the probability of a conflict. Further advantages are the increase of vision when parking. Ninety degree parking often encourages people to reverse into spaces. If this is perceived to be a problem then the adoption of 70–80° parking could reduce the proportion of reverse parking. A considerable problem associated with having the circulator run outside the main door of the shopping center is the introduction of double parking and the consequent reduction of the capacity and traffic flow along the circulator. Further, parking is a commonly used tool in traffic calming (Schlabbach, 1997), where the location of parking is used to slow traffic movement and discourage the use of particular roads.

An important principle in the design of parking facilities is the reduction of conflicts. Conflicts are related to accidents and to inconvenience. Their reduction will, therefore, increase the level of service of the parking facility. The capacities of the components of a parking system are largely independent of the speeds of the parking vehicles. Any effort to increase the speeds of cars on aisles or ramps are unlikely to benefit the parking operators; indeed, the park operators may feel that excessive speed is dangerous and attempt to reduce it. Hence, provided free-flowing conditions prevail, the parker obtains little benefit from increased speed.

The size of spaces, the angle of parking, and the width of parking aisles may also influence the availability and usability of parking systems. In Australia, the size of parking spaces is related to the vehicle dimensions. The base dimension incorporates critical conditions for maneuvering and stationary vehicles. It has been derived from field tests of the design vehicle and represents the minimum dimensions achieved using a skilled driver and providing zero clearance (Standards Australia, 1993). It provides the starting point for determining the size of a parking space. Clearances are then added to the base dimensions to take into account the parking turnover and type of land use served by the parking space. Clearly the standard vehicles used in the design process must be representative of the vehicle fleet and the type of vehicle using the parking area.

Illegal parking can directly influence traffic flow and safety in parts of the traffic system. The enforcement of parking policy is therefore an important aspect of traffic control. If enforcement is lax then the performance of the traffic system will be reduced and the movement of vehicles in the traffic system influenced.

5. Rural parking

Most parking problems occur in urban areas. However, in rural locations that have a high-speed environment and significant parking requirements, parking provision needs to be given special consideration. Where regular parking can be expected on rural highways (e.g., at rural industrial areas, local halls, theatres, or places of special interest), off-street parking should be provided. Where the parking demand is small it may be provided on street, but it should be separated from through traffic lanes by kerbed areas or guardrails and have properly designed entrances and exits (AustRoads, 1988). These include wayside stopping places, picnic and rest areas, scenic lookouts, truck parking areas, tourist information bays, and freeway service centers. At each of these facilities, the amount of parking, the size of parking bays, and the layout of the parking area in terms of parking bay arrangement and entry or exit driveways needs to be carefully designed to suit the particular usage and environment (McCluskey, 1987). The location of driveways needs careful consideration of the sight distance required, along with other geometric treatments to ensure that traffic can enter and leave the parking area safely and with minimum interference to through traffic.

6. A case study: Melbourne, Australia

Melbourne, Australia is a city of three million people. Its central city has many of the attributes of other large cities. In particular, there is a considerable growth in home unit development and a focus for formal recreational activity (e.g., theatre, art galleries, etc.).

Parking is generally provided by private enterprise in Melbourne. Private companies apply to the central council for permission to run a car park. The central city council has a traffic restraint policy that keeps parking to a reasonable limit. The council can therefore restrain the amount of parking to levels of 4 spaces per 1000 m^2. Public transport usage ranges between 40% and 50% and is higher than in other parts of the city region.

A dynamic parking guidance system was introduced in 1996 to supply information on the availability of parking and the route to it. This system required a constant traffic system to be introduced across all parking operators using the system. Parking operators were required to have consistent parking tariffs to reduce confusion of presentation of information on the parking guidance systems.

Swanson and Bourke Roads, the two major streets in the central area, are closed to parking and traffic. Clearways are used extensively to control parking and to maximize road capacity during peak times.

Private operators set parking pricing. On-street parking meters are used extensively and parking stations are generally located around the periphery of the urban area. Loading bays are used extensively to allow ready access to shops and offices.

Parking management, parking information provision, and parking systems design assist in providing a supportive environment for controlling traffic movement in Melbourne.

7. Concluding remarks

Parking controls can influence the level of transport usage, the environment, and access. They can enhance the quality of an urban area or discourage development. In order to maximize the benefits to the community, the management of parking should be integrated with transport and urban management schemes. This integration is becoming easier with the development of new technology to assist in the guidance of traffic and in the pricing of the transport and parking systems. Improved understanding of the impacts parking policy has on transport demand and traffic movements is required so that policies can be developed which support the goals of urban regions.

References

AustRoads (1988) *Guides for the traffic engineering practice, part II*. Sydney: AustRoads.

Barton-Ashman Associates Inc. (1983) "Shared parking demand for selected land uses", *Parking*, Autumn.

Burrage, R.H. and E.G. Mogren (1957) *Parking*. Saugatuck, CT: Eno Foundation for Highway Traffic Control.
Finstad, G.A. (1996) "Garages: The key to a successful transportation system", *Institute of Transportation Engineers Journal*, 66(5):38–42.
McCluskey, J. (1987) *Parking: A handbook of environmental design*. London: Spon.
McShane, M. and M.D. Meyer (1982) "Parking policy and urban goals: Linking strategy to needs", *Transportation*, 11:131–152.
Ministry of Planning and Environment (1987) "Melbourne metropolitan parking update study: Overview report".
Morall, J. and D. Bolger (1996) "The relationship between downtown parking supply and transit use", *Institute of Transportation Engineers Journal*, 66(2):32–36.
Schlabbach, K. (1997) "Traffic calming in Europe", *Institute of Transportation Engineers Journal*, 67(7):38–40.
Shoup, D.C. (1997) "Evaluating the effects of cashing out employer-paid parking: Eight case studies", *Transport Policy*, 4(4):201–216.
Shoup, D.C. (1999) "The trouble with minimum parking requirements", *Transportation Research: Policy and Practice*, 33A(7/8):546–574.
Shoup, D.C. and Wilson, R.W. (1992) "Employer-paid parking: The problem and proposed solutions", *Transportation Quarterly*, 46(2):169–192.
Standards Australia (1988a) "Manual of uniform traffic control devices: Parking controls", SA, Sydney, Australian Standard AS 1742.11.
Standards Australia (1988b) "Manual of uniform traffic control devices: General", SA, Sydney, Australian Standards 1742.1.
Standards Australia (1993) "Off-street parking", SA, Sydney, Australian Standard AS 2890.1.
Taylor, M.A.P., P.W. Bonsall and W. Young (2000) *Understanding traffic systems: Data, analysis and presentation*. Aldershot: Ashgate.
Trout, N.D. and G.L. Ullman (1997) "A special event park-and-ride shuttle bus success story", *Institute of Transportation Engineers Journal*, 67(12):38–43.
Urban Land Institute (1983) *Shared Parking*. Washington, DC: ULI.
Weant, R.A. and Levinson, H.S. (1990) *Parking*. Saugatuck, CT: Eno Foundation for Transportation.
Young, W. (1987) "Parking guidance systems", *Australian Road Research*, 17(1):40–47.
Young, W. (2000) "Modelling parking", in: D.A. Hensher and K.J. Button, eds., *Handbooks in transport 1. Transport Modelling*. Oxford: Pergamon.
Yousif, S. and Purnawan (1999) "On-street parking: Effects on traffic congestion", *Traffic Engineering and Control*, 41(9):424–427.

Chapter 25

ROAD VEHICLE DESIGN STANDARDS

MICHAEL C. CASE
Royal Automobile Club of Victoria, Melbourne

MAX G. LAY
Sinclair Knight Merz, Melbourne

1. Objective of road vehicle design standards

Vehicles are an integral part of any road transport system. They provide the key function of physical mobility, whether it be for the transport of people or their commodities. In the developed world, and many developing countries, it is motorized vehicles that provide the primary source of mobility. As such, they have become pivotal to the social and economic well-being of these countries and their people.

Unfortunately, motor vehicles are also the primary source of many of the detrimental outputs from transport systems that affect communities. Road trauma is an unfortunate result of the interaction of vehicles and the road network, with an estimated 125 000 road transport fatalities in 1999 in the 29 OECD nations alone (Organisation for Economic Co-operation and Development, 2000). Motor vehicles also significantly affect both human health and the environment through the emission of ozone-depleting substances, greenhouse gases, air toxins, and noise.

Vehicle design standards are intended to reduce the unwanted impacts of our transport systems, and to meet consumer and industry expectations. When complemented by the necessary regulatory framework, they provide a set of legal standards with which vehicles in a market must comply.

2. International design regulations

2.1. *The origins of design standards*

The need for design standards came about due to the increasing use of motor vehicles, and the hazardous impacts they have on communities. However, not all

Handbook of Transport Systems and Traffic Control, Edited by K.J. Button and D.A. Hensher

countries and jurisdictions have taken the same approach to the development and implementation of vehicle standards.

Research on early regulations shows that they were originally developed on a localized basis. That is, individual countries usually developed regulations independently of other countries. In the case of the U.S.A., Australia, and some of the European countries, vehicle regulation originated on a state-by-state basis, often resulting in a number of different standards within the same country. We see an example of this approach in the differing requirements in Europe for the driver's seating position. In the U.K. new vehicles are required to have the driver seated on the right, whereas vehicles sold new in mainland Europe are required to be left-hand drive.

In the past it was believed to be more appropriate to have the local jurisdiction responsible for matters relating to transport. As the interaction between many of the country regions was not strong, there was not really an issue of harmonization of vehicle standards. As a result, many individual countries today have regulations that reflect their own societal needs, traffic environment, local air quality, motor vehicle industry, legislative and legal systems, and cultural differences.

In the early 1960s, the impact of road trauma caused by motor vehicle crashes began to concern politicians with the result that, throughout the world, a concerted effort was made to upgrade the regulations covering vehicle safety. At the same time, the regulators began to understand that, due to the high mobility created by the automobile, regulations from neighboring states and countries needed to be harmonized to ensure that vehicles from one jurisdiction could travel to another without the need for temporary modification.

2.2. Different standards in different countries

North America, Europe, and Japan have traditionally accounted for the lion's share of the world's production and consumption of passenger vehicles. While local production in many developing countries is growing, manufacturers from these three economies built over 78% of the total of 58 million cars and trucks built worldwide in 1998, and accounted for over 77% of world sales (*Automotive News*, 1999).

In the U.S.A., the National Highway Traffic Safety Administration (NHTSA) administers the Federal Motor Vehicle Safety Standards (FMVSS). These standards have been mostly developed in the U.S.A., in response to local conditions, experience, and research. Canada uses a system of vehicle design standards based largely on the FMVSS.

In Europe, the United Nations Economic Commission for Europe (UNECE) administers the development of automotive safety and emissions regulations for adoption by its contracting parties, which includes representation from essentially all European countries.

Japan has also developed its own system of design standards. Although this has traditionally been a domestic activity, Japanese standards have often been based to some degree on either U.S. or European requirements.

These three economic powers have also served as the primary centers for vehicle standard rule-making in the world. Over time, the U.S.A. and Europe have developed very dissimilar systems for testing and certifying motor vehicles, while Japan's system utilizes elements of both the U.S. and the European systems. As the automotive markets in other countries have grown, their governments have created their own regulatory regimes, sometimes incorporating elements of other systems. While drawing on the supposed best aspects of other systems may sound like an appealing approach, the end result has been a proliferation of new regulatory procedures that resemble neither the U.S.A. nor Europe.

Australia's system of vehicle regulations provides a good example of one country's standards which have drawn on aspects of all three of the major automotive players. Sixty per cent of Australia's current vehicle design standards, known as the Australian Design Rules (ADRs), are aligned fully or partially with UNECE regulations; the remainder either are based on U.S. or Japanese standards, or are standards unique to Australia (Federal Office of Road Safety, 1999). The ADRs will be the basis for much of the discussion in this chapter.

2.3. Harmonization of international design standards

With vehicles increasingly being made and sold in a global market, there is a clear necessity to meet local customer demands, while at the same time standardizing common concepts and technologies as much as possible. The ability to design and build products that can be sold in different markets has become the goal of virtually every vehicle and auto part manufacturer. Conflicting and overlapping regulations impede that ability. They create barriers to trade and limit consumer choice, while adding unnecessary costs that may be borne by consumers.

The Australian Government and the automotive industry have identified vehicle standards harmonization with international norms as an essential element in trade facilitation. Different standards alone add 5–10% to the costs faced by the world's exporters entering the market for the first time (Federal Office of Road Safety, 1999). The total costs to the automotive industry due to the lack of internationally harmonized standards and conformity assessment arrangements are significant.

Many efforts currently exist to harmonize vehicle design standards internationally. The main mechanism for achieving this is Working Party 29

(WP29) of the UNECE. WP29 provides an international forum for development of vehicle standards, and its active membership includes not only European member countries, but also countries such as the U.S.A., Canada, Australia, Korea, South Africa, and Thailand.

Under the World Trade Organization rules, to which Australia is a signatory, only the regulations developed by the UNECE meet the definition of an "international" standard in the vehicle standards field (as opposed to national standards such as the Japanese or U.S. standards).

3. Developing design standards

3.1. Scope of vehicle design standards

Vehicle design standards are intended to set performance and design requirements for all road vehicles sold new in a country, whether they are manufactured in that country or imported. They cover those design aspects that affect a vehicle's safety, emissions, and security, but do not directly cover aspects such as build quality or esthetics.

Using the Australian system as an example, the list below shows the range of vehicle types that are usually covered by vehicle regulations (Department of Transport and Regional Services, 2000).

(1) motor cycle,
(2) passenger car,
(3) forward control passenger vehicle,
(4) off-road passenger vehicle,
(5) light omnibus,
(6) heavy omnibus,
(7) light goods vehicle,
(8) medium goods vehicle,
(9) heavy goods vehicle,
(10) very light trailer,
(11) light trailer,
(12) medium trailer, and
(13) heavy trailer.

Design standards are generally used for new vehicles prior to sale. However, many standards are also used to ensure in-service vehicles continue to comply with relevant standards and community expectations.

3.2. Identifying the need for a standard

A vehicle standard can be initiated in a number of ways. The call for a particular standard could come from manufacturers, industry experts, consumer groups, or government departments. It could be driven by an identified problem, or by a shortcoming in previous standards. But either way, it is essential that the body that oversees the management of vehicle standards answers the following questions favorably before a standard can be developed further:

(1) Will the standard be effective in improving road safety, emissions, or theft prevention?
(2) Do the forecast benefits outweigh the costs of developing, introducing, and enforcing the standard?
(3) Can manufacturers achieve the proposed standard, both technically and financially?
(4) Will the standard create any barriers to trade or restrict competition?
(5) Will the standard contradict an existing standard or relevant legal requirement?

3.3. Involvement of key stakeholders

For standards to effectively achieve the objectives of governments, manufacturers, consumers, and other relevant bodies, it is essential that these key stakeholders all be involved in the development of standards where appropriate.

Australia provides a good example of how key stakeholders are involved in standards issues. The Australian Transport Safety Bureau (ATSB), the government department that oversees vehicle design standards, develops the ADRs within a consultative process involving federal and state government departments and key industry stakeholders. The Australian Automobile Association (AAA) and other relevant bodies represent consumer groups in this process, while the Federal Chamber of Automotive Industries (FCAI) represents the vehicle manufacturers and importers. In accordance with the Trans-Tasman Mutual Recognition Act 1997, the New Zealand Government is also included in the consultative process.

The ATSB also works with agencies in the U.S.A., Japan, and Europe to ensure that Australia accesses the latest best international practices. This is consistent with the Commonwealth Government's policy to harmonize, wherever possible, with international standards.

Towards the end of the development process public comment is sought on a draft standard, ensuring that all key stakeholders in the standard being considered have been consulted. This process ensures that any new or modified vehicle design standard has the best possible chance of meeting the expectations of industry and consumers, and can effectively meet its objectives.

4. Enforcement of standards

4.1. Legislative support

It is difficult to enforce a design standard unless it is written into a country's legislation, making compliance with the standard mandatory. With regard to new vehicles, it is essential that this legal backing be provided to ensure that motorists are provided with vehicles that comply with the relevant standards.

For example, in the U.S.A. the NHTSA has a legislative mandate, under Title 49 of the U.S. Code, to issue Federal Motor Vehicle Safety Standards and Regulations to which manufacturers and importers of motor vehicles and components must conform and certify compliance with. In Australia, the ATSB has a legislative mandate in accordance with the Commonwealth's Motor Vehicle Standards Act (MVSA) 1989.

Often design standards are not enforced by law, but are a voluntary requirement. For example, an industry may require its member companies to comply with a code of practice. A voluntary standard is usually one that is not deemed to have a great enough safety or environmental implication to be covered by legislation, or might only involve in-service vehicles or aftermarket product.

4.2. Type approval systems

It would not be feasible to inspect every new vehicle sold onto a market for compliance with every regulated design standard. Not only would such an exercise be extremely time-consuming, but some standards require crash testing to determine compliance, which adds cost and complexity to the issue. To overcome these issues, most countries use a compliance mechanism called a "type approval" system.

Under a type approval system, only one vehicle representing the make and model of a particular vehicle is tested to each standard to demonstrate compliance. If the vehicle tested complies, then all others of the same design and manufacture are accepted as complying.

In Australia, the ATSB does not physically test vehicles for certification purposes. The manufacturers are responsible for ensuring compliance with the ADRs. The certification process requires the vehicle manufacturers to conduct the tests specified by the various ADRs. Manufacturers can conduct those tests at their convenience providing the tests are conducted according to ADR requirements. The tests can be done in test facilities associated with the manufacturer or in independent facilities located in Australia or overseas.

This highlights another benefit to be gained by the harmonization of vehicle standards. Where standards can be developed to be common with other countries, the cost of homologation can be significantly reduced, and thus competition in a market is not restricted by differing standards.

When the Administrator of Vehicle Standards is satisfied with the evidence provided that the vehicle complies with all applicable ADRs, the ATSB issues a Compliance Plate Approval (CPA) certificate. This is the authority to fit compliance labels to vehicles of the specified make and model. Fitting an identification label is a legal requirement. It indicates to a local registering authority and the consumer that the vehicle complies with the ADRs.

4.3. Compliance approval for low-volume vehicles

The high costs of homologating a vehicle to local design standards can provide a significant barrier to the sale of vehicle models expecting low-volume sales. It can restrict competition, reduce consumer choice, and prevent new vehicle technologies entering a market. Therefore, some jurisdictions have compliance procedures for low-volume vehicle models which do not require proof of compliance with every standard.

In Australia, this is covered by the Specialist and Enthusiasts Vehicle Scheme. This is a government policy that is implemented at the discretion of the Administrator of Vehicle Standards under Section 14 of the MVSA. The scheme operates with significant concessions compared with the regulation applied to full-volume manufacturers, but limits are placed on the numbers of vehicles any particular importer supplies to the market, and there are restrictions on the model types.

4.4. Conformity of production

It is common practice for the body overseeing vehicle design standards to perform an audit of companies supplying vehicles to a local market, to ensure that their company processes continue to comply with the relevant standards. In Australia, these are called Conformity of Production (COP) audits.

COP audits of production and testing facilities of major vehicle suppliers to the Australian market are performed by the ATSB to ensure the ADRs are complied with during the manufacture and supply of vehicles. They are intended to provide the Administrator of Vehicle Standards with reasonable assurance that manufacturers have adequate controls in place to ensure that the production of vehicles continues to conform with applicable ADRs and that the vehicles are built to approved specifications. They involve detailed examination of procedures followed through design, purchasing, and manufacturing processes.

4.5. *Vehicle recalls*

There are often cases where vehicles are sold to a market which do not meet the local design standards. Despite a vehicle model achieving type approval, it might be that an error in production or a faulty component results in a number of that vehicle model having a fault that causes non-compliance with a standard or standards. If the non-compliance is determined to be a safety issue which will or may cause injury, then the government body overseeing vehicle standards will require the affected vehicles to be recalled and rectified by the supplier.

Safety-related recalls are generally the only type enforced by law. Problems that do not affect the safe operation of a car are carried out at the discretion of the manufacturer or distributor, usually when publicity about the fault would be financially damaging. The steps of a recall procedure which are enforced upon a supplier by the relevant government body are:

(1) identify which batch of vehicles may have the safety defect;
(2) advertise this and/or write to owners; and
(3) rectify the safety defect free of charge.

In Australia, there is a voluntary code of practice developed by the FCAI for the conduct of safety recalls by its member companies, which include most vehicle manufacturers and importers. This recall system covers cars, motorcycles, trucks, buses, accessories, and individual parts, and was the first recall procedure in the world to be covered by a code of practice.

4.6. *In-service regulations*

Although most vehicle design standards are intended for new vehicles prior to their sale, many standards are also used by local jurisdictions to determine the "roadworthiness" requirements of in-service vehicles.

Often in-service vehicle requirements are implemented and enforced by the local jurisdiction or state, and not at a national level. This is certainly the case in both the U.S.A. and Australia, where roadworthiness requirements are the responsibility of state governments.

State governments usually use the new vehicle design standards as the basis for their registration and in-service performance management regimes. After all, these are what the vehicles were originally built to comply with. However, issues which may be unique to individual state and territories, such are air quality, road safety performance, road design, and local conditions, are taken into consideration when determining in-service standards.

Enforcement of in-service vehicle standards also varies between the jurisdictions. The variety of in-service inspection programs employed include compulsory annual vehicle inspections, inspection at change of ownership, and roadside inspections of vehicles suspected by police to not comply with in-service standards. As with the in-service standards, the method of enforcement is determined largely by local factors, and the decision over which approach may be most effective and cost-effective.

5. Implications for transport systems

Vehicle design standards can be clearly shown to provide benefits to the community. They are effective at reducing the hazardous impacts of transport systems on road users, the general public, and the environment.

The effectiveness of vehicle design standards in improving road safety and reducing environmental impacts can be demonstrated using Australian experiences.

5.1. Road safety

The increase in road fatalities in the 1950s and 1960s, as a result of greatly increased usage of motor vehicles, provided the catalyst for governments in Australia and elsewhere to introduce vehicle safety requirements.

Since the introduction of the ADRs in 1969 to coincide with the mandatory fitment of seat belts, Australia's road toll has been steadily decreasing. This trend has been sustained by a combination of factors. Road safety improvements can be attributed to initiatives such as drink–driving and speeding campaigns, improvements in road design, and the introduction of traffic management technologies, as well as improvements in vehicle design standards.

Monash University Accident Research Centre (MUARC) has considered improvements to vehicle design in isolation from other road safety activities by examining "crashworthiness" (the risk of injury when involved in a crash). As shown in Figure 1, MUARC found significant improvements in the crashworthiness of vehicle manufactured each year since 1964 (Newstead et al., 1999).

The probability of severe injury for drivers involved in a crash of the average vehicle manufactured between 1967 and 1996 has declined steadily. Figure 1 also shows when various safety-related ADRs were introduced. This "real-world" data highlights the fact that improvement in vehicle safety design, and the progressive

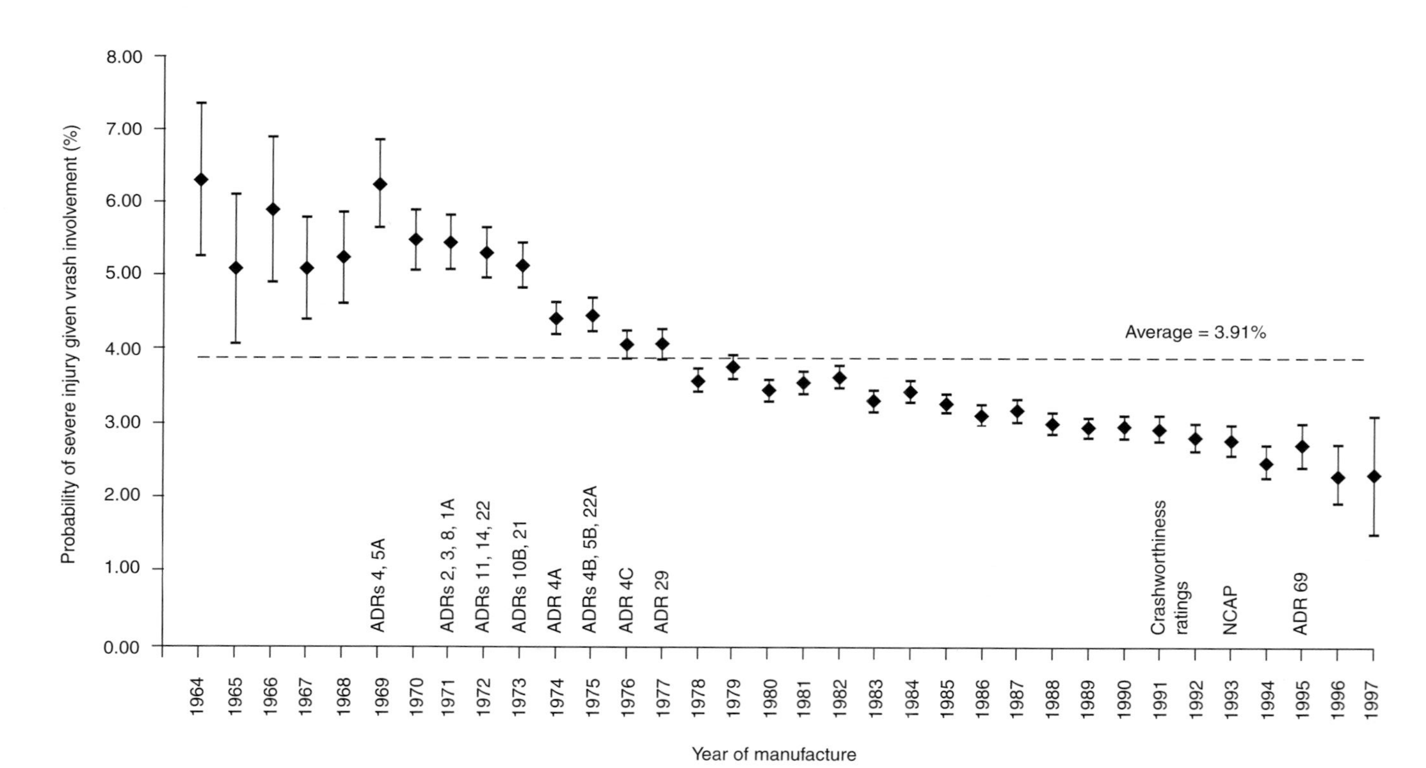

Figure 1. Crashworthiness by year of manufacture (with 95% confidence limits).

introduction of ADRs, has been effective in significantly improving the level of vehicle occupant protection provided in new vehicles.

5.2. Environment

Transport activities in total are the most significant contributor to ambient air pollution in Australia. Road vehicles, because of their numbers and their use throughout urban areas, are the dominant source of transport pollutants (Australian Academy of Technological Sciences and Engineering, 1997). In particular, motor vehicles have been shown to be the major source of emissions of carbon monoxide, hydrocarbons, oxides of nitrogen, and lead compounds in urban airsheds. Vehicle are also a significant source of noise emissions.

The control of emissions from motor vehicles is therefore a primary tool in the management of urban air quality and greenhouse emissions. Initiatives to reduce motor vehicle emissions are mainly aimed at two general areas: vehicle and fuel characteristics; and the way in which cities accommodate motor vehicles through, for example, transport and land use planning.

The main regulatory mechanism for control of motor vehicle emissions in Australia is embedded in the ADRs, which set limits for the following:

(1) noxious exhaust emissions or carbon monoxide, hydrocarbons, oxides of nitrogen, visible smoke, and particulates (ADRs 30, 36, 37, 70);
(2) evaporative hydrocarbon emissions (ADR 37); and
(3) noise emissions (ADRs 28, 39, 56).

Emission standards for new vehicles have been demonstrated to be an effective method for reducing the impact of vehicle emissions on air quality. Since their introduction in the 1970s, they have progressively been tightened to reduce the exhaust emissions of new vehicles. For example, in 1986 ADR 37 was introduced, which banned the use of leaded petrol in new vehicles and set emission limits which encouraged the use of catalytic converters, a proven device for reducing hazardous exhaust emissions.

Data from the Environment Protection Authority of Victoria shows that the city of Melbourne has experienced reductions in airborne lead, ozone, carbon monoxide, particulates, nitrogen dioxide, and sulfur dioxide since the early 1980s, despite a larger vehicle fleet (Environment Protection Authority, 1998). It is strongly suggested that increasingly tighter vehicle emissions standards have been a significant contributor to these improvements.

However, significant emission improvements can still be made. The planned harmonization of Australian emission ADRs with UNECE standards should be effective in realizing these desired improvements.

References

Australian Academy of Technological Sciences and Engineering (1997) "Urban air pollution in Australia", ATSE, Melbourne, inquiry report.

Automotive News (1999) *'99 market data book – May 1999*. Detroit: Automotive News.

Department of Transport and Regional Services (2000) "The Australian design rules for road vehicles – as at determination 1 & 2 of 2000", Commonwealth Department of Transport and Regional Services, Canberra.

Environment Protection Authority (1998) "Air emissions inventory – Port Phillip Region", Environment Protection Authority of Victoria, Melbourne.

Federal Office of Road Safety (1999) "Review of the Motor Vehicle Standards Act 1989", Chaired by the Federal Office of Road Safety, Canberra, MVSA Review Task Force.

Newstead, S., M. Cameron and C. My Le (1999) "Vehicle crashworthiness ratings and crashworthiness by year of manufacture", Monash University Accident Research Centre, Melbourne.

Organisation for Economic Co-operation and Development (2000) "Road deaths fell in OECD Countries in 1999, but injuries rose", OECD, Paris, news release, 20 April.

Chapter 26

JUNCTION DESIGN

ROGER N. BIRD
University of Newcastle

1. Introduction

The purpose of road junctions is to transfer road users from one traffic stream to another safely and efficiently. Despite this appearing to be a simple statement, junction design is complex. A junction is a component of a road network. It is not unusual for junctions to be the major source of delay and conflict within a network (Institution of Highways and Transportation, 1997). Conflict between road users at junctions can lead to accidents and is detrimental. Delay is also seen as detrimental as it leads to network inefficiency and driver frustration. Therefore junctions should be designed as an integral part of the network. This will affect the location, configuration, and detailed design of junctions.

It is good practice to avoid having junctions between roads of widely differing standards. For example, junctions on high-speed interurban roads should not be joined directly to minor suburban or rural roads. It may be necessary to provide short transitional highway links to allow drivers to adapt from one type of road to another in the interests of safety. A clear hierarchy of roads is often visible in new towns, illustrating the separation of different types of traffic. However, the role of a road may change over a period of years as land use and travel patterns change. Highway authorities should be aware of potential conflicts caused by such changes when assessing older road junctions and designing improvements, remedial works, or network improvements.

Junctions may also function in such a way as to discriminate between different types of road user or certain modes of transport. In most situations motorized traffic comprises the largest proportion of the traffic flows, and it is usual to design primarily for such users in order to minimize overall traffic delay. This imperative is no longer an overriding consideration for highway designers, as the political climate in many countries is now inclined towards restraint of motorized private transport, in favor of public transport and non-motorized transport. The latter usually means pedestrians and cyclists, although it may include rickshaws, animal-drawn carts, etc. The designer must be clear which goals the junction is expected to meet.

Handbook of Transport Systems and Traffic Control, Edited by K.J. Button and D.A. Hensher

2. Junction types

Road junctions can be divided into five main groups.

(1) *Uncontrolled junctions* – junctions at which there are no signs or road markings to indicate priority. In some road networks the majority of junctions are uncontrolled. Once there is sufficient traffic to warrant a designed junction, some form of priority or control will be included in the design.
(2) *Priority intersections* – junctions that are controlled by establishing that traffic on one road has priority over another. Signs and road markings indicate which vehicles should stop or give way.
(3) *Roundabouts* – junctions where traffic circulates around a central island and leaves at a chosen exit.
(4) *Signal-controlled junctions* – in which traffic streams alternately have priority as indicated by changing signals.
(5) *Grade-separated junctions* – in which conflicting traffic streams are separated by being at different levels.

Some junctions may combine elements of more than one group, e.g., a signalized roundabout with a flyover. This chapter does not deal with the design of traffic signals (see Chapter 34 of this book). Although uncontrolled junctions are not considered further in this chapter, some of the safety principles described here will apply. Where queues, delays, or accidents become unacceptable, some form of priority or control should be introduced.

3. Principles of design

Junction design usually focuses on two primary goals, viz., safety and efficiency. Safety is clearly the first priority, and should not be compromised in the interest of efficiency. However, caution must be also be exercised in the design of junctions where traffic restraint is a goal. A reduction of capacity at a junction may, in some cases, be seen as a desirable traffic restraint measure, but it may also create unnecessary hazards leaving the designer and highway authority in a compromised position should accidents arise. The following principles apply to any non-signalized junction.

3.1. Conflict

Conflicts at junctions arise as road users wish to use the same road space to execute different maneuvers, or the same maneuver at different speeds or on

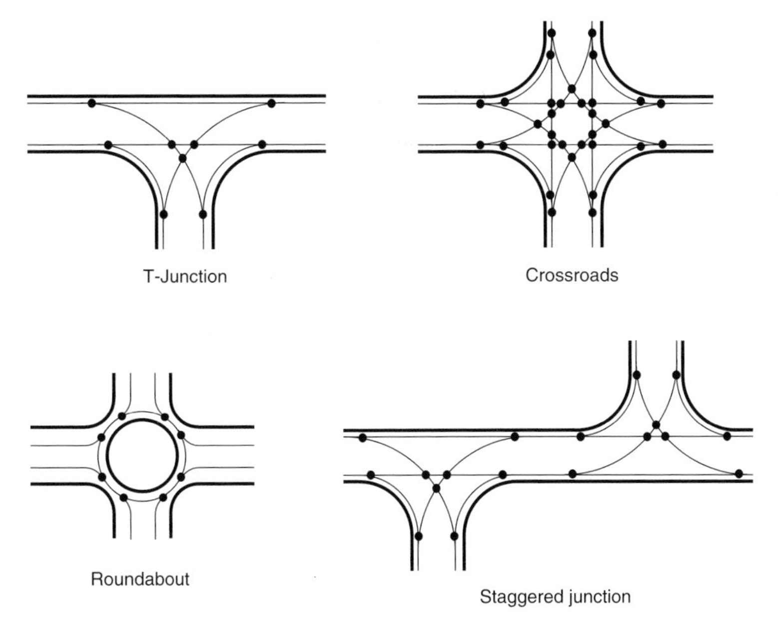

Figure 1. Conflict points at different types of junction.

differing paths. The form of junction controls these conflicts and should aim to reduce the risk of accidents to a minimum. Although many factors go together to cause a road accident, there is usually a human factor involved, where one or more road users have not seen or have incorrectly assessed the intended movements of others. Well-designed junctions minimize the possibility of such misunderstandings occurring or being severe enough to lead to collisions.

One way of achieving this is to consider each point where vehicle paths diverge, cross, or merge. These are known as conflict points, and have the potential to cause accidents. Selecting different forms of junction can reduce the number of conflict points. Figure 1 illustrates this point. Some layouts have fewer conflict points for the same number of arms. This has obvious advantages for users, and enables the designer to make these areas safer. The disadvantage is that each point will have more potential conflicts in a given time period. Further conflict points arise when non-motorized road users are included. The latter fall into two groups, those who mix with the motorized traffic on the main carriageway (e.g., cyclists), and those who use footways and need to cross the main carriageway. Adding these conflict points raises the issue of contrasting speeds and the associated risk caused by the speed/direction difference. The aim therefore is not necessarily to reduce the number of conflict points to a minimum, but to reduce the overall risk of conflict. The number of conflict points can of course be reduced by using dual carriageways or grade-separated interchanges or both. With full grade separation the only conflicts are merging and diverging.

It is good practice to identify the types of conflict and design accordingly. Merging, diverging, and weaving maneuvers are best carried out at low relative speeds and shallow angles (between 10 and 15°). This makes the maneuvers easier to carry out, and if accidents do occur they will be less severe. (This is particularly important on high-speed roads with grade-separated junctions.) Crossing maneuvers are best carried out at right angles. Achieving all these goals is not always possible, so attention should again be given to minimizing overall risk, by taking into account the number of road users and their vulnerability.

3.2. Traffic composition and priority

Junctions are usually designed for a particular level of traffic flow. Most highway authorities consider it uneconomic to cater for all expected flows without queues occurring. A series of design flows should be specified, knowing the likely frequency with which that flow will be exceeded, in the knowledge that on those occasions queues and delays are likely to exceed those predicted for the junction design. However, where there are tidal flows, typically on urban radial routes, flows may be significantly different between morning and evening peak periods. Designers must check their designs for each of these.

Local land use and the position of the junction within the network will define the likely traffic composition. The junction should be designed accordingly. Large numbers of heavy vehicles, buses, large goods vehicles, etc. will require appropriate corner radii for turning. Acceleration and braking distances may be affected, and gradients will have a more significant effect on heavy vehicles. The presence of non-motorized traffic must be considered and appropriate provision made.

3.3. Appearance

Good junction design must take into account how it will appear to each road user. People need to know what to expect at the junction, what shape or layout it is, where their chosen exit will be, and which lane they should be in while negotiating the junction. This requires clear and informative signs. The necessary information on signs will include textual information (e.g., names of destinations), symbols (e.g., standard nationally or internationally recognized symbols) and graphical elements (e.g., the layout of the junction). Signs should be designed to give information to each group of road users at a rate appropriate for the speed of travel. Care must be taken to avoid putting too much information on one sign,

Figure 2. Block-paved chevrons around a roundabout central island.

which cannot be read safely, or using too many intrusive signs. Some information can be given best by road markings, such as lane markings with appropriate direction arrows. At many roundabouts in the U.K. chevron signs have been replaced by patterned block paving (Figure 2). This serves both as a protective ring around the central island and as an effective means of indicating the edge of the island and the direction of circulation. As a result, fewer signs are needed on the roundabout island itself. Similar unobtrusive methods of assisting road users are to be encouraged. Further measures that can assist safety and efficiency are the use of street lighting and the installation of barriers to guide pedestrians to safe crossing points. When designing for pedestrians, allowance should be made for all users of footways, including low-speed wheeled vehicles, such as prams, pushchairs, and wheelchairs. Cyclists may share a footway, and in some countries it is accepted that young cyclists are safer using a footway than the main carriageway when no cycle track is available.

3.4. Environmental intrusion

Junctions are often intrusive to the surrounding area. The slowing down and speeding up of motorized traffic will create a concentration of vehicle engine noise and emissions. Street lighting is often installed at junctions in the interests of safety, and is regarded by many as intrusive. The necessary street furniture such as signs, pedestrian barriers, etc. can also be regarded as intrusive. These effects can be mitigated, but in most cases the necessary measures involve the use of land (e.g., to create visual barriers, noise fences, etc.). In urban areas this often not available, so decisions must be made about how much environmental intrusion is acceptable to meet the demands of road safety.

4. Priority intersections

4.1. Types

T-junctions are three-arm junctions where traffic traveling between two of the arms has priority over other movements. A crossroads has four arms where two routes cross each other. Priority is again usually given to the route with the highest through flow. A staggered junction has four arms, but with the minor arms offset from each other, effectively making two T-junctions on opposite sides of the major route. Priority junctions also exist with more than four arms. In such cases it is advisable to separate the converging roads into two junctions with a short link between, in order that road users are not confused by the layout, or have to look in all directions at once before being able to proceed safely. An alternative solution is to change the junction layout to a roundabout.

In most cases, priority junctions are laid out so that the maneuvers that have priority have the straightest alignment, so that speeds are maintained on the through route and delays are minimized. In some cases the dominant traffic movements travel round a corner, and a straighter route has to give way. In such cases priorities must be made clear by use of signs, road markings, and, if necessary, realignment of the road edges to emphasize the priorities.

4.2. Standards

The layout of any priority junction can be adapted to suit the level of traffic flows and the standard of road. On most lower-speed, urban, or lightly trafficked routes, a simple junction will suffice. However, as speeds and flows increase, it is advisable to provide more space to execute maneuvers, and, if necessary, to queue in safety. Traffic islands can be created using kerbs as physical constraints to channel through traffic, and to provide areas out of the through traffic streams where vehicles can accelerate, decelerate, or queue to turn. Guidance on the appropriate level of provision and physical dimensions will be available in national design standards. In some countries, an intermediate level of provision is that of ghost islands. These are road markings to channel flows in the same way as real traffic islands, but without creating physical barriers above the road surface.

4.3. Capacity

Normally national design standards give guidance on the layout of junctions for expected traffic flows. However, it is frequently necessary to assess older junctions or those which do not conform to, or are not covered by, standards. The formulae used for calculating the capacity of a priority intersection vary between countries. There are general principles that are common. Junction capacity is not usually calculated for the streams of traffic that have priority and do not need to give way. Similarly, capacity is not calculated for traffic that can leave a priority stream by making a nearside turn (i.e., left when driving on the left or right when driving on the right) and does not have to give way. The remaining traffic streams must find gaps in the priority streams in which to make their maneuvers. Some maneuvers will need coincident gaps in more than one stream. Different researchers and highway authorities have studied this subject and put forward various empirical relationships. These generally express the capacity of a particular movement as a function of the traffic volumes, angles, curve radii, sight distance, and road dimensions. The capacity is expressed in vehicles or pcu (passenger car units) per unit of time for each stream of turning traffic. It is usual to compare the capacity with the demand flow to obtain the ratio of demand to capacity (d/c ratio). This can be used with queueing formulae to predict the queue lengths and delays. These are more tangible measures of a junction's performance than the capacity itself. When the demand-to-capacity ratio is well below unity, queues and delays will be small. As the demand-to-capacity ratio approaches unity, queues will begin to increase. When demand exceeds capacity, long queues can be expected, which increase considerably. Such calculations are often complex and computer programs exist to make the process easier. Junction design software must be treated as a tool with which the designer achieves a safe, efficient, and economic design, not as a calculator that can give "the right answer." There is a danger in designing junctions to aim for a low RFC value for all movements without considering whether the junction geometry is too generous (which may encourage higher speeds), environmentally intrusive, or costly.

It is by no means certain that standards can always be met. Constraints may include land availability, topography, and the cost of realigning approach roads and associated structures. Compromises will therefore have an impact on the safety or the efficiency, or both, of the junction. Clearly the highest priority is to maintain safety, but when large queues begin to form, drivers may accept gaps which are too small, so there is an inevitable link between safety and efficiency.

4.4. Design considerations

A simple layout is usually the safest, so it is advisable wherever possible to use T-junctions. Where routes cross, it is desirable to offset the minor arms to form a staggered junction. Where land use constraints prevent this, particularly in urban areas, there may be no solution except a crossroads. At a staggered junction, minor-road traffic crossing the major route has to make two turns. It is preferable to make the offside turn first, followed by a nearside turn off the major route (i.e., right then left when driving on the left, or left then right when driving on the right). This will lead to fewer vehicles queueing to turn right off the major road. The concentration of conflicts at a crossroads means that a simple crossroad layout is only appropriate for the lightest traffic flows. In heavier flows alternative solutions should be investigated, viz., channeling with islands, offsetting the minor arms to form a staggered junction, signalizing, using a roundabout, or banning offside turns from the major route and providing a safer alternative route.

5. Roundabouts

5.1. Operation

Roundabouts are junctions where all traffic is channeled onto a one-way circulatory carriageway that joins all entries to the junction (Brown, 1995; Crowne, 1989). There are two distinct ways in which roundabouts operate, either with offside priority at each entry, or with unregulated flow. Early experience with roundabouts was in conditions of low traffic flows, and no priorities were defined. As traffic flows increased, it became necessary to align the carriageway edges and islands to encourage safe maneuvering. To achieve this, a traffic island deflected entering flow to be parallel with circulating traffic. Traffic could then change lane to stay on the roundabout or leave at the next exit. To accommodate increased flows, a roundabout needed to have sufficient length and width for the necessary weaving maneuvers to take place, leading to large, polygonal central islands. If such roundabouts have inadequate capacity, circulating traffic streams can obstruct each other so that traffic cannot leave the junction. In this way the complete junction can "lock." If the unregulated flow regime is maintained, the solution is to lengthen or widen the weaving lengths of the circulatory carriageway, thus making the junction large. An alternative approach that has been adopted in many countries is to use offside priority at each entry. This change has consequences for the operation and design of roundabouts. The basic principle is that all traffic that has entered the junction is, thereafter, unimpeded to its exit. Thus offside-priority roundabouts should not "lock". It also means that traffic can enter a "slot" in the circulating flow, and need have minimal conflict with other vehicles while passing round the roundabout. On some very large roundabouts, there may be weaving between circulating lanes. As the conflict maneuvers concentrate at the entries (see Figure 1), the long lengths of circulating carriageway are not necessary, but layout of the entries becomes more significant.

It is interesting to note that while some countries, notably the U.K., have made extensive use of roundabouts for many years, they are far from common in other countries. In mainland Europe there has been a slow increase in the use of roundabouts during the late 1980s and 1990s. They are used in many of Asia's large cities and in Australia and New Zealand. In contrast, in the U.S.A., roundabouts are still rare. Indeed the AASHTO (American Association of State Highways and Transportation Officials, 1994) guidance does not even mention them. Most countries that make use of roundabouts in heavily trafficked areas have adopted the offside-priority rule. However, a few countries, such as India, still have unregulated flow. Roundabouts are well suited to most busy junctions, and are generally more efficient than priority intersections for handling large quantities of turning traffic.

5.2. *Types*

Roundabouts can vary considerably in size. The smallest are mini roundabouts. These have no physical central island, but one created by road markings that can be overrun by large vehicles. These are used where there is insufficient space for a "normal" roundabout. It is usual to regard these only as a remedial measure for a priority intersection. The size of mini roundabouts means that there is often only one vehicle at a time in the junction area, and the rapidly changing priorities lead to similar behavior to "filter in turn" junctions, which are used in some parts of the world.

If there is enough space for about 25–30 m diameter of circulating carriageway (sometimes called the inscribed circle diameter, ICD), a physical island can be used. A central circular island can be created with sufficient width of carriageway to permit large articulated vehicles to negotiate the roundabout. The majority of roundabouts will fall into this group. The largest roundabouts are typically associated with grade-separated junctions and can be over 1 km in circumference. Designers of junctions of this size would do well to consider the use of smaller roundabouts and two-way link roads to achieve savings in time for road users and in cost of construction, as an alternative. Circulating speeds tend to increase as the size of roundabout increases, leading to an increased risk of accidents as road users try to assess gaps and enter the roundabout. Another form of large roundabout is a one-way gyratory traffic system, frequently found in urban areas. There is no clear dividing line to define when these should be considered as

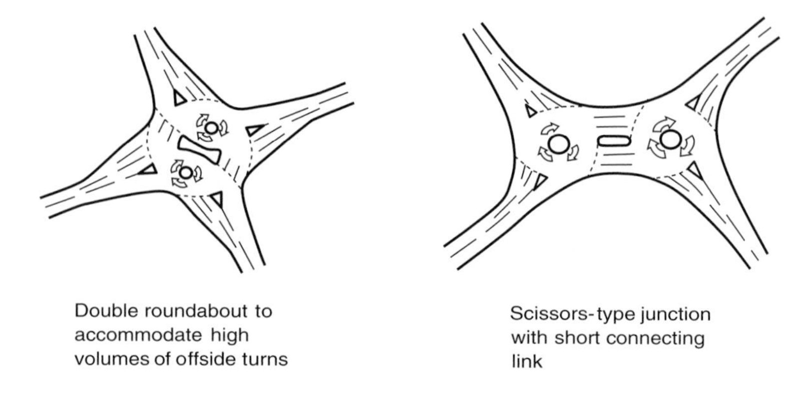

Figure 3. Double roundabouts.

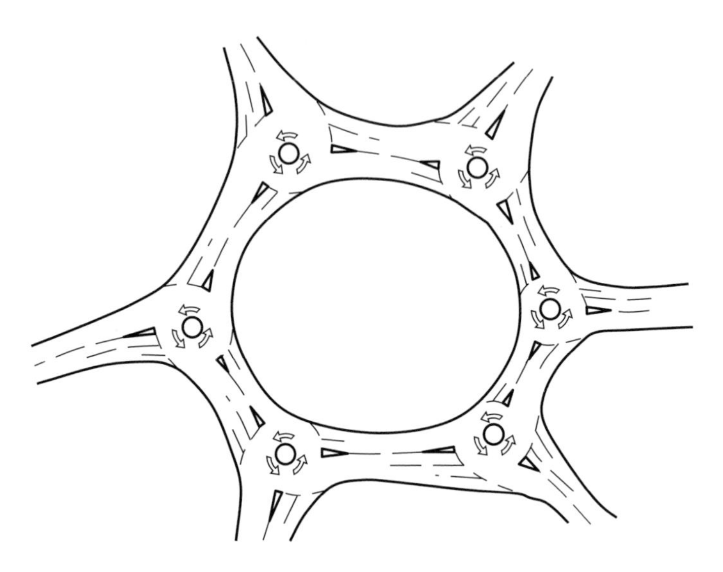

Figure 4. Ring junction.

roundabouts; each should be considered individually, and appropriate principles applied.

Various other types of roundabout have been used. If there is a particularly high proportion of offside turns at a four-arm junction, it may be appropriate to use a double roundabout (see Figure 3). A scissors-type junction may be suitable in some cases. Some large roundabouts have been found to operate more effectively as two-way rings with a three-arm junction at each entry (see Figure 4).

5.3. Capacity

The method of calculating the capacity depends on the way in which the roundabout operates. For unregulated roundabouts the capacity is assessed by calculating whether each section of the circulating carriageway can accommodate sufficient weaving maneuvers. An example of this method is the Wardrop formula, used for many years in the U.K. and still used in India, where offside priority has not been adopted. For offside-priority roundabouts, capacity is a function of the circulating flow past each entry and the geometry of the entries. Following the adoption of offside priority in the U.K. in 1966 (Ministry of Transport, 1966), various capacity studies were carried out. Kimber (1980) proposed a formula which has been widely used since then. It can be applied to any offside-priority roundabout, regardless of whether it was designed as such or is an unregulated roundabout simply re-marked for offside-priority operation. Using a formula of this sort is not straightforward, as the capacity of each arm is a function of the circulating traffic flow, which in turn is a function of the actual flow able to enter the other arms of the roundabout. Therefore each arm is dependent on the others, requiring an iterative procedure to balance the capacity calculations. Various computer programs exist to assist designers (e.g., ARCADY, Rodel, Robosign) (see Webb and Pierce, 1990).

Clearly, local driving patterns will affect junction capacity calculation. Studies in Switzerland by Tan (1991) have indicated that traffic leaving the roundabout on one exit has a significant effect on the capacity of the next entry if the length between entries is small. Appropriate formulae have been suggested for use there. In contrast, Troutbeck (1990) reports that studies in Australia have shown that leaving traffic has little effect on the next entry.

5.4. Design

As local practices in the use of roundabouts vary, only general matters will be mentioned. A guiding principle should be to slow traffic before entering the roundabout (Kay et al., 1992). This is particularly important for offside-priority roundabouts, as the entry is the main point of conflict. Traffic can be slowed by a variety of means, one of which is gentle deflection of the approach road so that the driver cannot see straight through the junction until quite close to the give way line. This will cause drivers to slow their vehicles until they are at a point where they can see that it is safe to enter the junction. Exit geometry should encourage an easy exit from the roundabout to keep the circulating carriageway clear with due regard to speeds. On unregulated roundabouts the main conflict is a weaving maneuver, so the streams of traffic should be aligned to be parallel, and, if possible, traveling at similar speeds at the start of each weaving section.

Roundabouts may not operate effectively in situations of heavy flow between two arms with few vehicles making any other turns. In such cases the dominant flows may prevent minor-road traffic from entering the roundabout. As with any junction, it is usual to design with peak-hour traffic in mind. Equally, the designer should allow for the opposing peak period when vehicle movements are likely to be reversed. Both cases should be checked, even if one peak is more intense than the other. The traffic flows may change between morning and evening, but the road geometry will not.

A potential disadvantage of roundabouts is that although they are efficient for motorized traffic, they present problems for non-motorized traffic (e.g., cyclists) which use the carriageway, and all users of footways who need to cross the carriageways. Research in The Netherlands (CROW-werkgroep "verkeerspleinen ii," 1993) has shown that the best form of provision for cyclists is a cycle track around the roundabout. In some cases cyclists have been allocated a circulating lane with signs to reinforce their priority over entering traffic. Controlled crossings on the entries and exits will aid footway users. On a roundabout entry a crossing is not inappropriate, as drivers are encouraged to slow down for entries. However, siting a crossing on a roundabout exit needs careful consideration. If it is too close to the roundabout it will cause traffic leaving the roundabout to stop at a point where it is accelerating away from the junction and where queues may block the roundabout itself. However, if it is further from the roundabout pedestrians will have a longer diversion to use it, and may cross at less safe locations. If space, budget, and environmental constraints permit, a subway or footbridge may be the best solution.

6. Grade-separated junctions

Grade-separated junctions are those where different traffic movements are accommodated by roads at different levels to allow conflicting traffic flows to cross each other unobstructed. Because of the cost of the necessary structures (bridges and tunnels), they are used almost exclusively on heavily trafficked roads.

6.1. Types

A junction may be fully grade separated with all crossing maneuvers accommodated at different levels. Alternatively, some maneuvers could be separated with the rest occurring at an at-grade junction. In its simplest form this could be a flyover or underpass taking a through-traffic movement over or under a busy junction. Grade-separated junctions therefore take many forms, depending on the traffic flows involved and the amount of space available.

6.2. Capacity

The capacity of a grade-separated junction will be assessed by looking at the various points where conflicts occur. If there is an at-grade junction as part of the interchange, it can be assessed as such. If queues are likely to form at an at-grade section of the junction, there should be enough space to accommodate the queues without affecting free-flowing high-speed roads. The main conflict points are the merges and diverges where main roads and link roads join and separate. Most national standards give guidance on how many lanes should be provided, how long they should be, and the angles at which they join. The usual parameters that govern choice will be the volume of traffic, proportion of heavy vehicles, design speed, and gradient. Some standards (e.g., in the U.S.A.), put emphasis on continuity of the number of lanes before and after junctions, and on the use of auxiliary lanes to allow for this. Some sections of roads at grade-separated junctions may carry significant amounts of weaving traffic. Each country will have a way of assessing how provision should be made for this.

7. Summary

A road junction needs to be designed for safe and efficient use by all road users. The designer's task is a matter of balancing these requirements with the needs and concerns of society at large, which can include environmental and political demands (AustRoads, 1988, 1993, 1995, 1999; Department of Transport (U.K.), 1994–2000). Care needs to be taken to ensure that the junction strikes the right balance between all these (often conflicting) requirements.

References

American Association of State Highways and Transportation Officials (1994) *A policy on geometric design of highways and streets*. Washington, DC: AASHTO.

AustRoads (1988, 1993, 1995, 1999) *Guide to traffic engineering practice*. Vermont South: ARRB.

Brown, M. (1995) *The design of roundabouts: A state of the art review*. Crowthorne: TRL.

CROW-werkgroep "verkeerspleinen ii" (1993) "Rotondes" [Roundabouts in The Netherlands], CROW, Ede, publication 79.

Crowne, R.B. (1989) *Interactive roundabout design*. Stafford: Rodel Software Limited and Staffordshire County Council.

Department of Transport (U.K.) (1994–2000) *Design manual for roads and bridges*. London: The Stationery Office.

Institution of Highways and Transportation (1997) *Transport in the urban environment*. London: IHT.

Kay, W.A., S.L. Irani and Katesmark S. (1992) *Advanced roundabout design with Robosign*. London: PTRC.

Kimber, R.M. (1980) "The traffic capacity of roundabouts", Transport and Road Research Laboratory, Crowthorne, Laboratory Report 942.

Ministry of Transport (U.K.) (1966) *Roads in urban areas*. London: HMSO.

Tan, J. (1991) "Entry capacity formula of roundabouts in Switzerland", in: *Proceedings of PTRC European Transport, Highways and Planning 19th Summer Annual Meeting*. London: PTRC.
Troutbeck, R.J. (1990) "Traffic interactions at roundabouts", in: *15th Australian Road Research Board Conference Proceedings*. Vermont South: ARRB.
Webb, P.J. and J.R. Pierce (1990) *ARCADY user guide*. Crowthorne: Transport and Road Research Laboratory.

Chapter 27

HIGH-OCCUPANCY ROUTES AND TRUCK LANES

DAVID PITFIELD and ROBERT WATSON
Loughborough University

1. High-occupancy vehicle lanes

1.1. Definition

The American Public Transportation Association defines high-occupancy vehicle (HOV) facilities as "an exclusive or controlled access right-of-way which is restricted to high occupancy vehicles at all time or for a set period of time." As O'Flaherty (1997) points out, a variety of HOV facilities are in operation throughout the world. They include busways or transitways on separate rights-of-way, exclusive lanes, and priority for HOVs at intersections. In North America, where such facilities are more widespread than elsewhere, HOV facilities, sometimes called commuter lanes, can be open to private car pools, van pools, motorcycles, and public buses. Elsewhere, including Britain, the emphasis is on bus usage and often access is also given to taxis and emergency vehicles and sometimes to cyclists and disabled drivers. Generally, HOV facilities can be categorized as "add-a-lane" or "take-a-lane." In the first case, new infrastructure is provided whereas in the latter case, existing infrastructure is restricted to HOV use.

1.2. Justification

In the absence of total growth in the number of vehicles, miles traveled, and number of trips taken, all HOV facilities (if taken up) serve the purpose of converting some trips from single-occupancy to multiple-occupancy vehicles and provide better facilities to vehicle trips that were already multiple occupancy. This can increase the person movement efficiency of a travel corridor and so will have an impact on congestion in general, on time delays to buses and so their reliability in particular, and on emissions and so the environment. Indeed, HOV facilities are one aspect to be considered for inclusion in any Green Transport Plan (GTP), which were first required of larger employers in southern California because of air

Handbook of Transport Systems and Traffic Control, Edited by K.J. Button and D.A. Hensher

quality regulations, and are now being implemented elsewhere, for example, in the U.K., as a result of the U.K. Government's 1998 White Paper *A New Deal for Transport: Better for Everyone* (Department of the Environment, Transport and the Regions, 1998), in The Netherlands, in Spain, in Germany, in France, and in Belgium. Particular companies promoting ridesharing as part of their GTPs in Europe are BASF in Germany, and Renault, L'Oreal, Charles de Gaulle Airport and Disneyland Paris, in France. The provision of HOV facilities might in itself encourage car sharing.

1.3. Location and trend

Pucher (1995) reports on 1993 data that demand management by HOV facility is practiced in over 20 cities in America, where express lanes for HOVs have been established on about 60 key arterial freeways. In addition a few cities have experimented with selective ramp metering on freeways to give priority to HOVs. This is in part because non-compliance with U.S. Environmental Protection Agency (EPA) standards requires urban areas, through state and local planners, to establish specific plans for discouraging single-occupant private car use and to assess the potential for pollution reduction.

By 1996, 75% of lane miles of HOV facilities in the U.S.A. were in six states: in California, Florida, Virginia, Washington, Texas, and Hawaii. About one-third of existing HOV projects and over one-half of proposed projects were in California.

By 1999, 30 cities had HOV schemes that were open, covering more than 120 interstate and state highways and other key arterials and totaling over 1200 lane miles. About 35% of these schemes are in California.

Schemes under construction or being planned, designed, or proposed would extend the lane miles by about 900 and entail new facilities in 59 cities, of which 15 are in California. Indeed, California has 32% of such new projects (American Public Transit Association, 1999).

This general expansion is against the background of a fall in U.S. car pooling (Ferguson, 1997). At the same time, New Jersey has decommissioned two of its HOV lanes and introduced legislation requiring a study of the effects of HOV lanes on air quality and traffic safety, and this has been followed by the effectiveness of HOV lanes being questioned by other states. Decommissioning has also been sought in New York, Minnesota, and Virginia. Proposed legislation in California requires regular assessment of the effectiveness of HOV lanes and for the conversion to mixed-flow lanes or high-occupancy toll (HOT) lanes if HOV lanes are found not to be effective. Further, there is also proposed legislation in Arizona, California, Georgia, and Minnesota to allow single-occupant alternative-powered low-emission vehicles or vehicles paying a toll to use bus HOV lanes.

Of Canadian cities, only Vancouver has either open or proposed HOV lanes and in the U.K., the Leeds HOV scheme is the first such scheme on an urban road in Europe (Figure 1). This experimental scheme, part of the European Increasing Car Occupancy project (ICARO), became permanent in November 1999. It aims to encourage car sharing and is a combined bus, cycle, and HOV car lane operated at peak times (Leeds City Council, 1999.)

Elsewhere, in many cities and on some arterial routes, the focus is on bus lanes. These are usually for restricted users for certain periods of the day. For example, a controversial bus lane has been opened on part of the U.K. M4 motorway between London Heathrow Airport and central London. This is reserved for buses, coaches, and taxis but has been famously violated by the British Prime Minister, Tony Blair. The U.K. Transport Research Laboratory reports journey time improvement for all traffic types and it is estimated that accident rates will fall by 20%.

1.4. Effectiveness and evaluation

An indication of the worldwide distribution of HOV facilities other than busways is indicated by the preponderance of U.S. empirical literature examining the phenomenon. It was noted above that there are recent moves to assess the effectiveness of HOV lanes. How have they been judged to be effective?

The study comparing HOV and HOT lanes used measures of delay and lane utilization. Studies have cast doubt on the relative effectiveness of HOV lanes by contrast to general-purpose lanes by examining vehicle miles traveled and emissions and so called into question Federal Government policy restricting funding for general-purpose lanes when air quality standards have not been met. Further studies have highlighted violations of minimum passenger occupancy requirements in car pools that result in extra delay, undermine the effectiveness of bypass lanes at metered on-ramps and call into question the integrity of such schemes in the mind of the public. In addition, assessment of safety at 16 HOV projects showed that the introduction of an HOV project tended to increase the facility accident rate based on vehicle-miles of travel but not on person-miles (Federal Highway Administration, 1979). In general, bus accident rates for freeway HOV projects were slightly higher than overall average freeway accident rates, and average bus accident rates for arterial street projects were many times higher than the average bus accident rates for freeway projects. A study in Virginia showed that allowing motorcycles to travel in HOV lanes had no adverse impact on motorcycle safety or on congestion. Surveillance and enforcement issues have also been studied and the use of video tape compared with enforcement officers.

2+
LANE

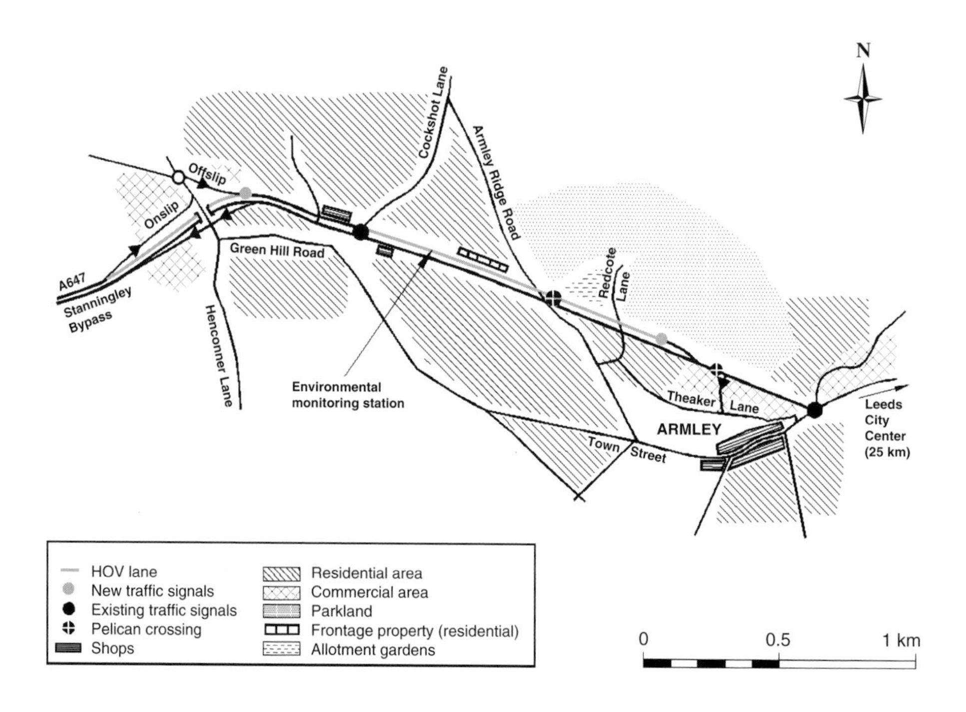
N
Offslip
Onslip
A647
Stanningley
Bypass
Green Hill Road
Henconner Lane
Cockshot Lane
Armley Ridge Road
Redcote
Lane
Environmental
monitoring station
Theaker
Lane
ARMLEY
Town
Street
Leeds
City
Center
(25 km)
HOV lane
New traffic signals
Existing traffic signals
Pelican crossing
Shops
Residential area
Commercial area
Parkland
Frontage property (residential)
Allotment gardens
0
0.5
1 km

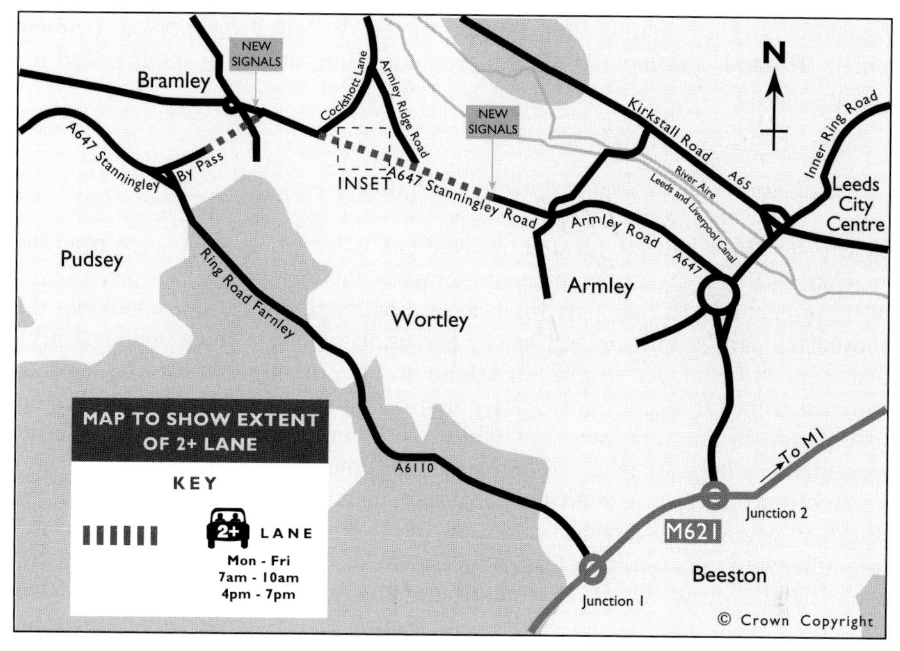

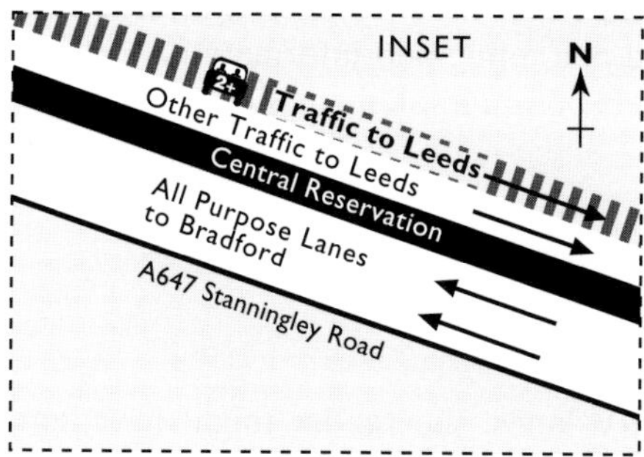

Figure 1. The Leeds 2+ HOV facility.

An evaluation of the Leeds experimental scheme showed that car pools have been formed as a result of the scheme and that car occupancy rates have increased. In addition, HOV lane users are reported as having saved time, as has the bus operator in the morning peak. Violation levels have been slight, and environmentally, noise reductions are noticeable and air quality improvements have resulted from reduced vehicle emissions.

Evaluation methods have been focused on by the U.S. Federal Highways Administration (Federal Highway Administration, 1996). A two-year study was designed to give quick-response procedures to planners and engineers for predicting and evaluating the impact of HOV lanes on person demand, vehicle demand, automobile occupancy, congestion, delay, air quality, and fuel consumption.

1.5. Marketing HOV lanes and the impact on car sharing

Clearly, for HOV lanes to be effective when aimed at car pools it is required that car sharing increases in the presence of such facilities. The European Union DG12 project ICARO on increasing car occupancy across Europe concludes that ridesharing can be encouraged by the provision of HOV lanes, amongst other measures, and evidence from Long Island and Leeds supports this. By contrast, evidence from Orange County shows that only the rate of car pooling among peak period commuters increased, whilst the formidable barriers to ridesharing prevented any increase when commuters at all times were considered.

Marketing efforts have been made in advance of the provision of HOV facilities, especially where these are to be located in areas where the local population is unfamiliar with such facilities, for example in the early 1990s in connection with the Long Island expressway, the first suburban HOV lane in New York State, which opened in 1994. The results of these marketing efforts gained interest among employers in the HOV concept, developed a constituency for HOV facilities on Long Island, and provided a co-ordinated marketing and informational program that could meet the needs of potential users before implementation.

2. Truck lanes

2.1. Definition

With this section following on from HOV lanes, the reader might have some expectation that "truck lanes" would refer to similar exclusive rights-of-way, but with use being restricted to trucks. However, truck lanes of this type are very much the exception. Typically the term "truck lanes" is used to describe the situation where trucks are restricted to a certain lane or lanes, but with other traffic permitted to use these lanes as well as the lanes from which trucks are excluded. Trucks covered by these restrictions are usually delimited by weight, or, more strictly speaking, mass (e.g., >4.5 t or 7.5 t) and, as would be expected, are defined as vehicles designed for the carriage of goods.

2.2. Justification for restrictions

Four reasons are put forward for the use of restrictions.

Improvement of traffic flow

In certain traffic conditions it may well be that the overall average speed of traffic can be improved by restricting trucks from using the "fast" lanes of multilane highways – this is most likely to be the case on rising grades, where overtaking trucks are traveling at a speed that is little higher than the vehicles being overtaken, but considerably slower than passenger vehicles traveling in the same lane. It should always be considered whether increasing average speeds is an adequate measure, as, for instance, it takes no account of the economic consequences of improving passenger vehicle journey time against worsening truck journey time.

Reduced accident rates

Reducing lane sharing between trucks and passenger vehicles and reducing the amount of lane switching by trucks may reduce accident rates and severity. Different fields of vision due to the different size and length of passenger vehicles and trucks means that there is a risk of small passenger vehicles not being seen by truck drivers: this is particularly an issue when trucks switch lanes to overtake. When collisions do take place between trucks and passenger vehicles, the difference in mass and strength of construction inevitably means that the occupants of the passenger vehicles are at considerable risk: statistics show that in serious accidents of this type 98% of fatalities are in the passenger vehicles (Insurance Institute for Highway Safety, 1995).

Public pressure

Even where there is not the empirical evidence to show that truck lanes will make a significant difference to traffic flow or safety, negative public perceptions of the impact of trucks on passenger vehicle speeds or highway safety, perhaps in the latter case through "high-profile" accidents reported in the press, can be used as a argument to introduce restrictive truck lanes.

Pavement wear

Truck lane restrictions have been used on a temporary basis to equalize pavement wear across different lanes of multilane highways. Permanent restrictions from using certain lanes potentially allow different design standards to be used for the lanes with possible savings in construction and maintenance.

2.3. Problems with restrictions

The economic effect on industry of slowing down trucks has typically not been considered, but needs to be.

Perhaps more important are concerns regarding the impact on safety. Two issues need to be taken into account:

(1) Restricting trucks to particular lanes can have the effect of producing a "moving barrier" (that is, a continuous stream of trucks in the "slow" lane) making it difficult and dangerous for passenger vehicles to get across from the fast lanes to exit the highway.
(2) Increasing the average speed of passenger vehicles has the negative effect (from a safety point of view) of increasing the differential between truck and passenger vehicle speeds. Specifically, there is the potential for more serious accidents through misjudgments of speed and fast-moving passenger vehicles can come into contact with slow-moving trucks.

There is also a possible additional complication for highway maintenance: pavement wear will be focused on the slow lanes with these lanes needing more frequent repair.

2.4. Categories of truck lane

The most basic form of truck lane is the "crawler lane," provided on hills to take slow-moving trucks off the main lane and hence enable passenger vehicles to continue to move freely (Figure 2). Similar is the "overtaking lane" or "passing lane" provided to enable faster vehicles to overtake. In both cases there may be no legal restriction on which lane is used by trucks or other vehicles, giving the opportunity to faster-moving trucks as well as other vehicles to overtake slower vehicles. Design considerations include whether vehicles in the "crawler lane" have to merge with overtaking traffic (as shown in the figure) or whether it is more appropriate (in particular safer) for the "crawler lane" to be made the main carriageway, with overtaking traffic having to merge into this lane (Morrall and Hoban, 1985; Mendoza and Mayoral, 1996).

The next category is where trucks are forbidden to use certain lanes on a multilane direction-segregated highway, such as an interstate highway or

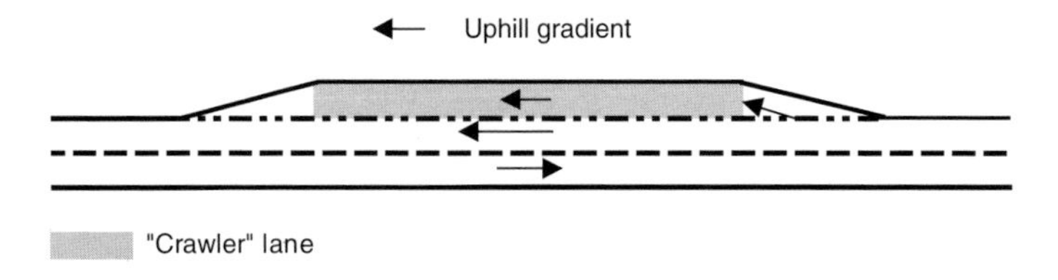

Figure 2. "Crawler lane."

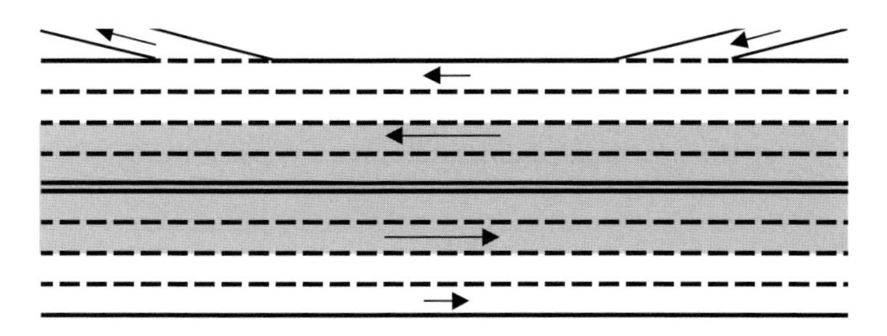

Figure 3. "Fast lane" restrictions – trucks banned from lanes shown shaded.

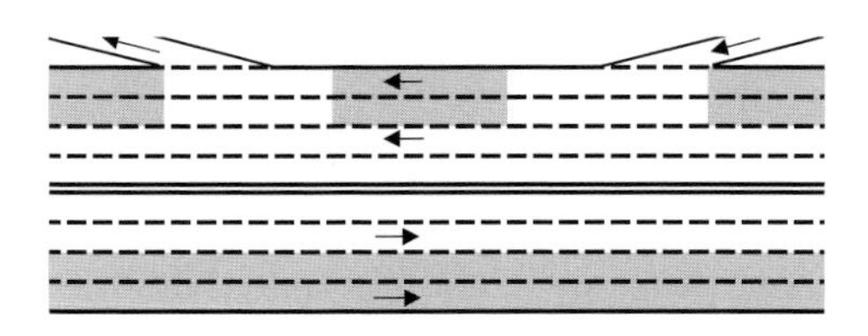

Figure 4. Long-distance trucks and "on–off" passenger vehicles predominate – trucks banned from lanes shown shaded.

motorway. Normally it will be the fast lane(s) that is restricted, enabling passenger vehicles to travel without being slowed down by trucks (Figure 3). The restriction may be for certain congested sections of highway or, as in the U.K., a national ban on trucks using the fast lane of motorways of three lanes or more (in each direction).

Consideration has been given (Hoel and Vidunas, 1997) as to which lanes are most appropriate for restriction. Particularly where truck traffic is long distance and passenger vehicles are using a highway for short "on–off" journeys, there is some attraction in restricting trucks to using lanes away from the exit/entry junctions (Figure 4); however, the safety implications of slow-moving trucks that do want to exit or join crossing lanes of higher-speed passenger vehicles need careful consideration. The position is also more complicated if exit/entry junctions are not always on one side of the highway – here conflicts between vehicles traveling at different speeds are inevitable and the introduction of truck lanes is most problematic.

Where volumes of traffic are so great as to justify additional construction and the land is available to introduce further lanes, it may be possible to achieve genuine separation of trucks and passenger vehicles, with lanes separated by barriers (Figure 5) or even a separate roadway on a different alignment. Design issues such as whether to put the truck-only lanes in the center or at the edge of the existing highway will need to be considered on the basis of local location-specific

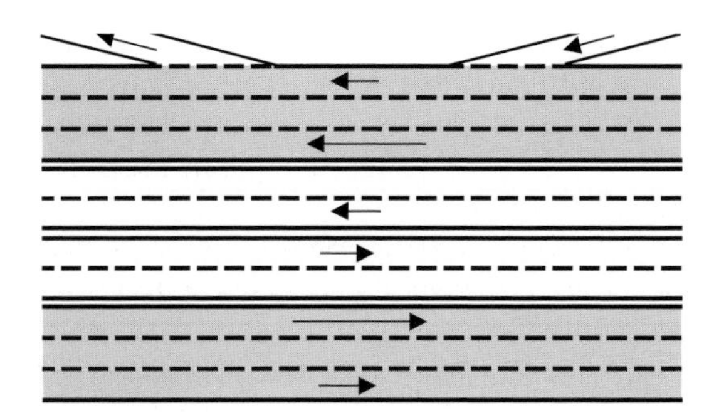

Figure 5. Long-distance trucks and "on–off" passenger vehicles predominate – trucks have separated lanes; banned from lanes.

volumes. The geometrics of the existing highway, availability of additional land, and grade separation of access and egress to the separate lanes are important considerations (see American Association of State Highway and Transportation Officials, 1994 for general guidelines on design).

2.5. *Evaluation methods and effectiveness*

Any scheme to introduce truck lanes should be assessed using a cost–benefit approach. Important costs to include in the assessment will be land purchase, the capital cost of physical work, and any additional maintenance that will be incurred. Benefits will be the travel time savings for the occupants of passenger vehicles and reduced accidents (e.g., though complete segregation). Potential negative benefits will be the economic effect of reduced average truck speed and any increases in accidents (e.g., through higher speed differentials).

The Federal Highways Agency in the U.S.A. has developed a software package "Exclusive Vehicle Facilities" (EVFS) which enables a range of options to be assessed from an economic viewpoint, producing net present values (NPVs) and a cost–benefit analysis. EVFS has been assessed by Hoel and Vidunas (1996), who conclude that the software is very valuable as a tool to make an initial assessment, but limitations in the range of options it can consider mean that some additional calculation may be required.

Truck lanes can be expected to achieve a positive NPV where congestion is evident, the highway passes through hilly terrain, and a significant proportion of the traffic is trucks. As traffic volumes grow further, additional restrictions are likely to be appropriate.

Short "crawler" or "overtaking" lanes are cost-effective ways of improving average journey speeds, especially on two-lane, two-direction highways; see, for example, Harwood et al. (1988).

Truck lane restrictions on interstate highways and motorways tend to reduce average truck speeds and hence have negative as well as positive effects. They must therefore be considered on a case-by-case basis. Even where the evidence of economic and safety benefits is not significant, public pressure is likely to lead to further restrictions being introduced.

Total separation of trucks requires significant investment, particularly at junctions, and, unless traffic volumes are very high, a positive net benefit may be hard to prove.

Finally, there may be an opportunity for synergy between HOV and truck lanes, with trucks being allowed to use HOV lanes, recognizing the economic benefits of efficient freight transport and also, pragmatically, using up some of the spare capacity often found in HOV lanes (Carson et al., 1996).

References

American Association of State Highway and Transportation Officials (1994) *A policy on geometric design of highways and streets*. Washington, DC: AASHTO.

American Public Transportation Association (1999) *1999 transit fixed guideway inventory*. Washington, DC: American Public Transit Authority.

Carson, J., F.L. Mannering, D. Nam and A. Trowbridge (1996) *The potential for freight productivity improvements along urban corridors*. Washington, DC: Federal Highways Administration.

Department of the Environment, Transport and the Regions (1998) *A new deal for transport: Better for everyone*. London: HMSO.

Federal Highway Administration (1979) "Safety evaluation of priority techniques for high occupancy vehicles", FHWA, Washington, DC: Report FHWA-RD-79-59, http://www.itsdocs.fhwa.dot.gov/jpodocs/repts_te/2yd01!.pdf.

Federal Highway Administration (1996) "Predicting high occupancy vehicle lane demand", FHWA, Washington, DC: Report FHWA-SA-96-073, http://www.itsdocs.fhwa.dot.gov/jpodocs/repts_te/2y701!.pdf.

Ferguson, E.F. (1997) "The rise and fall of the American carpool: 1970–1990", *Transportation*, 24:349–376.

Harwood, D.W., C.J. Hoban and D.L. Warren (1988) "Effective use of passing lanes on two-lane highways", *Transportation Research Record*, 1195:79–91.

Hoel, L.A. and J.E. Vidunas (1997) "Exclusive lanes for trucks and passenger vehicles on interstate highways in Virginia: An economic evaluation", Virginia Transportation Research Council, Charlottesville, VA.

Insurance Institute for Highway Safety (1995) "Large trucks", Insurance Institute for Highway Safety, Arlington, VA.

Leeds City Council (1999) "HOV lane information sheet", issue 5, http://www.leeds.gov.uk/lcc/highways/various/twop_inf.html.

Mendoza, A. and E. Mayoral (1996) "Design guidelines for truck lanes on Mexican two-lane roads", *Transportation Research Record*, 1523:91–98.

Morrall, J.F. and C.J. Hoban (1985) "Design guidelines for overtaking lanes", *Traffic Engineering and Control*, 26:476–484.

O'Flaherty, C.A., ed. (1997) *Transport planning and traffic engineering*. London: Arnold.

Pucher, J. (1995) "Urban passenger transport in the United States and Europe: A comparative analysis of public policies. Part 1. Travel behavior, urban development and automobile use", *Transport Reviews*, 15:99–117.

Chapter 28

URBAN TRAFFIC FLOW

ANTHONY D. MAY
University of Leeds

1. Introduction

Road space is a scarce resource, and traffic engineers and planners need to ensure that roads are able to accommodate as much traffic as possible, subject to safety and environmental constraints. In other words, we need to maximize the capacity of the road. But how is capacity measured, and what influences it? These questions lie at the core of traffic flow theory, a science which has attracted both theoretical and empirical analysis. In this chapter there is space only to introduce the concepts, the principal parameters, and some relationships which enable us to estimate capacity. We do this initially for an individual length of road, and subsequently for road networks. Further developments of these concepts are presented in a useful text by Gerlough and Huber (1975). Much of what follows is taken from an earlier textbook, which expands on the engineering implications (O'Flaherty, 1996). We acknowledge the publisher's permission to do this.

2. The principal parameters

2.1. The range of conditions

Every driver will have experienced the range of traffic conditions which can be experienced on the same length of road, and which are illustrated in Figure 1. When traffic is light, the road is relatively empty (Figure 1a) and drivers are free to choose their own speeds. As traffic increases (Figures 1b–d) drivers are more constrained by other vehicles, and less able to overtake; they are thus less able to choose their own speeds, and average speeds fall. As traffic levels increase further, traffic forms into platoons of slow-moving vehicles (Figure 1e), which may stop and start. Finally (Figure 1f), traffic levels become so great that queues form, and traffic may be at a standstill for considerable periods. These six conditions are

Handbook of Transport Systems and Traffic Control, Edited by K.J. Button and D.A. Hensher

Figure 1. Levels of service.

referred to in the U.S. *Highway Capacity Manual* (Department of Transportation, 2000) as levels of service A to F.

Which of them, though, represents the capacity of the road? Clearly Figure 1f has the most vehicles in it, but is not a desirable state of affairs.

2.2. Measures of quantity

There are in practice two ways in which the number of vehicles can be counted on a road. One is illustrated by Figure 1. One can photograph a length of road x, count the number of vehicles n_x in one lane of the road at a point in time, and derive a rate per unit distance. This measure is called the concentration of traffic (sometimes referred to as density) and is denoted by the parameter k (veh/m). Thus

$$k = n_x/x. \tag{1}$$

The second approach is to stand at the side of the road for a period of time t, and the number of vehicles n_t passing that point in one lane in that period, and derive a rate per unit time. This measure is called the flow of traffic (sometimes referred to as *volume*) and is denoted by the parameter q (veh/s). Thus

$$q = n_t/x. \tag{2}$$

Since one is usually concerned to get as many vehicles through a road as possible in a given time, it is q rather than k which one wishes to maximize, and capacity is then described as the maximum value of q, q_{m}. However, it is not immediately clear which of Figures 1a–f represents this condition.

2.3. Measures of quality

While traffic engineers and planners are concerned to achieve flows approaching capacity, at least at busy times, individual drivers will be most concerned about the quality of their journeys. Their main concern will be with the speed at which they can travel. Individual speeds (in m/s or km/h) are easy enough to measure, but traffic engineers and planners will be more concerned with average speeds. Here again there are two ways of measuring speed, and the distinction between them often causes confusion. The difference between the two methods can be illustrated by a simple example.

Three vehicles are recorded as taking 6 s, 8 s and 10 s to cover 100 m. Method 1 would calculate the average time, 8 s, and divide this into the distance. This gives an average speed of 100/8 or 12.5 m/s. Method 2 would calculate the individual speeds, 16.7, 12.5 and 10.0 m/s, and average them, as 39.2/3 or 13.1 m/s. Both are correct; they simply represent different ways of averaging. Clearly it is important to be consistent in the averaging method which is used. The two speeds can be related to the two different measures of quantity and separation. For example, suppose that a stream of traffic includes vehicles traveling at n different speeds $u_1, \ldots, u_n$ and that two different approaches are adopted to determine the overall average speed.

The first involves photographing a length of road on which there are k vehicles per unit length. Let the concentration of set i of vehicles with speed u_i be defined as k_i. The average speed of the k vehicles is referred to as the space mean speed, $\bar{u}_s$, since the average is of all vehicles in a given space (at one point in time). Then $\bar{u}_s$ is given by

$$\bar{u}_s = \sum_1^n k_i u_i / k. \tag{3}$$

The second involves recording all vehicles passing a point in a given time, i.e., q per unit time. Let the flow of set i of vehicles with speed u_i be defined as q_i. The average speed of the q vehicles is referred to as the *time mean speed*, $\bar{u}_t$, since the average is of all vehicles passing in a given time. Then? is given by

$$\bar{u}_t = \sum_1^n q_i u_i / q. \tag{4}$$

However, only one of these two speed averages is related to the parameters q and k.

3. The fundamental relationship

3.1. The relationship between concentration, flow, and speed

For any given stable traffic condition, the three parameters k, q, and u are directly related. This can be seen from the simple example in Figure 2, which considers a kilometer of road, on which all vehicles are traveling at the same speed. The parameters K, Q, and U are concentration in veh/km, flow in veh/h, and speed in km/h, respectively. By definition there are K vehicles, each with speed U, in the kilometer of road at any instant. If the flow is recorded at the end of the road, Q vehicles will pass per hour. The vehicle at the start of the kilometer will take $1/U$ hours to reach the end at a speed of U km/h. It will then be the Kth vehicle to pass the end of the road, and this will happen in K/Q hours. Thus

$$1/U = K/Q$$

or

$$Q = KU. \tag{5}$$

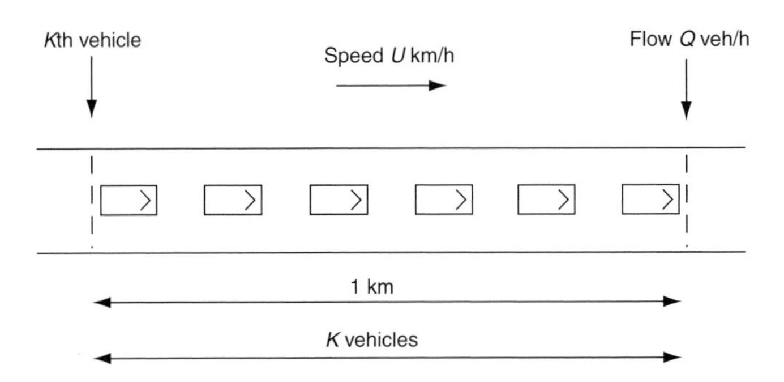

Figure 2. A simple derivation of the relationship between concentration, flow, and speed.

This is dimensionally correct, viz., veh/h = (veh/km)(km/h).

The same expression, using meters and seconds, is

$$q = ku. \tag{6}$$

3.2. *The shape of the fundamental relationship*

Whilst eq. (6) is important, it is still necessary to know the values of two of the parameters in order to calculate the third. In practice, on a given road, a given concentration is likely to give rise to a certain value of flow and a certain value of speed, subject always to the variations in driving conditions and driver behavior. This relationship can best be seen by considering the three parameters a pair at a time, as in Figure 3. In each of them there are two limiting conditions, which can be thought of by reference to Figures 1a and f.

In Figure 1a flow is very low (approaching zero), and so is concentration; the speed, which is likely to be at its highest, is referred to as the free flow speed u_f. In Figure 1f, the speed is zero and so, since the traffic is not moving, is the flow; however, the concentration is at its highest, and is referred to as the jam concentration k_j.

Figures 3a and 3b illustrate that as speed falls from u_f to zero, and as concentration rises from zero to k_j, flow first increases and later falls again to zero. It seems reasonable to expect that it only rises to one maximum, and this can be thought of as the capacity q_m. As flow increases towards capacity it is relatively easy to understand what is happening; as Figures 1a–1e show, vehicles increasingly disrupt one another, reducing the individual driver's ability to choose his/her own speed or overtake others. Speed thus falls and concentration increases. Beyond capacity it is less easy to explain what is happening. In practice such conditions are caused by queues from downstream conditions, perhaps a

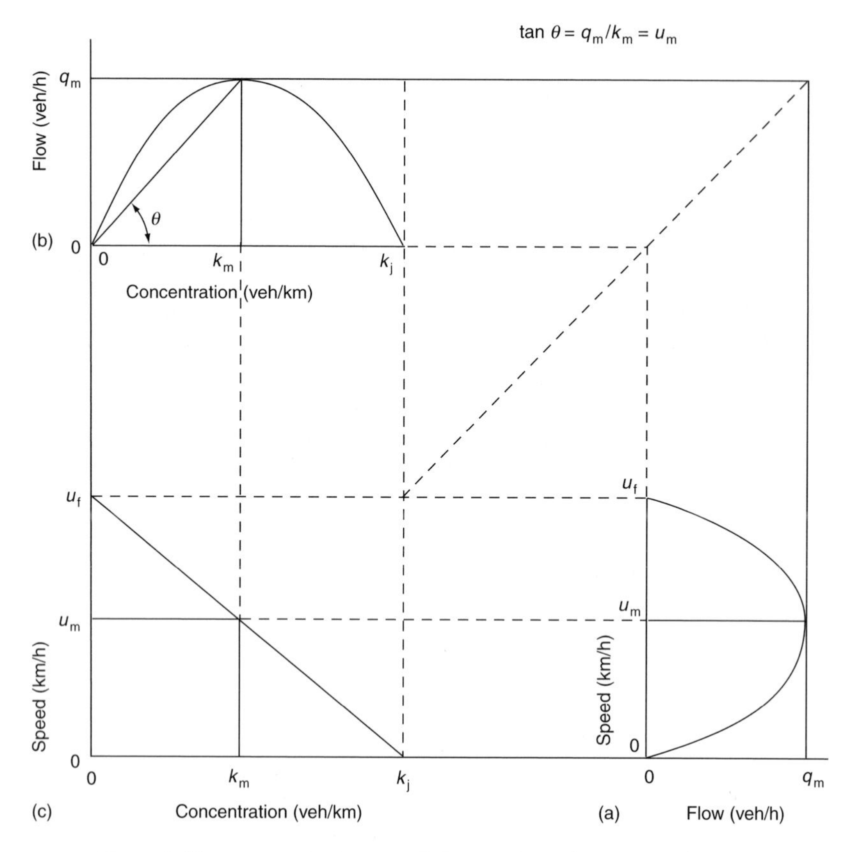

Figure 3. Flow–concentration, speed–flow, and speed–concentration curves.

junction, or an accident, or even a gradient or tight curve whose capacity is slightly lower. The queue leads to increased concentration; speeds fall further, and vehicles which cannot flow past the point join the queue as it stretches upstream.

3.3. Empirical relationships

Several analysts have attempted to fit relationships to observed data. This is most easily done with concentration and speed (Figure 3c), since the relationship is monotonic. However, care needs to be taken in manipulating data to fit such a relationship (Duncan, 1979). The most common relationship, shown in Figure 3c, is a linear relationship between speed and concentration, and this was first suggested by Greenshields (1934):

$$u = a + bk. \tag{7}$$

Using the two limiting conditions, it can be shown that this becomes

$$u = u_f(1 - k/k_j). \tag{8}$$

Other analysts have noted from empirical data that the relationship is not quite linear but slightly concave. They have suggested logarithmic and exponential alternatives (Greenberg, 1959; Underwood, 1961). Each of these relationships has been supported by theoretical analysis, primarily based on an analogy with fluid flow (Lighthill and Whitham, 1955) and the interpretation of car-following behavior (Herman et al., 1959). However, it is important to stress that traffic flow is not a wholly scientific phenomenon, but one which depends on the vagaries of driver behavior. Indeed, some authors have suggested that there is no reason why conditions above and below capacity should be part of the same relationship, since they arise in different ways (Edie, 1961; Hall and Montgomery, 1993).

3.4. *Estimation of capacity*

Subject to these caveats, it is possible to use eq. (8) to provide a theoretical estimate of the capacity of a length of road from a pair of observed data points. Figure 3b shows the relationship between flow and concentration which, combining eqs. (6) and (8), is

$$q = ku_f(1 - k/k_j). \tag{9}$$

The maximum value of q can be determined by differentiation:

$$\frac{dq}{dk} = u_f - 2u_f k/k_j, \tag{10}$$

which, when equated to zero, gives

$$k_m = k_j/2 \tag{11}$$

and, from eqs. (8) and (9),

$$u_m = u_f/2 \tag{12}$$

$$q_m = k_j u_f/4. \tag{13}$$

3.5. *The appropriate value of average speed*

Before applying this approach to estimating capacity, however, one needs to know which average of speed to use. This can be seen immediately by reference to eqs.

(3) and (4), in conjunction with eq. (6). Equation (3) can be further simplified to give

$$\bar{u}_s = \sum_{1}^{n} q_1/k = q/k. \tag{14}$$

Equation (4) cannot be further simplified in this way. Thus it is space mean speed which is related to flow and concentration in the fundamental relationship.

3.6. Practical capacity

Care is needed in using these empirical relationships to determine capacity in this way, since there are no scientific laws explaining the behavior of traffic. Moreover, in practice vehicles are not homogeneous, and practical analysis uses the concept of passenger car units (pcus) to describe the effect of vehicles of different sizes. Typical values of capacity (the highest flows recorded on a section of road) are in the range 1500–2500 pcu/lane h, and for jam concentration 150–200 pcu/lane km. Free-flow speeds are inevitably more variable, and depend on drivers' perceptions of the safety and acceptable maximum speed on the road. It is thus unwise to design for the capacity as calculated in eq. (13). British practice advocates design for maximum flows around 15% below the theoretical capacity, referring to this as practical capacity. U.S. practice uses the six levels of service shown in Figure 1, and recommends design for peak hour flows not exceeding level of service C on major interurban routes, and level of service D in urban areas.

3.7. Factors affecting link capacity

The capacities suggested above may apply to urban arterials with no frontage activity, but on most urban roads link capacities will be much lower than this. Many urban roads are not purpose-built, and have substandard widths or alignments. Most will operate two-way, and friction between the opposing streams will reduce capacity. Urban road space is often used for parking, and this will significantly reduce capacity both because it takes up space and because vehicles parking disrupt other traffic. Pedestrian activity will also reduce capacity, whether it is informal or managed through protected crossings, and there is some evidence that the type of frontage land use itself influences capacity. Finally, several traffic management measures are specifically designed to reallocate capacity to other users, including bus and cycle priorities, and traffic calming measures (which implicitly reallocate capacity to residents and pedestrians) (see Chapters 19 and 20).

More importantly, in urban networks, the link capacity will rarely be the determining factor. It is usually the downstream junction which determines the rate at which traffic can pass through an individual link, and the junction's capacity will often, in turn, depend on the other traffic flows entering it. This requires us to analyze the network as a whole, rather than simply its constituent links.

4. The characteristics of urban networks

4.1. Links and networks

So far we have considered an individual length of road, or link, operating on its own, and have implicitly assumed that its performance is not affected by the ability of traffic to leave the link (at the downstream end) or by queuing to enter the link (at the upstream end). In practice, in urban networks at least, life is not as simple as this.

The downstream end of any link will be a junction, which will largely determine its capacity. The principal types of junction control are

(1) priority junctions where the link has priority, which impose little restriction on capacity;
(2) priority junctions where traffic leaving the link has to give way to other traffic, for which capacity is critically dependent on other traffic;
(3) signaled junctions where the capacity is largely determined by the signal control; and
(4) roundabouts, where capacity is determined by the turning movements in the junction.

All of these are described more fully in Chapter 26.

At the upstream end, there will also be a junction, and its performance may be affected by the link. If the demand to enter the link exceeds the downstream junction capacity, a queue will form and, in the extreme, will spread into the upstream junction, reducing its capacity. In such conditions control of the upstream junction needs to focus on reducing the adverse impact on queues (Quinn, 1992). Conversely, it is possible that other traffic movements will have restricted the performance of the upstream junction, resulting in a flow much lower than capacity, and relatively high speeds.

All of these considerations make the analysis of urban networks far more complicated than that for individual links, and there have been far fewer attempts to consider them. As a generality, performance will depend on

(1) the shape of the network;
(2) the type of traffic control;

(3) the controls on individual links;
(4) the pattern of journeys through the network (through versus terminating; short versus long; radial versus orbital); and
(5) the overall level of demand.

4.2. Appropriate parameters

A first step in analyzing urban networks is to define the network equivalents of the link parameters flow (q), concentration (k), and space mean speeds ($\bar{u}_s$). While flow can still be measured as trips made in the network, some make longer journeys than others. A more appropriate parameter therefore is veh-km/h. Concentration can still be measured as vehicles on the network at any time. Speed needs to be averaged across the network, and this is most readily done by dividing the total travel in a network in a given time (veh-km/h) by the total time spent by vehicles during that period (veh-h/h).

However, an important distinction needs to be drawn between the "period" used to measure these parameters to determine the performance of the network, and that which is appropriate for measuring the costs experienced by drivers. The former can simply be based on observations during a specified period (say between 0830 and 0845) during which all travel distance and all travel time is recorded. Each vehicle is observed from its location at 0830, or the time and point at which it enters the network, to the time and point at which it leaves, or its location at 0845. This gives measures of veh-km performed during this period. A rectangular time–space domain, covering every activity throughout the network in the given time period, is the basis for the assessment.

However, individual drivers will experience conditions over shorter or longer periods than this. Short, uncongested journeys may be completed within 5 minutes; longer journeys in congestion could well take 30 minutes. At higher levels of congestion drivers may have to queue to enter the network. All of these need to be taken into account in determining the cost to users of a given network in a given state. Strictly, this means that each driver needs to be tracked from the time he or she enters (or starts queuing to enter) the network, thus giving measures of veh-km supplied, and veh-h supplied, for a given group of drivers starting in a given time period. The time–space domain for measurement is thus much less precise. A fuller specification of these distinctions can be found in May et al. (2000).

4.3. Performance and supply curves

Given the complexity of measuring either performance or supply, it is not surprising that there are few surveys of actual network conditions. The best

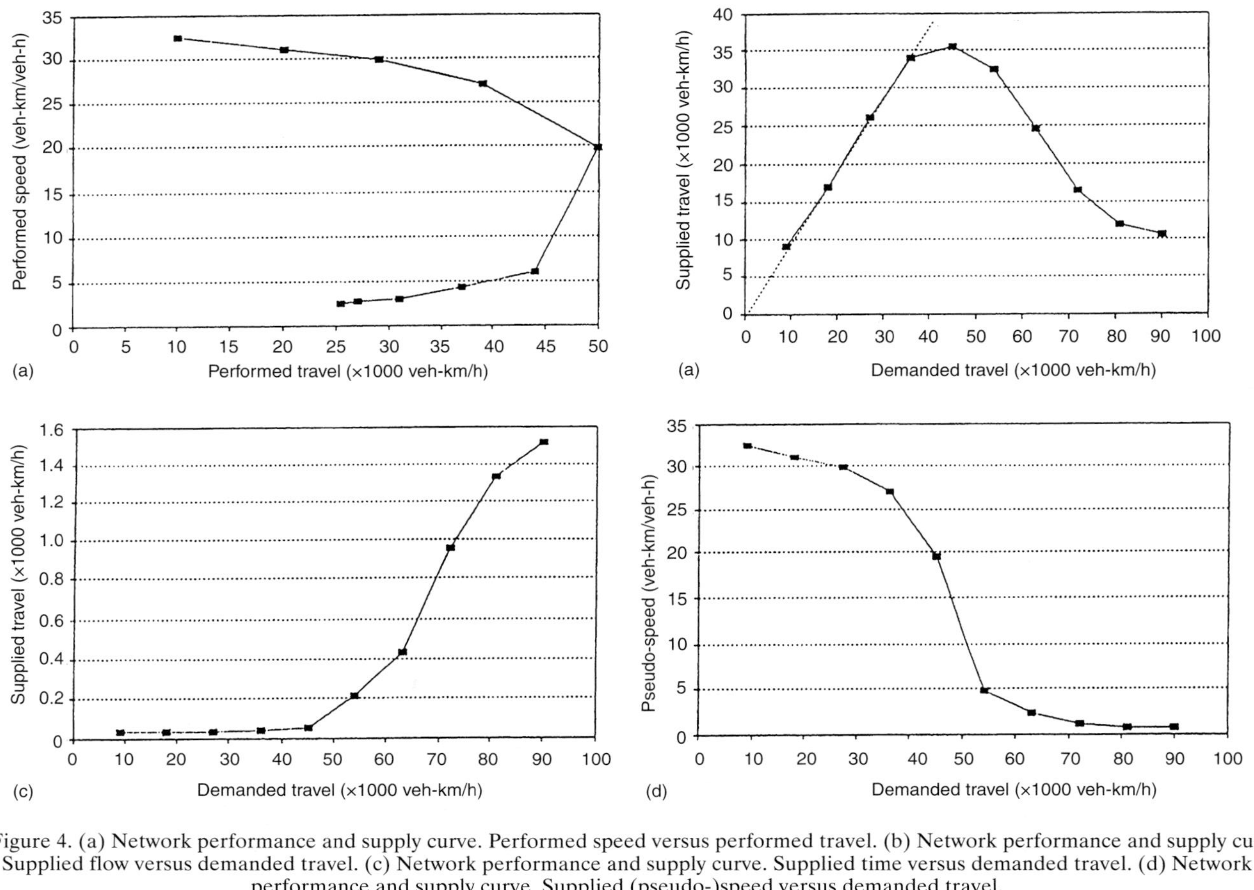

Figure 4. (a) Network performance and supply curve. Performed speed versus performed travel. (b) Network performance and supply curve. Supplied flow versus demanded travel. (c) Network performance and supply curve. Supplied time versus demanded travel. (d) Network performance and supply curve. Supplied (pseudo-)speed versus demanded travel.

documents are early work by Wardrop (1968) and subsequent studies by Harrison et al. (1986) and Ardekani and Herman (1987). However, all of these generate performance curves, which are of engineering interest, rather than the supply curves which are needed to understand the interaction between demand and supply.

May et al. (2000) have carried out an analysis by microsimulation of a grid and a ring radial network, taking a given matrix of journeys, which was factored to represent increasing demand. An example of their results for a 6 × 6 grid network is given in Figure 4. They show that

(1) speed–flow curves similar to Figure 3 apply on networks (Figure 4a); these describe the performance of the network;
(2) beyond "capacity," the travel which actually takes place may be reduced as demand increases further, because of blockages in the network (Figure 4b);
(3) the time costs of network use rise rapidly at higher levels of demand, largely as a result of longer queues in and on the approaches to the network (Figure 4c);
(4) as a result equivalent speeds ("pseudo-speeds" including the effect of queuing) fall rapidly at higher levels of demand (Figure 4d).

The same study also found that the shapes of these relationships were significantly affected by the shape of the network, the pattern of demand over the network, and the distribution of demand over time.

5. Summary

Any individual road will have a theoretical capacity, measured as the maximum number of vehicles which can pass through it per hour. As capacity is approached, speeds fall. Additional demand may then lead to increased queues and lower speeds. It is possible to develop relationships between flow, concentration (in vehicles per kilometer), and speed. However, these relationships are dependent on variations in driver behavior. In practice the capacity of urban roads will also be affected by road width and alignment, parking, and frontage activities, as well as by measures taken to manage and allocate capacity.

The performance of urban networks is more complex to analyze, and depends critically on the configuration of the network, the type of junction control, and the pattern of demand. Different parameters from those used to analyze individual roads are needed. There has been much less analysis of performance of urban networks, but it appears that relationships similar in form to those for individual roads may apply. However, these do not reflect the costs to individual users, for whom queuing and resulting travel times can increase rapidly at higher levels of demand.

References

Ardekani, S. and R. Herman (1987) "Urban network-wide traffic variables and their relations", *Transportation Science*, 21:1–16.

Department of Transportation (2000) *Highway capacity manual*. Washington, DC: U.S. DOT.

Duncan, N.C. (1979) "A further look at speed/flow/concentration", *Traffic Engineering and Control*, 20(10):482–483.

Edie, L.C. (1961) "Car following and steady state theory for non-congested traffic", *Operations Research*, 9:66–76.

Gerlough, D.L. and M.J. Huber (1975) "*Traffic flow theory: A monograph*", Transportation Research Board, Washington, DC, Special Report 165.

Greenberg, H. (1959) "Analysis of traffic flow", *Operations Research*, 7(1):79–85.

Greenshields, B.D. (1934) "A study of traffic capacity", in: *Proceedings of the 14th Annual Meeting of the Highway Research Board*. Washington, DC: Highway Research Board.

Hall, F.L. and F.O. Montgomery (1993) "The investigation of an alternative interpretation of the speed-flow relationship on U.K. motorways", *Traffic Engineering and Control*, 34(9):420–425.

Harrison, W.J., C. Pell, P.M. Jones and H. Ashton (1986) "Some advances in model design developed for the practical assessment of road pricing in Hong Kong", *Transportation Research*, 20A(2):135–144.

Herman, R., E.W. Montroll, R.O. Potts and R.W. Rottery (1959) "Traffic dynamics: Studies in car following", *Operations Research*, 7(1):86–106.

Lighthill, M.J. and G.B. Whitham (1955) On kinematic waves II: A theory of traffic flow on long crowded roads", *Proceedings of the Royal Society*, A229(1178):317–345.

May, A.D., S.P. Shepherd and J.J. Bates (2000) "Supply curves for urban road networks", *Journal of Transport Economics and Policy*, 34(3):103–345.

O'Flaherty, C.A., ed. (1996) *Transport Planning and Traffic Engineering*. London: Arnold.

Quinn, D.J. (1992) "A review of queue management strategies", *Traffic Engineering and Control*, 33(11).

Underwood, R.T. (1961) "Speed, volume and density relationships", in: B.D. Greenshields, H.P. George, N.S. Guerin, M.R. Palmer and R.T. Underwoods, eds., *Quality and theory of traffic flow*. New Haven, CT: Yale University Press.

Wardrop, J.G. (1968) "Journey speed and flow in central urban areas", *Traffic Engineering and Control*, 9(11):528–532.

Chapter 29

TRAFFIC FLOW CONTROL

LAURIE SCHINTLER
George Mason University, Fairfax, VA

1. Introduction

Traffic flow control manages the movement of people and goods at specific locations on roads, airways, and other modes of transportation using traffic control devices or measures. Often, the objective of traffic control is to expedite traffic, or to increase the capacity of a transportation facility; however, in some cases it is intended to impede or slow down traffic, at the entry point to a neighborhood for example. While traffic control may be as simple as placing a STOP sign at an intersection, it can also be a complex process, requiring the implementation of a range of traffic control devices and continual surveillance, communication, and prediction. In some countries like the U.S.A., Canada, and Japan, intelligent transportation systems (ITS), or the application of advanced technologies to transportation, are improving the ease with which these functions can be carried out and enhancing the effectiveness and efficiency of traffic control measures.

2. Surveillance and communication

The determination of traffic control measures requires some understanding of traffic conditions both before and after a strategy is implemented. Traffic engineers and planners typically monitor a range of road operating conditions, including the speed and travel time of vehicles, density of traffic, and delay. Volume-to-capacity (v/c) ratios are commonly used in the U.S.A. to measure conditions on road segments, and also at intersections. This ratio is essentially an indicator of capacity sufficiency, where a v/c ratio of 1 corresponds to that point where the road can not sustain any additional traffic. A ratio of 0.8 on a road segment, for example, indicates that the road can accommodate 20% more traffic before reaching traffic jam conditions. Based on these ratios, density and other parameters, six levels of service, A through F, are defined for traffic flow analysis.

Handbook of Transport Systems and Traffic Control, Edited by K.J. Button and D.A. Hensher

At the extremes, level of service A corresponds to free-flow speed and travel time and F relates to that point where jam density occurs.

There are several factors that can influence the operating conditions of a road or intersection. Aside from traffic control devices and measures, these include geometric conditions (e.g., alignment, speeds, lane width), traffic conditions (e.g., composition of traffic), and recurring and non-recurring events. Daily rush hour or periodic planned events like a concert are examples of recurring congestion, while random or irregular events such as incidents are considered to be non-recurring (McShane and Roess, 1990). In the U.S.A., almost 60% of freeway congestion can be attributed to occurrences of the latter type (Meyer, 1997).

Traffic conditions are monitored through a variety of mechanisms. Detectors wired by phone lines or computerized connections are commonly used to monitor traffic-volume characteristics at certain intersection approaches or freeway locations. There are three types of detectors: pressure-plate, magnetic-loop, and sonic. Only pressure-plate detectors can observe vehicles by axle; however, they have the disadvantage of being most vulnerable to deterioration in inclement weather (McShane and Roess, 1990). Closed-circuit video and aerial helicopters are also used to monitor traffic conditions and to detect incidents as they occur on freeways. A relatively new means for observing traffic, and in particular estimating vehicular speeds and point-to-point travel times, is to designate certain vehicles as probes using GPS satellite technology and transponders, or cellular technology. Use of this technique is still limited, though, due to privacy and technical issues.

Information collected via sensors, cameras, helicopters, and probe vehicles is ideally transmitted back to a transportation agency or traffic control center that has the authority to implement appropriate control strategies as needed. There are a number of technical and institutional issues related to communications. How much information should be transmitted? How often should the information be updated? How much information can be transmitted by the existing system? Who should have access to the information? Should the information be communicated back to travelers and if so, how and to what extent?

3. Prediction

Traffic control often requires forecasting, or an estimate of traffic parameters given some control strategy, to predict the performance of a traffic control system. There are three classes of models: microscopic, macroscopic, and mesoscopic. Examples of microscopic models are car-following theories, time–space diagrams, stochastic queuing analysis, and microsimulation. Macroscopic models include capacity analyses, speed–flow–density relationships, shock wave analysis, deterministic queuing models, and macroscopic simulation (May, 1990). Meso- scopic analysis combines microscopic and macroscopic elements.

Queuing models are used to estimate delay at a bottleneck given some level of capacity and the arrival and departure pattern at the facility. Computer simulation is "a numerical technique for conducting experiments on a digital computer, which may include stochastic characteristics, be they microscopic or macroscopic in nature, and involve mathematical models that describe the behavior of a transportation system over extended periods of real time" (May, 1990). Microsimulation models are based on theories of how vehicles maneuver through traffic. Car-following theories are commonly employed and they describe how one vehicle follows another. Development of a simulation model entails a number of steps, including the determination of model parameter settings (i.e., calibration) and the comparison of model output with actual observed conditions (i.e., validation). To date, the application of simulation models has been limited primarily to sections of roads, rather than to networks. This is changing, however, as data collection mechanisms and computing technologies improve.

4. Traffic control

Strategies vary in terms of the degree of traffic control they provide, the costs of implementation, the circumstances under which they should be applied, technological requirements, and institutional considerations (e.g., degree of agency or jurisdictional co-ordination required, privacy and liability concerns.) Many countries also have guidelines for the use of traffic control devices, and in particular road markings, traffic signs, and intersection signals (e.g., both Australia and the U.S.A. have a *Manual of Uniform Traffic Control Devices*). The principal purpose of these standards is to establish consistency and familiarity for the motorist and pedestrian.

4.1. Road markings and traffic signs

Road markings and traffic signs are common in many countries, and include those designed to regulate existing traffic laws, provide warnings to travelers about hazardous conditions, and offer guidance to motorists (e.g., directions or location of services). Although there are some signs and markings that are relatively common between nations, some differences do exist. In contrast to the U.S.A., Europe does not use yellow center and edgeline markings and in London, a yellow crosshatch at intersections is used to indicate where vehicles are not permitted to queue (Federal Highway Administration, 1998).

While traffic signs and road markings have tended to be a "low-technology" medium for control, this is changing. For example, some countries in Europe use variable speed limit signs, which display speeds that are dynamically adjusted based on traffic conditions (Federal Highway Administration, 1998). Australia has experimented with the concept of a variable message sign that displays the observed speed and license plate number of motorists exceeding the speed limit. The sign was tested at a selection of construction sites and after two weeks appeared to have a positive impact on speed enforcement. The percentage of speed violators decreased for both the weekday and the weekend (Ogden and Taylor, 1999). The city of Melbourne in Australia uses data collected via loop detectors to predict travel times, which are then displayed on a sign color-coded by congestion level (Ogden and Taylor, 1999).

4.2. Intersection signalization and co-ordination

Traffic signals can be an effective tool for mitigating congestion on surface streets. Many countries have guidelines for the use of traffic signals. In Australia, an intersection generally may not be signalized unless it has a minimum of at least 600 veh/h over any four hours of an average day on the major road and 200 veh/h or high pedestrian traffic on the minor road. Some exceptions to this rule are if the intersection has a high accident rate or it is part of a co-ordinated signal system (Ogden and Taylor, 1999).

In many countries, there are three modes of operation for traffic signalization: pretimed, semiactuated, and fully actuated. With pretimed signalization, the cycle length, phases, green times, and change intervals are predetermined. Different signal programs can be defined by day of the week or time of day based on anticipated traffic demand at the intersection. In fully actuated control, the signalization parameters change dynamically as traffic conditions at the intersection change. Sensors strategically placed at each intersection approach are required to monitor traffic demand. Semi actuated control is when only part of an intersection has actuated control (Transportation Research Board, 1995). This is appropriate when one street in the intersection bears the majority of the traffic through the intersection. An example of this situation would be an intersection located along a major arterial providing access to a residential area or commercial parking lot.

Signalization can be done locally (i.e., at only one or a few intersections) or in co-ordination with other intersections in the area. If implemented properly, signal co-ordination can significantly smooth traffic flow through an area, keep vehicles at a constant and ideal speed, encourage the formation of platoons which reduce headways between vehicles, and result in more efficient use of existing capacity. Effective co-ordination requires surveillance technologies (e.g., sensors) to monitor traffic conditions, communication devices, and computer software and algorithms to optimize signal parameters.

Traffic signal co-ordination systems have been successfully implemented in a number of cities. In Richmond, Virginia, for example, signal co-ordination resulted in a 9–14% reduction in delay and a 28–38% reduction in stops at intersections in the system (Meyer, 1997). The effectiveness of signal co-ordination can be impeded, however, if there is insufficient capacity, roadside parking, short signal spacing, heavy turn volumes at certain intersections (McShane and Roess, 1990), or even a lack of interjurisdictional co-ordination.

4.3. Ramp metering

Ramp metering, or freeway entrance-ramp control, is used to control traffic flow on freeways in countries such as the U.S.A., England, and The Netherlands. In the U.S.A. alone, there are over 2000 miles of freeway controlled by ramp meters (Kang and Gillen, 1999). There are generally four types of metering strategies: simple metering, demand-responsive metering, gap-acceptance metering, and pacer and greenban systems. The first type uses a fixed metering rate based on expected freeway traffic volume and some acceptable level of ramp traffic. Demand-responsive metering, on the other hand, adjusts its rate based on actual freeway conditions. Use of this strategy requires some surveillance of traffic or vehicular spacing, using sensors for example. With gap-acceptance metering, a green light is given when two vehicles with acceptable spacing between them are identified upstream. Due to technical requirements and only marginal benefits, this strategy is rarely utilized. Lastly, pacer and greenband systems are extensions of gap-acceptance metering and use lights to optimize ramp speeds based on the location of gaps in upstream traffic.

Ramp meters have been fairly successful in regulating freeway traffic. One of the earliest and most successful systems was initiated in Denver, Colorado during the late 1970s. Just a decade or so after ramp meters were installed on selected ramps, freeway speeds had increased by 58% and accidents decreased by 5%. A later study highlighted the importance of co-ordinating ramp signals, particularly under highly congested conditions. Minneapolis/St. Paul, Minnesota also has a successful ramp metering system, implemented as part of a larger freeway management system consisting of CCTVs, variable message signs, and highway advisory radio. Within the first 10 years of having parts of this system, average peak period vehicular speeds increased from 34 to 46 mph, and accidents fell by 38%. Along the M6 motorway in Great Britain, a combination of fixed-time ramp meters and variable message signs has led to a 3.2% increase in throughput, and in Zoetemeer, The Netherlands, there were similar improvements (see Kang and Gillen (1999) for more information on these and other ramp metering experiences).

While ramp meters can contribute to smoother and safer traffic flow, the ability to do so is also constrained by a number of factors. The effectiveness of ramp metering strategies is limited in areas where few alternative routes exist (e.g., frontage roads) (McShane and Roess, 1990) or where arterials feeding into the ramp have the potential to become highly congested (Meyer, 1997). Further, the experiences of Denver and other places demonstrate the importance of centralized control in ramp metering, particularly when a freeway is highly congested.

4.4. *Traffic advisory and information systems*

Traffic information systems can help to smooth traffic by providing travelers with information, which allows them to make better travel decisions regarding route and departure time, for example. Information can be communicated to travelers via a number of means including changeable message signs, the radio, television, the internet, highway advisory radio, cellular and landline phones, and citizens-band (CB) radios. Further, the information provided can range from periodic updates of general traffic conditions or the development of incidents to dynamic route guidance. Some automobile companies now offer in-vehicle advisory systems. For example, BMW's navigation system can give directions along with information on the location of gas stations, restaurants, etc. as requested. The benefits of such a system were demonstrated in the TravTek program, where 100 vehicles were equipped with in-vehicle monitors displaying traffic information. The system resulted in an approximately 37% reduction in wrong turns and a 33% decrease in stops (Meyer, 1997).

The effectiveness of traffic information systems is limited by several factors: the quality, relevance, and timeliness of information provided, the availability of alternative routes and other travel options, and the share of travelers who have access to information. Providing travelers with "perfect" or real-time traffic information is currently not feasible due to limitations in the coverage and capability of surveillance and communications technology. This is likely to change in the near future as technologies evolve.

4.5. *Limited access lanes*

Limited access lanes limit use to certain classes of vehicles (e.g., high-occupancy, truck, bus). The primary purpose of a high-occupancy vehicle (HOV) lane is to reduce congestion and improve capacity by diverting single-occupant motorists to carpooling. While HOV lanes appear to be benefiting and saving time for motorists who use them, utilization of some HOV facilities is still relatively low and the impact on overall freeway congestion and throughput mixed. High-occupancy toll (HOT) lanes are a relatively new phenomenon, which allow single-occupant motorists to use the HOV lanes at some monetary cost or toll. Two HOT lanes recently implemented in California show positive outcomes in terms of facilitating car pool formation.

4.6. Road pricing/electronic payment

Road pricing includes tolls and congestion (or value) pricing. Tolls can be fixed or tiered based on time of day (e.g., peak vs. off-peak) or day of the week. The primary purpose of a toll is to generate revenue for government or private-sector entities, although it may also have an impact on traffic flow by discouraging automobile use or use of a particular facility (e.g., a bridge). Congestion pricing on the other hand is more directly intended to reduce congestion and improve traffic flow. It is based on economics, and the notion that motorists should pay for the negative externalities or effects they have as a result of contributing to traffic. True congestion pricing would vary dynamically with changes in congestion, with higher prices imposed at more congested times. Road pricing can be particularly effective in areas where "bottlenecks" develop such as at a major bridge crossing. The impact that road pricing has had on congestion and traffic flow in areas where such schemes have been implemented is still mixed. Further, political, technical, and institutional issues still surround the implementation of a true congestion pricing program, where prices vary in real time (Hanson, 1995).

Electronic payment provides an alternative to cash, tickets, or tokens that can be used for tolls, transit fares, and parking. For toll collection or parking, vehicles can be equipped with an electronic device or tag that transmits vehicle and account information and charges the appropriate toll to that account. For public transit or parking, an individual may use a variety of technologies including prepaid accounts (debit cards), smart cards, or magnetic stripe cards, which could be multipurpose or dedicated to transportation activities. Electronic payment systems could be used in conjunction with congestion pricing systems.

4.7. Incident and emergency management systems

Accidents and other similar non-recurring events are a significant source of traffic congestion. An incident detection and management system uses surveillance, communications, and other technologies, along with institutional arrangements, to detect when and where an accident occurs, send out emergency professionals to respond to the accident, and inform motorists of the situation and reroute traffic if necessary. There are a variety of strategies that can be used to reduce detection and verification time, improve response time, improve site management, and improve motorist information (see Meyer (1997) for a full summary of these strategies). Massachusetts established a special *SP number, for example, which allows motorists with cell phones to report accidents or motor vehicle violations. With this system in place, incident waiting time decreased by some 40% (Meyer, 1997). While advanced technologies have been shown to enhance the effectiveness of incident management systems, institutional arrangements such as cross-jurisdictional, interagency co-ordination are also essential to the success of such systems.

An emergency management system can enhance traffic flow by providing immediate notification of an incident and triggering an immediate request for assistance. As a result, personal security on the road is increased and the time for an emergency vehicle to reach an incident is reduced. A range of capabilities are being tested and researched, including automatic collision notification, which transmits information regarding location, nature, and severity of the crash to emergency personnel; user-initiated signals for incidents like mechanical breakdowns or car jackings; and fleet management capabilities for emergency vehicles such as route guidance and signal priority.

4.8. Advanced vehicle control and safety systems (AVCSS)

Advanced vehicle control and safety systems can improve traffic flow and vehicle safety, and most elements of this system are classified by near-term reliance on self-contained systems within the vehicle. Some of these elements are longitudinal collision avoidance, lateral collision avoidance, intersection collision avoidance, vision enhancement for crash avoidance, safety readiness, and precrash restraint deployment. Automated highway systems (AHS) put suitably equipped vehicles under fully automated control, usually for some dedicated right-of-way. In theory, such systems should significantly enhance the capacity of a roadway by facilitating smooth and efficient movement of vehicles. The concept is still being tested and its implementation faces several barriers, including motorists' attitudes toward giving up control of their vehicle, and liability issues (McMillan and Sanford, 1998).

4.9. Traffic calming and street space management

In certain cases, the objective of traffic flow control is to slow down traffic with the underlying goal of improving an area's living environment. Traffic calming measures include, for example, speed bumps, lowered speed limits, the provision of bike lanes and adequate pedestrian rights-of-way, and street blockades. There are numerous examples of successfully implemented traffic calming programs, particularly in areas of Europe where the culture appears to be more tolerant of automobile restrictions and favorable toward non-motorized forms of transportation. In Germany for example, traffic calming measures in some neighborhoods resulted in a 20% reduction in accidents, a 50% reduction in severe accidents, and a significant decline in noise levels in these areas (Meyer, 1997).

References

Federal Highway Administration (1998) "The scan report: innovative traffic control practices", FHWA, Washington, DC, Report FHWA-RD-99-039.

Hanson, S. (1995) *The geography of urban transportation*. New York: Guilford.

Kang, S. and D. Gillen (1999) "Assessing the benefits and costs of intelligent transportation systems: Ramp meters", California PATH Research Report, UCB-ITS-PRR-99-19.

May, A. (1990) *Traffic flow fundamentals*. Englewood Cliffs: Prentice-Hall.

McMillan, B. and K.L. Sanford (1998) "Automated highway systems", *IEEE Journal*, Oct./Nov., 7–11.

McShane, W. and R. Roess (1990) *Traffic engineering*. Englewood Cliffs: Prentice-Hall.

Meyer, M.D. (1997) *A toolbox for alleviating traffic congestion and enhancing mobility*. Washington, DC: Institute of Traffic Engineers.

Ogden, K.W. and S.Y. Taylor (1999) *Traffic engineering and management*. Victoria, Australia: S&N.

Transportation Research Board (1995) *Highway capacity manual*, Washington, DC: TRB.

Chapter 30

TRAFFIC CONTROL DEVICES AND MANAGEMENT

KENNETH J. BUTTON
George Mason University, Fairfax, VA

1. Introduction

Traffic, irrespective of the mode involved, is the subject of a plethora of controls and is both directly and indirectly managed by outside authorities. It only takes casual observation to see this on the streets. These are filled with the architecture of management (directional signs, pedestrian rights of way, traffic signals, parking notices, roundabout deployment, priority markings, etc.). Equally, at airports, air traffic control towers do not have that name without a reason. Much less transparent, but equally important are the devices that are used to control other forms of traffic. Such controls are taken for granted, with most debates revolving around the detail rather than the acceptance of the need for such management of the system. Most of these controls and management affect transportation infrastructure use and most frequently, but not exclusively, that part of the infrastructure that is publicly supplied.

If traffic were not regulated in some form there would be at best higher costs and at worst chaos. This would embrace not only the transportation system itself but also the wider social effects of the system on such things as the natural environment and public health. As in all activities, there must be a code of enforceable law with effective policing.

It is as a result of simple pragmatism that this portfolio of control devices and management tools has emerged to tackle a variety of problems associated with traffic (Button, 1993). These tools may be of a physical kind but are often of a legal nature. Indeed, legal controls tend to dominate traffic behavior. Many of the devices and management techniques used are so common that they are taken for granted, e.g., driving on one side of the road, junction priorities, pedestrian crossing points, and one-way streets. In terms of hardware, traffic signals, information signs, and street maps (a powerful tool of management) are assumed standard parts of the road transportation system in its widest sense. There are parallels elsewhere. Air traffic control manages airspace using a range of institutional and technical approaches and there are numerous rules of the sea.

Handbook of Transport Systems and Traffic Control, Edited by K.J. Button and D.A. Hensher

Rail operations are strictly controlled. Much of the discussion here relates to road systems and the controls deployed to manage their use, but the issues that emerge generally have direct parallels for other modes. There are rights of way in the sky and in shipping lanes just as on urban streets.

Traffic control is not a new phenomenon. In Roman times there were bans on when wheeled vehicles could use the streets to limit noise nuisance. Priority was given to the official traffic, most notably that of the crown, in the Middle Ages in Europe. Somewhat later, in the U.K. there were rigidly enforced speed limits for the earliest motor cars determined by the speed at which a walker with a red flag could precede the vehicles. It was, however, with the advent of the large-scale use of motor cars and trucks at the beginning of the 20th century and the growth in air traffic during the second half that traffic control devices and management took on a major significance.

2. The scope

The perception of what constitutes traffic control devices and management and why they are used depends very much on perspective. Indeed, in many cases even those involved in traffic management are unaware that some of the most widely adopted practices (such as the side of the road on which people drive) represent a traffic control device. It is simply ingrained. From an intellectual perspective, any mechanism that directs traffic, be it institutional, informational, or physical, may be seen as part of traffic management. At a more pragmatic, operational level there are micro matters. In road traffic engineering terms this may involve the use of traffic lights, road markings, priority signs, and the like. For air traffic control there are informational devices such as radar and guidance systems (see Chapter 38). There are also controls over public transit, often extending to who provides and is responsible for the network that is being controlled and managed at a more micro level. At the most basic level, control devices and management can come down to the mechanical application of standards and rules such as those contained in the regularly updated *Manual of Uniform Traffic Control Devices* (U.S. Department of Transportation, 2001) and the *Highway Capacity Manual* (U.S. Department of Transportation, 2000) used by road traffic engineers in the U.S.A. Similar manuals exist in many other countries with the aim of ensuring that best practice and operational control tools are used.

Equally, the perceived motivation for deploying traffic control devices and management depends on the position one comes from. If one is a road traffic engineer or airport operator then the problems are those of containing congestion, safety, and environmental protection. From an economic perspective they are matters of economic efficiency. But the idea of efficiency can differ. Road traffic engineers think in terms of such indicators as speed–flow relationships. For privately owned trucking companies it is profit, and the control devices embody such things as computerized scheduling systems and tracking systems, often within a just-in-time framework (Taylor, 2001).

It also depends on the level at which the topic is addressed. For example, the economist sees the need for people to drive on the right (or left as the case may be) in terms of reducing transaction costs. Strict economic theory, following the arguments developed by the Nobel Prize winner Coase, maintains that if all property rights, be it to road space, the atmosphere, or rail track, were allocated then subsequent buying and selling of these rights would produce an optimal outcome. In practice, however, it is impossible in many cases to allocate rights in any meaningful way, and legal controlling devices are imposed. From a more pragmatic, and by necessity narrower local road traffic engineering perspective, congestion is just seen as an observable phenomenon that requires control and management. In this very narrow context traffic management may mean simply directing flows along the road network to reduce observable congestion. This embodies a set of localized instruments for traffic management.

Further, there is no generically correct set of traffic control devices or management procedures. In terms of traffic engineering, for example, much depends on the nature of the road network under review (a grid system common in the U.S.A. requires a different approach from the less structured networks found in most older European cities; the automobile-dominated streets of the U.S.A. require different priorities from the much more mixed-trafficked street of India), the motivations behind the policy, and the funds available to provide hardware, operational management, and policing. Thus even the traffic management manuals that are produced in most countries are often of little relevance in other places.

Since many of the individual issues concerned with road traffic control devices receive individual treatment in this handbook (e.g., traffic lights in Chapter 35, junction design in Chapter 26, traffic flow control in Chapter 29, route guidance systems in Chapter 33, and traffic calming in Chapter 20) the emphasis here is on much broader issues. This chapter looks at traffic control devices and management in much broader terms and concerns itself with larger issues of why they are needed, not only on roads but in terms of other modes, and at the fuller range of instruments that are used.

3. Motivations

Traffic control and management is seldom conducted for a single reason. There are usually several motivations behind any action, although frequently one has priority. The road traffic engineer is traditionally concerned with safety and efficiency, environmental, urban-economy, and equity issues using such instruments as signs, regulations, markings, and junction controls designed to manage road traffic. While some of these objectives may be familiar, but conceived differently, air traffic control inevitably puts a different emphasis on each, as do controllers of maritime activities. Further, the urban transportation planners and regulators may add to the depth of the problems by considering actions to influence the way that transportation services are delivered, e.g., the nature of the company or agency offering the service.

In this more generic setting, a fuller list of motivating factors includes:

(1) *Safety*. Transportation is an inherently dangerous activity (see Chapter 14). Transportation users are often not fully aware of the danger their activities pose, either for themselves or for others. Public actions such as very low speed limits in towns, traffic calming, statutory vehicle design features (including the fitting of seat belts to cars, and air bags), and safety barriers are aimed at controlling for this.
(2) *The improvement of transport co-ordination*. Because there are numerous users of transport services, inefficient use may result if their decisions are made independently. This may be an argument for one-way streets that funnel traffic and essentially create more capacity. In other cases there may be mechanisms that co-ordinate the way that public transport is supplied so that users can use a network of services.
(3) *The existence of high transaction costs*. Congestion is a problem because it leads to inefficient use of infrastructure. It is caused when, unlike most commodities and services, transportation infrastructure is not bought and sold in a market sense. One reason for this is that it is impractical to do so. While free markets may theoretically be capable of optimizing output, this may involve high transaction costs. Drivers confronting each other on a road could bargain as to who has right of way at junctions but simple rules, say giving priority to the first vehicle to arrive or forcing vehicles to use only one side of a street, are likely to prove more efficient. Air traffic control serves a similar purpose (see Chapter 38) as does the old rule of the sea that "steam gives way to sail."
(4) *Environmental considerations*. Transportation affects the environment in a variety of ways ranging from local noise nuisances, through transboundary problems of acid rain gas emissions, to contributions to the greenhouse warming effect. A very wide range of controls to contain adverse effects involve regulations over vehicle and infrastructure design and over their use, as well as more generic measures such as Sweden's carbon tax (Chapter 15).
(5) *The assistance of groups in "need" of adequate transport*. This embraces the notion that, for a variety of reasons including faults in the existing pattern of income distribution, effective demand is not an adequate guide to transport resource allocation and wider, social criteria should, therefore, be sought.

Examples of this include regulations committing bus services to provide services in the evenings or weekends that do not meet commercial criteria. Added to this, at a much more micro level, can be issues of junction priority. Junction priorities allow access to the main traffic stream to smaller streams from side streets that in terms of maximizing the overall flow would not be efficient – that is why in many cases where traffic light systems breakdown the overall flow actually increases.

(6) *The need to reflect the genuine resource costs of transport.* In the case of certain finite, non-renewable resources (for example, mineral fuels) the market mechanism may fail to reflect the full social time preference of society. The government may, therefore, intervene to ensure that the decision-maker is aware of the true shadow price. The low speed limits introduced on roads throughout the U.S.A. during the first oil crisis were to conserve fuel.

(7) *The containment of market distortions.* Some suppliers of transportation services can exercise monopoly power and limit traffic access – e.g., the owner of a bridge or important stretch of road. Public ownership and regulation of infrastructure is often deployed to contain this. At the other extreme, unregulated competition may affect the quality of service offered by such modes as trucks and buses and can have safety implications. Licensing and other management instruments may be used to determine which operators are permitted to join the traffic flow.

(8) *The integration of transport into wider economic policies.* Land use and transport are clearly interconnected and some degree of co-ordination may be felt desirable if imperfections exist in either the transport or the land-use market. For example, traffic management may favor movements into areas where development is being stimulated and pushed away from areas where development is not being encouraged.

4. Techniques

The previous section touched upon some of the instruments used to control and manage transportation systems and here we look at some of the more important ones in greater detail. While much of the economic activity of the developed world is controlled by market forces, as we have intimated this is often less the case in the transportation field. Market forces are important but allocating property rights is often so difficult that a considerable range of options now exist.

The effectiveness of individual instruments depends as much on the context and institutional setting in which they are deployed as on any intrinsic quality. They are seldom used in isolation but rather in packages (e.g., the use of trucks is generally controlled by weight, speed, access, and a variety of other measures that are operational at the same time.) Enforcement is a further issue and the effectiveness of control and management procedures depends as much on policing and penalties as it does on public acceptance. The cost-effectiveness of alternative measures can also influence selection of instruments (e.g., taxation measures may be preferred to command-and-control regulations if the same result can be achieved at a lower overall cost). There is also the consideration of how the measures fit in with wider economic, social, and technical considerations. Traffic routing regulations, for instance, may be inappropriate if they conflict with land-use objectives.

Who has the power to control and manage the transportation system varies considerably between countries. In many cases the central government has direct control, but local governments (embodying states in federal structures) have some responsibilities. In many cases local powers are tempered by central government either lightly (e.g., by the issuing of guidelines or "manuals") or more forcefully (e.g., by the withholding of finance if broad compliance with the direction of national policy is not taking place) to ensure a degree of overall co-ordination.

It is useful simply to provide a listing of the various policy instruments under the following general headings with some comments on each:

(1) *Laws and regulations*. Government, states, and local authorities have various powers of direct control over traffic. There has grown up an extensive body of law which, in effect, controls and directs the activities of both transportation suppliers and users. These regulations may be at the vehicle level (e.g., regarding engine emissions and size), at the user level (e.g., the need to wear a seat belt and not to have drunk more than a prescribed amount of alcohol), at the operational level (e.g., the requirement to drive on the right side of the road and under a certain speed), or at the infrastructure level (e.g., regarding the engineering design features of a road and the class of vehicle allowed to use it). They tend to vary considerably not only between countries (e.g., which side of the road you must drive on and whether you may use a cellular telephone when driving) but also between states and cities (e.g., permitted parking times and legal levels of alcohol consumption). These legal regulations can also segregate traffic, e.g., reserving freeways for the exclusive use of specified types of vehicle such as those with a prescribed level of occupancy. While many legal regulations over road use relate to the way the automobiles and trucks are used, there may also be laws governing the way public transportation operates. This may mean the banning of certain types of traffic (e.g., jitneys in many countries) or for operators to co-ordinate their services over specified routes (Viegas and Macarion, 1999).

(2) *Directional controls*. To some extent these act as indicators of legal requirements and embody such things as halt signs at the simplest

level to fully synchronized computer-controlled traffic signal systems (for more detail see Troutbeck, 2000). Regulations of this type can be complex, especially when they vary between places and at different times of the day.

(3) *User fees*. The regulation of traffic over many parts of the transport network is managed by charging users fees. These embody such things as tolls on roads, airport landing fees, and seaport charges. The aim in some cases is simply that of cost recovery rather than traffic control, as in the case of the toll rings around a number of Norwegian cities. In other cases, such as the Area Licensing System in Singapore and the temporally differentiated tolls on some French autoroutes, the fees have more direct traffic management objectives such as the containment of congestion. It should be emphasized that if this is done it is not the same as a measure aimed at internalizing externalities (see Chapter 7) but is rather a mechanism for the authorities to reduce congestion by a certain, non-market based-amount.

(4) *Taxes and subsidies*. The government may use its fiscal powers either to increase or to decrease the costs of various forms of transport or services and hence aggregate traffic flows. Such measures are very widespread in the transportation sector and take a multiplicity of forms. This may be for social reasons, such as subsidizing public transit or automobiles for designated groups (see Chapter 24), to shape the pattern of travel behavior (e.g., to switch people from the automobile to public transit), or for environmental reasons (e.g., lower taxation on unleaded gasoline).

(5) *Licensing*. The government may regulate either the quality or the quantity of the transportation provided by its ability to grant various forms of licenses to operators, vehicles, or services. The system of driving licenses also influences the demand for private transportation and who may operate it. The number of taxicabs and, ipso facto cab traffic, in a city is often regulated through licensing or (as in the case of London) their physical design and who may drive them is controlled in that way.

(6) *The physical design of a system*. Many U.S. cities are designed on a grid pattern to meet the demands of efficient traffic movement. This is not, however, the only approach. In the 1960s the Buchanan Report (U.K. Ministry of Transport, 1963) highlighted the potential for designing traffic systems in such a way that transportation objectives could be integrated with broader land use and environmental objectives (Simmons and Coombe, 1997). Much emphasis has been on separating transportation from activities that are for environmental and safety reasons sensitive to it; for example, these are the reasons that many ring roads are constructed around cities. More recently, the integration of physical traffic calming measures (e.g., speed humps, barriers at pedestrian-only streets, and chicanes) into road design has aimed at slowing traffic and making it less

intrusive in urban areas. Other physical controls involve junction design and the use of various forms of roundabouts.

(7) *Moral suasion*. This often supplements legal controls (e.g., it may be illegal to drink alcohol and to drive but this is often reinforced by "drink and do not drive" campaigns that are of a purely persuasive nature). In many instances moral suasion is of a weak form, usually being educational or the offering of advice on matters such as safety (for example, advertising the advantages of the wearing of seat belts), but it may be stronger when the alternative to accepting advice is the exercise, by government or others, of its powers (for example, the refusal of a license or withdrawal of a subsidy). In France, there was a successful weak-moral-suasion campaign in the 1980s to discourage people from all leaving for their summer vacations on the same weekend and as a result causing traffic chaos.

(8) *Direct provisions*. Local and central government are direct suppliers, via municipal and nationalized undertakings, of a wide range of transportation services. (They are also responsible for supplying a substantial amount of transportation infrastructure, notably roads, and supplementary services, such as the police.) This provision gives direct control over the management of significant parts of many transportation networks and the amount of traffic that can flow over them, the nature of that traffic, and its temporal distribution. Perhaps the clearest example of this is in air transportation, where public ownership of airports has allowed the authorities to control air traffic.

(9) *Provision of information*. The government, through various agencies, offers certain technical advice to transport users and provides general information to improve the decision-making within transport. Many of these services are not specific to transport (e.g., weather services), while others assist the transport sector more directly (e.g., maps and charts). At the local level street signs direct traffic flows. In some case these may be essentially neutral in their orientation and simply offer objective information, but in other instances the signing of streets or the ways roads are depicted on maps can influence the routes selected by drivers. In some case few signs are posted, to deter non-resident drivers from trying to navigate an area (San Francisco has a tradition of attempting this). The advent of electronic information channels adds an extra dimension to the way that information can be used as a directive device.

(10) *Research and development*. Government may influence the long-term development of traffic management through its own research activities. These are, in part, conducted by its own agents and, in part, through the funding of outside research. The impact of these activities is an increasingly important indirect tool of traffic control as more effort is made to adopt intelligent transportation systems. Given the high costs of R&D in this

area, and the need for uniformity of standards, there is a tendency for a major governmental presence in terms of the specific traffic management tools that are developed and subsequently deployed.

(11) *Policies relating to inputs*. Transport is a major user of energy, especially oil, and also utilizes a wide range of other raw materials and intermediate products. Government policy relating to the energy and other sectors can therefore have an important bearing, indirectly, on controlling the way transportation is supplied and used. Higher fuel prices, for example, affect the road traffic mix if they encourage more freight to switch from truck to rail.

In many cases there is no single instrument that is automatically seen as a mechanism of controlling traffic and its effects. Table 1, for example, gives a stylized listing of the types of control that can be applied to control the environmental implications of traffic. Selection of one or a combination of these measures will to a large extent depend upon their perceived effectiveness in meeting the environmental objective. It may also depend upon factors such as the costs of implementation, the public acceptance of the measure, and any secondary

Table 1
Policy instruments to control environmental impacts of motor vehicles

	Market-based incentives		Command-and-control regulations	
	Direct	Indirect	Direct	Indirect
Vehicle	Emissions fees	Tradable permits Differential vehicle taxation Tax allowances for new vehicles	Emissions standards	Compulsory inspection and maintenance of emissions control systems Mandatory use of low-polluting vehicles Compulsory scrappage of old vehicles
Fuel		Differential fuel taxation High fuel taxes	Fuel composition Pasing out of high-polluting fuels	Fuel economy standards Speed limits
Traffic		Congestion charges Parking charges Subsidies for less polluting modes	Physical restraint of traffic Designated routes	Restraints on vehicle use Bus lanes and other priorities

Source: Cabajo (1991).

adverse effects. Similar criteria are used to select controls for junction priorities, parking restraints, safety improvements, and so on. There are inevitably trade-offs that require judgments as well as technical inputs (Lave and Lave, 1999).

There are also wider institutional issues that can affect the effectiveness of various control devices. Institutional jurisdictions vary in the packages of traffic control instruments that they deploy; this can cause, at the very least, confusion for travelers who cross jurisdictional boundaries and can impose costs if different technologies are required. In terms of electronic information systems, for example, there is a need for technical uniformity or at least an interface between systems if the adoption of multiple devices is to be avoided.

5. Trends

There is an ongoing pattern of development in both the instruments available for traffic control and the way in which traffic is managed. In part this reflects a change in priorities on the part of transportation policy-makers. Considerably more emphasis is now placed on reducing the environmental implications of transportation. All modes of transportation have improved in terms of their environmental impacts but this has largely been due to technical developments (unleaded fuel, the catalytic converter, etc.) rather than changes in traffic patterns or intensity. In terms of road transportation there is now more emphasis on combining infrastructure investment with operational controls over traffic (e.g., in the form of traffic calming measures) to supplement these technical measures.

In terms of traffic congestion, traditional physical management measures are still available in some contexts to reduce the problem (e.g., in the air transportation context, improved air traffic control technologies), but in others, and most notably road transport, the scope for further relief using these tools seems limited. Fiscal instruments, such as congestion pricing (value pricing in the U.S.A.), are now being adopted in some areas and there are plans in other major urban areas to use them.

The ways in which traffic control devices and management are now moving are also being greatly influenced by the fostering of intelligent transportation systems (ITS) (see Chapter 31). The need in many countries, because of the high cost of infrastructure expansions and frequently expressed public concern about such expansions, has led to efforts to make better use of the infrastructure that is available. This is possible because more information is now becoming available about traffic conditions through developments such as the creation of geographical information systems (GIS) and as computer technology allows this information to be used to direct traffic flows and enforce restrictions. The need, therefore, is likely to be for more sophisticated management in the future and it seems inevitable that this will entail greater use of information systems to make it function efficiently.

Although there are new dimensions to it, the new technology acts largely to make existing traffic control devices more efficient. For example, in the road context, it allows for more effective co-ordinated sequencing of traffic lights, the collection of tolls with less disruption to traffic, and the provision of real-time information on traffic signs. In the air transport context, there is now the scope, because of geopositioning and accident avoidance systems, for allowing planes to route themselves without fear of collision and hence to optimize individual flight paths.

References

Button, K.J. (1993) *Transport economics*, 2nd edn. Cheltenham: Edward Elgar.

Cabajo, J. (1991) "Accident and air pollution externalities in a system of road user charges", World Bank, Washington, DC, informal working paper.

Lave, C. and L. Lave (1999) "Fuel economy and auto safety regulations: Is the cure worse than the disease", in: J. Gómez-Ibáñez, W.B. Tye and C. Winston, eds., *Essays in transportation economics and policy: A handbook in honor of John R. Meyer.* Washington, DC: Brookings Institution.

Simmons, D.C. and D. Coombes (1997) "Transport effects of urban land change", *Traffic Engineering and Control*, 38:660–665.

Taylor, S.Y. (2001) "Just-in-time", in: A.M. Brewer, K.J. Button and D.A. Hensher, eds., *Handbooks in transport 2. Logistics and supply chain management*. Oxford: Pergamon.

Troutbeck, R. (2000) "Modelling signalized junctions", in: K.J. Button and D. Hensher, eds., *Handbooks in transport 1. Transport modeling*. Oxford: Pergamon.

U.K. Ministry of Transport (1963) *Traffic in towns*. London: HMSO.

U.S. Department of Transportation (2000) *Highway capacity manual*. Washington, DC: USDOT.

U.S. Department of Transportation (2001) *Manual of uniform traffic control devices*. Washington, DC: USDOT.

Viegas, J.M. and R. Macarion (1999) "Legal and regulatory options to promote system integration in public transport", in: H. Meersman, E. Van de Voorde and W. Winkelmans, eds., *World transport research: Proceedings of the 8th World Conference on Transport Research*, vol. 1. *Transport modes and systems*. Oxford: Pergamon.

Chapter 31

INTELLIGENT TRANSPORT SYSTEMS

MICHAEL A.P. TAYLOR
University of South Australia, Adelaide

1. Introduction

Transport technology is undergoing enormous and rapid change. The transport systems so familiar to us today will be unrecognizable within the next decade or two. New technologies, collectively known as "intelligent transport systems" (ITS) or the variants "intelligent transportation systems" or "advanced transport telematics," are radically altering the ways in which transport systems perform. They will also fundamentally change the ways in which we provide transport infrastructure and services and the ways in which people use that infrastructure and those services.

What are ITS? There are many definitions. Some of these are shown in Table 1. They all have similar threads concerning the use of new technologies for the improvement of transport systems and the realization of economic and social benefits. Some of the definitions stress these impacts, whilst others are more concerned with identifying the relevant technologies.

Overall, the aim of using ITS should be to improve the links between the infrastructure, the vehicles, and users to make individual components, modes, and the transport system as a whole work more efficiently and effectively within the context of wider transport, urban and regional development, economic, and environmental policies. We can suggest the following potential benefits from applications of ITS:

(1) better utilization of infrastructure;
(2) improved traffic flow;
(3) better service from public transport;
(4) enhanced safety;
(5) lower-cost freight transport; and
(6) reduced environmental impact.

In addition, technology is changing the nature of work and having an impact on the locations of homes and workplaces, on the pattern of trips made on a regional

Handbook of Transport Systems and Traffic Control, Edited by K.J. Button and D.A. Hensher

Table 1
Some definitions of ITS

Source	Region	Definition
AustRoads	Australia	ITS cover the integrated application of modern computer, electronic, information, and communications technologies to improve all facets and all modes of transport and their linkages. ITS are essentially a diverse range of sophisticated tools that can reduce the environmental effects of transport, reduce congestion in a managed way, and improve transport efficiency, safety, and sustainability
ERTICO	Europe	ITS are the marriage of information and communication technologies with the vehicles and networks that move people and goods. "Intelligent" because they bring extra knowledge to travelers and operators. In cars, ITS systems help drivers navigate, avoid traffic holdups, and avoid collisions. On trains and buses, they let managers optimize fleet operation and offer passengers automatic ticketing and real-time running information. On the road network, ITS co-ordinate traffic signals, detect and manage incidents, and display information, guidance, and instructions to drivers
ITS America, Inc.	U.S.A.	ITS are a broad range of diverse technologies that holds the answer to many of our transportation problems. ITS comprise of a number of technologies, including information processing, telecommunications, control, and electronics. Joining these technologies to our transportation system will save lives, save time, and save money
ITS Australia, Inc.	Australia	"ITS" is an umbrella term referring to the application of informed technology to transport operations in order to reduce operating costs, improve safety, and maximize the capacity of existing infrastructure. "Freight management systems" refers to those applications that benefit operators and regulators of the freight transport industry
ITS Canada, Inc.	Canada	ITS, an emerging global phenomenon, are a broad range of diverse technologies applied to transportation to save lives, money, and time. The range of technologies involved includes microelectronics, communications, and computer informatics, and cuts across disciplines such as transportation engineering, telecommunications, computer science, financing, electronic commerce, and automobile manufacturing. The annual world market for ITS is estimated to be $24 billion by 2001 and $90 billion by 2011, and access to this market is vital to the transportation and related technology sectors

Table 1
Contd

Source	Region	Definition
U.S. Department of Transportation	U.S.A.	ITS are the integration of current and emerging technologies in fields such as information processing, communications, and electronics applied to solving surface transportation problems
VERTIS	Japan	ITS offer a fundamental solution to various issues concerning transportation, which include traffic accidents, congestion, and environmental pollution. ITS deal with these issues through the most advanced communications and control technologies. ITS receive and transmit information on humans, roads, and automobiles

network, on the information requirements of intending travelers, and ultimately on the need to travel itself. As travel habits change, there are direct implications for the role of the road system and the way that it should evolve to best serve community needs (Taylor and D'Este, 1997).

This chapter attempts to link the impacts of new technology to policy and planning concerns in the transport sphere, such as travel demand management (TDM), road safety, environmental impacts (especially air pollutant emissions), and road pricing. Table 2 lists some ITS applications with the potential for profound impacts on transport policy, planning, and operations.

2. Interactions and their significance

It is clear that there is a complex array of technologies capable of application to transport systems. How will it affect the way that we use these systems and the way that transport is managed? A common feature of many of the emerging technologies is the growing reliance on communications and information. The biggest impending change is in the amount of communication and the flow of information between the components of the road system. At present the components largely work in isolation but in future there will be much more interaction between them. Three broad components may be identified:

(1) *Roads and tracks* – all the fixed infrastructure and organizations supporting the road network. This includes the road pavement, rail tracks, guideways, bridges, underpasses, road furniture, signs, and vehicle detectors, and the authorities that build and manage a transport network, and any commercial enterprises that sell transport-related services.

Table 2
Some applications of ITS

ITS application		Description	Examples
ATC	Automatic traffic control	Systems to control and facilitate the movement of traffic and pedestrians using real-time adaptive co-ordination of traffic signals. Roadway elements are monitored to provide real-time information on traffic flows and speeds. The systems may provide priority for transit and emergency vehicles	FETSIM in California and the SCATS and SCOOT traffic control systems used around the world
AVC	Automatic vehicle control	Systems that use computers, in-road sensors, radar, and other devices to control the steering and speed of vehicles, without necessarily involving a human operator in some or all of the driving task	Automated freeway lanes trialled near San Diego, CA under the CALTRANS PATH project. Intelligent speed adaptation trials in Sweden
DIS	Driver information system	Systems that provide drivers with information about traffic conditions and congestion on a network. This information may be provided prior to departure or during the journey	Transguide system in San Antonio (Texas Department of Transportation)
EPM	Environment/ pollution monitoring	Systems that monitor and process measurements from a number of sensors in an area or along a road, and broadcast warnings and implement traffic control strategies in response to changes in pollution levels or environmental conditions	Warning systems for fog and ice are used in many countries
ETC	Electronic toll collection	Systems for the automatic payment of tolls or road user charges on roads or bridges without the need for vehicles to slow down. Typically vehicles are equipped with a "smart card" or are electronically tagged so that as the vehicle passes a collection point, the charge is deducted from the card or from an account	Many ETC systems have been installed on freeways in many countries. A most notable example is the electronic road pricing system in Singapore (Menon and Keong, 1998)

Table 2
Contd

ITS application		Description	Examples
FMS	Freight management system	Systems designed specifically to improve the reliability, service quality, and efficiency of freight transport, by greater control of fleet operations through vehicle tracking and identification, computerized scheduling, and two-way communications	Booz Allen and Hamilton (1998) describe a number of proprietary systems used by freight companies in the U.S.A.
INM	Incident management system	Systems that seek to rapidly detect incidents (such as crashes, vehicle breakdowns, or lane blockages) on a road and suggest rapid response plans so that the incident can be cleared	Many freeways in the U.S. and elsewhere are equipped with INM systems, e.g. Maryland's CHART system
PTI	Public transport information	Systems providing travelers with information about the availability of public transport services and that may recommend specific services for a given journey. The information may be provided before the trip or during the trip, and may be based on timetables or on real-time information about vehicle progression	There are a number of PTI systems currently in use in the U.K.
PTM	Public transport management	Systems that enhance the efficiency and safety of public transport vehicles, such as computerized timetabling, dispatch and rostering, fleet monitoring, and intelligent control systems	Computerized vehicle dispatch systems commonly used by taxi companies are a familiar example
ROG	Route guidance system	Systems that provide navigation and route guidance information and travel planning information, through media such as electronic map displays or synthesized voice instructions	Many hire car companies now provide in-vehicle navigation systems in rental cars
RSE	Road safety enhancement	Systems that help drivers avoid crashes and/or provide improved protection and emergency response if a crash occurs	Collision avoidance automatic braking systems, and automatic emergency assistance systems

(2) *Vehicles* – all the locomotives, carriages, vessels, craft, cars, motorcycles, trucks, buses, and bicycles driven on the roads.
(3) *Users* – all those who use the systems, as drivers, operators, passengers, customers and service recipients

2.1. Trackway–vehicle interactions

These interactions require the flow of information between the infrastructure and the vehicle with no direct intervention by the users. At present, this flow does not exist for the vast majority of users. However, this is a basic element of applications such as vehicle tracking and identification systems (PTM, PTI, FMS), electronic toll collection (ETC), advanced traffic control (ATC), and automated highways (AVC).

2.2. Trackway–user interactions

Technologies that involve communication between the road infrastructure and the user will increase in importance and sophistication. Historically, information was static and directed equally to all road users. Advisory speed signs on curves would be a good example. From now on the information will be increasingly dynamic and personalized. For instance, different vehicles (such as large trucks, passenger cars, and motor cycles) may be given different advice about the speed to (say) negotiate a curve. Information about the current status of roadways and traffic (DIS, INM, FMS), public transport (PTI), or environmental conditions (EPM) will be collected and synthesized by relevant authorities and passed on to system users.

2.3. Vehicle–user interactions

The level of information provided to the user by the vehicle will change rapidly, especially for road transport. Currently the only information we receive from the vehicle relates to its own performance (speed, engine revs, engine temperature, fuel gauge, etc.). Sensors and computer systems can be built into the vehicle to provide route guidance information to drivers (ROG) and provide safety (RSE) and environmental warnings (EPM).

2.4. Trackway–vehicle–user interactions

Technologies that involve communication from the roadway to the vehicle and then to the user provide another layer in the road operating system. The technologies include vehicle navigation systems that include real-time information (ROG), vehicle dispatch systems (FMS, PTM), and driver and public transport information systems installed in vehicles, at bus stops, or computer kiosks, or available through computer networks (DIS, PTI).

2.5. *Other interactions*

The new technologies will also enable each of the system components to collect and process more and richer information. By using a wider range of sensors and exploiting high-speed communications and computing, road authorities will have access to much more information about current traffic conditions. This information can then be used to react to incidents more quickly (INM) and adjust traffic lights to ease congestion (ATC) and reduce pollution (EPM). Vehicles will also become "smarter." Sensors and onboard computers will monitor the condition of the vehicle, the roadway, and the driver to optimize vehicle performance (PTM) and help to protect the driver (RSE).

3. Benefits

So far this chapter has considered the capabilities of the emerging technologies, their applications, and the way that the relationships between transport system components will change. But what will the technologies deliver in terms of benefits? Six types of benefits were identified in the introduction. These benefits will be delivered by combinations of technologies working together towards the overall aim of making transport systems better serve the community, in both urban and rural areas. Improved traffic control, better information about traffic conditions and transport options, better vehicles, and improved public transport will work together to make it easier and more efficient to use the transport network. ITS will also improve the utilization of existing infrastructure and can certainly reduce the need to provide more roads and road transport services. In addition, reduced environmental impact can result from more efficient usage of the road network by more environmentally friendly vehicles.

Quantification of the individual benefits that new technologies will deliver is still in its infancy, although there are a number of studies now emerging, e.g., Garrett (1998), Booz Allen and Hamilton (1998), Baum et al. (1999), and Kulmala and Pajunen-Muhonen (1999). In many instances the overall benefits are likely to be case-specific, not generic, and will depend on the characteristics of each city or region and the combination of technologies that are in use. In addition, many of the technologies are complementary and their full benefits will only be delivered

Table 3
Potential benefits from emerging technologies

Technology	Benefit					
	Better utilization of infrastructure	Improved traffic flow	Better public transport	Enhanced safety	Lower freight costs	Reduced environment al impact
Advanced traffic control	✓	✓	✓	✓	✓	✓
Route guidance	✓	✓	✓	✓	✓	✓
Driver information systems	✓	✓	✓	✓	✓	✓
Incident management	✓	✓	✓	✓	✓	✓
Electronic toll collection	✓	✓	✓	?		?
Automatic vehicle control	✓	✓	✓	✓		✓
Public transport information			✓			✓
Public transport management			✓			✓
Road safety enhancement				✓		
Freight management systems					✓	
Environment and pollution monitoring						✓

when they are all in place. Table 3 summarizes the likely benefits from each of the technology categories.

Garrett (1998) cites European research suggesting the following likely improvements in transport systems performance achievable by the year 2017, from the systematic implementation of ITS technologies:

(1) a 15% increase in survival rates from road crashes (RSE);
(2) a 50% reduction in road fatalities (RSE);
(3) 25% reductions in travel times (INM and DIS);
(4) savings of 40 h per traveler per annum through the use of ATC;
(5) 50% reductions in delays through improvements in public transport priority (ATC and INM);
(6) 25% reductions in freight costs through improved efficiency of freight movements and fleet operations (FMS), and
(7) a 50% decrease in pollution in city centers through the use of advanced traffic management systems (ATC and EPM).

There remains a need for ongoing independent examination of the magnitude and extent of potential benefits. For example, a recent German study (Baum et al., 1999) indicated that there may be optimal levels of ITS implementation below that of full implementation across (say) a vehicle fleet. Baum et al. considered the introduction of a system of tandem vehicles for long-distance freight traffic on freeways, in which a vehicle driven by a human operator is shadowed by one under automatic control (i.e., the driver is actually controlling two vehicles at once). An economic analysis indicated that, for the travel data used, the best result was for only 20% of vehicles to be equipped with this system.

Booz Allen and Hamilton (1998) provides a first substantial attempt at identifying and quantifying the system-wide benefits of ITS. This report focused on four key inputs to the assessment of benefits:

(1) specification of current and future costs of performance measures such as road crashes, congestion levels, pollutant emissions, and economic costs of transport;
(2) identification of ITS applications where sufficient information existed to permit quantification of safety, congestion, travel time, emissions, and fleet efficiency benefits;
(3) consideration of the potential levels of application of ITS within the national context, and
(4) assessment of the current level of and projected increases in ITS implementations over a given time period (1997–2012).

The Booz Allen and Hamilton review of international experience suggested a wide range of benefit–cost ratios associated with ITS applications, ranging from 1.4 (for a freeway management system (INM) in Kansas City, U.S.A.) to 62 for a

Texas state-wide traffic signal co-ordination program (ATC). Typical results were for ratios of between 1.5 to 3.5. On this basis they calculated that the net present value of likely net benefits of ITS deployments in Australia in the period 1997–2012 was estimated to be between U.S. $1.61 billion and U.S. $6.19 billion.

4. Developments around the world

The European Union, the U.S.A., and Japan have large-scale R&D programs aimed at developing and trialing ITS applications. The current total budget for these is about U.S. $0.6 billion per annum. In large measure this is a response to chronic traffic congestion coupled with environmental problems and resistance to building more and more roads. The international R&D programs cover the full gamut of applications as described above. The strength of the worldwide R&D programs and the rapid implementation of ITS technologies across the globe point to a number of significant economic, social, and community impacts that may test our existing systems.

The application of ITS technologies has the potential to deliver significant benefits to road users and road system administrators. However, the technology and its benefits cannot be considered in isolation from a wider community context. Inherent in any major technological change, there are range of social, legal, regulatory, and institutional issues. In terms of the impacts of ITS, these include: privacy; information ownership; legal liability; equity; uniform standards; and legislation and regulation.

These issues cannot be fully explored or resolved here but there is an opportunity to introduce the circumstances under which they arise, to look at how they will affect individuals and the community as a whole, and to raise some important questions that should be addressed by the community. Many of the community issues, especially those affecting individuals, will arise because ITS allows components of the road system to interact in new ways and information to flow between the components to an extent and in ways that have been hitherto impossible.

Current transport technology is largely impersonal and operates on the basis of averages and totals. Each vehicle, each driver and, each transit passenger is treated equally and the way that the system is presented and responds is the same for every user. In future this is likely to change. ITS will allow roads and vehicles to recognize individuals and to communicate and respond on a personalized basis. In some circumstances this interaction will occur automatically. There is also the capability to collect, integrate, and disseminate information collected over a wider region to provide an overview for strategic decision-making. The impact of new and emerging technologies as noticed by individual members of the community will be to change their perception of the system as follows:

$$\{\text{Impersonal, Manual, Local}\} \Rightarrow \{\text{Personalized, Automatic, Wide Area}\}$$

Most current road system technologies are impersonal, and manual to the extent that interaction between the user and the road system is largely under the user's conscious control. Current technologies also appear to operate on a small local area (single vehicle or intersection), even if this perception may be incorrect. For example, a red light is perceived to apply to that particular intersection, whether or not it is part of a co-ordinated traffic control area spanning a large area. Emerging technologies are allowing road, vehicle, and user interactions that relate to specific individuals and occur automatically, possibly without direct input by the user.

4.1. *Privacy and information issues*

The first group of community issues arises because ITS can personalize the road system to such an extent that it possible to automatically identify and capture information about the activities of particular road users. For instance, automatic vehicle identification allows individual vehicles to be identified and tracked. If the technology is used on public vehicles (such as buses, ambulances, and police) or commercial vehicles then monitoring can be justified on the basis of control and operational efficiency. Community support is probably guaranteed for such applications. In addition, automatic vehicle identification on a voluntary basis, such as in the freeway travel-time monitoring systems in some U.S. cities (e.g., Levine and McCasland, 1994), is also likely to achieve widespread acceptance. However, when private vehicles are identified and their movements traced then there are immediately issues of privacy and ownership of information. The major concerns in Western societies at present may rest more with the possible revelation of one individual's movements to other individuals associated with that person, rather than with a concern about a state or government spying on its citizens, but even these concerns should not be trivialized. The issues include:

(1) who owns the information?
(2) who has access to the information?
(3) what purposes can the information be used for?
(4) can road users retain their anonymity?

4.2. *Legal issues*

Privacy issues aside, there are other legal and legislative matters that need serious consideration in the implementation of new technologies, because they imply or require a large-scale change in the allocation of responsibility for transport behavior, especially for road networks. The ability to identify individuals can also be used as a mechanism for enforcement of regulations. Note, for instance, the strong interest in Europe at present concerning intelligent adaptive speed limits whereby a vehicle itself will be informed about the prevailing speed limit by a roadside device. Various alternative methods might then be applied to limit the speed at which the vehicle can travel or to inform the driver of the speed limit. Already there is discussion about the use and implications of such systems for speed limit enforcement (Malenstein, 2000).

Further, these fundamental changes in the ability to identify individuals and to remove some elements of vehicle control from individual drivers immediately lead to questions of legal liability. Identifying the vehicle and identifying the driver are two entirely different processes. Most technology is linked to the vehicle but vehicles can have many different drivers. Under current legal regimes the vehicle owner is usually responsible for the vehicle and liable for any offences committed by the vehicle. The onus is on the owner to identify the driver. This issue of the vehicle versus the driver will become more important as automatic vehicle identification technologies become more widespread.

Similar issues of responsibility and legal liability also arise where there is information flowing from the roadway to the driver or between the roadway and vehicle. DIS and AVC are examples. Who will be liable for an accident on an automated highway or for problems that arise if a driver follows directions provided by a driver information system? The current principle is that the driver is in control of the vehicle and cannot transfer any liability. This will need reexamination. Some ITS deployers may only design their systems to avoid anticipated risks, such as the identification of congested routes without the nomination of alternatives, to avoid possible liability issues. Nevertheless this strategy may of itself weaken the value of the information that is supplied, and thus lessen its impact on and credibility to travelers.

4.3. Equity and opportunity

Concerns have also been raised that the effects of ITS will be felt differently across the community. Not all members of the community have equal access to the technology or to the information and related services that it can provide. Will the greatest benefits accrue to those in the community with the greatest ability to pay? Similarly, not all members of the community are equally comfortable using information-based technologies. As with all new technologies, the rate of acceptance and ability to utilize the technology will vary within the population. This means that ITS technologies should be introduced in ways that are easy to use and accessible for the majority, with responsibility for wholesale changes to be accepted by government, as in the case of the introduction of electronic road pricing in Singapore, where the government paid for the retrofitting of the necessary hardware ("in-vehicle units") to all vehicles registered in the island republic (Menon and Keong, 1998). In other situations parallel systems may be required, such as the provision of roadside variable-message signage alongside the in-vehicle ROG and DIS systems available to some, so that all drivers are informed to some degree about downstream traffic conditions.

ITS may also produce changes in accessibility, travel behavior, and the pattern of traffic flows. For instance, driver information systems may encourage drivers to change their routes and patterns of trips. This has implications for the distribution of economic activity in the city. It will benefit some areas but equally will disadvantage others. This issue is explored further in the ITS chapter of the handbook on transport modeling in this series (Bonsall, 2000).

There are already many examples of the use of ITS around the world, and the pace of adoption of technological solutions to road system problems is sure to quicken. However, most applications of advanced transport technology are aimed at areas where road system usage is high, in terms of traffic congestion, heavily used public transport, and many trucks. This means that the technologies are probably being felt first and most extensively in the largest cities, where road system problems are most acute. Yet perhaps there are advantages, at least in social and political terms, in initial implementations of ITS such as electronic road pricing in smaller cities where the impacts may be felt less severely. The "toll rings" of the Norwegian cities may be good examples in this regard (Solheim and Assum, 1998).

5. Standards and institutions

Standards and institutional change relate to the creation of an environment in which the technologies can deliver maximum benefits with minimal negative impact on the community. The need for standards has already been established since without them there is a danger of having a random mix of incompatible technologies. This has immediate implications for the community in terms of ease of use and cost. It is inconvenient and costly to swap between different systems and to have more than one piece of equipment to do the same job. Likewise it is important that laws, regulations and institutional structures keep pace with changes in technology. Standardization does not require the adoption of a single technology at the expense of all other alternatives, but may be accomplished by the use of "open architectures" that allow alternative systems to work together.

There are also important issues relating to public acceptance and implementation. Many of the proposed technologies involve significant changes to the way that road users interact with the road environment. The success of these and other technologies largely depends on the level of public acceptance and willingness to utilize them. There is a danger that potentially beneficial technologies may not be adopted because they become associated with negative issues such as enforcement and invasion of privacy. There are also problems associated with managing the implementation of technological solutions. For many proposed systems, the full benefits will not be realized until a large percentage of the vehicle fleet is fitted with the required instrumentation or a large proportion of the road network is equipped with advanced instrumentation and signage. As a result, initial benefits may be small and there will be significant challenges associated with managing the transition period until full benefits are realized.

It is clear that ITS have direct implications for the community and community institutions. However, there are choices in terms of when and how the technology is introduced. This suggests that social impact analysis and community consultation should be integral components of the process of introducing new technologies into transport systems.

6. Conclusions

ITS are revolutionizing the world's transport systems. The imperative to make better use of existing road transport infrastructure and to improve the efficiency of road and rail transport operations suggests that advanced technologies will be increasingly used in planning, managing, and operating the state's transport system. Better practice may come by providing an environment to facilitate the application of ITS technologies which

(1) improve the efficiency of business;
(2) increase fuel efficiency and reduce and monitor noise and emissions;
(3) improve road safety;
(4) improve accessibility (demand-responsive transport and transport information systems); and
(5) improve travel data collection and monitoring systems.

In addition, attention is needed to the formalization, specification, and (where possible) quantification of the benefits to be obtained from the use of ITS technologies.

The developments in ITS are creating exciting opportunities but as we move to take advantage of these opportunities, we must not lose sight of the impact on the community and the broader legal and institutional issues, some of which have been introduced in this chapter. Further, at the broader level there is a need to adopt an integrated approach to ensure that uniform standards and regulations are developed and adopted.

References

Baum, H., W.H. Schulz, T. Geissler and M. Schulze (1999) "Methodological and empirical approach for the cost–benefit analysis of the CHAFFEUR system", in: *Proceedings of the 6th World Congress on Intelligent Transportation Systems*, Toronto, Paper 2022, CD-ROM. Toronto: ITS Canada.

Booz Allen and Hamilton (1998) "Intelligent transport solutions for Australia: Technical report", report to ITS Australia Inc., Canberra.

Garrett, A. (1998) "Intelligent transport systems – potential benefits and immediate issues", *Road and Transport Research*, 7(2):61–69.

Bonsall, P. (2000) "Information systems and other intelligent transport systems", in: Hensher, D.A. and K.J. Button, eds., *Handbooks in transport 1. Transport modelling*. Oxford: Pergamon.

Kulmala, R, and H. Pajunen-Muhonen (1999) "Guidelines for the evaluation of ITS projects", *Nordic Road and Transport Research*, 11(1):13–15.

Levine, S.Z. and W.R. McCasland (1994) "Monitoring freeway traffic conditions with automatic vehicle identification systems", *ITE Journal*, 64(3):23–28.

Malenstein, J. (2000) "Know your limits – a rejoinder", *Traffic Technology International*, April/ May, 15.

Menon, A.P.G. and C.K. Keong (1998) "The making of Singapore's electronic road pricing system", in: Proceedings of the International Conference on Transportation into the Next Millennium, Nanyang Technological University, Singapore.

Solheim, T. and T. Assum (1998) "Safety and the environment in road traffic – conflict or synergy?", *Nordic Road and Transport Research*, 10(1):18–19.

Taylor, M.A.P. and G.M. D'Este (1997) "The impact of emerging technology", in: D.C. Kneebone, ed., *Roads in the community. Part II: Towards better practice*. Sydney: AustRoads.

Texas Department of Transportation (2001) "TransGuide", www.transguide.dot.state.tx.us.

Chapter 32

TRANSPORTATION INFORMATION SYSTEMS

BRIEN BENSON
George Mason University, Fairfax, VA

1. Introduction

The first transportation information system used by human beings was likely a series of tree notches or rock piles indicating direction of travel. With the development of written language, road signs would display instruction in words and numbers. As human civilization and technology advanced, transportation information systems progressed in tandem. In particular, the telegraph and then the telephone became key technologies in communicating transportation information.

In the late 20th century computers brought the information technology revolution. The earliest application of information technologies to surface transportation in the U.S.A. was the electronic route guidance system (ERGS), a federally funded developmental project of the 1970s. ERGS, employing computers, and drawing on systems engineering concepts developed for NASAs Project Apollo, was designed to monitor traffic flow in congested areas and then, when appropriate, to reroute vehicles to less congested routes. Work on ERGS had been stimulated by the oil crises of the 1970s, and when oil decontrol measures adopted by the Reagan administration in 1981 brought a sharp decline in oil prices, interest in ERGS waned.

The application of electronics to surface transportation was revived as a federal policy objective by the 1991 Intelligent Vehicle-Highway Systems Act (IVHS Act). The act laid out an ambitious program of federal research in what would come to be called intelligent transportation systems (ITS), and also established federal leadership in creating a national plan for developing and deploying ITS. This legislation, and the follow-on 1997 Transportation Equity Act for the 21st Century (TEA 21), have provided $1 billion annually in federal funds for ITS. State and local governments in the U.S.A. spend roughly the same amount on ITS, and private industry, while restrained during the early research phase of ITS, is now becoming heavily involved in a variety of ITS technologies, as described below. This chapter will concern itself with surface transportation issues, since maritime and air transport have quite distinct histories, technologies, and issues. Also, in

Handbook of Transport Systems and Traffic Control, Edited by K.J. Button and D.A. Hensher

order to permit a focused discussion of issues and experiences, the chapter looks exclusively at intelligent transportation systems in the U.S.A., while noting that ITS is moving ahead briskly in both Europe and East Asia, particularly Japan.

2. Categories of intelligent transportation systems

The backbone of ITS is so-called "advanced traffic management systems," which, as the name suggests, include the full range of tools available to governmental traffic management agencies for controlling the flow of traffic on public thoroughfares (Sussman, 2000). These technologies include traffic signalization (including ramp metering and signal preemption for emergency vehicles and transit vehicles), electronic toll collection, incident management systems drawing on advanced technologies (for example, prompt identification of incident locations through a range of technologies), improved management of snow removal through automated fleet control systems, and quick response capabilities for emergency vehicles (police, ambulance, and fire) through the use of on-vehicle electronic maps which enable the driver to find the quickest route to an emergency. It is important to note that all such systems rely on some kind of traffic information collection.

Another major category of intelligent transportation systems is so-called "advanced traveler information systems" (ATIS) (Branscome and Keller, 1996). An important type of ATIS consists of systems built around traveler services, such as hotels, restaurants, tourist attractions, hospitals, and gasoline stations. These systems, sometimes called "electronic yellow pages," can typically be searched according to location requirements. Thus, for example, a traveler could request information about all Mexican restaurants in a certain town, or all gasoline stations within 5 miles of a certain freeway exit. ATIS traveler services are available on a variety of platforms, including handheld units, in-vehicle units, kiosks located at traveler rest points and popular tourist sites, and, increasingly, over the internet.

Such databases need to be updated regularly to be of maximum utility, and those that are part of a network can more easily be kept current than stand-alone databases, such as those found in handheld units. Its seems likely that, in coming years, more and more electronic yellow pages will be internet-based.

There are a number of information technologies which can enhance the usefulness of electronic yellow pages. The first is wireless communications. Thus, for example, after a traveler chooses from his/her electronic yellow pages a restaurant or hotel he/she would like to use, the traveler can call the facility on his/her cellphone for further information or to make reservations.

Another technology that can support electronic yellow pages is the electronic map, which is a geographic information system that contains road networks and that can be queried for directions. Thus, a traveler can instruct an electronic map to give him/her directions from his/her current location to any destination. Typically the traveler asks for the shortest route, but he/she may ask for the shortest route that, say, avoids freeways or has the minimum number of traffic signals. In some systems the traveler is able to ask for the quickest route, based on historical travel data, or even the quickest route based on current traffic conditions. The system then provides directions on a dashboard screen, or delivers them orally via a voice synthesizer. Electronic maps are now widely available, including, for example, Hertz's NeverLost system. One important use of the electronic map is to guide a traveler to a destination selected from an electronic yellow pages.

Yet another information technology that can supplement electronic yellow pages is automated vehicle location, or AVL. AVL is a system attached to a vehicle that continually updates the location of the vehicle. It does this either through communicating with a satellite that provides a point of reference (a global positioning system) or through "dead reckoning," whereby a vehicle's speed and turns are recorded by various sensors and then used to continually recalculate the vehicle's location. When AVL is available, an electronic yellow pages can provide a driver with information about a facility with reference to the driver's current location. Thus, for example, a traveler could query the database as to the availability of gasoline stations or Mexican restaurants within 5 miles of the vehicle's current location.

One specialized application of AVL is "Mayday" service, whereby a vehicle which has had an accident automatically transmits its location, together with a distress signal, to an operations center at some distant location. The operations center then dispatches an emergency vehicle to the site of the accident. Mayday is particularly useful in rural areas, where an accident leaving a driver unconscious might go unnoticed for hours and, by the time assistance arrived, the driver might be beyond help. This danger is especially acute where the vehicle leaves the road (for example, then striking a tree), or when the accident occurs at night.

Another category of advanced traveler information systems is those which provide current ("real time") traffic information. The backbone of most traffic information collection systems is the magnetic loop detector, a device embedded in pavement which records traffic flow through magnetic induction caused by vehicles passing over the detector. Loop detectors are typically spaced every ½ to 1 mile along a highway. Although loop detectors are the workhorse traffic sensors in this country and abroad, they have some important problems associated with them, which limit the accuracy of the information they provide.

First, they are inclined to malfunction because of the constant pounding they receive from traffic above, and a sizable percentage are often out of operation at any given time. Maintenance is difficult, because repair requires that the roadway be torn up, which can normally be done only during a few hours late at night when traffic is light. Even when functioning properly, loop detectors have their limitations. The spacing of the detectors means that several minutes may elapse before a traffic tie-up or easing of congestion is detected. And even when changes in traffic flow are detected, it is not necessarily obvious whether these are momentary or of more lasting significance. As a result, skilled personnel in a traffic information center are required to interpret the data from loop detectors, and often such personnel are not available.

Despite these problems, no cost-effective replacement technology has been found, though there has been extensive development and testing of alternative sensors, including infrared electronic eyes, acoustic sensors, and radar. These alternatives normally must operate from the roadside, which, among other problems, makes it difficult for them to track traffic conditions on individual lanes.

One technology in the developmental stage holding immense promise is the monitoring of cellphone traffic to determine the speed at which traffic is moving. Such a system, which relies on standard telecommunications infrastructure, including cellphone towers, monitors only the electronic energy from cellphone conversations; it does not monitor the actual conversation. Assuming that technical, financial, and privacy issues are solved, this technology has the potential to supersede loop detectors as the principal source of real-time traffic information. At present U.S. Wireless Corporation is the leader in the industry.

One useful way to correct for the deficiencies in loop detectors is the use of video cameras placed along the highway. Such cameras can typically be panned remotely by a traffic information center in order to view any location along the highway. A video feed is then sent to the center, where traffic experts can make a judgment as to the significance of the loop detector data.

Traffic information systems are normally supplemented with information about weather-related road conditions, such as snow or ice accumulation, and with information about road construction impacting traffic flow, in order to produce comprehensive reports on highway travel conditions. Such reports are used for traffic control purposes, including directing incident management teams to the site of an accident or rerouting traffic in case of a major accident that shuts down all lanes of traffic.

3. Information dissemination

Reports about highway travel conditions may also be disseminated to the traveling public, in a number of different ways. At present the most widespread technique is the variable message sign (VMS), sometimes called the changeable message sign (National Cooperative Highway Research Program, 1997). VMSs, placed overhead across roadways, can now be found in some three dozen of the U.S.'s biggest metropolitan areas. VMSs are normally used to alert motorists to congestion ahead, and typically display a message like "congestion 5 miles ahead" or "congestion between exit *xxx* and exit *yyy*." Lane closures are also often displayed in the case of incidents or where there is maintenance work. More advanced VMS systems, such as those in Atlanta, display travel time between major freeway exits.

Some questions about the optimal use of VMSs remain to be answered (Benson, 1996). Research has shown considerable driver resistance to the display of information about "recurrent" congestion, that is, rush-hour congestion that occurs regularly. Many drivers consider such information to be obvious, and therefore a distraction. On the other hand, for the out-of-town traveler or the occasional motorist, information about rush-hour congestion may not be obvious.

A second unanswered question is the use of VMSs for other than traffic information, such as announcements of public events, safety exhortations, and postings of the time of day. Research shows that travelers prefer to limit VMSs to the display of traffic information, but there is a rationale for including other public service information.

Finally, there is some question about the value of displaying travel times between major locations, such as freeway exits or intersections. The choice of locations is necessarily arbitrary, and may not be meaningful to many travelers, particularly those from out of town. In any case, research shows that many drivers prefer to know travel speeds rather than travel times. With all these caveats, however, the use of VMSs to display current traffic conditions is a major source for travel information.

The foregoing discussion has concerned fixed-location variable message signs. Another type of equipment is portable VMSs, which are smaller than fixed VMSs, and are located alongside highways. Portable VMSs have special uses, including notification of construction work ahead and notification of a driver's speed, calculated by use of radar. Such VMSs are typically left in place for a period of at least several weeks. Portable VMSs can also be used on an emergency basis, being towed to the location of a major accident and used for traffic control.

Variable message signs provide information visually. Alternatively, information can be provided orally. The most common such technology is the well-known radio "traffic-cast," available in nearly every American city, courtesy of Metro Traffic or Shadow Traffic. These for-profit firms provide local radio stations with free traffic broadcasts in exchange for air time, which they then sell to advertisers. MetroTraffic and Shadow Traffic collect traffic information from a range of sources, including aerial surveillance, monitoring of police and other emergency vehicle calls, and "probe" vehicles, or vehicles whose drivers, under contract, telephone information about traffic conditions where they are driving. Radio traffic broadcasts are limited in the number of corridors about which they can report, because of limits in broadcast time, and they often lack access to the most sophisticated systems for collecting and analyzing traffic information. But they have evolved a method of meeting the traffic information needs of a large portion of the nation's urban population without recourse to government funding, and they provide an impressive benchmark against which all the newer traffic information technologies may measure themselves.A second technology for providing traffic information orally is the highway advisory radio, or HAR. HARs are permitted only a small amount of power by the Federal Communications Commission, and so their range is normally limited to only a few miles. Typically a roadside sign calls motorists' attention to the fact that a certain type of information is currently available on HAR, and lists the station's frequency. Information may concern parking availability at a nearby airport, tourist attractions, and the like. On occasion HAR provides current traffic information when there has been a disruption in normal traffic patterns, and in these instances a variable message sign alerts the motorist when such information is available.

Thus far we have discussed well-established technologies for disseminating traffic information. With the advent of cable television, and then cellular phones, and most recently the internet, a new range of dissemination technologies has become available. Cable television currently broadcasts traffic information in two formats: as maps, and as video feeds from roadside cameras. Traffic-flow maps depict traffic conditions along freeways and major arterials, typically displaying slow traffic in red, moderate congestion in orange, and free flow in green. Video feeds typically are rotated among a number of key locations along major thoroughfares.

Cable TV is directed at the traveler who has not yet started his/her trip. Such pretrip information is not particularly helpful in the case of rapidly changing traffic conditions, but it can be very useful in the case of events likely to tie up traffic for at least an hour, such as a major accident, especially bad weather, or holiday weekend congestion. In such instances the traveler may decide to postpone his/her trip or take an alternative route.

Travelers already en route can use cellphones or pagers to learn current traffic conditions. The most important such use of cellphones is access to phone-in traffic information services, such as those provided in a dozen cities by the firm SmartRoutes. The driver uses a menu to select the route he/she is interested in, and a recorded message, regularly updated, then gives the driver traffic conditions along that route.

An alternative approach is a subscription arrangement, in which a traveler notifies a service in advance of what routes he/she is interested in. Whenever unusual traffic conditions develop along that route, the service notifies the traveler automatically over either his/her cellphone or a pager. Monthly service charges may run to about $50.

A recent decision by the U.S. Federal Communications Commission authorizes every state in the country to adopt "511" as a three-digit phone number offering traffic information, which offers the possibility of much greater utilization of phone-in systems. For example, in Cincinnati use of a traffic information phone-in number increased by 75% when the number was dropped from seven digits to three. Implementing this decision raises a number of difficult issues, including what agency in each state will manage the number, how the service will be financed, to what extent the service will accommodate long-distance calls, to what extent it will accommodate cellphone calls, and how it will be publicized (for example, should it be displayed on VMSs?). As of the year 2001 implementation has begun in seven states, Arizona, California, Kentucky, Michigan, Minnesota, Ohio, and Virginia.

The most recent technology to be brought to bear on traffic information is the in-vehicle internet connection, typically through a laptop computer. Traffic information websites are already widespread, presenting maps of traffic conditions, video feeds from roadside video cameras, and other corridor-specific information (Nowakowski, 2000). As use of the web increases, such sites could well become the primary in-vehicle source of traffic information.

All such traveler information systems based on real-time traffic information are designed primarily to assist travelers in cutting down on travel time. Yet the actual value to travelers of travel time savings remains a subject of substantial debate. A number of different theories of the value of travel time savings have been propounded, and one recent such study focused on the benefit of ITS-produced time savings suggests that the length of trip is the preeminent parameter in determining such value (Yang et al., 1999).

4. Commercial trucking

We turn now to the use of ATIS in commercial trucking, which is called "CVO" (commercial vehicle operations) by the ITS community. One of the earliest, and most effective, uses of ITS in the early 1990s was commercial vehicle automated fleet control systems. By using the automated vehicle location technologies described earlier in this chapter, trucking firms can keep track of the location of all their vehicles on the road, even when the fleets are spread throughout the nation. Such information has several applications. First, en route trucks that are not fully loaded can be instructed, through cellphones or pagers, to pick up newly placed shipment orders along their route of travel. Second, customers can be kept informed about the exact location of their shipment, permitting them to anticipate the shipment's exact time of arrival. Finally, automated vehicle location systems give management a clear picture of the exact routes and schedules being taken by drivers, permitting the tightening up of operating procedures.

CVO has also improved the efficiency with which state regulatory authorities can monitor commercial vehicles. Current law requires that trucks' licenses and safety credentials be reviewed at every state border, which in the past has meant that trucks must stop at every border crossing. ITS technologies now permit a truck to encode its license and safety credentials in an electronic transponder, which can be read electronically as the truck approaches a state border. Assuming the credentials are in order, the truck is permitted to cross the border without stopping. Using this technology, the HELP system along the west coast interstate highway corridor, running from the state of Washington through California and into the Southwest, and the I-75 Advantage program, running along I-75 from Georgia to Canada, now permit thousands of trucks daily to satisfy state border requirements without stopping.

Another regulatory requirement is that commercial vehicles be weighed at specified intervals along major highways, a rule arising from the fact that overweight trucks are more likely to tip over under certain conditions, and also are particularly damaging to pavement. Normally this requirement means that trucks must pull off the road and line up to be weighed on spring-loaded scales. However, ITS "weigh-in-motion" technologies, using piezoelectric sensors, permit such weighing to be done without the truck stopping, although the truck must pull off the road and slow down. Such technologies are beginning to be deployed around the country. One experimental technology would permit weight-in-motion without the truck needing to leave the main road or even slow down.

A final application of ITS to commercial trucking is the collection and distribution of traffic information along lengthy interstate corridors. Most traffic information systems are oriented to individual metropolitan areas, where the bulk of travel occurs, but interstate trucking can benefit from systems that collect and disseminate traffic information along corridors hundreds of miles long. Establishing such information systems presents major problems, including securing the co-operation of all relevant political jurisdictions, agreeing on quality standards and formats for the data, agreeing on an equitable method of financing the operation, and finding an overall manager. At present the I-95 Coalition, including over two dozen transportation authorities along I-95 from Richmond, Virginia, to Portland, Maine, has the most advanced interstate traffic information system in the nation.

5. Transit

We turn now to the use of ATIS in transit. Automated vehicle location systems, discussed above, have brought several benefits to transit. First, AVL permits much-improved fleet control. A major problem in managing transit fleets is the bunching that occurs when vehicles cannot keep to their schedule, because of delays due to traffic congestion or difficulties in loading and unloading passengers. AVL permits a fleet control center to identify when such bunching is occurring, and the center can then instruct the vehicles to speed up or slow down so as to relieve the bunching.

Another use of AVL is to protect the security of drivers and passengers. For example, in the case of an on-board robbery, assault or other incident, the driver can send an immediate distress signal to a control center, which, knowing the exact location of the vehicle, can then direct police, fire, ambulance, or other needed emergency support to the vehicle.

Finally, AVL systems can also be used to provide expected-time-of-arrival information to travelers waiting for a bus or train. Studies have shown that many transit travelers strongly dislike waiting for trains or buses, particularly during inclement weather, and the impatience and anxiety associated with waiting can be diminished if passengers are kept informed of exactly how much longer they must wait. Such time-of-arrival systems are beginning to be deployed.

Another use of information technologies to support transit is found in the increasingly sophisticated route-and-fare information systems available to travelers. A frequent complaint of transit users is the difficulty in learning what connections are needed to travel from origin to destination, and what the fare will be. Knowing the fare is particularly important in the frequent cases where exact change is required to board a bus. In the past such information has only been available through printed brochures, which are often hard to find and hard to read, or through phone-in information centers, which are often understaffed. Highly automated information systems now permit phone-in systems to provide route-and-fare information promptly. Such information can also be presented through kiosks placed in key locations, such as transit interchange stations.

A new application of ITS to transit is the VMS display of information about parking availability at nearby transit stations. Lack of parking is a major deterrent to the use of rail transit, and it is expected that, if automobile commuters are informed by VMS that a nearby transit rail station has parking available, a significant number of travelers may choose to drive to that station and continue their commute by transit rather than automobile. Such a system has been installed as a pilot project in the Maryland suburbs of Washington, DC.

Commuter ridesharing programs, often classified as a kind of transit, are also benefiting from ITS. Electronic databases of ridesharers are now fairly common around the country, and increasingly sophisticated databases and software are increasing the effectiveness of such systems. For example, one system searches for potential matches of ridesharers along the full length of a commute corridor, rather than using the traditional approach of looking for matches only among residents of a particular community. Another system aggressively seeks out downtown employers willing to participate in a ridesharing program, and then looks for matches among employees of companies located fairly close to one another.

6. The role of the public and private sectors

A critically important issue in ITS is the relative roles played by the public and private sectors. During the early, developmental stages of ITS the public sector and, in particular, the federal government were of fundamental importance in supporting research, development, and pilot deployments. Such state-of-the-art work as that on "hands-off-the-wheel" automated highways systems were funded almost exclusively by the federal government, as were such pilot projects as Florida's TravTek, an early 1990s demonstration in which onboard electronic route guidance systems were provided continually with updated traffic conditions via wireless communications.

As ITS evolved, public sector leadership increasingly shifted to states and localities, where the actual systems were to be deployed and maintained. Thus, in the year 2001, states and localities are estimated to spend on ITS somewhat more than what the federal government spends, which is $200 million annually. The influence of federal policy is gradually waning as states and localities pay a larger and larger percentage of the costs of deployment, and as the expertise necessary to evaluate and to plan for ITS projects is increasingly available at the state and local level. The emphasis in ITS at the state and local level has been on traffic management, including signalization upgrades, electronic toll collection, and state border electronic clearance of trucks.

The private sector, while in considerable measure taking a wait-and-see attitude towards ITS during the earliest stages, is becoming increasingly active, and a 1992 study forecasting that by the year 2020 the private sector would account for 80% of expenditures on ITS now seems quite prescient (IVHS America, 1992). This will largely be original equipment on automobiles, including electronic maps, radar-activated braking, sophisticated night vision devices, and the Mayday system described above.

One effort to bridge the gap between public and private sector programs has been federally sponsored "public-private partnerships." While in a handful of instances, such as Minnesota's GuideStar program and the state of Virginia's Travel Shenandoah traveler information program, genuine sharing of responsibility and risk has occurred, most such efforts have failed – primarily because of important institutional differences between the public and private sectors (e.g., Hall, 1999).

7. Conclusion

To summarize, then, during the last 10 years information technologies have dramatically altered transportation information systems, offering a raft of important new services to highway travelers, transit users, and truckers. Some of these services are already widely available, while others are just beginning to be deployed. It seems likely that improvements in transportation information systems during the next 10 years will be at least as great as those during the past 10 years. The private sector will increasingly be the driving force behind such changes.

References

Benson, B.G., (1996) "Motorist attitudes about content of variable-message signs", *Transportation research record*, No. 1550, *Human performance, driving simulation, information systems, and older drivers*. Washington, DC.: National Academy Press.

Branscome, L.M. and J.H. Keller (1996) *Converging infrastructures: Intelligent transportation and the national information infrastructure*. Cambridge, MA: MIT Press.

Hall, R.W. (1999) "Institutional issues in traveler information dissemination: Lessons learned from the TravInfo field operational test", *ITS Journal*, 5(3):3–38.

IVHS America (1992) *Strategic plan for intelligent vehicle-highway systems in the United States*. Washington, DC: IVHS America.

National Cooperative Highway Research Program (1997) *NCHRP Synthesis 237: Changeable message signs: A synthesis of highway practice*. Washington, DC: Transportation Research Board.

Nowakowski, C., P. Green and M. Kojima (2000) "How to design a traffic information website – a human factors approach", *ITS Quarterly*, 8(3):41–51.

Sussman, J. (2000) *Introduction to transportation systems*. Boston: Artech House.

Yang, H., O.S. Ma and S.C. Wong (1999) "New observations in the benefit evaluation of advanced traveler information systems", *ITS Journal*, 5(3)251–274.

Chapter 33

ROUTE GUIDANCE SYSTEMS

DONNA NELSON and PHILIP J. TARNOFF
University of Maryland, College Park, MD

1. Introduction

Until recently, route guidance systems were an insignificant element of the intelligent transportation systems (ITS) market. These systems were primarily used by the more sophisticated small-parcel delivery services, trucking firms, and taxi services. With the expansion of the internet and the growing popularity of in-vehicle electronics, route guidance systems are experiencing explosive growth. Current forecasts anticipate that these systems will become a major economic force in ITS. As they become increasingly integrated with the roadside infrastructure, route guidance systems will also have a major influence on transportation management.

1.1. Definition and alternative forms of route guidance

Route guidance is the provision of information to travelers to facilitate their selection of a path from their origin (or current location) to their destination.

There are many alternative forms of route guidance, including:

(1) In-vehicle route guidance vs. out-of-vehicle guidance. In-vehicle guidance provides information to travelers while they are en route to their destination. Out-of-vehicle route guidance (designated as stand-alone systems in this chapter) may be as simple as a paper map, or it may consist of internet-provided routing consisting of turn-by-turn instructions.

(2) Autonomous route guidance systems vs. systems fully integrated with the infrastructure. Autonomous route guidance is performed by a software-based system that performs its calculations using a static map database. It does not take into consideration prevailing traffic and roadway conditions. The fully integrated form of route guidance provides information based on the influence of prevailing traffic conditions. This latter form of route guidance is known as dynamic route guidance.

Handbook of Transport Systems and Traffic Control, Edited by K.J. Button and D.A. Hensher

(3) Operator-based route guidance vs. automated systems. Many vehicle manufacturers offer operator-based route guidance. Motorists desiring routing information call operators from their vehicles. The vehicle's location is automatically transmitted to the operator, who then provides aural directions to the motorist. Automated systems do not rely on a human operator, but use computer software and map databases to automatically calculate and present directions to the motorist.

As indicated in Table 1, different systems use a variety of criteria to provide routing information. Currently, the majority of systems rely on shortest-path criteria for provision of routing information, an approach that frequently leads to unsatisfactory results. As indicated in the "future" categories of this table, it is likely that additional criteria will be added as these systems become more sophisticated.

To a certain extent, increasing the number of options available to the user as well as the reliability of the routing information will be a function of the accuracy of the data available to the route guidance system from the infrastructure. It is not possible to calculate reliable travel times or costs for alternative routes without information related to vehicle speeds, toll costs, and other critical information. Use of route guidance to optimize traffic flow in a corridor will become increasingly important as the market penetration (percentage of equipped vehicles) increases.

2. Stand-alone systems

Stand-alone systems include any system that (ideally) is not utilized in a vehicle while the vehicle is in motion. Stand-alone systems tend to be used by travelers prior to their departure (pretrip planning) or by dispatch operators located in a fixed dispatch center. Primarily, the general public uses pretrip planning, while dispatch operations are used by commercial services such as trucking companies, taxi services, and delivery services. The paper map is the original stand-alone route guidance system. The advent of accurate computer mapping and the popularity of the internet have increased the use of computers as replacements for the paper map.

2.1. *Internet-based systems*

Many internet-based route guidance systems are available such as Mapquest, Yahoo! Maps, and MapBlast!. A recent search identified more than a dozen such systems. The majority of existing systems use shortest path (or distance) as the

Table 1
Route guidance criteria

Criterion	Comments
Shortest path	The most common routing criterion
Shortest time	Link travel times are essential for reliable routing
Maximum use of interstates	Available in many systems
Minimum use of interstates	Available in many systems
Minimum cost	Takes tolls into account but not actual travel cost. Not available in most systems
Tourist attractions	Vacationers make use of this feature, which would provide them with options for scenic routes or other tourist features
Optimum traffic management	Future – this feature will be important to the system manager (not the motorist) when high market penetration is experienced
Optimum mode split	Future – this form of route guidance will become more useful when accurate comparative travel times and costs are available. It will provide information on alternative modes as well as alternative routes

single criterion for providing routing. Most systems provide significant amounts of supplementary information such as hotels, restaurants, gas stations, and golf courses. MapBlast! even provides locations of speed traps along specific routes.

Figure 1 provides a sample of the Mapquest output. As shown in the figure, users are provided with both street maps and turn-by-turn directions. In addition, users may request return routing, which may be particularly critical in the presence of one-way streets and complex freeway ramps. Directions include the name of the route, miles of travel to the next turn, and direction of the next turn. This output is representative of the available internet-based systems.

Another category of information available on the internet is indirectly related to routing. This information includes the status of the roadways in various geographical areas, including congestion, average speeds, incidents, and construction. Status information is provided by public agencies responsible for operating the highway system and by private firms such as SmartRoutes. Firms that have traditionally provided traffic reports over commercial radio, such as Metro Traffic and Shadow Traffic (both of whom have now merged), are also providing internet-based traffic reports, with plans for major expansions in this area.

With the exception of information provided by regional organizations such as the I-95 Corridor Coalition, this information tends to be restricted to urban areas and is of little value to the long-distance traveler. The information may be provided in the form of textual descriptions of incidents, maps with changing colors representing speed ranges, or closed circuit television images such as those shown in Figure 2.

FASTEST ROUTE | SHORTEST ROUTE | AVOID HIGHWAYS

DIRECTIONS	DISTANCE
1: Start out going South on 50TH AVE towards LEHIGH RD.	0.0 miles (0.1 km)
2: 50TH AVE becomes LEHIGH RD.	0.1 miles (0.1 km)
3: Turn RIGHT onto CORPORAL FRANK S SCOTT DR.	0.1 miles (0.1 km)
4: Turn RIGHT onto PAINT BRANCH PKWY.	1.0 miles (1.6 km)
5: Turn RIGHT onto BALTIMORE AVE/US-1.	2.1 miles (3.4 km)
6: Take the I-95 N/BELTWAY NORTH/I-495 N ramp towards BALTIMORE/SILVER SPRING.	0.2 miles (0.4 km)
7: Merge onto CAPITAL BELTWAY.	19.8 miles (31.8 km)
8: Take the VA-267 TOLL W exit, exit number 12B, towards DULLES AIRPORT.	0.4 miles (0.6 km)
9: Merge onto VA-267 W (Portions toll).	4.9 miles (7.8 km)
10: Take the exit on the left towards DULLES AIRPORT.	0.1 miles (0.2 km)
11: Merge onto DULLES AIRPORT ACCESS RD.	8.4 miles (13.5 km)
TOTAL ESTIMATED TIME: 46 minutes	**TOTAL DISTANCE: 37.0 miles (** 59.6km)

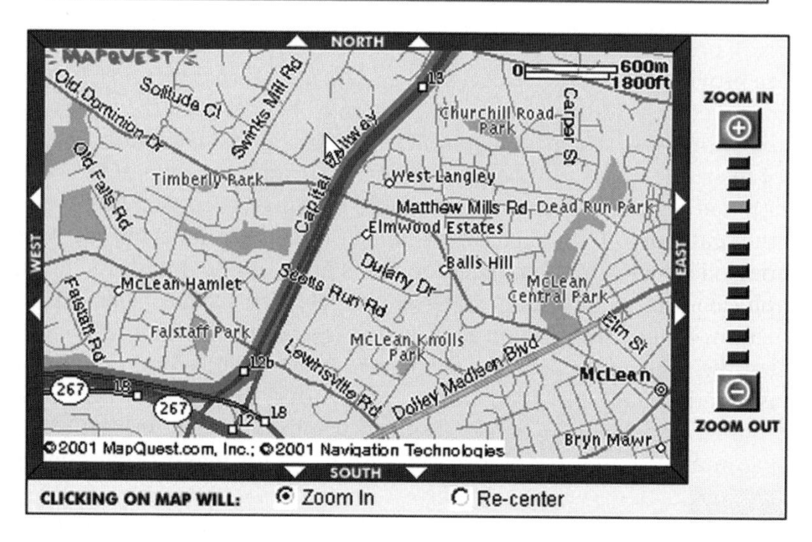

Figure 1. Sample internet-based route guidance with map and driving directions.

Figure 2. Sample closed circuit television image from a website.

Many innovative approaches are being taken to the display of roadway status on the internet. Two representative examples include:

(1) The Royal Automobile Club (RAC) of the U.K. (http://www.rac.co.uk/) enables users to enter their travel preferences and other details into personal travel profiles. A personalized update is then provided to the user on request. Other services have extended this concept from a "pull" operation (where the user pulls information from the internet) to a "push" operation in which the system automatically pushes information to the user when an incident occurs.

(2) The Subway Navigator website (http://www.subwaynavigator.com/) can provide optimum routes for subway use in more than 70 cities throughout the world. This well-designed website provides information in English or French. In addition to routing information, it provides estimated travel times and route maps for each system.

To date, internet providers of routing information have tended to rely on static information as the basis for routing information. This approach has been taken because of the absence of comprehensive traffic flow information from public agencies, and the high cost of installing equipment capable of collecting such data throughout an urban area. Obviously, internet-based route guidance would be most valuable if the two forms of information (static routing and roadway status) were to be combined into a single package. In this form, routing would take into consideration current roadway conditions. The development of this integrated capability is inhibited by the following two factors:

(1) Availability of adequate roadway status information for the long-distance traveler; and
(2) The duration between the time that the travelers receive the routing information at their origin and the time that they depart for the trip.

As the public agencies collect more extensive traffic flow information and access to the internet from the vehicle becomes more common, it is likely that this integration will occur.

2.2. *Dispatching systems*

Much of today's route guidance software began with the development of systems for trucking companies, small-package delivery services, and taxi services. These dispatcher-based systems rely on information provided by the dispatcher using route guidance software installed at the dispatch center. Route guidance software is used both to provide assistance to the unfamiliar driver, and to optimize the pickup and delivery assignments.

This type of software tends to be very specialized. Obviously, the requirements of a taxi company that picks up and drops off a single passenger without intermediate stops are quite different from those of a small-package delivery service whose operations are optimized by the number of intermediate stops that can be made on a single route. Other customers for this type of software include transit operators, school bus fleets, and mail services. Here again, the types of pickups and deliveries are very different for each service.

Commercial route guidance software is offered by a number of firms, such as GIRO Enterprises, who specialize in public transit, postal delivery, and school bus operations, the Lightstone Group, MicroAnalytics, Roadnet Technologies, Inc., and RouteSmart Technologies. Information related to the services offered by each of these firms is provided on their internet sites.

Typically, the capabilities provided include:

(1) Route optimization and scheduling for both personnel and delivery vehicles;
(2) Planning tools;
(3) Scheduling and information management tools that track the productivity of individual drivers;
(4) Dynamic addition of new stops to routes while deliveries are in progress, in response to calls for service; and
(5) Information to customers regarding pickup times, location, and scheduling of deliveries.

These route guidance systems have been shown to improve the efficiency of most delivery services. For example, the Lightstone Group (Samuel, 1998) reports that its system has saved the Dearborn Regional Clinical Laboratories in the Detroit area $250 000 per year because that organization managed to reduce its van fleet from 22 to 19 and reduce the number of drivers by five.

Most of these services, like their internet counterparts would benefit from the availability of up-to-date, accurate, real-time traffic information. Time is money to commercial carriers, and knowledge of unexpected congestion would significantly improve their efficiency.

3. In-vehicle systems

A quiet revolution is currently under way within the automotive industry. Through the availability of low-cost computing and increasingly sophisticated display technology, automobile manufacturers throughout the world are introducing in-vehicle electronics at a rapid rate. Route guidance systems are at the heart of these systems.

Two fundamental forms of route guidance currently exist. Although there is no universally accepted designation for these forms, they will be denoted as operator-assisted route guidance and autonomous route guidance for the purpose of this discussion.

3.1. Operator-assisted route guidance

In the U.S.A., General Motors' On-Star and Ford's Rescu provide operator assisted route guidance. Both European and Japanese manufacturers offer similar systems. A generic block diagram of the operator-assisted system is shown in Figure 3.

The operator-assisted system requires the motorist to pay an annual subscription fee to the automobile manufacturer or a third-party provider.

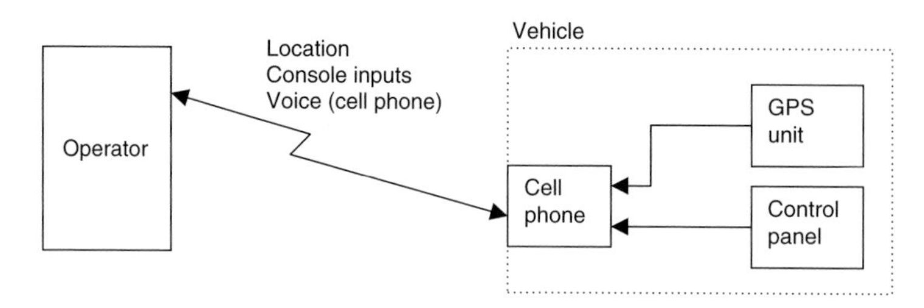

Figure 3. Operator assisted route guidance.

The operator provides personalized responses to the motorist's requests for information. While routing information is a fundamental element of the service provided by these systems, extensive ancillary services may also provided. In addition to routing information, they might include:

(1) nearest restaurants, hotels or tourist attractions;
(2) stolen-vehicle tracking;
(3) automatic dispatch of emergency services in the event that an airbag inflates, and no call is received from the motorist;
(4) dispatch of roadside repair services when a request is received from the motorist;
(5) response of law enforcement services in the event that the hijack alarm button is depressed;
(6) concierge services such as restaurant reservations or theater tickets.

All communication between the operator and the motorist are by voice and are performed using the cellular telephone. The simplicity of the phone is an asset of this form of system.

To date in the U.S.A. the operator-assisted systems have proven more popular than the autonomous systems. According to Russell Shields (1998), chairman of Navigation Technologies, development of this approach was encouraged by Motorola, a company with a strong interest in selling communications products, working in co-operation with Ford. Craig Lyman (1998), vice president of sales and marketing for Etak's consumer division, offers an alternative view. He feels that the operator-assisted systems attract consumers because of their security features, and that American consumers are not as interested in route guidance systems as their European and Japanese counterparts. He supports this statement with the relatively slow growth of the navigation (route guidance) market in the U.S.A. through 1998. Recent information published by the Consumer Electronic Manufacturers Association, however, indicates that a growing number of in-vehicle navigation systems are being sold in the U.S.A. as customers become aware of them and of the utility they offer.

3.2. Autonomous route guidance

Autonomous route guidance systems provide all of the routing and supplementary functions using equipment installed on the vehicle. No communication with remote operators is required. The information is contained on compact discs that are purchased by the motorist for specific geographic areas. For example, the 48 continental states in the U.S.A. are included on five CDs. Motorists are offered a subscription to a service that periodically provides new CDs with updated information. These systems are available from all vehicle manufacturers in Europe, Japan, and the U.S.A.

The autonomous route guidance system includes a GPS unit, a processor, a CD player, and a display. The GPS unit continuously provides the vehicle's location to the computer, which tracks the vehicle's position by co-ordinating the GPS information with the map data contained on the CD. When routing information is requested, the processor calculates the best path from the vehicle's current location to the entered destination. Most systems offer a variety of options, including shortest path, all interstates, or no interstates. The selected route is presented to the motorist in the form of turn-by-turn instructions that are presented both as visual and as audio outputs. Future systems are likely to employ DVD technology for map storage to increase storage capacity and the speed of operation.

Since they are self-contained, autonomous systems avoid reliance on uncertain cellular telephone contacts with remote operators. They also avoid the need for a continuing subscription payment to continue the service. However, these systems are not able to offer the range of features provided by operator-assisted services, since it is generally not possible to summon law enforcement and emergency services without subscribing to a separate service. In addition, the operator-assisted services are more likely to offer the most up-to-date information. A few vehicle manufacturers include an option of in-vehicle navigation, combined with "mayday" emergency services.

U.S. sales of autonomous systems have been outpaced by sales in Europe and Japan. However, as indicated by the projections of Table 2 (McManus, 1998), the market is anticipated to grow significantly by 2005.

This projected growth will depend on suppliers' ability to improve the ease of data entry, increase the functionality, and reduce the cost of autonomous systems.

3.3. Dynamic route guidance systems

Many research projects are currently under way throughout the world into a concept known as dynamic route guidance. Dynamic route guidance is defined as the provision of routing instructions that are automatically altered in response to changes in traffic and roadway conditions. This technology will be an important adjunct to ITS technology, in that it will facilitate the ability to relieve traffic conditions in the presence of major incidents. Examples of these projects include the CLEOPATRA project at the University of Leeds in the U.K., and the Dynamic Traffic Assignment (DTA) project sponsored by the U.S. Federal Highway Administration.

Dynamic traffic assignment requires the accurate projection of traffic conditions during the times that the motorist will traverse the assigned route. This requires an understanding of the area-wide traffic conditions under the unusual and unpredictable circumstances of an incident. Since incidents can take many forms, from a highway accident, hazardous spill, or special event to extreme weather conditions, this is a difficult task.

As the market penetration of equipment with dynamic-traffic-assignment features increases, the problem of forecasting traffic conditions is further compounded due to the need to estimate the impact of multiple vehicles being diverted over the same route. In this latter case, equity becomes an issue since it may be necessary to divert some motorists onto routes that are less desirable than others in order to maintain a balance of traffic on all available routes within a corridor.

Dynamic route guidance systems offer the potential to provide significant benefits to both individual drivers and the overall transportation system. Theoretically, motorists would be guided around incidents while the roadway system was being used in an optimum fashion. While these benefits are tantalizing, the implementation of this capability will require a fully integrated system that includes comprehensive surveillance of roadway conditions and co-ordinated dissemination of this information to motorists through a variety of route guidance devices, as well as through the media and infrastructure-based displays such as dynamic message signs and highway advisory radio. It also requires the development of new analytical tools for projection of traffic conditions.

Table 2
Market for location and guidance systems

Market	Units sold 1997	Units sold 2005
U.S.A.	10 000	129 million
Western Europe	50 000	188 million
Japan	2.5 million	17 million

Table 3
Performance of TravTek vehicles with respect to non-TravTek vehicles, due to traffic incidents

Measure of performance	Under recurring congestion (%)	No initial congestion (%)
Average trip duration	–12	–11
Average trip length	–4	–3
Vehicle stops	–18	–4
Fuel consumption	–10	–10
HC emissions	–10	–12
CO emissions	–4	–5
NO_x emissions	–3	–5
Accident risk	6	13

While the required vehicle surveillance capabilities are not currently available in the U.S.A. or Europe, Japan is currently exploring the practical application of this technology. The Vehicle Information and Communications System Center (VICS) led by the National Police Agency has installed a comprehensive communications and surveillance infrastructure that has permitted the development of a number of operational systems in the Tokyo area. Plans exist for the expansion of this service to Osaka and other major urban areas.

3.4. Benefits of in-vehicle systems

The most comprehensive evaluation of the benefits of route guidance was performed in connection with the TravTek project sponsored by the Federal Highway Administration during the early 1990s. This project included the instrumentation of 200 vehicles in the metropolitan Orlando, Florida area with route guidance capabilities to evaluate specific system architectures as well as the potential benefits of route guidance to the motoring public. Motorists in TravTek-equipped vehicles were provided with dynamic routing that took the presence of accidents and other forms of non-recurring events into account. The TravTek evaluation included an evaluation of the specific system, supported by traffic simulation studies to permit an estimation of the network-wide effects of the system for different market penetration levels. The most significant of the findings of this study were the simulated benefits to motorists in vehicles with TravTek features. These benefits, summarized in Table 3, show improvements in all areas except for safety. The small estimated impact on accidents for TravTek-equipped vehicles is the result of routing on surface streets, with their higher accident rates in the presence of freeway accidents.

4. The future of route guidance

It is clear that route guidance is an emerging technology that is still in its infancy. However, it offers significant potential to reduce motorist frustration and improve the efficiency of the highway system.

4.1. Future challenges

Much work needs to be done. The issues that must be addressed include:

(1) *Improved mapping accuracy.* High mapping accuracy is an essential requirement for route guidance systems. It is estimated that there are between 2 million and 10 million roadway sections for the U.S.A. alone. Each section has approximately 150 data elements associated with it. Thus, it is necessary to maintain more than one billion database elements. These elements are constantly changing. Continuing emphasis must be placed on maintaining an extremely high level of accuracy for these database elements.

(2) *Displays.* A number of alternative displays are currently available for route guidance systems. These include CRT displays, audio displays, and heads-up displays in which the image is projected in front of the windshield of the vehicle. None of these displays is completely satisfactory. Work must continue on the development of improved display technology that can be rapidly comprehended by the motorist and reduces the potential for driver distraction.

(3) *Dynamic traffic assignment.* The real benefits of route guidance, both to the motorist and to the overall transportation system, will occur when routing takes traffic and roadway conditions into account. This will require the establishment of a close partnership between the public and private sectors. Improved prediction and optimization algorithms must also be developed.

If these problems can be solved, route guidance systems will achieve widespread acceptance and significant market penetration.

4.2. Vision for the future

Wireless internet is becoming increasingly popular. Many handheld and in-vehicle products are becoming available that provide convenient access to this powerful medium. As the internet moves into the vehicle, the range of services (including route guidance) that can be accessed by the motorist will be extensive. Real-time interfaces of in-vehicle devices such as GPS units, alarms, and airbags with the internet are already feasible. With this type of interface, the alternatives described in this chapter will disappear. It will be possible to obtain internet-based routing information from sites such as Yahoo and Mapquest in the vehicle. These sites will be able to sense vehicle location automatically and provide en-route guidance. They will be able to supplement this information with hotel, restaurant and reservations services that are already available on these sites. In addition, the motorist will have access to all of the other information available on the internet.

With the availability of these expanded services, the popularity of route guidance as one of a number of information services will become explosive. However, it is equally likely that the internet-based devices available from the information technology industry will replace the specialized route guidance devices currently being offered by vehicle manufacturers. If these projections are correct, the industry is likely to see significant technology and structural changes in the future.

References

Lyman, C. (1998) "Taking turns", *ITS International*, 17:61–62.
McManus, S. (1998) "Locating markets", *ITS International*, 17:58–59.
Samuel, P. (1998) "Route leader's flying start", *ITS International*, 15:65–66.
Shields, R. (1998) "Shields of vision", *ITS International*, 15:67–69.

Chapter 34

MODELING TRAFFIC SIGNAL CONTROL

MIKE SMITH, JANET CLEGG and ROBERT YARROW*
University of York

1. Introduction

1.1. An overview of this chapter

In this chapter we consider the modeling of traffic signal control. This is an enormous subject with very many complications. So as to allow a self-contained account which is specific enough to be informative, we have made many choices which might well appear as oversimplifications; without the simplifying choices the chapter would have been much longer, much more complicated, and much less informative.

The chapter seeks to reduce the distance which currently lies between traffic signal control modeling and transport planning modeling. For this reason the chapter begins by giving a practical example of the way traffic signal control may be used to deliver public transport benefits.

Then we emphasize the simplest traffic signal control model, the continuous model of a single junction operating on a fixed-time basis first studied by Webster (1958), and show how this model may be used to "optimize" fixed-time signal timings but only with reference to standard delay-minimizing aims and only by making the assumption that flows are essentially unchanged by any control changes. Next we outline Miller's (1965) discrete model, with individual vehicles, and show how Miller's strategy for controlling a single junction arises from this model. Once again, Miller considered only standard delay-minimizing aims and made the usual assumption that flows are essentially unchanged by any control changes. Networks of signalized junctions are then considered, and the basic principles of TRANSYT, SCOOT, and SCATS are outlined. The optimization procedures within all three of these systems embody delay-minimizing aims and make the assumption that flows are essentially unchanged by any control changes made.

*We are very grateful for the help of Rahmi Ackelik and Peter Lowrie. We are also grateful for EPSRC support for traffic signal control research over many years.

Handbook of Transport Systems and Traffic Control, Edited by K.J. Button and D.A. Hensher

Finally, the chapter describes an approach to signal control modeling which should permit arbitrary transport aims or targets to be addressed by using signal control adjustments, and should permit reasonable allowance to be made for travelers' responses to control changes. Using this model, transport planning targets may be pursued using traffic signal controls.

1.2. Traffic signal control and transportation management

There are very many models of traffic signal operation and a vast literature concerning traffic signal control in real life. Much of this literature deals with hard-edged technical issues (including hardware specification and, these days, standards for communication protocols, for example) but it also extends through to "softer" policy issues (including the time required for pedestrians to cross a road and how to best provide this time, for example).

Throughout the traffic signal control literature, however, there has been a separation from the wider concerns of holistic transport management. This separation has become increasingly clear as transport management has moved toward the forefront of political and social debate; it is clearly seen by comparing models used by transport planners and signal control engineers. Transport planning models usually cover a large geographical area and invariably envisage alternative futures, in which transport flows change substantially as a consequence of different planning assumptions. On the other hand, traffic signal control models usually cover a small geographical area and traffic flows are assumed fixed.

This separation is illustrated in this series on modeling transportation; signal controls are scarcely mentioned in those chapters which embrace transportation management. The separation has arisen for three main reasons: (i) there is a genuine conflict between the short and the long term, (ii) the-short term issues are more immediately and obviously pressing when viewed by an engineer with short-term control facilities ready to hand, and (iii) there are currently theoretical difficulties in the way of signals taking proper account of long-term targets. Here we discuss these issues, which are overlapping.

A conflict between short-term goals and long-term goals

If controls are sought which seek to encourage different routes or modes to be used in the future, so as to *reduce* congestion in the future, then the immediate effect will almost certainly be an *increase* in congestion on those routes and modes now being used. This conflict would seem to be unavoidable, and is rendered powerful by the lack of clear evidence that long-run demand management works in practice. Some evidence that signal control may be used to achieve long term-aims is available as a consequence of the MUSIC project; see, for example, Clegg et al. (2000). "MUSIC" stands for "Managing transport USIng traffic flow Control and other measures." The final report on the MUSIC project is available on the Web (Clegg, 2001).

The short-term is more pressing

The signal control engineer lives on the front line and so is continually seeking to control unpredictable minute-to-minute and hour-to-hour situations using the best technology available to him/her, designed for those timescales. Immediate concerns predominate. This leaves little time or peace to consider wider transportation planning or management concerns, which focus on the effects of policy over weeks, months, or years.

A long-term view requires theoretical tools which do not yet exist

To consider meeting long-term targets using traffic control is hard. Even if all practical issues are (for the moment) ignored, the problem of designing signal control strategies which meet long-term needs is yet to be solved.

The separation between signal control and long-term planning has, however, not been total; for example traffic signals play a large part in many bus priority schemes, and such schemes are often designed with a long-term view in mind. Such applications of signal control to bus priority are against the grain of signal controls more generally implemented. They are *exceptions*.

1.3. Public transport priority using traffic signal control

Public transport priority using traffic signals currently takes essentially two main forms: *queue relocation* and *selective vehicle detection*. These two techniques for giving buses priority may be combined.

Queue relocation

Here traffic signals are utilized so as to relocate vehicle queues upstream to where the queues may be bypassed by public transport on bus-only lanes. There are also many special implementations, designed using local engineering judgment, which seek to achieve low congestion on certain bus routes or low traffic flows along certain roads by holding non-bus traffic back; the best-known and longest-running of these is the Bitterne scheme in Southampton, U.K., which has for decades metered traffic joining a bus route so as to reduce congestion on the bus route.

Figure 1 illustrates the basic idea behind such queue relocation schemes; this particular arrangement has been implemented within the city of York to reduce delays to buses entering the city along an important radial road. New signals (called presignals as they are upstream of the congested signalized junction being "assisted") and a bus lane have been installed where the radial road is wide.

Signal timings at the downstream junction and the presignal are carefully adjusted so that almost all queuing occurs at the presignal and almost none at the downstream junction. The presignal may be thought of as a "tap" which meters traffic appropriately into the downstream junction, so that little or no queuing occurs there.

Figure 2 shows the effect of implementing this queue relocation scheme as measured on street "before and after." Here the performance measure adopted was the number of vehicles queuing at the signal-controlled junction downstream from the presignals. Vehicles queuing at the downstream junction delay buses, whereas the relocated vehicles do not delay buses as the bus lane bypasses these. Thus the difference between the "before" and "after" numbers of vehicles queuing at the downstream signal-controlled junction is a measure of the reduction in queuing delay felt by buses.

Selective vehicle detection

Within an existing traffic control system, with or without queue relocation such as that above, the presence of buses may be detected. The traffic signals may then accord buses priority at existing traffic signals. Care has to be exercised here if the signals are already successfully optimized for general traffic: just giving buses quick priority and then returning to the optimized scheme may cause high delays to general traffic. Furthermore, these higher delays may later interfere greatly with the bus traffic whose benefit is being pursued! Much effort has been spent on designing systems which seek to circumvent these problems.

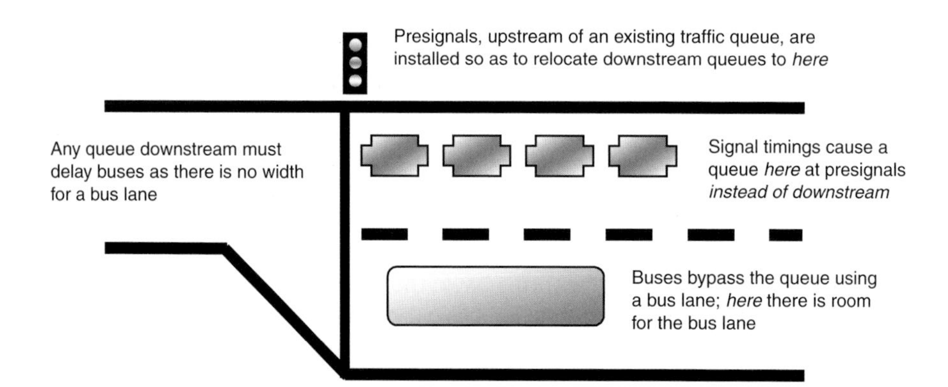

Figure 1. Diagram showing the operation of a queue relocation scheme in the city of York, U.K. Traffic flows from right to left. The implementation was part of the MUSIC project.

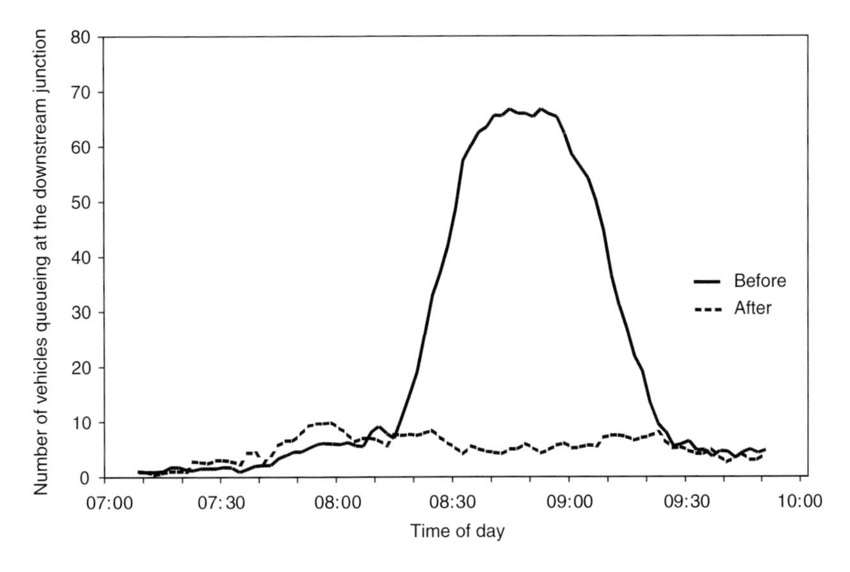

Figure 2. Number of vehicles queuing at the downstream signal-controlled junction.

1.4. *Traffic signal control within transportation management models*

The above exceptional use of traffic signals to manage the car/bus interaction suggests using traffic signal controls more uniformly to achieve multimodal transport management objectives. In order to do this, traffic signal controls must become part of transport management and transport planning models. As soon as this is done new opportunities and problems arise.

On the one hand, the vast number of new (signal) control variables should permit planning and management objectives to be met which could not previously be met, and should also allow far more discriminating numerical targets to be met far more precisely than heretofore. On the other hand, the vast number of new (signal) control variables renders the optimization of these variables far harder and the number of reasonable "options" to be considered becomes astronomical.

1.5. *New targets*

The need to connect traffic management and planning models and traffic control models is now being made more urgent, in the U.K. at least, because of the introduction of new targets. For example, *A new deal for transport* (Department of the Environment, Transport and the Regions, 1998) proposes that transportation should provide:

(1) cleaner air,
(2) a cut in the stranglehold of traffic in city centers,
(3) more quality places to live, and
(4) easier and safer walking and cycling.

How is signal control to help in providing these things? In this chapter we will outline the standard approach to signal control and then we will outline a new approach aimed at meeting new targets in line with the proposed provisions listed above.

2. Background to current traffic signal control

Traffic signals (with a movable arm, like signals which now control trains) were introduced in 1868 outside Westminster Abbey in the U.K. They were operated by a policeman and the objective was to separate conflicting traffic streams so as to avoid collisions. Collision avoidance has been a main purpose of traffic signal control ever since. The first automatic traffic signals were installed in Detroit in the 1920s. The first vehicle-actuated traffic signal was installed in Baltimore in 1928. (I am grateful to Underwood (1996) for some of this information.)

2.1. Standard traffic signal control strategies

Existing, standard traffic signal control strategies usually seek to do the best (usually interpreted as minimizing a combination of total travel time and stops for vehicles or travelers) for the observed local traffic pattern assuming that this pattern will not change. This is natural if a solution to the problems of just the next few seconds or minutes is being sought. However, this assumption that traffic patterns are fixed (which is very reasonable over short time scales) means that existing strategies do not generally, for example, try today to encourage tomorrow's traffic away from bus routes likely to be congested tomorrow, because in choosing the signal controls today there is a presumption that the traffic pattern will not change. It is clearly not reasonable to assume that tomorrow will be just as bad as today whatever traffic signal controls are chosen. as a "solution" to a current problem will thus be excluded.

An important exception to this rule is in the implementation of public transport priority. Here there is the intent to make public transport quicker with the aim of increasing patronage or reducing the decline of patronage. Here then a longer view is taken, but this longer view is against the grain of the rest of traffic signal control.

A wide-ranging review of current traffic signal systems is given by Wood (1993).

3. Continuous modeling of fixed-time isolated signals

Signals at a single road junction today are operated in a wide variety of ways and the simplest is the fixed-time method. In this method certain sets of lanes are called "stages." The stages are sets of lanes approaching the junction which are given green simultaneously. In the simplest case the stages are such that traffic approaching the junction along distinct lanes in the same stage follows paths which do not intersect, and so lanes in the same stage may safely be given green simultaneously.

If there are K stages at a junction the signal display changes from green to stage 1 to green to stage 2 to green to stage 3, ..., to green to stage K to green to stage 1, and so on. The method of changing display is said to be "fixed time" if the periods of time when the signal is green for stage k are all of the following form (where t_k is the time after some reference time and is measured in seconds):

$$t_k \text{ to } t_k + \tau Y_k,$$
$$\tau + t_k \text{ to } \tau + t_k + \tau Y_k,$$
$$2\tau + t_k \text{ to } 2\tau + t_k + \tau Y_k,$$
$$3\tau + t_k \text{ to } 3\tau + t_k + \tau Y_k, \ldots$$

and so on. That is, the time period during which stage k is green consists of repeated periods of time of constant length, τY_k seconds, which "come around" regularly (every τ seconds). τ seconds is called the cycle time of the signal (usually, $\tau \leq 120$), τY_k is the green-time for stage k and t_1 may be thought of as the offset associated with the junction. Different junctions will have their own t_1. Taken together, these describe the way in which the fixed-time evolutions at the separate junctions are related.

The simplest reasonable model of traffic signal control assumes that the above changes of green display, from one stage to the next, take some fixed time η seconds, say, during which the display is red for all those lanes not in both stages (to allow traffic which has just had priority in one stage the time to clear the junction before "new" conflicting traffic flows having priority in the next stage are allowed to flow). This time of η seconds may be called an "intergreen." We shall suppose that each lane is given green for exactly one period of time per cycle; this time then comprises the sum of the green-times of all those stages containing that lane together with the sum of the intergreens between those stages. In this simple model we are assuming that each intergreen equals η seconds so the proportions $Y_1, Y_2, \ldots, Y_K$ of the cycle time for which the stages $1, 2, 3, \ldots, K$ are given green add up to $1 - K\eta\tau$ (In this chapter we consider only this simplest case, as beyond this there are technical complications which do not affect the principles to be considered.)

Thus, given a signal-controlled junction, the signals are said to be operating fixed-time if the variables above, $\tau, Y_1, Y_2, \ldots, Y_K$, are fixed as time passes. In this case it would be natural to seek the "best" values of these variables. This has always been the overwhelming reason for modeling traffic signals (even with very complicated models) – to find approximations to the "best" signal timings.

The first thoroughgoing modeling study of isolated traffic signals was by Webster (1958) and this was followed up by Webster and Cobbe (1966), which has become a standard reference on fixed-time signal control. Webster developed a formula (now known as Webster's formula) for the delay imposed on a single traffic stream by a traffic signal operating on a fixed-time schedule, determined by *settings* τ, Y_1, Y_2, ..., Y_K, and used this formula to justify a simple rule for calculating fixed-time settings which approximately minimizes the total of all the delays experienced by all vehicles passing through the junction in a unit time. This study involved (i) theoretical modeling, (ii) simulation to validate the delay formula, and then (iii) optimization to estimate a "reasonable" signal setting. In the many studies which have succeeded this pioneering work, all three elements have also played a part.

A more rigorously correct and complete treatment of the problem of optimizing fixed-time signals has since been given by Allsop (1971), still using Webster's delay formula. Allsop's method sometimes, particularly for junctions where there are lanes which belong to separate stages, leads to timings which are rather different from Webster's. The principle behind Allsop's method is classical optimization and is outlined below.

Both Webster and Allsop make the central assumption that the traffic flows through the junction are given as fixed average flow rates along each lane approaching the junction. By "fixed" here we mean that the average flow rates are unaffected by the timings chosen. Innocuous though this assumption appears to be, particularly for "isolated" junctions, it has become the single greatest barrier to the effective design of traffic signal timings to achieve reasonable targets today.

3.1. Delay formula, routes, lanes, and stages

In order to optimize the signal control variables above we need an objective function; in this signal control context the objective function has usually been chosen to be the total rate of delay D at the junction. (The units of D are seconds per second, so D is dimensionless.) In order to minimize D we require a formula which estimates the average medium-run delay d_i to vehicles on a single lane caused by traffic signals operating on a fixed-time basis. (This leads to a corresponding formula for the total rate of delay D.) An example of such a formula is Webster's formula. This was designed to estimate the average delay per vehicle (in seconds per vehicle, say) as a function of (i) the average traffic inflow rate along that lane (x_i vehicles per second), (ii) the proportion of time the lane is given green (y_i), (iii) the saturation flow of the lane (s_i vehicles per second), and (iv) the cycle time (τ seconds). The formula is

$$d_i(x_i, y_i) = 9/20\{\tau(1 - y_i)^2/(1 - x_i/s_i) + x_i/[s_i y_i(s_i y_i - x_i)]\}. \tag{1}$$

The flow along lane i, x_i above, is the sum of the flows along all routes to which the lane belongs. Thus

$$x_i = \sum\nolimits_{\text{those routes } r \text{ such that } r \text{ contains lane } i} X_r. \tag{2}$$

Also, the green-time awarded to lane i, y_i above, is the sum of those green-times awarded to the stages to which the lane belongs, together with the sum of those intergreens between the stages containing lane i. Thus

$$y_i = \sum\nolimits_{\text{those } k \text{ such that stage } k \text{ contains lane } i} Y_k + [(\text{number of stages containing lane } i) - 1]\eta. \tag{3}$$

The second term of eq. (1) is close to a well-known formula, derived by Pollaczek and Kintchine, for the average delay experienced by a Poisson stream of traffic being served by a server with identical service times equal to $1/s_i y_i$ seconds. The first term approximates the additional delay due to the stop–start nature of traffic signal control.

In this chapter we shall consider just the Y_k as control variables: the cycle time is to be fixed. Again, this is so as to be as simple as possible. We are thinking of congested urban networks and in such networks the cycle time is often fixed at the maximum allowed.

4. Continuous modeling and optimization of fixed-time isolated signals for fixed flows

In this section we outline the standard method of optimizing traffic signal green-times at a single junction operated on a fixed-time basis. We shall utilize the simplest possible reasonable model, based on that just presented, and follow Allsop (1971). The junction will have just two lanes approaching it, lane 1 and lane 2, with saturation flows s_1 and s_2, and just two stages, comprising lanes 1 and 2, respectively. Since there are just two approach lanes to the signal and just two stages,

$$y_1 = Y_1 \quad \text{and} \quad y_2 = Y_2.$$

In this case we write $d_1(x_1, Y_1)$ for $d_1(x_1, y_1)$ and $d_2(x_2, Y_2)$ for $d_2(x_2, y_2)$.

4.1. Stages and stage green-times

As above, we shall suppose that stage 1 is given green for τY_1 seconds and stage 2 is given green for τY_2 seconds during each cycle of τ seconds. Of course, we then have

$$Y_1 + Y_2 = 1\text{–}2\eta/\tau.$$

Here the intergreen whenever the signal changes is η seconds and this time becomes lost to all lanes. So the time lost over a cycle, which embraces two signal changes, must be 2η seconds. If $m\tau$ seconds is the minimum green-time for each approach during one cycle then we also have $Y_1 \geq m$, $Y_2 \geq m$.

4.2. *Supply feasibility*

Let the flows x_1, x_2 be fixed. We suppose that the stage green-times are adjusted so that the fixed traffic flows can get through the junction. Thus we suppose that $x_1 < Y_1 s_1, x_2 < Y_2 s_2$.

4.3. *Constraints on the stage green-times*

For this two-approach junction the constraints on the green-time vector $\boldsymbol{Y}$ are thus

$$x_1 < Y_1 s_1, \quad x_2 < Y_2 s_2, \quad Y_1 + Y_2 = 1 - 2\eta/\tau, \quad Y_1 \geq m, \quad Y_2 \geq m.$$

4.4. *Optimizing the signal settings for fixed flows*

We are supposing that the vector of lane flows $\boldsymbol{x}$ and the cycle time τ are both fixed. It is easy to check that Webster's formula $d_i(x_i, y_i)$ is decreasing and strictly convex in y_i. Thus if we let D be the sum of all the delays experienced at the junction per second,

$$D(\boldsymbol{x}, \boldsymbol{Y}) = x_1[a_1 + d_1(x_1, Y_1)] + x_2[a_2 + d_2(x_2, Y_2)]$$

and is (for a fixed flow vector $\boldsymbol{x}$) a convex function of the vector $\boldsymbol{Y} = (Y_1, Y_2)$ of the stage green-times. There are many well-known ways of determining a feasible vector $\boldsymbol{Y}$ which minimizes the convex function D subject to the linear feasibility constraints on $\boldsymbol{Y}$.

One way is as follows: begin with any feasible stage green-time vector $\boldsymbol{Y}$. Determine the marginal costs

$$\partial D/\partial Y_1 \quad \text{and} \quad \partial D/\partial Y_2.$$

These $\partial D/\partial Y_i$ will be negative, as the delay d_i will be smaller if Y_i is larger. They give the degree to which D changes when the stage green-time vector $\boldsymbol{Y}$ changes. $-\partial D/\partial Y_1$ and $-\partial D/\partial Y_2$ may be thought of as pressures P_1 and P_2 on the stages 1 and 2, respectively. Increasing the green-time for that stage under most pressure is natural.

Suppose

$$P_1 = -\partial D/\partial Y_1 < -\partial D/\partial Y_2 = P_2.$$

Then a swap of time δ *from* the less pressurized stage 1 *to* the more pressurized stage 2 will reduce D (by about $\delta[\partial D/\partial Y_1 - \partial D/\partial Y_2]$) if δ is small. (It is reasonable to specify the swap rate more precisely and also more generally as $-(P_1 - P_2)$.)

As green-time is swapped toward stage 2, D must decline and under natural conditions $\boldsymbol{Y}$ approaches the stage green-time vector $\boldsymbol{Y}^*$ which minimizes D for fixed X. *At* this stage green-time vector $\boldsymbol{Y}^*$ it is impossible to swap further according to the above rules (otherwise D could be reduced further). Under natural conditions, at the optimum $\boldsymbol{Y}^*$,

$$P_1 = -\partial D/\partial Y_1 = -\partial D/\partial Y_2 = P_2.$$

Here the pressures on the two stages are equal and D is minimized.

4.5. *More general policies*

It is possible to build other objectives into a signal-setting policy by letting $P_1(\boldsymbol{x}, \boldsymbol{Y})$ and $P_2(\boldsymbol{x}, \boldsymbol{Y})$ be other functions of flow and green-time, not necessarily $-\partial D/\partial Y_1$ and $-\partial D/\partial Y_2$. We still think of $P_1(\boldsymbol{x}, \boldsymbol{Y})$ as the "pressure" on stage 1 and $P_2(\boldsymbol{x}, \boldsymbol{Y})$ as the pressure on stage 2. For reasonable pressures, if $P_1(\boldsymbol{x}, \boldsymbol{Y}) < P_2(\boldsymbol{x}, \boldsymbol{Y})$ it makes sense to swap green-time from the less pressurized stage 1 to the more pressurized stage 2; so it makes sense to swap green-time from stage 1 to stage 2 at a rate $-[P_1(\boldsymbol{x}, \boldsymbol{Y}) - P_2(\boldsymbol{x}, \boldsymbol{Y})]$ just as before. Our new control policy is now determined by $P_1(\boldsymbol{x}, \boldsymbol{Y})$ and $P_2(\boldsymbol{y}, \boldsymbol{Y})$. So we call this movement of the signals "policy *P*." Policy *P* clearly depends on the form chosen for $P_1(\boldsymbol{x}, \boldsymbol{Y})$ and $P_2(\boldsymbol{x}, \boldsymbol{Y})$. For any natural *P* the ensuing adjustment of $\boldsymbol{Y}$ is just as straightforward as it was with the delay-minimization policy.

The precise green-time adjustment above gives the rate of change of Y_1 as $P_1(\boldsymbol{x}, \boldsymbol{Y}) - P_2(\boldsymbol{x}, \boldsymbol{Y})$ and the rate of change of Y_2 as $P_2(\boldsymbol{x}, \boldsymbol{Y}) - P_1(\boldsymbol{x}, \boldsymbol{Y})$. Under natural conditions this swap rule leads to a $\boldsymbol{Y}$ vector which equalizes $P_1(\boldsymbol{x}, \boldsymbol{Y})$ and $P_2(\boldsymbol{x}, \boldsymbol{Y})$ and minimizes the maximum of $P_1(\boldsymbol{x}, \boldsymbol{Y})$ and $P_2(\boldsymbol{x}, \boldsymbol{Y})$.

Policy *P* is clearly "copied" from the delay-minimization policy above. See Smith and van Vuren (1993) for an exploration of the ideas implicit in this observation.

There are three natural choices for $P(\boldsymbol{x}, \boldsymbol{Y}) = (P_1(\boldsymbol{x}, \boldsymbol{Y}), P_2(\boldsymbol{x}, \boldsymbol{Y}))$ which stand out. These are (as there are just two stages)

$$P_1(\boldsymbol{x}, \boldsymbol{Y}) = -\partial D/\partial Y_1 \quad \text{and} \quad P_2(\boldsymbol{x}, \boldsymbol{Y}) = -\partial D/\partial Y_2,$$
$$P_1(\boldsymbol{x}, \boldsymbol{Y}) = x_1/s_1Y_1 \quad \text{and} \quad P_2(\boldsymbol{x}, \boldsymbol{Y}) = x_2/s_2Y_2,$$
$$P_1(\boldsymbol{x}, \boldsymbol{Y}) = s_1d_1(x_1, Y_1) \quad \text{and} \quad P_2(\boldsymbol{x}, \boldsymbol{Y}) = s_2d_2(x_2, Y_2).$$

The first choice makes policy *P* identical to the delay-minimization policy, as we have already seen. The second makes policy *P* into an "equisaturation" policy: there is no change in Y when the pressures are equal, and this happens when both lanes are equally saturated. The third choice makes policy P into a policy called "P_0" (Smith, 1980): this policy tends to favor lanes with high saturation flows – wider roads – and, under natural assumptions, maximizes the capacity of a general network.

5. Discrete modeling and optimization of fixed-time isolated signals for fixed flows

Delay formulae such as that quoted here assume some random variability in traffic flow, but are not able to deal with flows which change substantially, as they do in the peak period. Now while it is clearly reasonable to model signal green-times as fixed – we do after all control these in real life – we must allow for substantial variation in traffic flows – we do not control these. So we shall now suppose that traffic flows are variable. We shall also here suppose that arrival traffic is modeled as a sequence of individual vehicle arrivals rather than as a smooth flow. If the flow is variable it is clear that a fixed-time signal may, for some time periods, give too much time to a minor road joining a major road (if, for example, there is no traffic on the minor road for appreciable periods). It is thus natural to suggest that the presence of a side-road vehicle should be used to trigger a period of side-road green. This simple discrete model, with little in the way of quantification, led to the introduction, and prevalence for many years in the U.K. at least, of *vehicle* actuated VA signals. In many implementations of VA signals the side-road green was triggered by a single vehicle crossing a detector.

Miller (1965) introduced greater discrimination and quantification in the above discrete model which demonstrated that the total rate of delay at a junction would be reduced, compared with the standard VA case, if the side road was only given green when a number of vehicles had already accumulated on the side road, because switching display incurs intergreens and so switches should only occur when it has become "worth it." He developed a strategy for changing the display which compares the effect of changing the display immediately with the effect of changing the display in h seconds, $2h$ seconds, $3h$ seconds, and so on, taking account of the time lost when a signal display is changed. The display is changed as soon as an immediate change incurs least total delay. A model is used to assess the hypothetical effects of different alternatives, and the answers depend on assumptions about future arrivals. Miller made simple assumptions here.

Miller compared the performance of this system with a VA system using computer simulation; he regarded fixed-time settings for single junctions as "scarcely worthy of comparison." In Miller's simulation results, Miller's method reduced delays (compared with the VA case) by a maximum of about 40%; dropping to zero as the flow approached the junction capacity.

Miller suggested that his strategy could be implemented within a computer-controlled system and calculated that (compared with the VA case) the expense would be justified if the rate of delay saving achieved was at least two vehicle-hours per hour per junction for the peak period; or, equivalently, that a reduction of two vehicles queuing at each junction could be achieved.

This method of optimizing signals, by comparing the likely consequences of an immediate as opposed to a delayed change of display, has since been adopted in several studies. Miller's strategy was driven to a logical extreme by Robertson and Bretherton (1974), who determined the optimum control of a single intersection when the whole sequence of future arrivals is accurately known. Miller also considered the problem of linking signals to benefit traffic in a network, and briefly considered the offsets and cycle times likely to be beneficial.

6. Network modeling and control with fixed route flows and lane flows

The simplest way to model a network of signal-controlled junctions is by modeling each of these junctions separately and choosing green-times which minimize delay at each, using one of the methods initiated by Webster, Allsop, or Miller (say) for each single junction, as outlined above. We briefly indicate a natural modification which involves "whole network" variables and allows junctions with more than two approaches. For simplicity we suppose that the network has flow from a single origin to a single destination. New variables, additional to those specified previously, are as follows: X_r = the flow on route r, and T is the total flow from the origin to the destination.

The stages now refer to the whole network and here we will suppose lost times and minimum green-times to be zero for clarity. The variables X_r, Y_k, x_i, y_i are related according to eqs. (2) and (3) above. Then the constraints on the problem may be expressed as feasibility constraints as follows.

6.1. Feasibility constraints

The (route flow, green-time) pair $(\boldsymbol{X}, \boldsymbol{Y})$ is said to be feasible if

$$\sum X_r = T, \quad \sum Y_k \quad \text{and} \quad x_i \leq y_i s_i \quad \text{for all lines } i.$$

Much traffic control modeling is done under this assumption of a rigid demand where the total travel T does not depend on travel cost.

6.2. Total travel cost or time and the standard traffic control problem

We suppose that the cost of traversing lane i is, for given (x_i, y_i), to be $a_i + d_i(x_i, y_i)$ where a_i is constant and d_i is given by eq. (1) above. Then the cost or time of traversing route r is naturally to be

$$C_r = \sum_{\text{those } i \text{ such that lane } i \text{ belongs to route } r} [a_i + d_i(x_i, y_i)],$$

where the x_i and the y_i are as defined previously in terms of $\boldsymbol{X}$ and $\boldsymbol{Y}$. A standard traffic control problem may then be stated as follows: for fixed $\boldsymbol{X}$ and so fixed $\boldsymbol{x}$, minimize

$$\sum_r X_r C_r = \sum_i x_i [a_i + d_i(x_i, y_i)],$$

subject to the feasibility constraints listed above. Note that the lane flows x_i and the route flows X_r are regarded as fixed in this problem. (Also, offsets are not considered here.)

6.3. *Network effects*

Still assuming that the route flows and lane flows are fixed, it is nonetheless clear that nearby traffic signals should be controlled together in a co-ordinated way. Indeed computer tests of Miller's method have shown that while the method did give advantages at a single junction, the bulk of the gains were lost when two adjacent junctions were controlled in the same way. (In fact Miller did seek to extend his method so as to allow for the delay at the next junction.)

These issues have led to methods of choosing green periods along an arterial which cause a "green wave" to match the average speed of traffic, allowing the bulk of the traffic to traverse the arterial without stopping. See, for example, Morgan and Little (1964). The objective of these systems is to maximize the "bandwidth" along the arterial; this is (roughly) the length of a green period which can pass throughout the whole arterial at a natural speed and subject to natural constraints. Since, usually, there are two directions, a weighted sum of two "bandwidths" is minimized. The problem is purely geometrical, and traffic flows are not accounted for, so for a large bandwidth to translate into benefits, the traffic flows must not be too large. Recently a heuristic approach to achieving co-ordination in a whole network has been outlined by Stamatiadis and Gartner (1999).

Allsop (1968) calculated the offsets which minimize total delay in a network controlled by fixed-time signals. Minimization of total delay has always been a central objective of traffic control, and other objectives, like maximizing bandwidth, have usually been utilized as simplified surrogates standing in for this central objective.

In the above work the traffic flows are essentially steady. Further complications arise when dynamic or peak-period problems are considered. One of the first people to address such issues was Gazis (1964). He considered the optimum control of a system of oversaturated intersections in a dynamic setting. Here the inputs to the network, again fixed (independent of the control parameters chosen), are given as smooth curves; traffic accumulates behind saturated bottlenecks, and discharge rates are maximal (any "extra" holding back of traffic is not permitted). In a dynamic context the "minimize total travel time" objective tends to lead to unfair consequences as minor flows are held up so as to reduce delay for major flows. (Fairness has typically not been well treated within signal control modeling and this may well have had unforeseen consequences when control systems have been applied in practice.)

6.4. Other extensions of these ideas

The ideas just introduced in this chapter may be developed in many directions, some rather different from those listed above. For example Akcelik (1994, 1996) and Akcelik et al. (1999) have shown how greater realism may be introduced into the optimization of signal timings and also how strict optimization has to relaxed in order to allow for practical difficulties which arise as soon as real networks are considered. This work shows how far the approach outlined here can be developed in order to deal with real problems, and also shows how simplified the approach taken here really is.

Nash (1996) and Lowrie (1996) give descriptions of signal control techniques for isolated junctions and networks. These provide a clear and fairly complete statement of Australian signal-setting practice, allowing for theoretical issues and implementation issues.

6.5. TRANSYT (TRAffic Network StudY Tool)

The most famous method of setting signals in a road network is the TRANSYT method (Robertson, 1969). The ideas lying behind the TRANSYT model and optimization are similar to the ideas lying behind the model and optimization of Gazis above. However, TRANSYT is usually run as if traffic was essentially repetitive, as at the peak of the peak period.

TRANSYT is a model of traffic signals in a network operating on a fixed-time basis and includes facilities for optimizing the signal timings. TRANSYT requires a detailed model of the road network, including the geometry of all relevant traffic signals, saturation flows at stop lines, travel times along lanes, significant bottlenecks with their saturation flows, and major sources and sinks of traffic. TRANSYT also requires traffic flows and these are to be as close as possible to those actually occurring on the network during the time period being studied, or the average of these flows. In the model input flows are constant over periods of time. These and the entire modeling period may be chosen by the user to suit the network being modeled.

The central elements of the TRANSYT model are the flow profiles. These are essentially histograms with an integer flow in vehicles per hour (*y* axis) plotted against time (*x* axis). For given signal timings throughout the time period the model manipulates the flow profiles so that, as time passes, a natural evolution occurs. At each junction the time-varying inflow pattern to a downstream lane is calculated as a sum of relevant outflows from upstream lanes and for each lane output flow profiles are determined from input flow profiles, mainly by shifting the profile by the time taken to traverse the lane. (Dispersion along a lane and also random additional delays at junctions are represented by further adjustments.)

The model will then estimate the flows, delays, and stops over the study period. The performance index (PI), a weighted sum of travel times and stops on all lanes, is then calculated. This value of the PI is associated with the fixed signal timings.

For optimization, the traffic model above changes the fixed-time signal timings systematically by a small amount, and evaluates the PI for each change. Beneficial changes are kept and the procedure continues until there are no small changes which benefit the PI.

The optimization relies on various assumptions embedded in the model; the main one is probably that the route flows are unchanged as the signals are optimized.

6.6. SCOOT (Split Cycle Offset Optimization Technique)

SCOOT is a responsive traffic control system for optimizing network performance on line; see Hunt et al. (1981, 1982). The method was developed by TRL, Ferranti, GEC, and Plessey over several years of research. It is installed in at least 35 towns in the U.K.

For selecting offsets between junctions, the principles of SCOOT are similar to those of TRANSYT (minimize the PI), but green-time at a junction is divided so as to minimize the maximum of a modified "degree of saturation" at each junction. The cycle time is chosen so as to reduce the maximum "degree of saturation" to below 90% if this is possible. The model optimization is performed quickly and the calculated signal changes are implemented on line. By working continuously, SCOOT seeks just small but frequent signal-setting changes.

Following a change in traffic flows SCOOT changes green-times fairly rapidly to deal with the new flows, but then steps back toward the previous green-times so as to not overshoot; and so it is carefully damped.

6.7. SCATS (Sydney Co-ordinated Adaptive Traffic System)

SCATS was developed by the Roads and Traffic Authority of New South Wales in the early 1970s. It is now installed in over 40 towns throughout the world, including Australia, New Zealand, China (and Hong Kong), Malaysia, the Phillipines, Indonesia, the U.S.A., and Ireland.

The system provides three basic modes: Adaptive, Flexilink, and Isolated. In the Flexilink mode signals do not respond to changing flows; the signals follow precalculated optimized fixed-time plans and so co-ordination is maintained. In the Isolated mode all relevant signals function on their own, either following a fixed-time plan or responding to local detectors.

The Adaptive mode is the most advanced. This has two "levels": strategic and *tactical*. Strategic control is the top level and continually estimates the most appropriate signal timings (cycle times, green-times, offsets) for the conditions prevailing over the whole network. The signal timings implemented by this centrally computed, responsive, network-wide control are then subject to tactical control at each intersection; this provides local responsiveness and allows for rapid response to rapid changes in local conditions. As with SCOOT frequent small changes are made so as to follow the traffic conditions.

SCATS facilitates the provision of active public transport priority. Such a system is operational in Melbourne with over 470 signals. Trams in Melbourne are fitted with transmitters which allow them to be detected and accorded priority at traffic signals. Congestion caused by the application of tram priority is rapidly recognized by the SCATS system, which implements automatic compensation.

6.8. Other real-life strategies

There are very many other urban traffic control strategies and behind each one is a signal control model and an optimization procedure. These models all differ to a greater or lesser degree but they all have aspects which are similar to the models described earlier in this chapter in more detail. We mention three by name: OPAC (Gartner, 1983), PRODYN (Henry and Farges, 1989), and MOVA (Vincent and Young, 1986).

These strategies all seek to optimize, using mainly standard delay criteria, and they all assume that traffic flows are not influenced by the signal control actions taken.

6.9. Modeling assessments of standard signal control policies when choices are variable

As we have already suggested, it is hard to design practical traffic signal control models and strategies which take correct account of drivers' route choices. This issue will be outlined below. It is, however, comparatively easy to assess network performances of traffic control strategies, allowing for future route choices, using a computer model. See van Vuren and Van Vliet (1992) and Ghali and Smith (1994a,b) for a comparison of delay-minimization, equisaturation and P0 using assignment models in which these strategies have been implemented.

7. Signal control modeling and optimization for new targets with variable flows

As soon as we seek signal controls which achieve targets while taking account of travelers' future choices the optimization becomes difficult – even in our simple setting. Perhaps the simplest assumption that we can make is that drivers' route choices will tend to restore the equilibrium state. Thus we must consider the traffic equilibrium condition introduced by Wardrop (1952).

7.1. *Wardrop equilibrium on a fixed set of routes*

Here we suppose that the vector of stage green-times Y and so the vector y of lane green-times are fixed. Then the triple $(\boldsymbol{X}, \boldsymbol{Y}, \boldsymbol{M})$ is an equilibrium on the specified set R of routes if

$$C_r = \sum_{i \text{ such that lane } i \text{ belongs to route } r} [a_i + d_i(x_i, y_i)] = M \quad \text{for } r \text{ in } R,$$

$$C_r = \sum_{i \text{ such that lane } i \text{ belongs to route } r} [a_i + d_i(x_i, y_i)] \geq M \quad \text{for } r \text{ not in } R,$$

$$X_r = 0 \quad \text{for } r \text{ not in } R.$$

M (seconds) is the minimum cost or time needed to travel from the origin to the destination.

7.2. *Optimization subject to equilibrium*

Let $Z(\boldsymbol{X}, \boldsymbol{Y}, \boldsymbol{M})$ be a measure of the total disbenefit arising if the route-flow pattern is $\boldsymbol{X}$, the stage green-time pattern is $\boldsymbol{Y}$ and the cost-to-destination pattern is M. We wish to solve the following problem: minimize $Z(\boldsymbol{X}, \boldsymbol{Y}, \boldsymbol{M})$ subject to the condition that $(\boldsymbol{X}, \boldsymbol{Y}, \boldsymbol{M})$ is an equilibrium on a fixed set R of routes.

The need to solve this problem was emphasized by Allsop (1974). There has been much research on this: see, for example, Fisk (1984), Yang and Yagar (1995), Yang (1996), Smith et al. (1996), Chiou (1997), and Clegg et al. (2001). The equilibrium constraints here are non-convex for Webster's delay formula and indeed for almost all possible delay formulae.

However, if we make two simple assumptions these two constraints can be made to be convex by relaxing the problem somewhat. The first assumption is that, as far

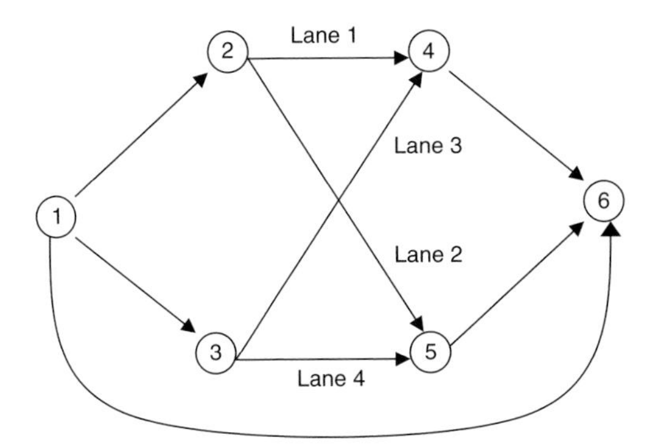

Figure 3. A simple example network.

as calculating delay is concerned, a red period of an appropriate length has the same effect on delays as a real vehicle. The second assumption says that the signal control facilities are very discriminating: starting at any feasible lane green-time vector, any single-lane green-time may be separately diminished. While these are important assumptions here, they are actually not likely to be very seriously wrong.

We exploit the new assumptions as follows. Introduce a new "bottleneck delay" b_i (which will normally become equal to $d_i(x_i, y_i)$) and consider the new constraints

$$C_r = \sum\nolimits_{i \text{ such that lane } i \text{ belongs to route } r} (a_i, b_i) = M \quad \text{for } r \text{ in } R,$$

$$C_r = \sum\nolimits_{i \text{ such that lane } i \text{ belongs to route } r} (a_i, b_i) \geq M \quad \text{for } r \text{ not in } R,$$

$$X_r = 0 \quad \text{for } r \text{ not in } R,$$

$$d_i(x_i, y_i) \leq b_i \quad \text{for all } i.$$

The first three constraints here are always convex as they are linear, and the last is convex if we make our two assumptions. So in this case these constraints are all convex. If the objective function (say Z) to be reduced is convex too, and this is a reasonable assumption, then the problem becomes a problem of minimizing a convex function subject to convex constraints. Clearly we would probably need to solve several of these for various choices of the set R of "active routes."

We now give some example results of following these ideas, using the simple network illustrated in Figure 3. This network was first considered by Yang (1996); it has one origin (node 1) and one destination (node 6) and a total of five possible routes. The network also has two signal-controlled junctions (at nodes 4 and 5). The detailed data for this network are given in Clegg et al. (2001).

The problem is to find signal timings which minimize total travel cost subject to user-equilibrium behavior by travelers. We have followed Yang and calculated timings for a range of rigid demands: in our case we have calculated timings for each rigid integer demand from 1 to 328 (the greatest possible demand for this network, given the maximum throughput of each lane for given green-times and the strict capacity constraints). We have compared these "optimal" timings, which are very long-winded to calculate, with P0 timings, which are quick and easy to calculate.

In Figure 4 the bold curve shows the "optimum" green-times for lane 4 and the fainter curve shows the green-times for lane 4 generated by P0. For this network the easy-to-determine P0 green-times are, for both junctions, very close to the hard-to-determine optimum green-time when flow is high.

8. A signal control strategy classification

It seems natural to fit traffic signal control strategies into one of the following three classes: fixed-choice local strategies, fixed-choice network strategies, and variable-choice network strategies, according to their properties (see Table 1).

8.1. *Fixed-choice local strategies*

Some signal control strategies utilize local traffic flows to "drive" the strategy; on the assumption that car route choices and mode choices are fixed. These may naturally be called fixed-choice local strategies.

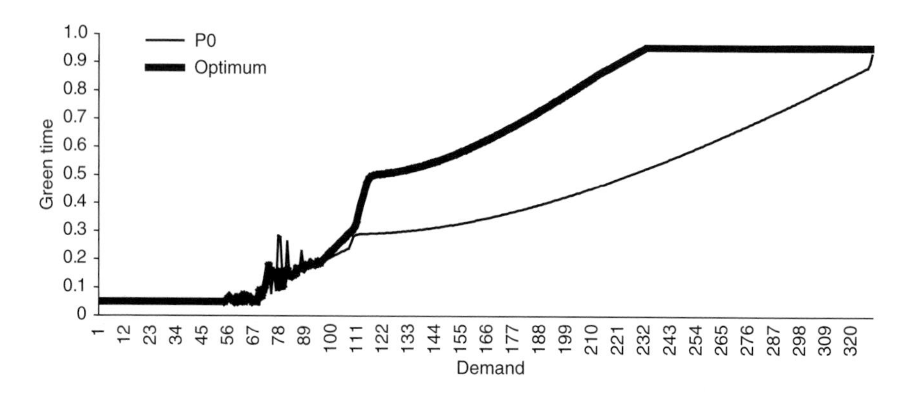

Figure 4. Green-times for lane 4.

Table 1
Traffic control strategies fitting into the three strategy classes

Class of strategy	Idealized strategy	Implemented strategy
Fixed-choice local	Delay minimization, equisaturation	MOVA
Fixed-choice network	Delay minimization, equisaturation	TRANSYT, SCOOT, SCATS
Variable-choice network	P0, bilevel optimization	MUSIC

8.2. *Fixed-choice network strategies*

Strategies which use more distant "network" traffic flows, such as TRANSYT or the SCOOT strategy, utilize network-wide information to drive a control strategy. But, still, actions taken on one day are (usually) concerned with just that day – any consequent rerouting tomorrow is not considered in deciding today's timings. Optimization ignores future route and mode changes: the strategies are fixed-choice network strategies. It is natural to view these strategies as short-term strategies since the assumption of fixed choices is more correct over a timescale of seconds and minutes than over a timescale of days, weeks, and months.

8.3. *Variable-choice network strategies*

Strategies in this class adjust controls over time aiming to minimize (or reduce) a network objective, while taking account of future routing changes and future congestion, so they are variable-choice network strategies. It is natural to view these strategies as long-term strategies since the assumption of variable choices is more correct over a timescale of days, weeks, and months than over a timescale of seconds and minutes.

8.4. *Less standard or newer strategies*

These are the strategies in the bottom row of Table 1. P0 maximizes overall network capacity in certain natural models while allowing for route choices and also uses only local data as if route choices were fixed (see Smith, 1979). Bilevel optimization seeks to find optimal signal timings taking account of travelers' responses as described earlier in this chapter (Smith et al., 1998; Clegg et al., 2001). The MUSIC method is described by van Vuren et al. (1998), Clegg et al. (2000), and Ghali and Smith (1994b).

9. Conclusion

This chapter has considered modeling traffic signal control from a variety of viewpoints. The way modern traffic signal optimization models have arisen, from models of fixed-time signals, through Miller's theoretical on-line "computer-controlled" model and the real-life TRANSYT fixed-time network model to the real-life SCOOT and SCATS rapid-response models, has been outlined.

The way in which signal control has for many years been divorced from transport management and planning has been indicated and a way of integrating traffic control and traffic management has been hinted at.

Finally, we have given a classification of signal control strategies which may be helpful.

The subject has attracted the attention of very many people over very many years and the survey provided here can only seek to provide a flavor of the ideas involved.

References

Akcelik, R. (1994) "Estimation of green times and cycle time for vehicle actuated signals", *Transportation Research Record*, 1457:63–72.

Akcelik, R. (1996) "Progression factor for queue length and other queue-related statistics", *Transportation Research Record*, 1555:99–104.

Akcelik, R, M. Besley and R. Roper (1999) "Fundamental relationships for traffic flows at signalised intersections", ARRB Transport Research Limited, Vermont South, Australia, Research Report ARR 340.

Allsop, R.E. (1968) "Selection of offsets to minimise delay to traffic in a network controlled by fixed-time signals", *Transportation Science*, 2(1):1–13.

Allsop, R.E. (1971) "Delay-minimising settings for fixed-time traffic signals at a single road junction", *Journal of the Institution for Mathematics and its Applications*, 8(2):164–185.

Allsop, R.E. (1974) "Some possibilities for using traffic control to influence trip distribution and route choice", in: *Proceedings of the 6th International Symposium on Transportation and Traffic Theory*. New York: Elsevier.

Chiou, S.-W. (1997) "Optimisation of area traffic control subject to user equilibrium traffic assignment", in: *Proceedings of the 25th European Transport Forum*, Seminar F, 2:53–64.

Clegg, R.G., A. Clune and M.J. Smith (2000) "Traffic signal settings for diverse policy goals", Proceedings of the European Transport Forum, Seminar K.

Clegg, J., M.J. Smith, Y. Xiang and R. Yarrow (2001) "Bilevel programming applied to optimising urban transportation", *Transportation Research B*, 35: 41–70.

Department of the Environment, Transport and the Regions (1998) *A new deal for transport: Better for everyone*. London: The Stationery Office.

Fisk, C.S. (1984) "Optimal signal controls on congested networks", in: J. Volmuller and R. Hammerslag, eds., *Proceedings of the 9th International Symposium on Traffic and Transportation Theory, Delft*. Utrecht: VNU Science Press.

Gartner, N.H. (1983) "OPAC: A demand responsive strategy for traffic signal control", *Transportation Research Record*, 906:75–81.

Gazis, D.C. (1964) "Optimum control of a system of oversaturated intersections", *Operations Research*, 12::815–831.

Ghali, M.O. and M.J. Smith (1994a) "Comparisons of the performances of three responsive traffic control policies, taking drivers' day-to-day route-choices into account", *Traffic Engineering and Control*, 35:555–560.

Ghali, M.O. and M.J. Smith (1994b) "Designing time-of-day signal plans which reduce urban traffic congestion", *Traffic Engineering and Control*, 35:672–676.

Henry, J.J. and J.L. Farges (1989) "PRODYN", in: *Proceedings of the 6th IFAC–IFIP–IFORS Symposium on Transportation*. Oxford: Pergamon.

Hunt, P.B., D.I. Robertson, R.D. Bretherton and R.I. Winton (1981) "SCOOT – a traffic responsive method of coordinating signals", Transport and Road Research Laboratory, Crowthorne, U.K., Report 1014.

Hunt, P.B., D.I. Robertson and R.D. Bretherton (1982) "The SCOOT on-line traffic optimisation technique", *Traffic Engineering and Control*, 23:190–192.

Lowrie, P.R. (1996) "Signal linking and area control", in: K.W. Ogden and S.Y. Taylor, eds., *Traffic engineering and management*. Melbourne: Monash University.

Miller, A.J. (1965) "A computer control system for traffic networks", *Proceedings of the 2nd International Symposium on the Theory of Road Traffic Flow*. Paris: OECD.

Morgan, J.T. and J.D.C. Little (1964) "Synchronising traffic signals for maximum bandwidth", *Operations Research*, 12:896–912.

Nash, D. (1996) "Traffic signals – design and analysis", in: K.W. Ogden and S.Y. Taylor eds., *Traffic Engineering and Management*. Melbourne: Monash University.

Robertson, D.I. (1969) "TRANSYT: A traffic network study tool", Road Research Laboratory, Crowthorne, U.K., Report LR253.

Robertson, D.I. and R.D. Bretherton (1974) "Optimum control of an intersection for any known sequence of vehicle arrivals", in: *Proceedings of the 2nd IFAC–IFIP–IFORS Symposium on Traffic Control and Transportation Systems*. Amsterdam: North Holland.

Smith, M.J. (1980) "A local traffic control policy which automatically maximises the overall travel capacity of an urban road network", *Traffic Engineering and Control*, 21:298–302.

Smith, M.J. and T. van Vuren (1993) "Traffic equilibrium with responsive traffic control", *Transportation Science*, 27:118–132.

Smith, M.J., Y. Xiang, R. Yarrow and M.O. Ghali (1998) "Bilevel and other modelling approaches to urban traffic managment and control", in: P. Marcotte and S. Nguyen, eds., *Proceedings of Equilibrium and Advanced Transportation Modelling*. Boston: Kluwer.

Stamatiadis, C. and N.H. Gartner (1999) "Progression optimisation in large scale urban networks: A heuristic decomposition approach", in: A. Ceder, ed., *Proceedings of the 14th International Symposium on Traffic and Transportation Theory, Delft*. Oxford: Pergamon.

Underwood, R.T. (1996) "Development of the traffic engineering profession", in: K.W. Ogden and S.Y. Taylor, eds., *Traffic engineering and management*. Melbourne: Monash University.

van Vuren, T. and D. Van Vliet (1992) *Route choice and signal control.* Aldershot: Avebury.

van Vuren, T., I. Routledge and M.J. Smith (1998) "MUSIC: Putting the 'M' into UTMC", *Engineering and Control*, April:1–5.

Vincent, R.A. and Young C.P. (1986) "Self-optimising traffic signal control using microprocessors – the TRRL MOVA strategy for isolated intersections", *Traffic Engineering and Control*, 27(7/8):385–387.

Wardrop, J.G. (1952) "Some theoretical aspects of road traffic research", *Proceedings of the Institution of Civil Engineers*, 2(1):235–278.

Webster, F.V. (1958) "Traffic signal settings", Department of Transport, HMSO, London, Road Research Technical Paper 39.

Webster, F.V. and B.M. Cobbe (1966) "Traffic Signals", London: HMSO.

Wood, K. (1993) "Urban Traffic Control, Systems Review", Transport Research Laboratory, Crowthorne, U,K., Project Report 41.

Yang, H. (1996) "Equilibrium network traffic signal setting under conditions of queueing and congestion", in: Y.J. Stephanedes and F. Filippi, eds., *Applications of advanced technologies in transportation Engineering. Proceedings of the 4th International Conference*. American Society of Civil Engineers.

Yang, H. and S. Yagar (1995) "Traffic assignment and signal control in saturated road networks", *Transportation Research*, 29A:125–139.

Chapter 35

RAILWAY SCHEDULING

ROBERT WATSON
Loughborough University

1. Definitions

Railway scheduling is the process by which the "demand" for rail transport (passenger and freight) is brought together with "supply side" constraints (such as available infrastructure capacity, rolling stock, and staff) to produce timetables and resource plans that meet the demand at an appropriate level of cost. This process is also known as "train planning."

Timetables show how trains travel over time and usually take the form of "tables" or "time–distance graphs."

Resource plans map rolling stock and staff to the trains that are in the timetables, taking into account all the operational, legal, and union rules that need to be applied.

Schedulers or planners are the railway personnel who put together timetables and resource plans. For very small railways (for instance a city-center-to-airport rail link) there may be just one or two schedulers, responsible for the whole process; major railroads and national railways will have up to several hundred schedulers, with different groups of staff responsible for timetabling and resource planning and individual staff specializing in particular tasks within the process.

2. Time horizons

Railway scheduling is undertaken at different times for different reasons.

2.1. Strategic planning

This is where changes to the infrastructure are being considered and "what if" questions are being asked, typically looking 2–10 years ahead. "Back of an envelope" assessments are undertaken (e.g., what train service could be operated

Handbook of Transport Systems and Traffic Control, Edited by K.J. Button and D.A. Hensher

if some extra trains were leased or what train service could be operated if extra tracks were provided?). As these ideas become firmer, detailed timetables are produced to assess the likely performance of revised infrastructure layouts (perhaps with additional platforms or a new passing loop to let fast trains overtake slow trains) and to confirm the additional rolling stock required to operate a proposed future timetable.

2.2. *Tactical planning*

This is scheduling over a time horizon where the infrastructure tends to be fixed, but the mobile resources (rolling stock and staff) can be varied in quantity, quality, and intensity of operation. Tactical planning is often split into "long-term planning" and "short-term planning." Long-term planning produces the timetables and resource plans that are to be in operation for typically up to a year in the future; short-term planning makes the changes to this plan that are always needed to a greater or lesser extent to cope with supply or demand fluctuations from a few weeks to a few days ahead of operation (an example of a supply fluctuation is a shortage of train crew due to sickness; examples of demand changes are statutory holidays and special occasions such as major sporting events). The long-term plan is produced over a number of months and, particularly for passenger railways, has to be completed some weeks or months before the new timetable comes into operation, to allow for publication of new timetables for passengers and transmission to reservation and customer information systems.

2.3. *Operational planning and control*

However good the tactical planning process, real-time perturbations are an inevitable feature of transport operations. Rescheduling ultimately takes place "real time" to accommodate last-minute changes in demand (usually from freight customers), train failures and delays, infrastructure reliability problems, and staff sickness. Train makeup, car routing (decisions about which freight cars go on which trains), and yard operation decisions are now made.

3. Stages in the process

Scheduling in all three time horizons defined above follows a similar high-level process, as set out in Figure 1 and described below.

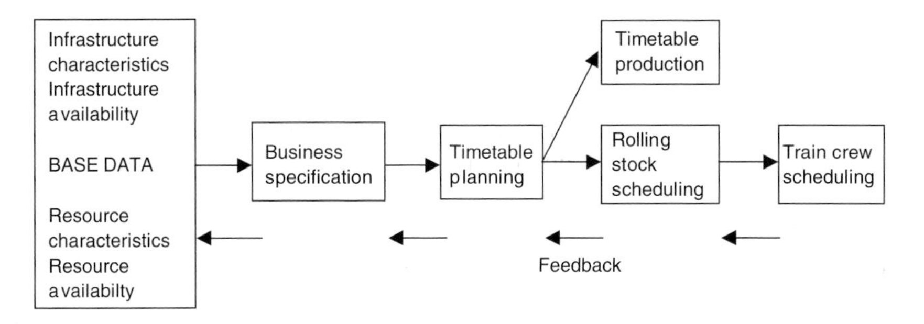

Figure 1. Overview of railway scheduling process.

3.1. Base data

The process starts by initially defining:

(1) Infrastructure characteristics (number of tracks, where trains can pass each other, speeds trains can travel, etc.). The timetable planning process works to a simplified view of the railway. This simplified view (often known in the U.K. as the "planning geography") works at a level of detail that matches with the variables at play during planning. Hence all junctions and stations are included in the planning network created, but individual switches and signals are not.

(2) Infrastructure availability (in practice often a statement of when the infrastructure is "unavailable"). This indicates the constraints placed on when trains can run, usually because the infrastructure is being maintained. This will typically be stated in the form of a maintenance "window" – a period of hours on a daily, weekly, or occasional basis when trains are banned. In more limited circumstances, it may also not be possible to run trains because signaling staff or other key operating staff are not on duty.

(3) Resource characteristics – whether particular rolling stock can run on particular routes (limitations include gauge, curvature, weight, or signaling interference) and the performance characteristics of particular rolling stock on particular routes (in particular the "timings": the time it takes particular rolling stock to travel over every leg of the planning "geography").

(4) Resource availability (rolling stock numbers, numbers of staff, locations, etc).

3.2. Business specifications

It is normal for several, potentially conflicting business specifications to be produced – these specifications come from the differing requirements of customers of the railway. Within unified state railways, specifications come from the different business units – usually some combination of international, intercity, suburban, regional, and freight; for railways where there are a number of separate train operating companies competing for access to the infrastructure, each will provide its own requirements. In either case high-level train service specifications (sometimes called "service plans") will be produced, indicating the general level of service required (e.g., number of trains per hour, and the type and stopping pattern of these trains). These specifications take into account the level of resources (particularly rolling stock) available.

3.3. Timetable planning (sometimes known as timetable development or, in the U.K., access planning)

The train service specifications are passed to the timetable planners, whose task is to produce timetables that are "conflict free" (so that if the timetable was worked to exactly in practice, no train would be delayed by any other).

Figure 2 provides an overview of the timetable development process; each stage is now described.

Creating detailed train schedules

The times provided in the business specifications are turned into detailed schedules, accurate to fractions of a minute, taking account of the details of the infrastructure and characteristics of the trains.

Previous timetables

Previous timetables provide valuable input to the schedule creation task (it may be that some schedules are identical; alternatively some may be the same, but offset in time). Previous timetables may also be of value to the conflict resolution stage – these solutions may be applicable again with, perhaps, minor changes.

"Graphing" schedules

The next stage towards producing a timetable is to draw the schedules on a time–distance graph (using either a computer package or graph paper and pencil). Once all schedules are drawn on the graph, it is possible, by eye, to judge whether there are "conflicts" between train schedules – where there is not sufficient distance and time between trains on the same track going the same direction ("headway"), traveling in opposite directions on single track, or traveling on tracks which cross ("junction margins"). A simplified example of timetable

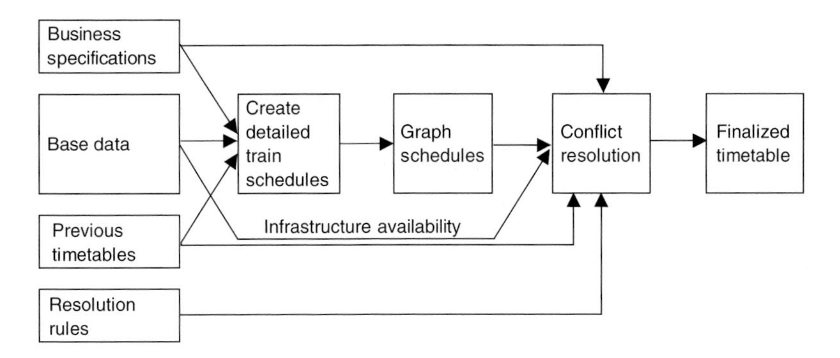

Figure 2. Timetable planning process.

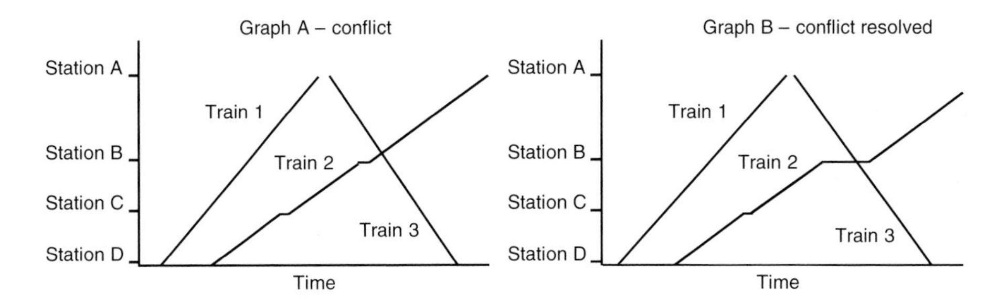

Notes
Single track with passing loops at stations
Train1: fast train non-stop from station D to station A
Train2: slow train stopping at stations C and B
Train3: fast train non-stop from station A to station D
In graph A train 3 conflicts with train 2 on the single track between A and B – conflict resolution required
In graph B train 2 is held at station B to "cross" train 3

Figure 3. Simplified timetable graphs.

graphing is given in Figure 3. For a more detailed exposition see Ford and Haydock (1992).

An alternative approach is to print out the schedules in tabular form and to check for "headway" or "junction margin" deficiencies by comparing the times of schedules at key locations.

Conflict resolution

Once conflicts have been detected, it is necessary to "resolve" them. In the U.K. the jargon "flexing" has come to be used to describe the process whereby the planner moves the schedules on the graph to achieve a conflict-free timetable, whilst, as far as practicable, achieving the requirements in the business specification. Inevitably, on busy routes compromises are required; very skillful planners are required to achieve satisfactory "paths" (the term used to describe a train schedule once it has been put on the graph) to meet all the business requirements.

Resolution rules

There are rules about what represents a satisfactory solution – not only must the basic infrastructure headway and junction margin rules be obeyed, but a set of rules will apply regarding the extent to which it is acceptable to, say, increase the overall journey time of an intercity train to give a reasonably fast schedule for a freight train (practice varies, but typically in Europe, for instance, it is not acceptable to increase the journey time of a passenger train at all to improve the journey time of a freight train). In some circumstances, these rules may be formalized in a document that has been agreed between the infrastructure company and the train operators (in the U.K., for instance, these generally applicable rules are included in a document called the *Access Conditions*).

Finalized timetable

Once the resolution process has been completed, the timetable can be fed into the rolling stock scheduling process and also to the timetable publication process.

3.4. Rolling stock scheduling

All the services in the timetable have to be allocated to rolling stock "diagrams." A diagram is a listing of the services that a notional item of rolling stock will undertake during a day. It is constructed by "associating" (linking) the end of one service with the start of another to form a continuous string. There are rules regarding associations that have to be followed (e.g., the end of one service must be in the same location as the start of the next, unless an "empty stock" or "relocation" service is added, and the start of the next service to be linked must be at least a certain number of minutes later than the end of the previous service). The rolling stock schedule is complete once all the services have been allocated to diagrams: it is then possible to be certain how much rolling stock is required to operate the timetable (which might be different from the input resource availability).

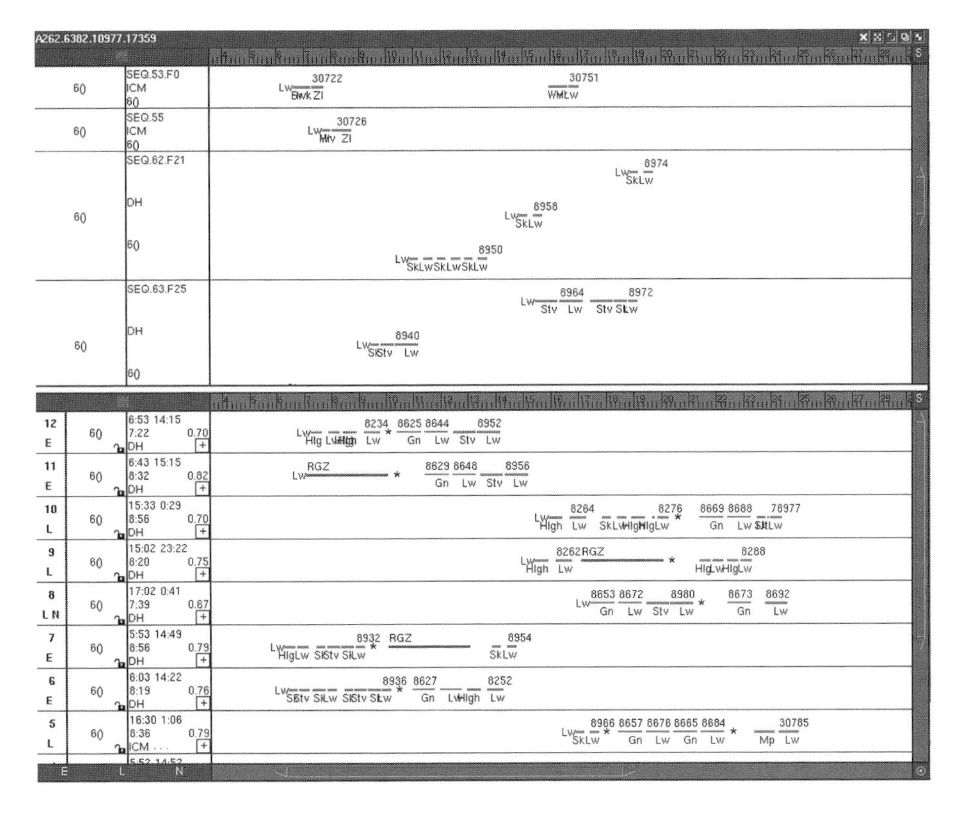

Figure 4. Train crew diagrams in preparation (screen shot from CREWS software, copyright Siscog).

3.5. *Train crew scheduling*

Once the rolling stock plan has been developed, it is then possible to overlay this with train crew schedules (often known "train crew diagrams and rosters"). All the rolling stock diagrams have to have matching train crew diagrams (although this will usually not be a "one to one" match), taking into account how many crew are required, the various rules regarding train crew working hours and knowledge, and various ancillary tasks that have to be performed (such as reporting for duty and being briefed on any particular safety or operational issues – known as "signing on," – handing in cash at the end of the shift, training, etc.). The second part of the train crew scheduling task is to produce rosters. This involved putting the diagrams together in sequences to produce "links" such that when a named member of staff is allocated to that link, it provides for him or her a series of diagrams that make up a working week (or number of weeks) that matches with the various rules on working hours. When put together, the overall roster must be made up of links that cover all of the work to be undertaken. If this scheduling work has been done as part of tactical planning, once the roster has been put together and, usually, agreed with the staff representatives or union, it is then possible to allocate people to the links. This is typically done some weeks in advance (unlike rolling stock, which is often only allocated to diagrams "on the day"), to give employees some knowledge of their forward working hours.

Figure 4 gives an example of a train crew scheduling task in progress. The top half of the figure shows rolling stock diagrams that have still to be allocated to a train crew diagram; the bottom half shows train crew diagrams that have already been constructed (e.g., 12E is a completed "early turn" (morning) diagram). Time of day runs from left to right (hours over 24 are on the next day, e.g., 25 is one o'clock in the morning); numbers such as 30772 and 8234 represent the train diagrams to be "covered" by that train crew diagram; the * indicates a meal break.

3.6. Timetable production

Once timetable development is complete, documentation can be produced for passengers (known in the U.K. as the "public timetable") and staff (a "working timetable" giving much more detail than the public timetable).

4. Key issues

As will be understood from the foregoing, railway scheduling is a complex process. Timetable planning in particular is complicated for railways, primarily because railways have a "single degree of freedom" – forwards and backwards (whereas road-based transport has two degrees of freedom and air three). All "overtaking" of trains or "meeting of trains" (on a single track), and "crossing of trains" (at junctions) have to be planned in detail if delays are not to result. Railway scheduling is critical to the effective utilization of railway infrastructure and in the planning of future infrastructure developments, with schedulers expected to produce robust plans that optimize the use of network and resources whilst delivering solutions in a reasonable time.

4.1. Need to deliver solutions in a reasonable time

In general, railways are becoming more commercially aware and wish to move more rapidly to meet the needs of their customers.

At the strategic planning stage, detailed timetable development and simulation (see Section 6) projects which look at the likely performance of a revised infrastructure layout and revised timetables produce timetables at a very detailed level – often with train times planned down to when they would pass individual signals for the whole route being examined. This process can take many months. As business growth makes the need for greater capacity more pressing, timescales for these assessments need to be reduced.

Within tactical planning, the long-term planning process typically starts 12 to 15 months in advance of the commencement date for the timetable – i.e., before the previous timetable has even commenced. There is, unsurprisingly, considerable pressure to reduce these timescales, to permit assessment of the effectiveness of a timetable in practice before planning the next one and to be more responsive to changing customer needs.

A further issue for "mixed" railways (that is, railways that carry both passengers and freight) is that the nature of logistics and freight distribution is such that the tactical planning takes place much closer to the train running, with timescales measured in weeks, days, or even hours, rather than the months ahead that passenger trains need to be planned. This creates an uncomfortable mismatch between planning horizons, with the widely held view in passenger-train-dominated railways that freight suffers some disadvantage, having to accept "what is left" after the passenger services have been planned.

4.2. Need for robust solutions

"Robustness" (by which is meant the extent to which a timetable can accommodate minor perturbations due to, for instance, locomotive or infrastructure failures without substantial "knock on" delays to other services) is often only tested when the timetable becomes operational, which means that there is a risk that there will be substantial delays should the timetable or resource plan not be capable of coping with the inevitable problems that occur. Where infrastructure management has been split from train operation, this can have serious financial implications for the infrastructure operator if there is a "performance regime" in place with the train operators, whereby poor performance due to defects in the timetable results in penalty payments to the train operators. Although a considerable length of time is taken over producing a long-term plan, the complexity of the task does lead to imperfections – indeed there are examples of substantial "short-term planning" changes being necessary in the weeks after a new long-term plan has been put into effect, to deal with particular performance problems caused by deficient planning.

4.3. Need for solutions closer to the optimum

The time required for producing a workable solution means that a wide variety of timetable options cannot be considered as part of the tactical planning process. Typically in the U.K., for instance, no radically different options are considered – in the time available the task is simply to produce a timetable that is operationally feasible and meets most of the commercial aspirations of the train operators. This prevents sensible discussion about the merits of different timetables and can lead to timetables being implemented that are short of what could have been achieved in terms of meeting train operators' aspirations. This can mean that business is turned away because the planners cannot find enough paths, with revenue implications for the infrastructure operator as well as the train operator. Similar problems apply to resource planning. In practice, it is currently very often only possible to look towards solving each stage of the process separately (often with the objective only being to produce an operable solution in the time available).

Scheduling, then, is in practice predominantly sequential, with the feedback loops shown on Figures 1 and 2 little used. As discussed, achieving an optimum solution requires account to be taken of the interaction of the different stages of the business process. In theory this could be achieved by seeking to optimize across the whole process using some sophisticated operations research technique. However, currently, optimization across the whole process is unachievable due to the complexity of the problem. The practical alternative is feedback from one part of the process to others, sometimes called "iteration." For instance, if a particular timetable results in inefficient use of train crew then it will be necessary for "feedback" to be given from the train crew scheduler to the timetable planner, with sufficient time in the process to allow the timetable planner to take account of this feedback and develop an improved timetable.

5. Complexities

There are a number of complexities that make the delivery of an improved process difficult:

(1) the complexity of the railway scheduling problem, with the need to consider infrastructure and resource efficiency alongside robustness and time taken to produce a solution;
(2) the complexity and interlinked nature of national railway networks, leading to the timetable planning process of necessity requiring a national solution, but with the scale of the task requiring the development of a solution in a

reasonable time to be divided between a number of different people and often between different locations;

(3) the congested nature of many rail routes, either through infrastructure rationalization to reduce cost or, especially more recently, growth in traffic volumes, leading to timetable planners having difficulty finding solutions that meet all the business requirements, if indeed such a solution exists at all;

(4) the additional complexity that has been introduced to the processes by the current trend towards separation of infrastructure management and train operation, leading to no one organization or individual being in a position to decide between different compromises if no solution meets the requirements of all the companies involved;

(5) the relatively limited software support available to railway schedulers.

6. Software

It is perhaps surprising that scheduling tasks are often performed largely without significant software support.

"Data management" packages are the primary tools used by schedulers to store and manipulate the core input data to the scheduling process (this includes infrastructure data, performance characteristics about the trains to run on that infrastructure, and rules that apply to the deployment of staff), and sophisticated graphical interfaces are used to manually adjust train times to achieve the business required, and eliminate any "conflicts" between trains.

Simulation packages that take timetables developed using these packages and assess them for robustness are also available. These packages take a timetable and overlay this with a number of possible perturbations (perhaps points failures or train breakdowns) and assess how much overall delay might be caused. Timetables where there is little "knock on" delay caused by a failure are robust; timetables where a lot of incremental delay occurs need further work or, perhaps, indicate that the infrastructure is being used rather too close to capacity for a robust timetable to be produced. Typically simulation packages require very detailed infrastructure maps, down to individual turnouts/switches and signals, and are hence complex and time-consuming to set up and run. They are frequently used as "strategic planning" tools to validate infrastructure improvements for robustness and are also used, for the same purpose, where major restructuring of a timetable is intended. Typically they are run for a single station or just a few miles of infrastructure. Time constraints mean that these packages are rarely used as part of the tactical planning process or to examine "whole route" effects during strategic planning.

Work is now under way to develop and implement software that can provide much greater support and, gradually, should enable better schedules (both in terms of robustness and efficient use of resources) to be produced in less time. Bussieck et al. (1997), Caprara et al. (1997), Cordeau et al. (1998), and Ferreira (1997) provide useful summaries of these developments (and also provide summaries of some stages of the scheduling process), covering timetable planning, crew and rolling stock scheduling, freight car routing, yard models, car management (all focused on a freight-dominated North American/Australian-style freight railway operation), and train dispatch (real-time operation).

7. Conclusions

Railway scheduling is central to the effective utilization and development of any railway system. A complex series of time-consuming activities are undertaken to match demand for rail transport with available resources. Railway schedulers face a number of challenges to the timely delivery of effective, robust timetables; in time the introduction of new software should substantially improve the schedules that can be produced.

References

Bussieck, M.R., P. Kreuzer and U.T. Zimmermann (1997) Discrete optimisation in public rail transport", *Mathematical Programming*, 79:415–444.

Caprara, A., M. Fischetti, P. Toth, D. Vigo and P.L. Guida (1997) Algorithms for railway crew management", *Mathematical Programming*, 79:123–141.

Cordeau, J.-F., P. Toth and D. Vigo (1998) A survey of optimization models for train routing and scheduling", *Transportation Science*, 32:380–404.

Ferreira, L. (1997) "Planning Australian freight rail operations: An overview", *Transportation Research Part A*, 31:335–348.

Ford, R. and D. Haydock (1992) "Signalling and timetabling", in: N.G. Harris and E.W. Godward, eds., *Planning passenger railways*, Glossop: Transport Publishing.

Chapter 36

PUBLIC TRANSPORT SCHEDULING

AVISHAI CEDER
Technion-Israel Institute of Technology, Haifa

1. Introduction

There is a saying by George Bernard Shaw: “The person who behaves sensibly is my tailor. He takes my measure anew every time he sees me. All the rest go on with their old measurements.” A good public transport scheduling program is one which has a tailor for frequent updating. However, only a few public transport organizations actually update their scheduling data or are aware of the benefits they may gain by this updating. This chapter provides an overview of old and new methods for efficient task scheduling. This overview is accompanied by some examples used as an explanatory device for the prescribed methods.

Public transport scheduling covers in essence four major modes of operation – airline, railway, bus, and passenger ferry – as illustrated in Figure 1. While the scheduling processes of buses, railways, and passenger ferries are similar, airline scheduling has some special features, not covered in this chapter. Nonetheless, in what follows there is a brief reference to the latter.

1.1. Airline scheduling

Airline scheduling is different from that of buses, railways and passenger ferries because the timetable is developed first and the routing afterwards. Airline scheduling comprises four basic elements: timetable development, fleet assignment, aircraft routing, and crew scheduling. Ideally all four elements should be solved simultaneously, but this is intractable even with modern computer aids. A good overview of airline scheduling problems and solutions can be found in Yu (1998).

When airline scheduling, special attention is given to the crew scheduling problem. That is, a given flight schedule is to be covered by a set of crew itineraries, rotations, or pairings, each consisting of a connected sequence of flights (legs) that begins and ends at a certain airport, usually called the home

Handbook of Transport Systems and Traffic Control, Edited by K.J. Button and D.A. Hensher

base. In this case the total crew time is to be minimized subject to a number of constraints on the itineraries. The traditional crew scheduling model is the set partitioning problem (described below in Section 4.2 for bus, railway, and passenger ferry crew scheduling). A recent survey of airline crew scheduling can be found in Barnhart et al. (1999). One interesting approach to the solution of this problem is an optimization tool using an artificial neural network, reported by Lagerholm et al. (2000).

1.2. Bus, railway, and passenger ferry scheduling

The bus, railway, and passenger ferry operational planning process includes four basic components, usually performed in sequence: (1) network route design, (2) setting timetables, (3) scheduling vehicles to trips, and (4) assignment of drivers (crew). It is desirable for all the four components to be planned simultaneously to exploit the system's capability to the greatest extent and to maximize the system's productivity and efficiency. However, this planning process is extremely cumbersome and complex, and therefore seems to require separate treatment of each component, with the outcome of one fed as an input to the next component. In the last 20 years, a considerable amount of effort has been invested in the computerization of the four components mentioned above, in order to provide more efficient, controllable, responsive schedules. The best summary, as well as the accumulated knowledge from this effort, has been presented at the second to the eighth International Conferences on Public Transport Scheduling, and appears in books edited by Wren (1981), Rousseau (1985), Daduna and Wren (1988), Desrochers and Rousseau (1992), Daduna et al. (1995), Wilson (1999), and Voss and Daduna (2001).

This chapter focuses on three components of scheduling, namely: timetabling, vehicle scheduling, and crew scheduling, while assuming that the public transport network is unchanged. The third scheduling component is usually divided into creating crew duties and creating crew rosters (rotation of duties among the drivers). A functional diagram of these scheduling elements appears in Figure 2.

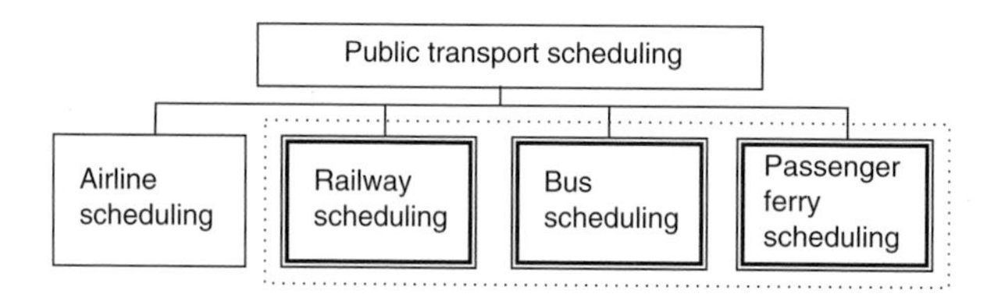

Figure 1. Four major modes of public transport with emphasis on the similarities between bus, railway and passenger ferry.

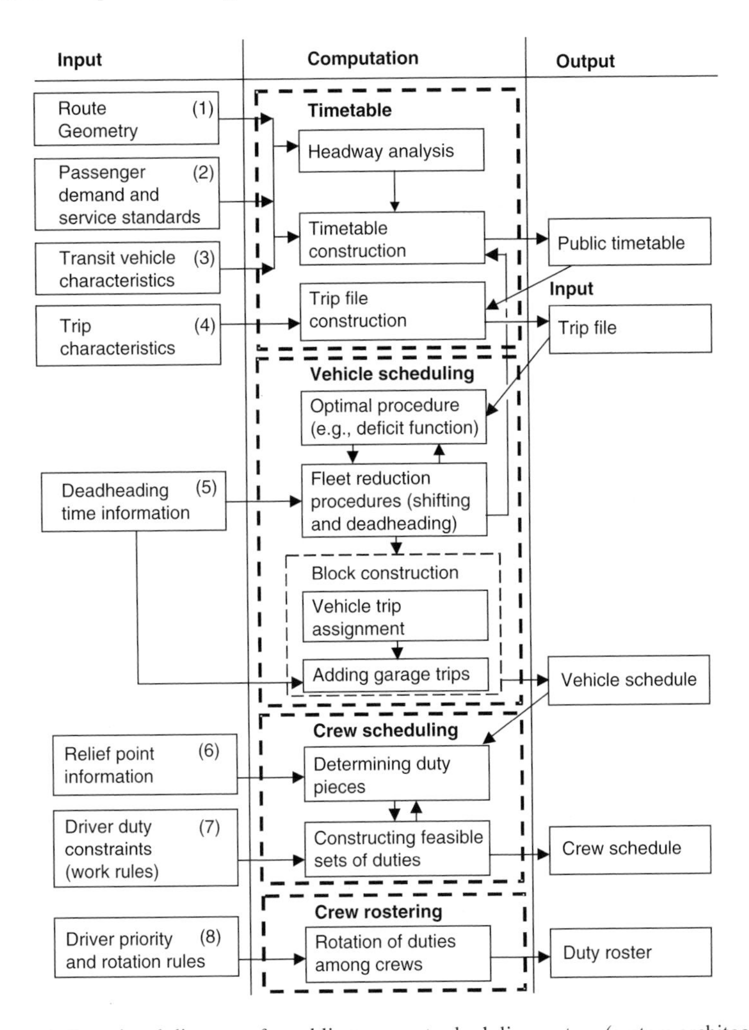

Figure 2. Functional diagram of a public transport scheduling system (system architecture).

The timetable component in Figure 2 is aimed at meeting the general public transportation demand. The demand varies during the hours of the day, with the day of the week, from one season to another, and even from one year to another. This demand reflects the business, industrial, cultural, educational, social, and recreational transportation needs of the community. It is the purpose of this component to set appropriate timetables for each transit route to meet the variation in the public demand. Determination of timetables is performed on the basis of passenger counts, and must comply with service frequency constraints.

The purpose of the vehicle scheduling component in Figure 2 is to schedule vehicles to trips according to given timetables. A transit trip can be planned either to transport passengers along its route or to make a deadheading trip in order to connect two service trips efficiently. The scheduler's task is to list all daily chains of trips (some deadheading) for a vehicle, ensuring the fulfillment of the timetable requirements and the operator requirements (refueling, maintenance, etc.). The major objective of the task is to minimize the number of vehicles required. The purpose of the crew scheduling component in Figure 2 is to assign drivers to the outcome of vehicle scheduling. This assignment must comply with some constraints, which usually are dependent on a labor contract. Finally, the crew rostering component in Figure 2 usually refers to priority and rotation rules, rest periods, and drivers' preferences. Any transit company, which naturally wishes to utilize its resources more efficiently, has to deal with problems presented by various pay scales (regular, overtime, weekends, etc.), and by human-oriented dissatisfaction. All the components in Figure 2 are very sensitive to internal and external factors, a sensitivity which could easily lead toward an inefficient solution. In Figure 2 the items under the heading "input" are numbered, and their general description is listed below; one must bear in mind that their values differ by time of day and by day of the week.

(1) *Route geometry*:
   (i) route number;
   (ii) nodes, stops, and timepoints on a route;
   (iii) pattern and sequence of nodes on a route.
(2) *Passenger demand and service standards*:
   (i) passenger loads between adjacent stops on a route;
   (ii) load factor, desired number of passengers on board the transit vehicle;
   (iii) policy headway, the inverse of the minimum frequency standard.
(3) *Transit vehicle characteristics*:
   (i) vehicle type;
   (ii) vehicle capacity;
   (iii) running time, vehicle travel time between stops and/or timepoints on a route.
(4) *Trip characteristics*:
   (i) trip layover time (maximum and minimum);
   (ii) trip departure time tolerances (maximum departure delay and maximum departure advance).
(5) *Deadheading time information*:
   (i) list of garages, name, and location;
   (ii) list of trip start and end locations;
   (iii) deadheading times from garage locations to each trip start location (pull-outs);

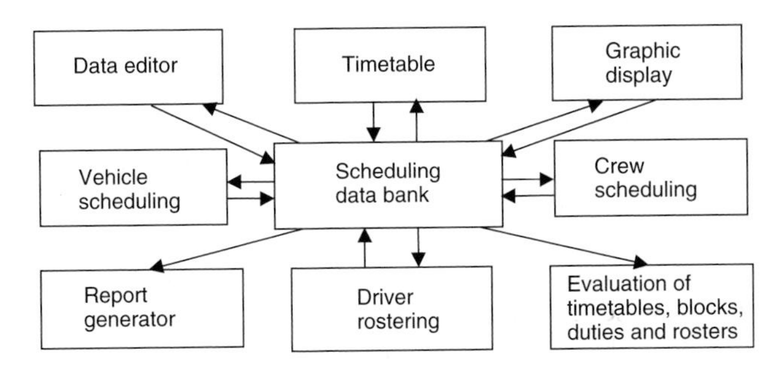

Figure 3. Overall view of a public transport scheduling system.

(iv) deadheading times from trip end locations to garage locations (pull-ins);
(v) deadheading time matrix between all trip end and start locations.

(6) *Relief point information*:
(i) relief point location (stops, trip start and end points, garages);
(ii) travel times between the relief points.

(7) *Driver duty constraints* (*dependent on a labor contract*):
(i) type of duty (early, late, split, full, tripper, etc.);
(ii) duty length (maximum spread time);
(iii) number of vehicle changes on duty;
(iv) meal breaks;
(v) duty composition;
(vi) other work rules.

(8) *Drivers' priority and rotation rules*:
(i) list of drivers by name and type (e.g., part-time, full-time, seniority);
(ii) driver priority or equality rules;
(iii) workday on and off pattern.

The complexity involved in the public transport operational planning process has challenged researchers to develop automated computerized procedures, which has led to a number of software packages being available on the market. An overall view of such a piece of software is illustrated in Figure 3. It is worth mentioning that the evaluation module of such a software package should be based on an external input related to cost coefficients, and performance criteria. The cost coefficients include vehicle cost (fixed and variable), crew cost (fixed and variable), service benefit, and other costs. The performance criteria include measures of passenger service, measures of vehicle and crew schedules, measures of duty rosters, and other criteria.

Section 2 of this chapter summarizes some methods for constructing timetables, while Sections 3 and 4 provide a basic overview of vehicle scheduling and of crew scheduling and rostering, respectively. These sections emphasize certain methodologies that are used in the example in Section 5 for crystallizing and clarifying the process of public transport scheduling. Finally, Section 6 presents some concluding remarks.

2. Timetables

2.1. *Mathematical programming approaches*

Mathematical programming methods for determining frequencies and timetables have been proposed by Furth and Wilson (1981), Koutsopoulos et al. (1985), and Ceder et al. (2001). The objective in Furth and Wilson (1981) is to maximize the net social benefit, consisting of ridership benefit and wait time saving, subject to constraints on total subsidy, fleet size, and passenger loading levels. Koutsopoulos et al. (1985) extend this formulation by incorporating crowding discomfort costs into the objective function and treating the time-dependent character of transit demand and performance. Their initial problem comprises a non-linear optimization program relaxed by linear approximations. Finally, Ceder et al. use mixed integer programming and heuristic procedures for constructing timetables with maximum synchronization, that is, maximization of the number of simultaneous arrivals of vehicles to connection stops.

2.2. *Passenger-data-based approach*

A public transport timetable is commonly constructed for given sets of derived frequencies. The basic criteria for the determination of frequencies are to provide adequate space in vehicles to meet passenger demand, and to ensure a minimum frequency (maximum-policy headway) of service. Ceder (1984) described four methods for calculating the frequencies. Two are based on point-check (counting the passengers on board the transit vehicle at certain points), and two on ride-check (counting the passengers along the entire transit route). In the point-check methods the frequency is the ratio of the passenger load at the maximum load point (either the load for the whole day or the load in each hour) to the desired occupancy or load factor. In the ride-check methods the frequency is the ratio of the average or restricted-average passenger load to the desired occupancy. The average load is determined by the area under the load profile (in passenger-km) divided by the route length (km), and the restricted average is higher than the average value, in order to ensure that in a certain percentage of the route length the load does not exceed the desired occupancy. This desired occupancy (or load factor) is the desired level of passenger load on each vehicle, in each time period (e.g., number of seats).

In a follow-up study Ceder (1986) analyzed optional ways of generating public timetables. This analysis allows one to establish a spectrum of alternative timetables, based on three categories of options: (a) selection of type of headway, (b) selection of frequency determination method for each period, and (c) selection of special requests. In category (a) the headway (time interval between adjacent departures) can be equal or balanced. Equal headway refers to the case of evenly spaced headways, and balanced headway to the case of unevenly spaced headways but with even average passenger loads at the hourly maximum-load point. In category (b) it is possible to select, for each time period, one of the four frequency determination methods (two point-check methods, and two ride-check methods) mentioned above, or a frequency specified by the scheduler. In category (c) it is possible to request clock headways (departure times that repeat themselves each hour, easy to memorize) and/or a certain number of departures (usually for cases with limited resources).

The outcome of these analyses is a set of optional timetables in terms of departure times of all vehicles at all specified timepoints, using passenger load data. Each timetable is accompanied by two comparison measures, which are used as an evaluation indicator in conjunction with resource saving. The first measure is the total required vehicle runs (departures) and the second is an estimate of the minimum required fleet size at the route level only.

3. Vehicle scheduling

3.1. Objectives

This second component in Figure 2 determines, in an efficient manner, the construction of chains of trips or blocks (the vehicle schedules). The main objective of this function is to construct vehicle blocks while either (a) using the existing number of vehicles (while minimizing the total deadheading kilometers and disruption to the timetable) or (b) minimizing the number of vehicles required to carry out the schedule (the trip schedule).

In addition, the assignment of vehicle chains to garages should be determined in an efficient manner. The attainment of these objectives can be carried out through the interaction of the following subfunctions:

(1) trip characteristics study;
(2) deadheading-trip construction;
(3) intertrip deadheading-trip insertions;
(4) timetable shifting;
(5) vehicle trip chaining;
(6) garage chain assignment.

3.2. Exact-solution approaches

The problem of scheduling vehicles in a multidepot scenario is known as the multidepot vehicle scheduling problem (MDVSP). This problem is complex (NP-hard) and considerable effort has been devoted to solving it in an exact way. Reviews and descriptions of some exact solutions can be found in Desrosiers et al. (1995), Daduna and Paixao (1995), Löbel (1999), and Mesquita and Paixao (1999).

An example formulation of the MDVSP is as follows:

$$\text{objective function:} \quad \min_{y} \left\{ \sum_{i=1}^{n+1} \sum_{j=1}^{n+1} c_{ij} y_{ij}, \right. \tag{1}$$

where i is the event of ending of a trip at time a_i, j is the event of starting a trip at time b_j, and

$$y_{ij} = \begin{Bmatrix} 1, & \text{ending is connected to start} \\ 0, & \text{otherwise} \end{Bmatrix}.$$

For $i = n + 1$, then $y_{n+1,j} = 1$ if a depot supplies a vehicle for the jth trip. For $i = n + 1$, then $y_{i,n+1} = 1$ if after the ith trip end, the vehicle returns to a depot, and $y_{n+1,n+1}$ is the number of vehicles that remain unused at a depot.

The cost function c_{ij} takes the form

$$c_{ij} = \begin{Bmatrix} K; & i = n+1; & j = 1, 2, \ldots, n \\ O; & i = 1, 2, \ldots, n; & j = n+1 \\ L_{ij} + E_{ij}; & i, j = 1, 2, \ldots, n & \end{Bmatrix}, \tag{2}$$

where K is the saving incurred by reducing the fleet size by one vehicle, L_{ij} is the direct deadheading cost from event i to j, and E_{ij} is the cost of the idle time of a driver between i and j.

This formulation, which appears in a similar form in Gavish et al. (1978), covers the chaining of vehicles in a sequential order from the depot to the transit routes alternating with idle time and deadheading trips, and back to the depot. This is a zero–one integer programming problem that can be converted to a large-scale assignment problem. In addition, the assignment of vehicles from the depots to the vehicle schedule generated in the above chaining process can be formulated as a "transportation problem" of the type known in all of the operations research literature.

Löbel (1999) uses a branch-and-cut method for the MDVSP with the generation of upper bounds and the use of Lagrangian relaxations and pricing. Mesquita and Paxiao (1999) compare in this problem, a linear relaxation based on a multicommodity network flow approach.

3.3. A heuristic deficit function approach

The following is a description of the deficit function approach described by Ceder and Stern (1981), for assigning the minimum number of vehicles for a given timetable. A deficit function is simply a step function that increases by one at the time of each trip departure and decreases by one at the time of each trip arrival. Such a function may be constructed for each terminal in a multiterminal transit system. To construct a set of deficit functions, the only information needed is a timetable of required trips. The main advantage of the deficit function is its visual nature. Let $d(k, t, S)$ denote the deficit function for terminal k at time t for the schedule S. The value of $d(k, t, S)$ represents the total number of departures minus the total number of trip arrivals at terminal k, up to and including time t. The maximal value of $d(k, t, S)$ over the schedule horizon $[T_1, T_2]$ is designated $D(k, S)$. Note that S will be deleted when it is clear which underlying schedule is being considered.

If the set of all terminals is denoted as T, the sum of $D(k)$ for all $k \in T$ is equal to the minimum number of vehicles required to service the set T. This is known as the fleet size formula. Mathematically, for a given fixed schedule S,

$$D(S) = \sum_{k \in T} D(k) = \sum_{k \in T} \max_{t \in [T_1, T_2]} d(k, t), \tag{3}$$

where $D(S)$ is the minimum number of buses required to service the set T.

When deadheading (DH) trips are allowed, the fleet size may be reduced below the level described in eq. (3). Ceder and Stern (1981) described a procedure based on the construction of a unit reduction DH chain (URDHC), which, when inserted into the schedule, allows a unit reduction in the fleet size. The procedure continues inserting URDHCs until no more can be included or a lower bound on the minimum fleet in reached. The lower bound $D_m(S)$ is determined from the overall deficit function, defined as

$$g(t, S) = \sum_{k \in T} d(k, t, S),$$

where

$$D_m(S) = \max_{t \in [T_1, T_2]} g(t, S).$$

This function represents the number of trips simultaneously in operation. Initially, the lower bound was determined to be the maximum number of trips in a given timetable that are in simultaneous operation over the schedule horizon. Stern and Ceder (1983) improved this lower bound to $D_m(S') > D_m(S)$ on the basis of the construction of a temporary timetable S' in which each trip's arrival time is extended to the time of the first trip that may feasibly follow it in S.

The deficit function theory was extended by Ceder and Stern (1982) to include possible shifting in departure times within bounded tolerances. Basically, the shifting criterion is based on a defined tolerance time, with a maximum advance of the scheduled departure time of the trip (early departure) and a maximum allowed delay (late departure). Once a terminal k is selected, the algorithm searches to reduce $D(k)$ by shifting departure times (if allowed). Then all of the $d(k, t)$ values are updated. When no more shiftings are possible, the algorithm searches for a URDHC from the selected terminal while considering possible blending between DH insertion and shiftings in departure times.

In the fixed-schedule problem, the algorithm terminates when $D(S)$ is equal to the improved lower bound. In the variable-schedule problem (when shifting is allowed), the algorithm also uses this comparison, and if $D(S)$ is equal to the improved lower bound, the URDHC procedure (with shiftings) ceases and the shifting-only mode is applied. If the latter results in a reduction of $D(S)$, the URDHC procedure is again activated. The process terminates when $D(S)$ cannot be further reduced. Finally, all of the trips, including those that were shifted and the DH trips, are chained together to construct the vehicle schedule (blocks). One common rule for creating the chains is "first in first out" (FIFO). The FIFO rule simply links the arrival time of a trip to the nearest departure time of another trip (at the same location), and continues to create a schedule until no connection can be made. The trips considered are deleted and the process continues.

4. Crew scheduling and rostering

4.1. Overview

The last two components of the public transport operational planning process (shown in Figure 1) are the assignment of drivers to carry out the vehicle schedules and the rotation of duties among drivers. The purpose of the assignment function is to determine a feasible set of driver duties in an optimal manner. The criterion for this determination is based on an efficient use of manpower resources while maintaining the integrity of any work rule agreements. The construction of the selected crew schedule is usually a result of the following subfunctions:

(1) duty piece analysis;
(2) work rule co-ordination;
(3) feasible-duty construction;
(4) duty selection.

The duty piece analysis function divides or partitions each vehicle block at selected relief points into a set of duty pieces. These duty pieces are assembled in the feasible-duty construction function. Other required information is the travel times between relief points, and a list of relief points designated as required duty stops and start locations.

Theoretically, each relief point may be used to split the vehicle block into new duty pieces. Usually, it is more efficient to use one or more of the following criteria for the selection of which relief points to include:

(1) minimum duty piece length;
(2) select a piece so that the next relief point selected is as close as possible to the maximum duty part time (maximum time before taking a break);
(3) only a few (say two) relief points in each piece;
(4) operator decisions.

In order to utilize any crew scheduling method, a list of work rules to be used in the construction of feasible driver duties is required. The work rules are the result of an agreement between the drivers (or their unions) and the public transport company (and/or public authorities).

The determination of different feasible sets of duties may be based on, for example, one or more of the following performance measures:

(1) number of duties (drivers),
(2) number of split duties,
(3) total number of changes,
(4) total duty hours,
(5) average duty length,
(6) total working hours,
(7) average working time,
(8) number of short duties,
(9) costs.

Once the set of duties is established, it is common to group them into rosters. A roster is defined as a periodic duty assignment which guarantees that all the trips are covered for a certain number of consecutive days (which can be a week, a month, or any other period). Commonly a roster contains a subset of duties covering 6 or 7 consecutive days (called weeks). The length of a roster is typically between 30 days and 60 days (5 to 10 weeks). The usual rostering problem is to find a feasible set of rosters to cover all the duties in the minimum number of weeks, implying the minimization of the number of crews required, as explained in detail in Caprara et al. (1999).

It is worth noting that for railways the vehicle scheduling component is obviously not important, and instead it is required to construct a work schedule of the train crews (drivers and conductors together).

4.2. Mathematical approaches

Crew scheduling and rostering are widely treated in Wren (1981), Rousseau (1985), Daduna and Wren (1988), Desrochers and Rousseau (1992), Daduna et al. (1995), Wilson (1999) and Voss and Daduna (2001). Specific detailed studies can be found in, for example, Bodin et al. (1983), Carraresi and Gallo (1984), Bianco et al. (1992), Caprara et al. (1999), and Freling et al. (1999).

The basic formulation of the crew scheduling problem is one of zero–one integer linear programming, called a set partitioning problem (SPP). In this SPP the objective is to select a minimum set of feasible duties such that each task is included in exactly one of these duties:

$$\min \sum_{q \in Q} c_q x_q \tag{4}$$

such that

$$\sum_{q \in Q(j)} X_q = 1, \quad \text{for all } j \in J, \tag{5}$$

$$x_q = (0, 1), \quad \text{for all } q \in Q. \tag{6}$$

where c_q is the cost of duty $q \in Q$, Q is the set of all feasible duties, and $Q(j) \in Q$ is the set of duties covering task $j \in J$. A binary zero–one variable x_q is used for indicating if duty q is selected in the solution or not. Constraint (6) ensures that each task will be covered by exactly one duty. An easier way to solve the crew scheduling problem is to relax constraint (6):

$$\sum_{,q \in Q(j)} x_q \geq 1, \quad \text{for all } j \in J. \tag{7}$$

Equation (7) represents a new problem called a set covering problem (SCP). This SCP is usually solved first and the solution is then changed to handle the SPP by deleting overlapping trips. This deletion process involves changes in the duties considered and a crew member who is assigned to such a duty will make the trip as a passenger. Freling et al. (1999) explain that such a change affects neither the feasibility nor the cost of the duties considered.

A mixed (heuristic and exact) approach to solving the crew scheduling problem, mentioned in Bodin et al. (1983), is to

(1) generate all feasible pieces of work (to be derived from the vehicle blocks),
(2) establish an interval piece cost (based on piece characteristics and past experience),
(3) create an acyclic network from each block, where nodes are the relief points and arcs represent the cost of each feasible piece in that block,

(4) solve the shortest-path problem in order to establish the best (minimum-cost) pieces, and
(5) solve a matching problem while using two or three legal piece combination, and, if the solution is not feasible, reiterate part of the process by updating the piece combination cost and redo the shortest path (step 4) until the solution is feasible and satisfactory.

Part of this shortest path-approach is exhibited in the example problem in the next section.

Freling et al. (1999) provide several approaches for integrating the vehicle and crew scheduling problems. They believe, though can not prove, that a combined formulation, with heuristics for the column generation (piece coverage) procedure, results in better solutions, especially for situations when it is not allowed for the crew to change vehicles during a duty. More consideration of integration and of mathematical treatments appear in Voss and Daduna (2001).

Finally, the duty rostering problem in different public transport agencies and countries is also well covered in Voss and Daduna (2001). It is apparent that the duty rostering rules and priorities are dependent on the location and agency.

5. Example

The example in this section is used as an explanatory device for describing the basics of the scheduling procedures and considerations. Table 1 contains the necessary information and data for a 3 hour example of a transit line from A to B and B to A. Point B can be perceived as the central business district that attracts the majority of the demand between 6 and 9 a.m. There are 14 and 8 departures from A to B and B to A, respectively. The average observed maximum load on each trip, the service and DH travel times, the desired occupancies, and the minimum frequency are all shown in Table 1.

In order to construct a balanced-load timetable, the data shown in Table 1 were used for running the accumulative-load procedures described by Ceder (1986). The balanced-load timetable is based on even average loads on board the public transport vehicles as opposed to even headways (see Section 2.2). Given that the maximum load is observed at the same stop for each direction, the accumulative-load procedures determine the new departure times shown in Figure 4 for the direction B to A, together with the results for the direction A to B, which are also shown. There are 14 new departures for the direction A to B that are based on desired occupancies of 50 and 65 passengers. There are seven new departures for the direction B to A in Figure 4.

Once the departure times are set at both route end points, the vehicle scheduling component can be integrated into the two-direction timetables. First,

Table 1
Given data for the example problem (a)

Time	Departure time at the route departure point		Average observed maximum No. of passengers on board the vehicles		Travel time including layover time (min.)				Desired occupancy (passengers) on each vehicle	
					Service		Deadheading			
	A→B	B→A	A→B	B→A	A→B	B→A	A→B	B→A	A→B	B→A
6–7 a.m.	6:20 6:40 6:50	6:30 6:45	15 30 47	22 38	60	50	40	35	50	50
7–8 a.m.	7:05 7:15 7:25 7:30 7:40 7:50	7:10 7:25 7:45	58 65 79 90 82 62	52 43 59	75	60	45	40	65	50
8–9 a.m.	8:00 8:10 8:20 8:35 8:50	8:25 8:40 8:55	75 68 55 80 70	23 51 28	70	60	45	40	65	50

Note: (a) Minimum frequency: 2 vehicles per hour, for all hours both directions.

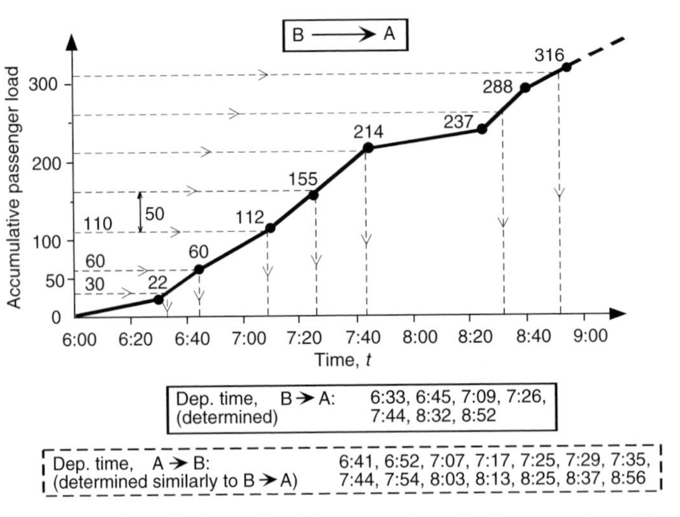

Figure 4. Determination of balanced load departure times for the example problem, direction B–A, with results also for direction A–B.

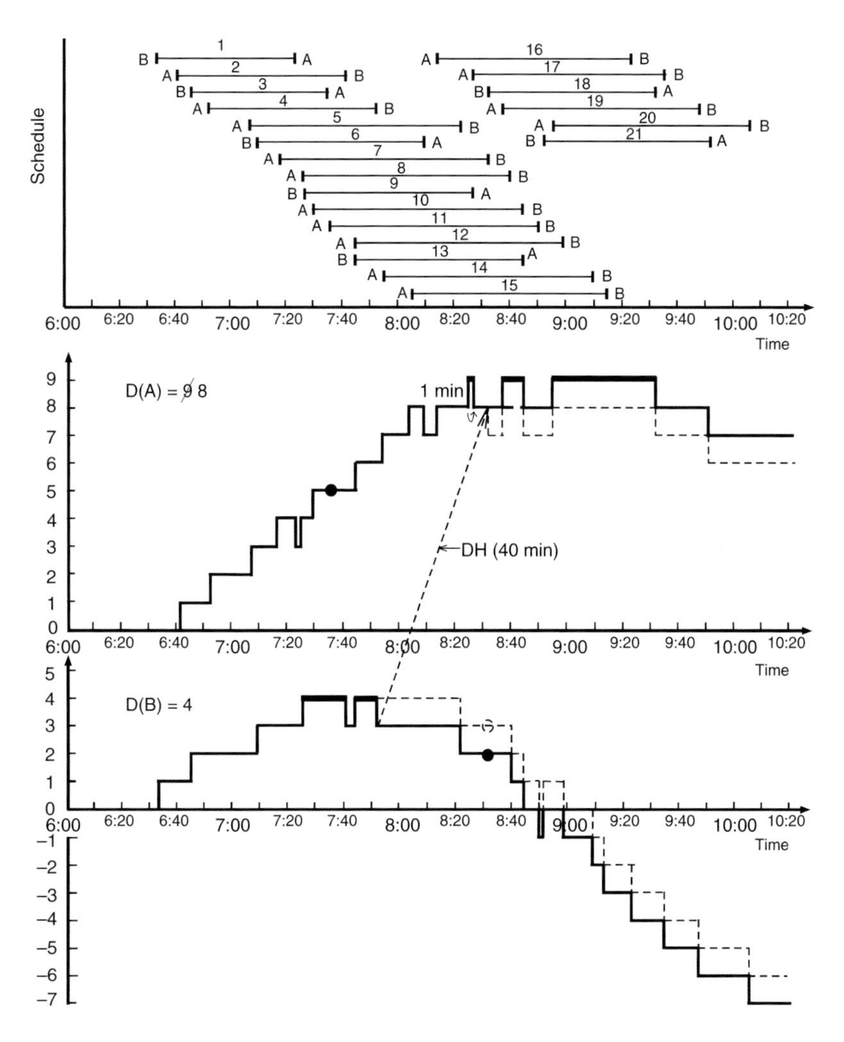

Figure 5. Deficit function analysis for the example problem.

two deficit functions (DFs) are constructed at A and B, as shown in Figure 5. These DFs are based on the schedule of 21 trips (14 A–B, 7 B–A), presented with respect to their travel times in the upper part of Figure 5. Second, the DF theory leads to saving one vehicle at $d(A, t)$ through a shifting of trip 17 by one minute forward (late departure), and inserting a DH trip from B to A (7:52 to 8:32 a.m). The total fleet required is then $8 + 4 = 12$ vehicles.

Once the deficit function procedure has been performed, the final efficient schedule can be set both for balancing the passenger loads and for the minimum fleet size required. The timetables at the route's end points appear in the upper part of Table 2, and the blocks in its lower part. The blocks (vehicle schedules) are constructed from the timetable using the FIFO rule. The first block, for example, starts with trip 1 which is linked with its first feasible connection at A, trip 8 (7:23 links to 7:25), and with trip 21 at B (8:40 links to 8:52).

Table 2
Timetable and vehicle schedule (blocks) of the example problem (see Figure 5)

A→B			B→A		
Trip No.	Departure time	Arrival time	Trip No.	Departure time	Arrival time
2	6:41	7:41	1	6:33	7:23
4	6:52	7:52	3	6:45	7:35
5	7:07	8:22	6	7:09	8:09
7	7:17	8:32	9	7:26	8:26
8	7:25	8:40	13	7:44	8:44
10	7:29	8:44	DH (a)	7:52	8:32
11	7:35	8:50	18	8:32	9:32
12	7:44	8:59	21	8:52	9:52
14	7:54	9:09			
15	8:03	9:13			
16	8:13	9:23			
17	8:26 (b)	9:36			
19	8:37	9:47			
20	8:56	10:06			

Block number	Trips in block (in sequence, via FIFO)
1	1–8–21
2	2–13–20
3	3–11
4	4–DH–19
5	5–18
6	6–16
7	7
8	9–17
9	10
10	12
11	14
12	15

Notes:
(a) Inserted deadheading trip.
(b) Trip 17 was shifted by 1 minute.

Block No.*	Segment (trip No. or combination of trips in parentheses)	Segment cost	Shortest-path solution (best pieces emphasized)
1	B-A (1)	4	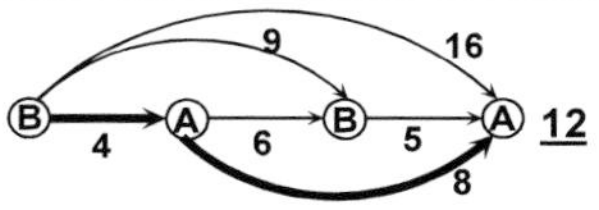 12
	A-B (8)	6	
	B-A (21)	5	
	B-A-B (1-8)	9	
	A-B-A (8-21)	8	
	B-A-B-A (1-8-21)	16	
2	A-B (2)	3	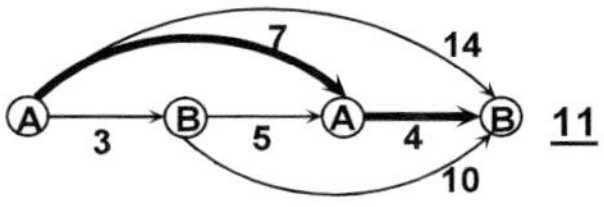 11
	B-A (13)	5	
	A-B (20)	4	
	A-B-A (2-13)	7	
	B-A-B (13-20)	10	
	A-B-A-B (2-13-20)	14	
3	B-A (3)	4	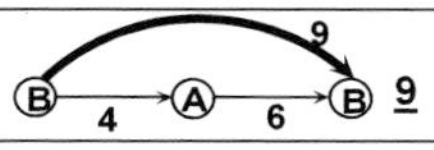 9
	A-B (11)	6	
	B-A-B (3-11)	9	
4	A-B (4)	4	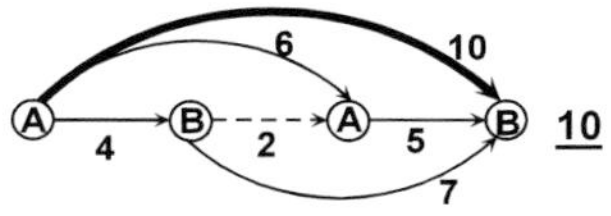 10
	B-A (DH)	2	
	A-B (19)	5	
	A-B-A (4-DH)	6	
	B-A-B (DH-19)	7	
	A-B-A-B (4-DH-19)	10	
5	A-B (5)	3	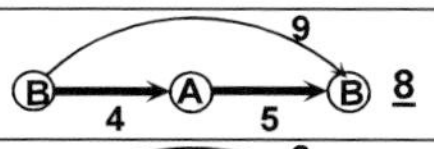 8
	B-A (18)	5	
	A-B-A (5-18)	9	
6	B-A (6)	4	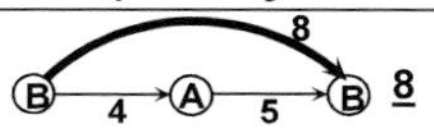 8
	A-B (16)	5	
	B-A-B (6-16)	8	
8	B-A (9)	3	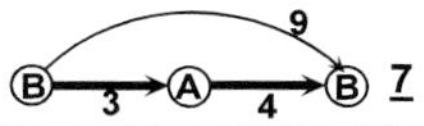 7
	A-B (17)	4	
	B-A-B (9-17)	9	

*Refers to blocks with more than one trip

Figure 6. Partitioning of each block into minimum-cost pieces.

The blocks appear in Table 2 are now subject to a crew scheduling procedure, like the one illustrated below for the sake of clarity. Given that both A and B are relief points, each block can be considered as a driver schedule or be partitioned into alternative pieces covering all possible combinations. Figure 6 shows how to partition the blocks into the combinations of pieces. In addition, each piece is assigned an internal piece cost based on the piece characteristics (e.g., time a day, arrival and departure locations, type of vehicle required) and past experience (how much is the cost of a similar piece in past crew schedules?). The right-hand column of Figure 6 contains the acyclic network of each block, representing all the possible pieces and their internal costs. It is then possible to apply a shortest-path algorithm such as the known Dijkstra algorithm (see any operations research literature on networks). The results of the Dijkstra procedure are emphasized in Figure 6 along with the minimum piece cost required to cover the whole block. These results are illustrated in Figure 7 for each block. Given that each piece is

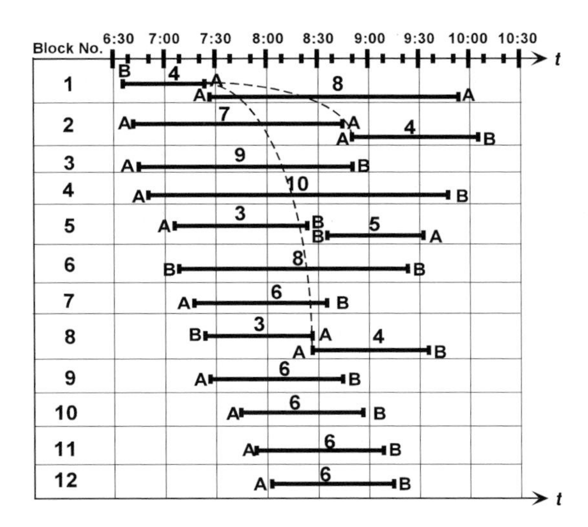

Figure 7. The results of partitioning the blocks into best pieces.

eligible to be covered by one driver, then Figure 7 shows that the 12 blocks require 16 drivers with a total cost of 95. However, there are more possibilities for matching some pieces if it is both legal (from the labor agreement perspective) and can reduce the cost. Taking into account the deadheading times in Table 1 as a measure for moving between A and B or B and A, there are only two possible matchings to examine: to connect the first piece of block 1 either with the second piece of block 2 or with the second piece of block 8, as shown in Figure 7. Both possible matchings require no a deadheading time. Usually, if the best (minimum-cost) feasible matching results in a cost reduction (less than 95) then this matching is selected, and the number of drivers will be reduced to 15 in this case.

This simplified example demonstrates some of the complexities involved in public transport scheduling. Furthermore, on top of the combinatorial problems inherent in the scheduling tasks, there are human dissatisfaction issues related to the crews that deserve attention and make this undertaking even more cumbersome.

6. Concluding remarks

In practical public transport scheduling, schedulers attempt to create timetables and to allocate vehicles and crew in the most efficient manner possible. However, these scheduling tasks are time-consuming and exacting, requiring the services of imaginative and experienced schedulers. The overview and methodologies presented in this chapter suggest that most of the scheduling tasks can be performed automatically or in an interactive human–computer mode.

The adoption of new practical scheduling methods will undoubtedly produce more efficient timetables and vehicle and crew schedules, which ultimately will result in savings in operational cost. Moreover, it will then be possible to better match public transport demand, which varies systematically by season, day of the week, time of day, location, and direction of travel, with the resultant public transport services. There is a saying by François Gautier: "More important than the quest for certainty is the quest for clarity." Many elements in the implementation stages of public transport scheduling are uncertain, mainly due to traffic congestion, vehicle breakdowns, and crew behavior. It is therefore important to emphasize the clarity of the scheduling processes in order to allow for immediate adjustments and smoothing of interferences.

References

Barnhart, C., E. Johnson, E. Nemhauser, and P.H. Vance (1999) "Crew scheduling", in: R.W. Hall, ed., *Handbook of transportation science*. Dordrecht: Kluwer Scientific.

Bianco, L., M. Bielli, A. Mingozzi, S. Riccardelli and M. Spadoni (1992) A heuristic procedure for the crew rostering problem", *European Journal of Operational Research*, 58:272–283.

Bodin, L., B. Golden, A. Assad and M. Ball (1983) Routing and scheduling of vehicles and crews: The state of art", *Computers and Operation Research*, 10:63–211.

Caprara, A., M. Fischetti, P.L. Guida, P. Toth and D. Vigo (1999) "Solution of large scale railway crew planning problems: The Italian experience", in: N.H.M. Wilson ed., *Computer-aided scheduling of public transport*. Berlin: Springer.

Carraresi, P. and G. Gallo (1984) A multilevel bottleneck assignment approach to the bus drivers' rostering problem", *European Journal Operational Research*, 16:163–173.

Ceder, A. (1984) Bus frequency determination using passenger count data. *Transportation Research*, 18A(5/6):439–453.

Ceder, A. (1986) "Methods for creating bus timetables", *Transportation Research*, 21A(1):59–83.

Ceder, A. and H.I. Stern (1981) "Deficit function bus scheduling with deadheading trip insertion for fleet size reduction", *Transportation Science*, 15(4):338–363.

Ceder, A. and H.I. Stern (1982) "Graphical person–machine interactive approach for bus scheduling", *Transportation Research Record*, 857:69–72.

Ceder, A., B. Golany and O. Tal (2001) "Creating bus timetables with maximal synchronization", *Transportation Research* B (in press).

Daduna, J.R. and A. Wren, eds. (1988) *Lecture notes in economics and mathematical systems*, vol. 308. *Computer-aided transit scheduling*. Berlin: Springer.

Daduna, J.R. and J.M.P. Paixao (1995) "Vehicle scheduling for public mass transit – an overview", in: J.R. Daduna, I. Branco and J.M.P. Paixao eds., *Computer-aided transit scheduling*. Berlin: Springer.

Daduna, J.R., I. Branco and J.M.P. Paixao, eds. (1995) *Lecture notes in economics and mathematical systems*, vol. 410. *Computer-aided transit scheduling*. Berlin: Springer.

Desrochers, M. and J.M. Rousseau, eds. (1992) *Lecture notes in economics and mathematical systems*, vol. 386. *Computer-aided transit scheduling*. Berlin: Springer.

Desrosiers, J., Y. Dumas, M.M. Solomon and F. Soumis (1995) "Time constrained routing and scheduling. in: M.O. Ball, T.L. Magnanati, C.L. Monma and G.L. Nemhauser, eds., *Handbooks in operations research and management science*, vol. 8. *Network routing*. Amsterdam: Elsevier.

Freling, R., A.P.M. Wagelman and J.M.P. Paixao (1999) "An overview of models and techniques of integrating vehicle and crew scheduling. in: N.H.M. Wilson, ed., *Computer-aided scheduling of public transport*. Berlin: Springer.

Furth, P.G. and N.H.M. Wilson (1981) "Setting frequencies on bus routes: Theory and practice", *Transportation Research Record,* 818:1–7.
Gavish, B., P. Schweitzer and E. Shlifer (1978) "Assigning buses to schedules in a metropolitan area", *Computers and Operations Research*, 5:129–138.
Koutsopoulos, H.N., A. Odoni and N.H.M. Wilson (1985) "Determination of headways as function of time varying characteristics on a transit network", in: J.M. Rousseau, ed., *Computer scheduling of public transport 2*. Amsterdam: North-Holland.
Lagerholm, M., C. Peterson and B. Soderberg (2000) "Airline crew scheduling using potts mean field techniques", *European Journal of Transportation Research*, 120:81–96.
Löbel, A. (1999) "Solving large scale multiple-depot vehicle scheduling problems", in: N.H.M. Wilson, ed., *Computer-aided scheduling of public transport*. Berlin: Springer.
Mesquita, M. and J.M.P. Paixao (1999) "Exact algorithms for the multi-depot vehicle scheduling problem based on multicommodity network flow type formulations", in: N.H.M. Wilson, ed., *Computer-aided scheduling of public transport*. Berlin: Springer-Verlag.
Rousseau, J.M., ed. (1985) *Computer scheduling of public transport 2*. Amsterdam: North-Holland.
Stern, H.I. and A. Ceder (1983) "An improved lower bound to the minimum fleet size problem", *Transportation Science*, 17(4):471–477.
Voss, S. and J.R. Daduna (2001) *Lecture notes in economics and mathematical systems. Computer scheduling of public transport*. Berlin: Springer.
Wilson, N.H.M., ed. (1999) *Lecture notes in economics and mathematical systems*, vol. 471. *Computer-aided scheduling of public transport*. Berlin: Springer.
Wren, A., ed. (1981) *Computer scheduling of public transport 2: Urban passenger vehicle and crew scheduling*. Amsterdam: North Holland.
Yu, G., ed. (1998) *Operations research in the airline industry*. Dordrecht: Kluwer Academic.

Chapter 37

SEAPORT TERMINAL MANAGEMENT

BRIAN THOMAS
B J Thomas Consultancy Services, Cardiff

1. The terminal concept

Specialization in shipping has been the most significant development in sea transport in recent times, driven by spectacular growth in international trade. Very large dry and liquid bulk carriers have appeared, and there has been a remarkable transition from conventional to what are termed unitized means of carriage in the general cargo trades. This has involved palletization, packaged timber, neo-bulk, barge carriers, and roll-on roll-off vessels, but it has been the arrival of the cellular container vessel which has produced nothing less than a revolution in the way goods are packaged and transported around the world.

Containerization has radically altered ocean transportation, changing trading patterns, ship routing and itineraries, ship design and size, cargo-handling equipment and operations, inland transport and freight terminals, commercial practices and customs procedures, employment and working practices, and information and communication systems. But containerization's greatest impact has been on ports and the way they accommodate container carriers and handle their cargoes. Seaports have had to construct dedicated container terminals (operated on an exclusive, preferential, or common-user basis) and invest in specialized handling equipment to provide fast and efficient services to ship operators (Thomas, 1994).

In conventional general cargo trades, the functional unit of a port is the berth, its activity centered on the quayside at which the ship moors. The traditional emphasis of the port's service is mainly directed at the ship and its owner. With the arrival of bulk and unitized trades, with their mechanized and automated facilities designed to transfer cargo as quickly as possible between the ship and inland transport, the emphasis moved away from the quayside to include also the landside activities. The name "terminal" was introduced, implying an integrated activity providing services not just to the shipowner but also to importers, exporters, and transport operators. The name "terminal" is now applied to facilities which provide a comprehensive range of specialized services to users. If a terminal is to work smoothly and

Handbook of Transport Systems and Traffic Control, Edited by K.J. Button and D.A. Hensher

efficiently, its activities must be fully integrated and co-ordinated, this requires a high degree of management control, a unified structure.

The main benefit of containerization to the ship operator has been the economies of scale. By building larger vessels, designed expressly to take containers, operators reduce the cost of transporting each container. However, these specialist vessels are extremely expensive to build and to operate and so ship operators have found it essential to reduce drastically the time these vessels spend in port. They do this by demanding high daily container-handling rates from terminal operators and secondly, and more seriously for many ports, by reducing the number of ports of call. Whereas conventional break-bulk general cargo ships call at several ports (perhaps six or seven) at each end of the trade route, these large container ships call at no more than two or three ports and some companies limit the calls to just one port at each end of the route. This has had a profound influence on the geography of sea transport and has resulted in a division of container ports into three categories as follows.

Pivot ports, hub ports, or load center ports, which are served by the specialist large container vessels and which take two forms. In the "hub and spoke system" the hub port is used to trans-ship containers from the mainline vessels to surrounding ports in smaller vessels; in the relay or interlining hub port, containers are interchanged between mainline vessels operating in different services (referred to as strings or loops); this method is used in many cases to link ship operators' east–west and north–south trades. Feeder ports, which gain access to the main container transport routes through smaller, short sea vessels, known as feeder vessels, by means of which their container trades are transshipped to and from the pivot ports; and direct call ports, which have sufficient local trade to merit direct calls by large vessels but are not used to transship containers to other ports in the region.

This division of ports is already widespread in the Far East, south-east Asia, the Mediterranean, north-west Europe, the Middle East, and the Caribbean, and is rapidly expanding in west Asia, and Central and Latin America. It is important to realize, however, that a single port may illustrate all three functional categories for different services. Whether or not a port becomes a pivot port for a ship operator has a striking effect on its business and on the terminal facilities and resources needed to handle vessels and the cargo generated as a result of this.

The current rationalization taking place in the shipping industry through the policy of consolidation and the formation of alliances and partnerships, the increases in size of vessels operating on the major routes, and ship operators' decisions to integrate their east–west and north–south services are likely to see a further polarization of container ports and the emergence of global, regional, and subregional hub ports in the future. This will require ports selected as global or regional hubs to make huge investments in the construction of very large terminals (so-called mega-terminals) and the provision of equipment and other resources to handle the high throughput this status will generate.

2. Terminal design and layout

In response to ship operators' demands for high container-handling rates, ports have had to build more specialized types of terminals, some adapted from existing berths but most designed and constructed as new facilities (Thomas, 1987). The most striking feature of a modern container terminal is its sheer size. Early terminals developed in the 1970s have grown from 10–15 hectares to over 75 hectares for modern terminals acting as global or regional hub facilities and handling large quantities of trans-shipment containers. Longer quays have to be constructed to accommodate the largest generation of vessels, now up to 350 meters in length, and the depth of water alongside has increased to 16 meters with predictions of 18 and possibly 21 meters required in the future.

The essential terminal facility is the quay at which the vessel berths to discharge and load its cargo. A length of about 300 meters is considered about right for a "single terminal unit," one that takes one large container vessel at a time. A "two-unit terminal" would have a length of about 500 meters and a three-unit terminal 750 meters. However, major terminals have several kilometers of quay length and are able to accommodate several mainline and feeder vessels at the same time. The quay must be wide enough to accommodate the large quayside gantry cranes that serve in most terminals for loading and discharging ships, to provide space for containers to be landed, and for container-moving equipment to maneuver.

Behind the quay is an extensive storage area – the container yard. It takes up about 60–70% of the total space of the terminal, and is used primarily to stack containers awaiting onward movement. On a large terminal, the container yard may accommodate over 10 000 containers at a time. It is set out in well-marked and numbered blocks, linked by roadways and aisleways along which vehicles and equipment travel. Some blocks are reserved for exports (normally those to the seaward side of the yard) and others for imports and empty containers (normally to the landward side). Some stacking areas are set aside for special containers: refrigerated containers, containers carrying overheight and overwidth cargoes, containers carrying dangerous cargoes, and so on.

In those terminals where a large proportion of cargo arrives as, or is collected as, separate consignments of break-bulk packages, it was common for a container freight station (CFS) to be provided within the terminal. However, the modern trend is to locate this facility outside the terminal and often outside the port to avoid the handling of break-bulk cargo, to reduce handling operations, and to release more space for container storage purposes. It is in the CFS that inbound containers are unpacked and the separate consignments of cargo are stored awaiting collection, and where export consignments are gathered and packed into empty containers before they are moved to the container yard ready for shipment. There are also areas set aside for various examination functions, for example customs and port health examination of containers and their contents.

The interchange area is that part of the container terminal to which access is permitted for road vehicles, to deliver and collect containers. There are two distinct types of interchange, depending on the type of container transfer equipment used at the terminal. In one arrangement the interchange is a separate, marked-out area of the surface. Containers are brought to, or taken from, road vehicles parked at slots at the interchange by the transfer equipment. The alternative is for the interchanges to be a series of lanes running along one side of each storage block. Road vehicles are permitted to drive into the container yard and to take and collect their containers at positions alongside the stacks. Movement of containers into and out of the terminal is controlled at a gate facility, where documentation, security and inspection procedures are attended to. For containers arriving or leaving by rail, a rail reception/dispatch terminal or railhead may be provided within the terminal or just outside it.

Administrative offices are provided within the terminal, where staff engaged in planning, administrative, and documentary activities are housed. The expensive and heavily used equipment requires regular maintenance and repair so a maintenance facility in the form of a suite of workshops is provided on the terminal. A parking area is also provided near the workshop for equipment not in use.

Container terminals have to provide a range of ancillary facilities not directly concerned with the movement or storage of containers but essential to their safety and security. There must be security fencing surrounding the entire terminal area – the terminal is a customs area – and lighting covering all areas so that work can continue safely and securely through the hours of darkness.

3. Terminal operations

The previous section provided a broad idea of the terminal's facilities and activities and illustrated some of the complexity of container terminal operations. In reality, they are more complex and demanding than that outline suggests, for two reasons: they must be carried out quickly and safely, and they usually take place simultaneously (Thomas and Roach, 1991). For example, for inbound containers at any one time there may be containers being discharged from the ship; being moved from the quayside to the container yard; being stacked in the container yard to await collection; being collected by road vehicles, rail wagons, or inland waterway; being shifted in the stacks to gain access to others being collected; and being moved to the CFS for unpacking and, after unpacking the empty containers being moved to the stacks to await owners' instructions.

For outbound containers we might see the receipt of containers as they arrive at the terminal by road, rail, or inland waterway; stacking of received containers in the container yard to await vessel arrival; movements of empty containers from the stacks to the CFS for packing with cargo; movement of packed containers from the CFS to storage in the container yard; shifting of containers in the stack, as they are assembled in the correct sequence for loading; transfer of containers from the container yard to the quayside; and the loading of containers onto the ship. Other terminal movements may include the transfer of containers from the container yard to the customs and port health facilities for examination, or to the dangerous-goods examination area; the arrival of empty containers from outside the terminal (e.g., from inland container depots (ICDs) or importers' premises), either for storage or to be shipped by sea as empties; and the dispatch of empty containers from the stacks to exporters' premises or to ICDs, for packing with exports.

This complex pattern of container movements is complicated further by special treatment of refrigerated containers, those with hazardous cargoes, those that are overheight, or overwidth, and so on. Altogether, on a large terminal, there may be many thousands of container movements a day. Managing this would not be possible without a well co-ordinated and controlled series of systems and subsystems, allowing the terminal to operate smoothly and efficiently. There are five main operational systems (assuming the terminal has a CFS), in fact:

(1) the ship operation,
(2) the quay transfer operation,
(3) the container yard operation,
(4) the receipt/delivery operation, and
(5) the CFS operation.

The ship operation consists of the movement of containers between the quayside and the vessel. The quay transfer operation consists of the movement of containers between the quayside and the container yard. The container yard operation is concerned with container storage and is largely an inactive "operation" – just the holding of the container in safety and security until it is ready to be moved, for dispatch or loading. In practice, there may be equipment assigned solely to the container yard, lifting containers to and from quay transfer equipment, in which case stacking and unstacking can be considered to be components of the container yard operation. The other, and in many ways the most significant, components of the container yard operation are the range of in-terminal movements (e.g., the movements of containers to/from the CFS, customs, and port health examination areas). The receipt/delivery operation consists of two distinct, linked subsystems. For example, for an inbound container leaving the terminal by road, there is first movement of the container from its stacked position in the container yard to an interchange (either alongside the stack or near the gate) and its landing on the vehicle. The road vehicle is then driven to the other sub-system, the handling of the road vehicle through the gate (container inspection, door seal checking, and the documentary procedures). For an outbound container the documentary and inspection procedures at the gate come first. There are equivalent operations for containers arriving and departing by rail and inland waterway.

All these activities that make up the various terminal operations are not independent of each other, of course. They are all closely interrelated and interdependent, and must be carefully co-ordinated if the terminal is to operate effectively and efficiently. If the activities get out of step with each other – one operation running slower than the others, for example – then one operation can interfere with and possibly delay another. The operations will be out of balance, and container-handling performance will be poor. Clearly, it is vitally important for those planning and supervising terminal operations to understand the interrelationships of the various operations and to recognize the dangers and symptoms of the component operations going out of balance. The important thing to remember is that demand (i.e., the quantity of work required) for a particular operation varies from hour to hour and from day to day. So continual adjustments have to be made between those components, moving resources from one to another as demand varies and as loading/discharging, in-terminal movements, and receipt/delivery activities increase and decrease in intensity.

To assist in this operation, terminals have invested heavily in computer hardware and software and developed extensive communication systems. It is true to say that modern terminals are systems driven, and great developments are taking place in the application of information technology and e-commerce to improve planning and control procedures and to raise terminal efficiency.

4. Container-handling systems

It should be clear that the performance and efficiency of a container terminal depend heavily on its handling equipment (Thomas, 1994). Indeed, the presence and activity of very large, fast-moving equipment is a characteristic of the container terminal. A considerable variety of equipment is used for handling containers in ports, and terminal design, layout, and operations differ accordingly. At the quayside, containers can be handled to and from the ships by jib cranes, multipurpose cranes, gantry cranes, and even mobile cranes. Quay transfer may be handled by tractors towing trailers, by straddle carriers, or by heavy-duty lift trucks ("front-end loaders"). In the container yard, stacking and unstacking may be carried out by straddle carriers, yard gantry cranes, or a variety of lift-truck designs, while receipt/delivery operations may also involve those equipment types, as well as tractor–trailer systems.

In operational terms, it is usual (and useful) to think in terms of container-handling systems, each a combination of equipment types working together to perform the shoreside handling function. Six systems can be conveniently distinguished, three pure and three combination systems, as follows.

4.1. Pure equipment systems

The chassis system, in which containers are both handled and stored on "over-the-road" chassis or terminal trailers, which are moved around the terminal by heavy-duty tractor–trailer units.

The straddle carrier direct system, in which all quay transfer, container yard, and receipt/delivery movements are performed by straddle carriers.

The lift-truck system, which is operated quite effectively in smaller container terminals. The truck – whether front-end loader or reach–stacker – lifts, transfers, and stacks containers at all stages of the terminal operation.

4.2. Combination systems

All other container systems (in fact, probably the majority of them) can be considered as combination systems, using various mixtures of the five types of transfer equipment as follows.

The straddle carrier relay system, in which straddle carriers are responsible for in-yard stacking and unstacking, while quay transfer and other movements are performed by tractor–trailer sets or other equipment.

The yard gantry system, where the container yard is equipped with rubber-tired, rail-mounted or overhead gantry cranes for stacking/unstacking, with tractor–trailer units used for quay transfer and other movements.

Combination systems, which are various "hybrid" combinations of straddle carriers, yard gantry cranes, and other equipment, with more than one type of stacking equipment in use at a time, each carrying out a function to which it is best suited.

Those, then, are the container-handling systems currently in use. The choice of equipment will influence the design and layout of the terminal, the stacking patterns, and the operating procedures. On the other hand, the land area and configuration available for the terminal may restrict the range of equipment from which the system can be selected.

5. Terminal organization

Now let us turn to the organization of the terminal. Specialization in shipping has brought great and visible changes to the landscape of ports and terminals. What is not so immediately apparent is that changes just as great have been necessary in administration, managerial, operational, and maintenance systems.

For historical reasons, the management structure of a conventional general cargo berth is fragmented. The port authority is responsible for berthing the vessel, but shipboard cargo handling is often contracted to a stevedoring company, acting as the agent of the shipowner. Quay transfer may be performed by another cargo-handling organization, while the transit storage sheds and open storage yards are normally under the control of the port authority. So, on its journey through the berth, a consignment of cargo is handled by several organizations, with different objectives, management structures, and working practices. Co-ordination of their activities is often poor, with weak management control and duplication of administrative procedures. This outdated organizational structure has no place in a modern container terminal, where a simple, speedy, and cost-effective transfer of cargo between transport systems is essential.

In order to properly plan, co-ordinate, and supervise the high-speed and complex flow of containers through the terminal, it is essential for the terminal to be controlled by a single entity or organization, responsible for all the activities within the terminal area. "Unity of command" is the key to effective terminal management, whether achieved by public or private investment and ownership, by the owner of the terminal managing all its activities itself or leasing the terminal to a single operator to manage.

Once the terminal organization has been decided, its management structure must be established. This should have its own corporate identity, best achieved by forming a new operating company with its own board of directors and its own clearly set out objectives, powers, and responsibilities. The management team will administer and control all day-to-day activities on the terminal through delegation to departmental heads; quick decision-making is essential.

The organizational structure of the terminal is based on departments (each under the control of a head of department), with agreed job titles and specifications, and with clear lines of authority and communication. The number of departments, and the details of their functions and responsibilities vary from terminal to terminal but there are usually four main departments – operations, engineering, systems, and administration and finance. If the terminal is completely independent, and located separately from any other port facilities, there will be in addition a marine department, looking after vessels' access to the terminal, pilotage, tugs, other craft, and so on.

The operations department is the terminal unit directly concerned with handling cargo, and its main division or subdepartment (often called the "traffic" department) has that primary function. If there is an on-terminal CFS, that might be managed through a separate division.

The engineering department is primarily a service unit for the operations department, providing all the maintenance services and facilities to keep the operations department functioning (Thomas, 1989). It is usually made up of separate civil, mechanical, and electrical engineering divisions and, in the case of an isolated, self-sufficient terminal, a marine division as well. The importance of plant and equipment to efficient operations has placed much greater emphasis on this function within terminals in recent years.

Modern terminals rely heavily on the use of computers and information technology. The systems department (alternatively known as the information systems department) is responsible for provision of all electronic information systems, computer services for tracking and control of containers, and communication equipment and facilities to all departments and personnel.

The administration and finance department acts as the terminal's secretariat and provides personnel and welfare services to all departments. It is responsible for the preparation and presentation of company accounts, provision of banking, insurance, legal, and other professional services, invoicing and billing of all terminal activities and services, and the maintenance of company administrative and financial procedures and practices.

6. Terminal management

Merely investing enormous sums in terminal construction, equipment, and information systems is no guarantee of an efficient and commercially successful operation (Thomas, 1994). The performance and profitability of the terminal depend to a very large extent on the quality of its managers, supervisors, and workforce. The keys to terminal efficiency are all "people-dependent" and based on the principles of good operations planning, strict container control, close supervision, continuous and reliable information systems, and good communication.

6.1. Operations planning

A characteristic of every efficient container terminal is that every activity is planned in detail in advance, nothing is left to chance or to last-minute off-the-cuff decisions. Effective planning is achieved by establishing a planning unit or planning cell within the operations department. The functions of the planning unit are to plan all terminal activities: ship planning, yard storage, CFS movements, receipt/delivery activities, etc., so that the necessary labor, equipment, and materials are always available to meet current needs.

Despite the detailed planning operating problems will arise, and in practice managers and supervisors must be prepared for the unexpected. Vessels may not arrive on schedule or may arrive in bunches, putting berthing facilities under pressure; road traffic congestion may slow down the receipt and delivery of cargo; as a consequence storage areas may become congested and equipment shortages or breakdowns may disrupt operations. Weather delays may also affect operations. For these and a host of other unpredictable reasons, operations may not proceed as planned and terminal managers must take immediate action to solve the problem. So, terminal management systems must be flexible and versatile so that managers can quickly adapt plans and put their contingency measures into immediate action.

6.2. Container control

Given the complexity and speed of terminal operations, the essential requirement is that the movements of containers are closely controlled, to prevent errors and to ensure efficient use of terminal resources. Appropriate instructions must be issued at exactly the right moment, so that each handling and documentary activity is carried out at the right time and place. Container control is the system which integrates all the operational subsystems and components, through the preparation and issue of a series of printed or written instructions. The basic rule is that a container is never moved unless a written instruction has been prepared and issued for the move.

6.3. Supervision of operations

The third principle is supervision of operations. All terminal operations must be closely supervised, to ensure that they run efficiently. Those with supervisory authority and responsibility are extremely important people within the terminal organization, and have a range of vital functions.

Supervisory employees are responsible for the deployment of resources at the beginning of the shift. They then make sure that all activities are performed how and when they should be, as the shift progresses. They ensure that terminal rules and regulations are strictly followed, to prevent unsafe practices. Supervisory employees must respond appropriately and immediately to emergencies and put contingency plans into action, redeploying resources (labor and equipment) as required. Supervisors must at all times promote teamwork and motivate the workforce, encouraging them to work effectively and efficiently.

6.4. Information systems

Comprehensive, reliable, and up-to-the-minute information on the location of containers, on progress on operational plans, and on resource allocation is essential if effective control is to be maintained over terminal operations. The terminal information system is at the core of all management and supervisory control, and prompt, accurate record-keeping is of crucial importance to operations, as well as to all administrative and financial activities on the terminal. The information system is essential for keeping track of activities, but is also important for monitoring performance of the various terminal operations and ensuring that the terminal is operating efficiently and cost-effectively – setting targets for each unit of the terminal and checking that the target performance is achieved.

6.5. *Communications*

The final management principle of an efficient terminal is good communications between all units and employees. At a formal level, this is achieved through regular meetings and briefing sessions (at the beginning and end of shifts, for example, when operational activities are handed over from one supervisor to another), and through the distribution of all relevant documents to concerned parties. At an informal level, good communications are characterized by effective telephone, radio, and personal contacts during the work periods, so that everyone who needs to know is constantly informed about the state of operations.

Through these five management and supervisory principles, effective control and co-ordination of terminal activities can be maintained, and an efficient terminal operation can be achieved.

7. Terminal safety and security

Seaport terminals are dangerous places to work in and there is a high risk of accidents, particularly for those involved in cargo operations. There is no doubt that the most important single factor of risk is that presented by the numerous heavy, fast-moving container-carrying and container-lifting machines. Other important danger factors are the sheer size of modern terminals, the mixture of container-carrying equipment, road vehicles, and pedestrians in the working areas, and the need for 24 hour working in all weather conditions. However, over 80% of all injuries are actually caused by unsafe and avoidable actions on the part of those working there.

To encourage safe working and to make visits by vehicle drivers, ships' crews, and others safe, the terminal must establish a "safety culture." The safety message must be built into every aspect of terminal design, operation, and activity, so that it becomes second nature to everyone in and visiting the terminal. The safety culture is achieved through setting up an effective safety organization, establishing a clear company safety policy, setting company rules and regulations, laying down safe working procedures, publishing the rules, regulations, and procedures as a safety handbook or manual, and providing safety training for all employees.

The final topic of seaport terminal management is the safety and security of the containers held in the safekeeping of the terminal within the container yard. The terminal is entirely responsible for, and accepts liability for, all the containers and their cargo while they are held within the terminal's perimeter fence. The terminal accepts responsibility for containers when they are received at the quay and the gate, and that responsibility remains until it is transferred to a vehicle driver, rail official, or barge master (in the case of inbound containers) or the ship's master (in the case of containers leaving by sea). All terminal employees must be constantly aware of this and exercise "due diligence" at all times in looking after them. If a container or any of its cargo is damaged or stolen while in the terminal's safekeeping, the terminal operator has to pay damages and/or compensation for that loss. So security is vitally important, and every employee has a responsibility in maintaining container and cargo safe custody against the risks of theft, damage, and deterioration.

8. Conclusion

This chapter has demonstrated that a modern container terminal is an enterprise where a bewildering variety of different activities take place at the same time. The main purpose of these activities is to transfer goods in containers, as quickly and efficiently as possible, between inland and maritime transport. The container terminal has a central role in the international transport of goods – it is an essential link in the transport chain. The efficiency with which the terminal carries out its function has a very significant impact on the speed, smoothness, and cost of the transportation of cargo from exporter to importer, and everyone employed in the terminal has a part to play in making the terminal work as efficiently as possible. This chapter has provided a basic understanding of the role of the terminal in that transport chain.

References

Thomas, B.J. (1987) *Manual on container terminal development policy*. Geneva: United Nations Conference on Trade and Development (UNCTAD).

Thomas, B.J. (1989) *Management of port maintenance*. London: Her Majesty's Stationery Office.

Thomas, B.J. (1994) *Container terminal operations*. Geneva: International Labour Organisation.

Thomas, B.J. and D.K. Roach (1991) *Operating and maintenance features of container handling systems*. Washington: World Bank.

Chapter 38

AIR TRAFFIC CONTROL SYSTEMS

GEORGE L. DONOHUE
George Mason University, Fairfax, VA

1. Background

Air transportation refers to the movement of people and material through the third dimension, usually in heavier-than-air vehicles. These vehicles range from 400 pound (182 kilogram) powered parachutes transporting one person 25 miles (46 kilometers) to 800 000 pound (364 000 kg) jumbo jet aircraft transporting 350 passengers 9000 miles (17 000 kilometers). In fact, jet aircraft have now been designed to the point that they can connect virtually any two points on earth non-stop in much less than a day.

The invention and development of the jet aircraft in World War II have led to the use of aircraft as a major mode of both domestic and international transportation. Since 1960, the year that the U.S. Department of Transportation began collecting statistics, the air mode of transportation has grown over four times faster in passenger traffic and seven times faster in tons of cargo than any other mode of transportation in the U.S.A. (the other modes of transportation have grown at roughly the rate of growth of gross domestic product). The International Civil Aviation Organization states that more than one third of all international cargo by value was shipped by air in 1998. It should come as little surprise that the technical and physical infrastructure is feeling the strain of this sustained growth rate.

In the U.S.A., there are over 10 000 aircraft in commercial service at the turn of the century. Roughly 60% of these aircraft are powered by high-bypass-ratio fanjets, the rest are powered by either jet or piston-driven propellers. The fanjet aircraft prefer to fly above 30 000 feet in altitude, whereas the propeller aircraft prefer to fly below 24 000 feet. Aircraft flying above 12 000 feet are usually pressurized due to the lack of adequate oxygen for passenger comfort and/or survival. The U.S.A. operates approximately 40% (i.e., operational rate) of the world's commercial air transportation. In addition, the U.S.A. uses aircraft for private transportation to a considerable extent.

Handbook of Transport Systems and Traffic Control, Edited by K.J. Button and D.A. Hensher

There are over 150 000 registered private aircraft in the U.S.A., with over 600 000 registered pilots. On any given day, there are over 5000 aircraft in the air (between the hours of 10:00 and 22:00) under positive separation control by the Federal Aviation Administration (FAA) air traffic control (ATC) system. Of this amount, approximately one third are involved in private transportation. There are also approximately three times this amount of private aircraft in the air that are not under FAA positive control. Europe operates an air transportation system that has approximately 65% of the operational rate of the U.S. system, but with very little private air transportation activity. Africa, South America, and Australia operate a considerable amount of private air transportation in addition to commercial air transportation because of the large intercity distances and lack of substantial ground transportation infrastructure.

There are roughly 15 000 U.S. military aircraft that come under operational control of the FAA, a civil agency, while in transit to military operational areas. This is somewhat unusual since many countries use a separate military ATC system to control their military aircraft in partitioned airspace. In time of war, the FAA can come under operational control of the US Department of Defense.

In the U.S.A., the ATC system equipment is owned and operated by the federal government. Increasingly, governments are turning this function over to private or government-owned corporations so that the rapid changes in technology (and subsequently required increased access to investment capital) can be incorporated into the ever-changing ATC systems that will be required in the future. Federally operated systems are typically financed by a combination of user taxes and general tax fund contributions. The newly privatized (e.g., Canada) or quasi-privatized systems (e.g. Germany, Australia, and New Zealand) are totally supported by user fees and have access to private financial capital.

2. Government services and responsibilities

Before the end of World War II, in 1944, the International Civil Aviation Organization was formed as part of the United Nations to regulate international civil aviation. There are approximately 180 member countries at the beginning of the 21st century. Each member country must have a civil aviation authority (CAA) to provide communications, navigation, surveillance, and air traffic management (CNS/ATM) services to internationally accepted standards. For the U.S.A., this agency is the FAA. In addition to the provision of CNS/ATM services, each country must provide aircraft safety oversight for the certification of aircraft airworthiness and aircraft operation. Until recently, these two functions (CNS/ATM and safety oversight) have been supplied by the same government agency. Since 1990, there has been a trend to privatize (through different means, ranging

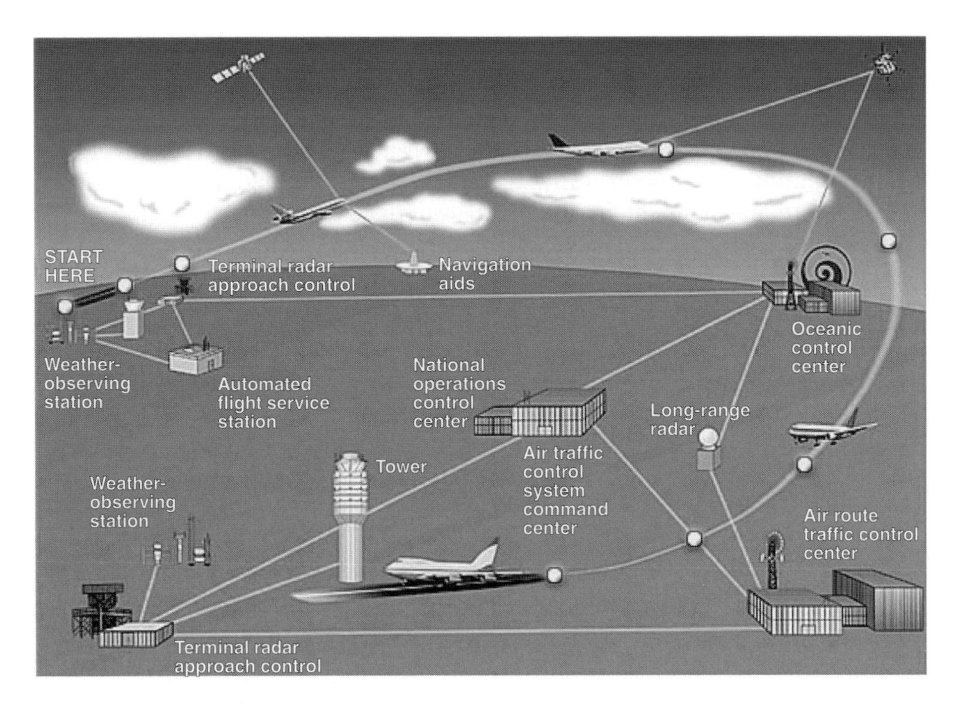

Figure 1. A pictorial representation of the many components and subsystems that exist in the current U.S. National Airspace System (NAS); this is typical of a modern air traffic management system (source: Federal Aviation Administration, 1999).

from wholly owned government organizations to complete privatization) the provision of CNS/ATM services and retain government safety oversight.

The CNS/ATM function has evolved from the provision in the 1920s of primitive navigation and communications services to a highly computerized ATM system with central flow control management (CFCM) or traffic flow management (TFM) utilizing space-based communications and navigation equipment as shown in Figure 1. With the advent of radar in World War II, the surveillance function was added to the CAA's provision of services in the late 1950s. The physical limitations of radar at that time set the aircraft separation standards that are still in use today. These separation standards (in conjunction with the number of runways that are available) set the maximum operational capacity that the air transportation system can support. These separation standards are typically 5 nautical miles (9 km) in high-altitude airspace (i.e., above 18 000 feet) and 3 nautical miles (6 km) in low-altitude airspace (typically within 60 nautical miles (110 km) of an airport). Airspace that does not have radar surveillance must maintain procedural separation using aircraft onboard navigation position fixes and ATC communications. These separation standards usually exceed 60 nautical miles (110 km) and are used in oceanic airspace and in undeveloped countries that lack radar services.

The radar physical properties that dictate these standards are beam width and sweep rate. In practice, the aircraft are routinely maintained at 7 to 30 miles (13 to 56 km) separation due to air traffic controller cognitive workload limitations. A typical controller can maintain situational awareness on four to seven aircraft at a time. When airspace sector loading exceeds this amount, controller teams work to maintain aircraft separation. These teams can be as high as three controllers per sector. In the U.S.A. there are over 730 sectors, and in Europe there are over 460 sectors. The number of sectors that are available to high-density airspace in the U.S.A. and Europe is limited to the number of communication channels that are available to the CAA. The number of communications channels available is dictated by the technical efficiency with which the allocated radio spectrum is utilized. The radio spectrum is allocated and controlled by the International Telecommunications Union (ITU), also a United Nations charter organization.

Unlike the U.S.A., where the FAA operates the entire airspace ATC (from airport tower to upper-altitude airspace), Europe has formed a central, trans-European organization called EUROCONTROL to co-ordinate the national provision of ATC services and operates the central flow management unit and upper airspace over central Europe. Both the U.S.A. and Europe have experienced considerable delays in introducing new technology over the last 20 years due to the difficulty of developing complex computer software that can demonstrate extreme levels of safety. This software is needed to greatly increase the level of ATC automation in order to reduce air traffic controller workload. Aircraft manufacturers have been much more successful in introducing computerized flight management systems (FMS) that reduce pilot workload and provide onboard aircraft collision avoidance systems (ACAS). Much of the increase in air transportation safety over the last 20 years has been attributed to the introduction of these aircraft automation systems.

3. A typical flight

Whether it is a commercial flight of a B777 for a major airline or a private pilot flying his/her own airplane, a flight begins with the flight planning activity 1 to 6 hours before the flight. FAA regulations require the pilot in command, or his/her designated representative, to check all factors (especially weather) that may affect the safety of the flight, including the status and availability of navigation aids and the status of the runways at the destination airport. For flight under instrument flight rules (all commercial flights essentially fly under instrument flight rules) a flight plan must be filed with the country's civil aviation authority, and this information is entered into a computer that is connected to the entire international air traffic control system. An air operations center (AOC), which begins a planning and replanning dialogue with the ATC system throughout the flight, performs this function for a major airline.

Prior to the actual flight, the pilot in command will check over the critical systems of his/her aircraft and make a final check of the weather at the origin and destination airports, as well as the weather over the entire route of the intended flight. Once the pilot deems it is safe to conduct the flight, he/she will contact the ATC clearance delivery authority via VHF radio (all future communications with ATC are via VHF radio unless the pilot is over the ocean out of line-of-sight to a VHF antenna, in which case satellite or HF communications will be used) to receive clearance to begin the flight and ensure that all information has been properly entered into the ATC computer system. Upon delivery of clearance, the pilot will contact ATC ground control to request taxi clearance and instructions to proceed to the end of the appropriate runway for take off. At the runway end, the pilot contacts the ATC tower to request clearance to take off on the designated runway.

After the ATC tower controller authorizes the aircraft to depart on the designated runway, the pilot contacts ATC departure control (located in a control facility known as a TRACON, or terminal radar control facility) upon liftoff. The TRACON facility may be located many miles from the airport and will typically handle several airports simultaneously. The ATC controller in this facility will guide the pilot throughout the climb phase to a cruise altitude of over 30 000 feet (or approximately 10 kilometers). At altitude, the pilot is turned over to an ATC controller at an en-route center. In the U.S.A., there are 20 en-route centers, which will handle traffic from approximately six to ten TRACONs (as a measure of comparison, the U.K. and Australia have two en-route centers). As the aircraft travels through a multitude of en-route sectors and centers, he/she is continually transferred to different controllers on different VHF frequencies. Upon approach to the airport and at top of descent, the pilot is handed to the ATC approach controller in the TRACON that controls the destination airport airspace. The approach controller assigns a runway for landing and a landing sequence to the arriving aircraft. Within about 10 miles of the airport, the pilot contacts the ATC tower controller and is guided to aircraft touchdown. After clearing the active runway, the pilot contacts ATC ground control for taxi instructions to the gate. When the aircraft is shut down, the active flight plan is terminated and the computer file is closed. Throughout the flight, a typical commercial aircraft of a major airline is in constant communication with the airline AOC via a VHF digital data link.

In the event that that the ATC system should experience an equipment failure, backup VHF radios are available to instruct the aircraft to separate using procedural separation techniques. In addition, since 1990 in the U.S.A., large passenger-carrying aircraft have used onboard, computerized collision avoidance equipment referred to as aircraft collision avoidance systems (ACAS). This type

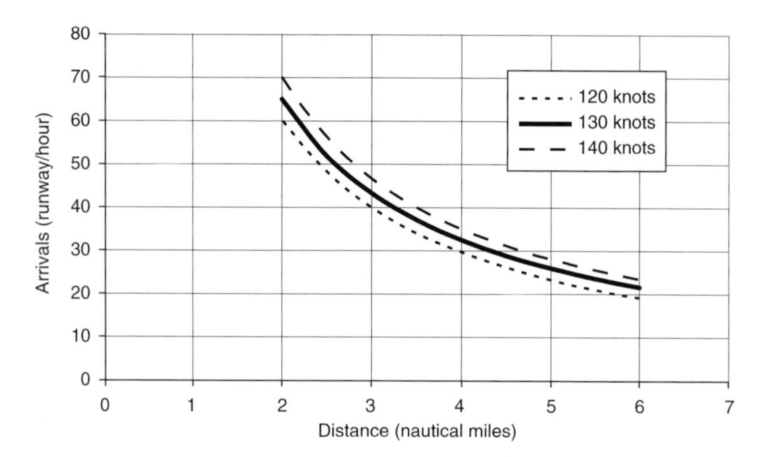

Figure 2. Runway arrival rate per hour as a function of aircraft spacing. Decreasing arrival spacing from an average of 4 nautical miles to 3 nautical miles increases the maximum capacity of the system by 34% (source: Donohue, 1999a).

of system has matured over the last 10 years and ACAS are planned to be introduced as standard equipment in aircraft throughout the world by 2005.

Except for the aircraft-to-AOC VHF digital data link, this communication and control system is virtually unchanged from its development over half a century ago and is reaching its capacity limit. Considerable international debate is ongoing on how this system will evolve over the next 15 years.

4. System capacity and delay

Most of the world allocates air transportation routes through government agencies. In the U.S.A., prior to 1978, the Civil Aeronautics Board (CAB) controlled the allocation of routes that commercial air carriers could provide. In 1978, the U.S. government deregulated the air transportation industry and allowed economic forces to shape the air transportation network. This system evolved very quickly to a hub-and-spoke network. At the beginning of the 21st century, there are approximately 60 hub airports in the U.S.A. owned and operated by local municipal governments, with a maximum capacity of about 40 million operations per year. Current forecasts predict that the future demand for air travel will significantly exceed supply, and delays will increase over the foreseeable future (Air Traffic Action Group, 1996; EUROCONTROL, 1999a,b; Donohue and Shaver, 2000). Decreasing aircraft separation in the final approach to a runway from an average of 4 nautical miles between aircraft to 3 nautical miles (Figure 2) could increase this capacity in the U.S.A. to over 53 million operations per year.

This increased capacity could be achieved by migrating from the use of radar surveillance to the use of aircraft broadcast Global Positioning System (GPS) satellite navigation fixes over a wireless digital data link (this is referred to as automatic dependent surveillance – broadcast, or ADS-B) (Donohue, 1999a). The unaugmented GPS position accuracy is better than 15 meters with a 1 second update rate. This capacity increase cannot be realized, however, without a change in the en-route separation procedures used by air traffic controllers, due to human cognitive workload limitations (Donohue, 1999b). Today, in both the U.S.A. and Europe, at least 2 minutes of delay per aircraft can be attributed to saturation of the en-route sectors. At individual high-density airports, these average delays can be as high as 10 minutes per aircraft at airport capacity fractions (*cf*) of over 0.9. Queuing theory would predict that airport delays will be proportional to $cf/(1 - cf)$. Increasingly, a central flow control function is being used to institute ground delay programs to anticipate these delays and hold aircraft on the ground at the point of origin rather than in the air at the point of destination. In the U.S.A., these delays are frequently triggered by a weather event at one or more of the hub airports.

There are four main actors that control the utilization of the air transportation system:

(1) the CAAs in the provision of regulations and aircraft separation/flow control standards and services;
(2) the airlines in their utilization of aircraft enplanement capacity/advanced avionics and the utilization of ATM information in their air operations centers;
(3) the airport operators in their provision of airport infrastructure, and
(4) the private aircraft operators in their provision of advanced aircraft avionics and non-interfering airspace utilization.

In order for the capacity and quality of service to increase in the 21st century, each of these players will have to make substantial capital investments in new equipment and significantly revise its operational procedures.

References

Air Transport Action Group (1996) "European traffic forecasts 1980–2010", ATAG, Geneva, www.atag.org/ETF/Index.htm [accessed 18 August 1999].

Donohue, G.L. (1999a) "A simplified air transportation system capacity model", *Journal of Air Traffic Control*, April–June, 8–15.

Donohue, G.L. (1999b) "A macroscopic air transportation capacity model: Metrics and delay correlation", presented at: Advanced Technologies and their impact on Air Traffic Management in the 21st century, Capri.

Donohue, G.L. and R. Shaver (2000) "United States air transportation capacity: Limits to growth: Parts I and II", presented at: Transportation Research Board 79th Annual Meeting, Washington, DC.

EUROCONTROL (1999a) "Performance review commission first performance review report", EUROCONTROL, Brussels, Report 1.

EUROCONTROL (1999b) "Performance review commission special performance review report on delays (January–September 1999)", EUROCONTROL, Brussels, Report 2.

Federal Aviation Administration (1999) *The national airspace system architecture, NAS version 4.0.* Washington, DC, FAA.

Chapter 39

AIRPORT SLOT CONTROLS*

ATA M. KHAN
Carleton University, Ottawa

1. Introduction

Many industry analysts have predicted that air passenger traffic is likely to double in the next 10 to 15 years. The growth in air travel, liberalization of market access by many commercial airlines, and proliferation of airline hubs will continue to generate high demand for airport slots (International Civil Aviation Organization, 2000). An airport slot is a scheduled day and time (usually defined within a 15 or 30 minute period) for an airline flight to arrive at or depart from an airport (International Air Transport Association, 2000a). The demand for airport service by airlines continues to exceed the available capacity at major airports around the world. Given that airport capacity at many sites cannot be increased due to environmental, political, and physical constraints, airport congestion has been on the rise. Owing to the interconnectedness of the air transportation network, congestion at an airport has adverse impacts on other airports.

Airport authorities and airlines are continuing to develop measures for overcoming or ameliorating situations of insufficient capacity. However, there is a growing list of sites where demand by airlines to initiate or increase commercial operations cannot be met because of lack of airport capacity. Consequently, there is pressure to utilize available capacity more efficiently. The balance between demand for service and available capacity can be achieved through airport slot controls. The establishment of the total number of slots and the allocation of slots to airlines are complex subjects involving supply side studies of airport capacity as well as demand management. This chapter describes how to use airport slot controls as a mechanism to balance demand and supply at congested airports.

2. Airport components and capacity

The interconnected functional components of an airport are used for serving aircraft, passengers, and freight. These include air traffic control (ATC), runways,

*This chapter is based on research sponsored by the Natural Sciences and Engineering Research Council of Canada.

Handbook of Transport Systems and Traffic Control, Edited by K.J. Button and D.A. Hensher

taxiways, aprons for parking aircraft, gates, terminal buildings, terminal curbs and parking, and land access (Figure 1). The challenge for airport planners, airport management, and airlines is to ensure that the various components offer users an acceptable level of service, do not become bottlenecks, and at the same time exhibit cost-effectiveness.

The capacity of an airport is the amount of passengers and cargo which can be accommodated in a given period of time (e.g., an hour). Since the various components of an airport are interconnected, airport capacity can be assessed by examining the combination of airside and landside facilities. Although in the past the capacity of the air traffic control subsystem was not an issue at most airports, it is becoming relevant to study it in combination with runways and other airside components, owing to growing air traffic demand. Airport capacity can be adversely affected by external factors, such as environmental restrictions. A number of airports have noise-related curfews. If two airports that are distant from each other have curfews, the available time for a flight to operate between them (i.e., the "window") is further narrowed (International Civil Aviation Organization, 2000).

The capacity of the runway subsystem is expressed in terms of the number of aircraft operations (landings or takeoffs) which can be performed safely per hour. It is shaped by such factors as the number, configuration, and physical characteristics of the runways, the surrounding area, the altitude, the type and mix of aircraft involved (larger aircraft may mandate greater separation), and air traffic control capabilities. The technological features of the air traffic control system, as well as en-route navigation, can directly affect runway capacity. The taxiway capacity is rarely an issue. The apron and gate capacity, defined as the number of aircraft that can be served per hour, is affected by aircraft size, the load factor, the nature of services required, and the efficiency of ground crew.

Terminal capacity is the amount of passengers and cargo which can be served in a given period of time. The type of passenger or the passenger mix can influence the throughput rate. Logically, international passengers who have to clear customs and immigration require more time and space than domestic passengers. The capacities of the remaining components of the airport, namely terminal curbs, parking, and land access have to be checked as well so as to avoid bottlenecks. Matching airside and landside subsystem capacity to the demand for use of

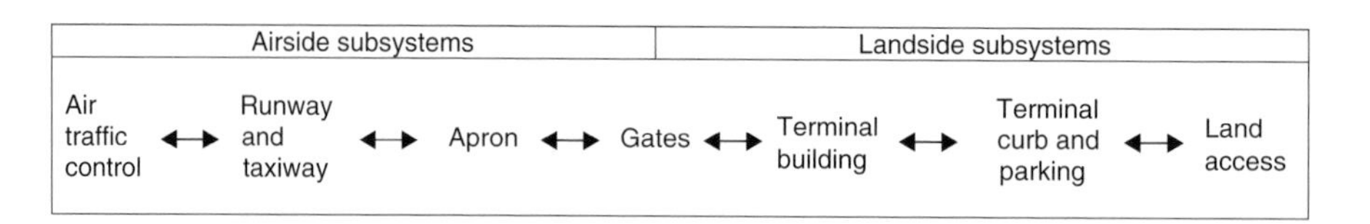

Figure 1. Functional components of an airport.

Table 1
The concept of an airport slot

Airport slot: 16:45 to 17:00
Airline A flight arrivals: 16:45, 16:48, 16:53, 16:58
Airline B flight arrivals: 16:50, 16:55
Airline A flight departures: 16:52
Airline B flight departures: 17:00

the airport requires careful planning and analysis. Additionally, co-operation of interest groups belonging to aeronautical authorities, airports, and airlines is essential for making adjustments, if required (Airports Council International and International Air Transport Association, 1996; International Civil Aviation Organization, 2000).

In the study of airport capacity and congestion, a key variable is the "peak traffic." Airports have to cope with traffic patterns reflecting underlying demand for air services and air carrier practices. These can result in peaking of traffic, which can fully utilize airport capacity. There are, of course, seasonal peaks as well, such as summer. At major hub airports, the airline scheduling practice of hub-and-spoke operations, which requires passengers to change planes, results in a series of peak periods throughout the day. It is a requirement of the design process that facilities should be able to serve a peak-period traffic representing the design year's busy-period conditions. Also, the expectation is that the various facilities (particularly landside facilities) should be able to cope with a usage level higher than the design peak level for a number of hours/periods during the year.

3. Airport slots

The use of an airport by an airline is expressed in terms of an airport slot. As noted earlier, a slot is the scheduled time of arrival or departure available for an aircraft movement on a specific date at an airport. A slot allocated to a flight of an airline will take into account capacity constraints at the applicable subsystems of the airport, e.g., apron, gates, terminal, etc. Although airline schedules have to be stated in more precise terms airport slots may be allocated in terms of a time period, such as 16:45 to 17:00, owing to variability in flight times, unavoidable delays etc.. In practice, a number of flight arrivals/departures may be assigned a common slot, although each will appear in the respective airline's schedule at a specific time (Table 1). The number of flight arrivals/departures that can be permitted within a slot is a function of airport capacity.

On the basis of permission to operate in an airport slot, airlines can develop their schedules. This process takes into account time to taxi to and from gates, landing/takeoff time, and estimated en-route time. An assumption is made that an ATC operation time will be made available as close as possible to the time necessary for the flight to operate on schedule. An ATC slot, which is the takeoff or landing time of an aircraft, is assigned by the relevant ATC authority to make optimum use of available capacity at points en route or at the destination airport. Therefore, there is a need to closely co-ordinate airport slots and aircraft arrivals/departures.

In practice, different airlines permitted to operate within a specific slot (e.g., 16:45 to 17:00) may show their scheduled flight departures at the same time for commercial or operational reasons. Frequently, flights are scheduled to depart on the hour (e.g., 17:00). Predictably, this practice at busy airports exacerbates peaking and often results in aircraft queues on taxiways waiting for takeoff clearance. At major busy airports, airline schedules may have up to 20 aircraft scheduled to depart at the same time. This results in a "scheduled delay."

4. Challenges of adjusting capacity to demand

Airport authorities carry out capacity analyses in order to determine the ability of the airport to serve projected demand and to identify problem spots. A thorough capacity analysis must examine the critical subsystems of the airport in question in terms of capacity deficiencies, and identify the possibilities of removing the capacity constraints through infrastructure or operational changes. Associated planning activities include estimates of the time and cost required to resolve the problems.

A fully interactive computer program known as CAPASS (airport terminal capacity assessment) and the U.S. Federal Aviation Administration (FAA) Advisory Circular AC 150/5060-5, "Airport capacity and delay," are commonly used by airport authorities and consultants (Airports Council International and International Air Transport Association, 1996; International Air Transport Association, 2000b). Likewise, a number of simulation models are available for the study of the flow of aircraft and passengers through the applicable components of the airport. The simulations provide realistic and dynamic evaluation of the complex interactions of the system.

Supply side approaches to increase capacity are as follows:

(1) build new airports;
(2) expand and/or improve existing airports;
(3) improve air traffic control capabilities with new technology and procedures; and
(4) enhance techniques and resources for passenger and cargo facilitation.

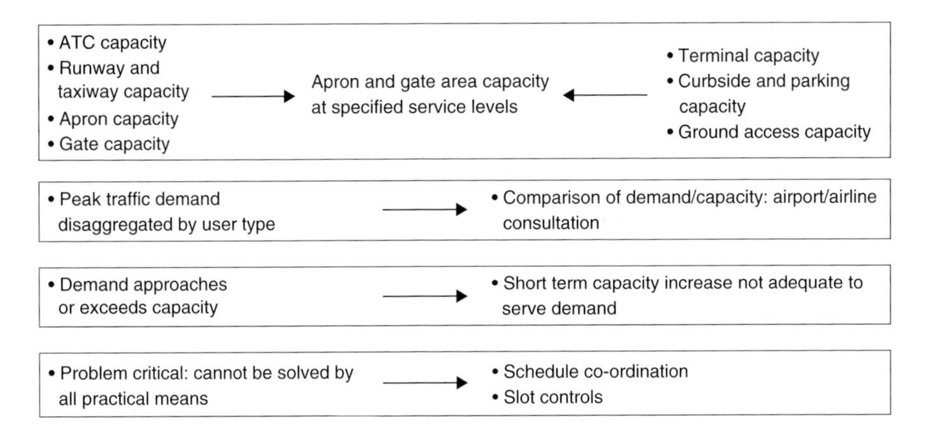

Figure 2. Various stages of the study of demand vs. capacity (adapted from International Air Transport Association, 1990).

In recent years, new airports have been built and new facilities have been added to existing airports. For example, in the 1990–1999 period, at capacity-constrained sites, three new airports, twelve terminals, and two runways were built. This expanded capacity placed some airports in the category where demand could be met without congestion. However, in the case of airports where capacity was added during early part of the 1990–1999 period, growth in traffic absorbed much of the capacity increase (International Civil Aviation Organization, 2000).

Improving ATC capabilities and enhancing passenger and cargo facilitation result in increased throughput for existing facilities. ATC improvement results in an increased number of movements that can safely be accommodated on a runway. This in turn increases the number of airport slots available. Thus governments and airlines are implementing programs to improve ATC services, especially in the U.S.A., Canada, and Europe. A recent notable development is the endorsement of the "gate-to-gate" air traffic management strategy by transport ministers at the European Civil Aviation Conference (ECAC) in January 2000 (International Civil Aviation Organization, 2000). Likewise, for passengers and cargo, the use of facilitation measures such as machine-readable passports, visa waiver programs, preinspection, and electronic cargo clearance increases the throughput of passengers and cargo of the existing facilities.

Adjusting capacity and/or demand to arrive at a balance is a complex and multistage endeavor (Figure 2). The first stage calls for an assessment of the capacities of all subsystems of the airport. Owing to the importance of the apron and gate area, capacity checks are made at specified service levels. The next stage is the examination of peak traffic demand disaggregated by user type. This leads to demand/capacity comparison and airport/airline consultations. The third stage comes into play when demand approaches or exceeds capacity. Following the assessment that a short-term capacity increase is not adequate for the demand to be served, the balancing of demand moves to the final stage. Here, all practical supply side and demand side measures are studied. If it becomes clear that the problem is critical and cannot be solved by any practical means, then a schedule co-ordination and slot control process is initiated.

5. Managing access at capacity-constrained airports

Over 120 airports around the world are capacity-constrained since the demand by air carriers for slots exceeds the available supply. If demand for air transportation continues to increase, other airports will experience capacity limitations and will not be able to serve any more aircraft movements (Abeyratne, 2000). Capacity shortages may occur during certain periods or in certain seasons (such as summer). In the most severe cases, demand can exceed capacity during all hours the airport is open. Thus, the severity of a capacity constraint can vary widely among airports. Likewise, different measures to deal with different situations can be required. On the assumption that supply side measures cannot be relied upon to alleviate shortages of capacity, demand side measures can be initiated (Box 1).

The issue of managing airport access (commonly known as the slot issue) is not new. For example, according to a 1989 report of the General Accounting Office (U.S.A.) (Mead, 1989), airport takeoff and landing slots required to gain entry at the four airports covered by the Federal Aviation Administration's high-density rule were not readily available for new entrants in the airline market. Thus, this practice was regarded as a barrier to competition in the airline industry.

The Airports Council International (ACI) and the International Air Transport Association (IATA) have jointly developed guidelines for airport capacity/

Box 1
Demand management measures for airports

- Use of IATA schedule co-ordination conferences and procedures
- Use of regulatory policies and practices (mainly for international services at airports):
  - annual limits on number of aircraft movements or passengers
  - allow new or expanded traffic rights when these can be accommodated at the airport(s)
  - negotiate access to slots bilaterally in advance
  - application of a policy of reciprocity
  - using noise rules as a constraint on airport access by airlines
- Diversion to alternate airports within the same area
- Market-oriented measures (e.g., peak-period pricing; buying, selling and leasing slots; slot auctions)

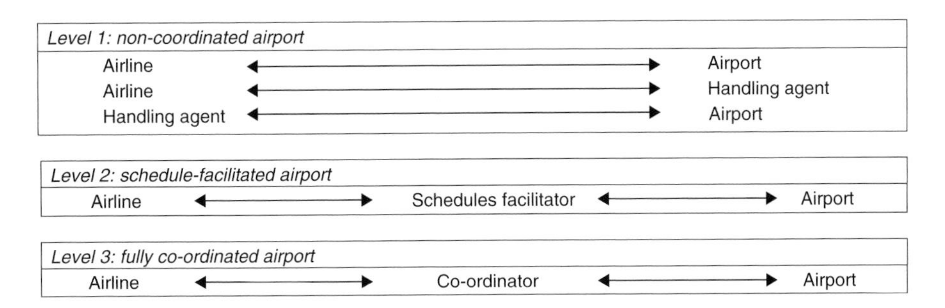

Figure 3. Levels of airport activity (adapted from International Air Transport Association, 2000b).

demand management to assist airport operators and airlines in addressing the peaking problem, and in maximizing the utilization of their facilities and other resources (Airports International and International Air Transport Association, 1996). Among these is the encouragement of the use of IATA schedule co-ordination conferences and procedures (described in detail in a later section of this chapter).

Regulatory policies and practices can be used for demand management purposes. Annual limits can be mandated on numbers of aircraft movements or numbers of passengers to be served. The level of demand that can be allowed has to match the airport capacity. According to regulatory policies, new or expanded rights cannot be permitted without expansion of slots. Bilateral negotiations to gain access to slots can be allowed. For international services, a policy of reciprocity can be followed. The availability of slots to airlines can be constrained by noise rules. This would take into account windows of operation between two airports with noise controls.

In major regions, excess demand that cannot be served at the subject airport can be diverted to existing or planned new airports. An area authority responsible for all airports can facilitate this measure.

The use of market mechanisms for managing demand can be studied. In the case of peak-period charges, the same charge is levied on all aircraft arriving or departing during the peak period. On the other hand, some market-oriented measures such as buying, selling, and leasing slots may not support broader policies on air transportation related to equity or political considerations. For example, access for new entrants or foreign carriers may be hindered by their inability to compete with the major airlines for expensive slots. Likewise, international air service agreements may be difficult to accommodate if slots are obtainable through the market mechanism. This subject is covered in more detail in Section 7.

6. Concept of slot controls

Airlines have to get approval of their schedules from the airport authority. Three broad categories of airports can be defined from the perspective of schedule clearance (Figure 3). The Level 1 airports have adequate capacity to meet user demand and are therefore non-co-ordinated in terms of schedule clearance. The Level 2 airports are those where demand is approaching capacity and therefore require schedule facilitation. This necessitates co-operation of various parties in order to avoid reaching an overutilization situation. The Level 3 airports are fully co-ordinated since demand exceeds capacity during the relevant period. It is not possible in this case to resolve the problem through voluntary co-operation between airlines. Even after consultation with all the parties involved, it is not possible to handle slot shortages.

The aviation industry, airport authorities, and other interests have to recognize and deal with the realities of airport congestion. The fully co-ordinated capacity-constrained airports are under pressure to maximize the use of available capacity and minimize the impact of congestion. Formal procedures are used to allocate available capacity and co-ordinate schedules.

7. Mechanism for slot allocation

The IATA schedule co-ordination conferences and procedures are the most widely used mechanism for managing scarcity of airport capacity around the world. At present about 200 airports participate in schedule co-ordination efforts. This system can enhance airport use by:

(1) serving as a convenient and transparent mechanism for over 260 participating airlines;
(2) enabling airlines to adjust their schedules on a worldwide basis by applying for or trading slots at the fully co-ordinated airports concerned;
(3) combining the efforts and expertise of all co-ordinators and airlines involved, at one of the IATA's biannual scheduling conferences; and
(4) helping to identify additional slots by appointing scheduling and operation experts to all co-ordinated airports, and assisting in capacity analyses at airports aiming for the fully co-ordinated airport status.

Table 2 shows the components of the IATA slot allocation process. According to this process, independent co-ordinators are assigned to each co-ordinated airport. These appointments are made following consultations with airport management, airlines, and their representative organizations. The co-ordinators are thoroughly experienced in airline and airport operations so as to facilitate schedule adjustments. Presently, there are 13 independent co-ordinators in the

IATA scheduling system. Additionally, there are five governmental agencies, which also serve as airport co-ordinators.

Schedule co-ordination conferences discuss airline schedules. These conferences are attended by all co-ordinators and are held twice each year, about four months before the start of the summer and winter scheduling seasons. Among other data required, airport capacity limitations applicable for the season under discussion are made known to the conferences by the appropriate authorities in consultation with airlines.

Airlines provide co-ordinators with schedule clearance requests about three weeks before the conference that cover the arrival and departure times required at the airports of interest. Following the collation of this information, the co-ordinator identifies periods in which slot requests exceed declared airport capacity. Airline requests of lower priority are offered the nearest alternative timings available. This enables demand at highly peaked periods to be reduced to match capacity.

The principle of historical precedence accords priority to a slot request if the airline operated at the same time the previous year. After historical slots have been allocated at conferences, 50% of the remaining slots available are first allocated to new entrants. The European Union has introduced a regulation which requires that a certain percentage of the newly available slots must be allocated to new-entrant carriers. The "use-or-lose rule" works as a safeguard against overbooking slots. These criteria are attractive for reasons of continuity of service, equity for those who were there before, the stability and continuity needed for planning and long-term investment, the opportunity for newcomers to become regulars in one year, and wide acceptance. The request of a developing-country carrier, particularly for slots required for starting a new service, could be given similar treatment to that accorded to new-entrant carriers in the group. The International Civil Aviation Organization (ICAO) secretariat has developed some preferential criteria for airport access by developing-country carriers (International Civil Aviation Organization, 1999).

A number of other criteria have been proposed and sometimes used. However, these are not accepted universally. These include priority for large over small aircraft, long over short routes, international over domestic service, and passenger over freighter services, and frequency capping, etc. The adoption of each criterion involves some groups receiving preference over others. The acceptance of any of these proposals requires careful consideration in the context of worldwide acceptability.

At the conference, schedules are fine-tuned, mainly through bilateral discussions between airlines and co-ordinators regarding alternative slots offered. Also, airlines can exchange slots offered or accepted. Given that a schedule change at one airport must affect one or more other airports, the conferences provide the best forum in which all such repercussive changes can be quickly and efficiently processed.

The end result of the conference is that the airlines secure firm schedules, which are regarded as the best compromise between what is requested and what is offered. The entire process is based on consensus. It is recognized to be flexible, fair, and open. It is rare that airlines leave the conference believing that the results have been inequitable.

8. Proposals for change

Proposals have been made to change the slot allocation system. The U.S. Government first floated the concept of the market mechanism in 1979. However, it was short lived due to international governmental opposition to paying a second time for rights already negotiated. Also, there was opposition to this idea due to conflict with the ICAO recommendation that airport user charges should be cost-related. At present, there is some renewed interest in the reliance on a market mechanism for allocating airport slots. It has some support in economic organizations such as the Organisation for Economic Co-operation and Development (OECD). A number of fairly detailed proposals have been advanced. For example, the concept of a secondary market in airport slots has been suggested in the European Commission.

According to the market mechanism, airlines are allowed to purchase, sell, and lease airport slots. The market value of slots would prevent hoarding and would give an incentive to airlines to use slots efficiently so as to recover their cost. The market mechanism would enable airlines to change slots quickly for the purpose of starting new services or exiting from unprofitable ones. However, concerns have been raised on competition grounds that the market mechanism would enhance the dominant position of larger airlines, which would be in a position to buy up the available slots and exclude potential competitors. In the U.S.A., there is actual experience with the market mechanism at four airports. While there were many buyers, there was a lack of sellers in this market. At the outset of the process in 1986, there was considerable activity in slot sales, purchases, and leases. Two years later, the number of slot exchanges dropped to one sixth of the total. There has been, however, a steady increase in the number of slots leased by airlines. From a base of less than 1% in 1986, the percentage of leased slots grew to 4.6% in 1993 (International Civil Aviation Organization, 2000a).

Further study and analysis of the various facets of the market approach to the slot issue are required (International Air Transport Association, 2000). From a legal perspective, Article 15 of the Chicago Convention requires that charges for use of an airport should not be higher for foreign aircraft than for national aircraft. This implies a single price for a slot. This rule would be difficult to implement, given that the market mechanism may result in different charges for national and foreign aircraft using the airport. If slots continue to be in short supply due to difficulties in adding facilities, the airlines are likely to recover the high price for slots from consumers in the form of higher fares. If the market mechanism were to be used, clear criteria would be required for application of competition law to slot auction, sale, purchase, or lease in order to avoid anticompetitive charges. It has been argued that if slots are auctioned, the process has to be transparent and non-discriminatory. The use of anticompetition laws could prevent undue concentration and abuse of a dominant position.

There are issues associated with the hybrid approach based on both the market mechanism and government regulation of slot allocation. There is actual experience with this approach in the U.S.A. International services that are committed under international obligations are guaranteed slots close to requested times. In the case of severe capacity shortages, domestic slots are withdrawn by lottery. Holders of domestic slots are allowed to buy and sell slots. Experience to date suggests that this hybrid approach has led to strengthening of the position of dominant airlines, and not of new entrants.

Suggestions have been made to maintain the structure of the present system. It has also been suggested that procedures and priorities could be modified by law. A weakness of this proposal is that resorting to law could be a roadblock to the present ability of co-ordinators and airlines to quickly resolve conflicting requirements. Experience indicates that conflicts can be resolved in a spirit of mutual co-operation within the time available before schedules must be published and flown. In its 1989 review of civil aviation policy, the U.K. Civil Aviation Administration (CAA) stated that the IATA slot allocation system works and better alternatives have not been developed (U.K. Civil Aviation Authority, 1995).

9. Conclusions

Slot controls at capacity-constrained airports are an effective means to create a balance between demand for airport use and supply of slots. The IATA slot allocation system is widely accepted. Improvements to the existing system can be brought about on the basis of fine-tuning the criteria for slot allocation, covering the treatment of new entrants, stage length, international vs. domestic flights, aircraft size, and airlines of developing countries. There is a renewed interest in market-oriented measures for demand management. However, these require further study and analysis.

All interest groups have recognized that detailed capacity analysis, network simulations, and studies of schedules and operations have the potential to squeeze out additional slots for the capacity-constrained airports.

Future IATA schedule co-ordination conferences should include the ATC components in schedule evaluation. The proposed schedules should be simulated, and schedule delay and potential bottlenecks on the ground and in the air should be identified. The necessary alterations to the schedules should be sought in order to mitigate the risks of congestion and its impacts.

References

Abeyratne, R.I.R. (2000) "Consequences of slot transactions on airport congestion and environmental protection", *Journal of Air Transportation World Wide*, 5(1):13–36.

Airports Council International and International Air Transport Association (1996) *Guidelines for airport capacity/demand management*, 3rd edn. Montreal/Geneva: ACI/IATA.

International Air Transport Association (1990) *Airport capacity/demand management*. Geneva/Montreal: IATA.

International Air Transport Association (2000a) *Schedule coordination services, airline schedule coordination & airport congestion*, www.IATA.ORG/SKED/About.htm. Montreal/Geneva: IATA.

International Air Transport Association (2000b) *Worldwide scheduling guidelines*, www.IATA.ORG/SKED/About.htm. Montreal/Geneva: IATA.

International Civil Aviation Organization (1999) "Policy and guidance material on the economic regulation of international air transport", 2nd edn., Appendix 3, ICAO, Montreal, Doc 9587.

International Civil Aviation Organization (2000) "Study on the allocation of flight departure and arrival slots at international airports", Appendix B, ICAO, Montreal, ANSConf-WP/11.

Mead, K.M. (1989) "Barriers to competition in the airline industry", General Accounting Office, Washington, DC, statement before the Sub-Committee on Aviation, Committee on Commerce, Science, and Transportation, United States Senate.

U.K. Civil Aviation Authority (1995) "Slot allocation: A proposal for Europe's airports", CAA, London.

AUTHOR INDEX

SUBJECT INDEX